# Introduction to
# Evolutionary Biology

**Fatik Baran Mandal**
Bankura Christian College
Bankura, West Bengal, India

**OXFORD & IBH PUBLISHING CO. PVT. LTD.**
New Delhi

Oxford & IBH Publishing Company Pvt. Ltd.
113-B Shahpur Jat, *Asian Games Village Side*
New Delhi 110 049, India

*Fax:* (011) 4151 7559
*Email:* oxford@oxford-ibh.in

ISBN 978-81-204-1779-3

Printed at Chaman Enterprises, New Delhi.

# Foreword

Two core concepts of the Biology are the making and the breaking; otherwise called the evolution and the extinction. Now evolution is not only the attainment of the complexity from the simplicity, perhaps it is the attainment of perfection in the structure, organization, behaviour and other characteristics of the organisms. The history of science shows the changing concepts of the evolution in various times. Undoubtedly, the Evolutionary Biology has been emerged as a discipline of tremendous potential for the well-being of human and the society.

This book provides a brief introduction to the basics of the evolutionary theory, some of the key evidences of evolution, the analysis of fossils and its implications, mechanisms of the evolution, co evolution and mimicry, and the evolution beyond biology. The discourse on the evolution of primates and our own species, the fact of becoming human deserves special mention. In short, the author has attempted to sketch the evolutionary biology in 11 chapters along with the glossary for important terms, summary, and references for each chapter.

*On the Origin of Species by Means of Natural Selection*, the famous book of Charles Darwin was published in the year 1859. The theory of evolution has drawn the efforts of many scientists for about 200 years. Research throughout the globe has produced powerful, multi-stranded theory of the evolution. Understanding modern evolutionary biology requires learning the language of the evolution. That living species can change over time and produce new kinds of species occupies the central position in the process of the evolution. All organisms share a common origin. The fossil records, the remains of life-forms were preserved for a long. Comparing these fossils with one another and with living forms provide evidence of change in life forms over time, or the evolution.

Darwin was interested in explaining the pattern of distribution of living organism. Species distribution patterns suggest the change over space. Before Darwin, the fossil record was mostly unknown. Many of those concerned with biology did not see the pattern of distribution of living species as evidence for past change. The Darwinian theory of evolution by natural selection emphasize that variation is the ground condition of life. Darwin suggested the driving force as the struggle for existence. In a later edition of *On the Origin of Species*, he used a phrase coined by Herbert Spencer and described the outcome of the struggle for existence as "survival of the fittest."

Evolutionary research could be divided into 2 major sub-fields. Microevolution concentrates on short-term evolutionary changes within a given species over few generations. Macroevolution focuses on long-term evolutionary changes, especially the origins of new species and their diversification across space and over millions of years. Macro evolutionary events which span many generations and growth and decay of many ecological settings are summarized in the geological time scale. Micro evolutionary studies in evolutionary biology have been made possible by the modern evolutionary synthesis. The modern synthesis defined a species as a reproductive community of populations that occupies a specific niche in nature.

The author, Dr. Fatik B. Mandal well-known for his 5 books has attempted his level best to cater to the needs of the students in a succinct way. I trust that his untiring and sincere efforts would have significance to the students and the scholars.

The Rt. Rev. Dr. Probal Kanto Dutta
Bishop, Durgapur Diocese
Church of North India
Durgapur, West Bengal
INDIA
Chairman Governing Body
Bankura Christian College

# Preface

Evolutionary concepts appeared in early Greek writings in the works of Thales, Anaximander and Aristotle. However, the glimpses of the theory of evolution emerged in the mid 19th century. The study of the reproductive cells became possible after the invention of the microscope. Lamarck for the first time presented an evolutionary theory. It included the inheritance of acquired characteristics as the operative force of evolution and was discredited for many years. Buffon suggested the influence of use and disuse in molding the organs of vertebrate animals.

In 1859, Charles Darwin published the theory of natural selection. Darwin's idea altered the view of our place in the universe. Darwin argued that all individuals struggle to survive, some get advantage of better chance of survival due to their small, heritable differences than individuals lacking these beneficial traits. Such individuals have the higher fitness. The useful traits become common in the population as their offspring survive more and finally become the norm. Harmful traits are eradicated quickly. Natural selection creates population which is highly suited to its environment. When individuals compete for limited resources, they go through the ecological selection. The useful traits are subjected to sexual selection and increase organism's chance of reproduction. Sexually selected traits make a male more attractive to the females.

Over many generations, the slow evolutionary change called anagenesis, change one species to evolve into another. However, most new species form in a speciation; a process Darwin called the "mystery of mysteries". Darwinian evolution is a slow process. However, much of the fossil record indicates long periods of stasis. Niles Eldredge and Stephen J. Gould argued that even communities start to change called the theory of punctuated equilibrium. Species struggle among themselves to survive and most become extinct over time. During the 20th century, E. Mayr, J.B.S. Haldane, J. Huxley, and T. Dobzhansky combined Darwinian evolution with the emerging knowledge of genetics to establish the modern synthesis.

The incidence of different traits in a population is driven by natural selection and genetic drift. Now, evolution is defined as the change in the frequency of alleles in populations over time. New traits are introduced into populations by gene flow from other populations, or by mutation. Most mutations are the neutral. Others are harmful, for example causing the inherited diseases like cystic fibrosis. Rarely mutations can lead to the beneficial new traits.

The central concept of the life's work of Darwin is the evolution by natural selection. Evolution explains adaptations, complexity and the existing biodiversity. It has significant bearing to medical sciences, agricultural sciences, and behavioural sciences and to understand the world. The various evidences for evolution is abundant, ever increasing and interconnected. Various thinkers have advocated the idea that various species are descended from the common ancestor. Simply, the evolution is the history of life and now it is the unifying force in biology. It ties fields like genetics, microbiology and palaeontology. In co-evolution, the evolutionary history of 2 species or groups of species gets intimately intertwined. Evolution is thus a beautiful and crucial concept for the human well-being. It is one of most significant revolutions in human thought. It solves problems. Evolutionary theory develops models to solve problems, revolutionizes the way of viewing the world and makes predictions. Such theory is applicable to fields like economics, linguistics, sociology, anthropology, psychology, and cosmology. Such theory is used for explanations for the origin of life, development of language, development of matter and universe, development of morals, culture, religion, behavior, and capitalistic economic models.

Random mutations, asteroid impacts and such other events have influenced the species evolution. Long sequences of DNA reveal more similarities and differences among species than found in their anatomy. Biologists could develop a general, predictive theory of the evolution and population dynamics of pathogens and their hosts, especially for rapidly evolving organisms like HIV and for rapidly migrating host species like modern humans. Analyses of genetic variation in resistance to pathogens in humans and other hosts are also needed for the well-being of the society.

Epigenetics refers to changes in phenotype or gene expression caused by mechanisms other than changes in the underlying genetic material. Epigenetics is applied only to developmental processes of organisms. This idea was non-controversial as all the cells of multicellular eukaryotes are genetically identical. Cells have different structures and functions as the genetic material is expressed variously in different cells. Development in multicellular eukaryotes proceeds by successive up- or down-regulation of the genes of given cells producing different structures and functions in an organism without changing its genetic material. All this is non-controversial.

The study suggests that epigenetic effects are not restricted to intra-individual phenotypic changes at the level of individual cells. Some significant phenotypic changes in adult organisms are not correlated with underlying genetic changes. Some phenotypic changes in adult organisms are caused by changes in expression of particular genes, rather than changes in genes themselves. Some purely phenotypic changes are heritable from parents to offspring and generate considerable controversy within the scientific community.

**Fatik Baran Mandal**

# Acknowledgements

The author wishes to express his sincere thanks to his family members without whose support this work would not have been completed. The author has collected figures used in this book from various sources particularly from the Wikipedia and Wikimedia .org. The author wishes to provide credits to the following links for various figures used in this book.

http://en.wikipedia.org/wiki/Comparative_genomics#mediaviewer/File:A_genome_alignment_of_eight_Yersinia_isolates.png

http://en.wikipedia.org/wiki/Drosophila_melanogaster#mediaviewer/File:Drosophila_melanogaster_-_side_%28aka%29.jpg

http://en.wikipedia.org/wiki/Hybrid_%28biology%29#mediaviewer/File:Liger.jpg

http://en.wikipedia.org/wiki/Hybrid_%28biology%29#mediaviewer/File:Zeedonk_800.jpg

http://en.wikipedia.org/wiki/Speciation#mediaviewer/File:Punctuated-equilibrium.svg

http://en.wikipedia.org/wiki/Speciation#mediaviewer/File:Speciation_modes.svg

http://en.wikipedia.org/wiki/Extinction#mediaviewer/File:Bufo_periglenes2.jpg

http://commons.wikimedia.org/wiki/File:Stegosaurus_Senckenberg.jpg

http://commons.wikimedia.org/wiki/File:Berlin_Diplodocus.jpg

http://en.wikipedia.org/wiki/Extinction#mediaviewer/File:Edwards%27_Dodo.jpg

http://en.wikipedia.org/wiki/Extinction#mediaviewer/File:Ectopistes_migratoriusMCN2P28CA.jpg

http://en.wikipedia.org/wiki/File:Extinction_intensity.svg

http://en.wikipedia.org/wiki/Bacterial_taxonomy#mediaviewer/File:Haeckel_arbol_bn.png

http://en.wikipedia.org/wiki/Gene#mediaviewer/File:Human_genome_by_functions.svg

http://en.wikipedia.org/wiki/Gene#mediaviewer/File:Gene. http://en.wikipedia.org/wiki/DNA#mediaviewer/File:DNA_Structure%2BKey%2BLabelled.pn_NoBB.png.

http://en.wikipedia.org/wiki/DNA#mediaviewer/File:Maclyn_McCarty_with_Francis_Crick_and_James_D_Watson_-_10.1371_journal.pbio.0030341.g001-O.jpg

http://en.wikipedia.org/wiki/Genetics#mediaviewer/File:Genetic_code.svg

http://en.wikipedia.org/wiki/Genetics#mediaviewer/File:Eukaryote_tree.svg

http://en.wikipedia.org/wiki/Genetics#mediaviewer/File:Ecoli_colonies.png

http://en.wikipedia.org/wiki/Reproductive_isolation#mediaviewer/

http://en.wikipedia.org/wiki/Sexual_conflict#mediaviewer/.

http://en.wikipedia.org/wiki/Reproductive_isolation#mediaviewer/File:Drosophila_speciation.

http://en.wikipedia.org/wiki/Evolution#mediaviewer/File:Speciation_modes_edit.svg

http://en.wikipedia.org/wiki/Ring_species#mediaviewer/File:Ring_species_seagull.svg.

http://en.wikipedia.org/wiki/On_the_Origin_of_Species#mediaviewer/File:Darwin_divergence.jpg

http://en.wikipedia.org/wiki/Outline_of_physical_science#mediaviewer/

http://en.wikipedia.org/wiki/Cloud#mediaviewer/File:Lowcloudsymbols.gifLow cloud weather map symbols: Includes low and upward-growing vertical.

http://en.wikipedia.org/wiki/Rock_%28geology%

http://en.wikipedia.org/wiki/Fossils#mediaviewer/File:Index_fossils.gifExamples of index fossils

http://en.wikipedia.org/wiki/Fossils#mediaviewer/File:CambrianRusophycus.jpgambrian

http://en.wikipedia.org/wiki/Coacervate#mediaviewer/File:Coacervats.JPG

http://en.wikipedia.org/wiki/Origin_of_life#mediaviewer/File:Blacksmoker_in_Atlantic_Ocean.jpg deep sea hydrothermal vent

http://en.wikipedia.org/wiki/Co-evolution#mediaviewer/File:Yuccaharrimaniae.jpgA flowering yucca plant that would be pollinated by a yucca moth

http://en.wikipedia.org/wiki/Ecological_speciation#mediaviewer/File:Gasterosteus_aculeatus.jpg

http://en.wikipedia.org/wiki/Fossil#mediaviewer/File:Microfossils.JPG

http://en.wikipedia.org/wiki/Animal#mediaviewer/File:AnimalsRelativeNumbers.png

http://en.wikipedia.org/wiki/Trace_fossil#mediaviewer/File:Cheirotherium_prints_possibly_Ticinosuchus.JPG

http://en.wikipedia.org/wiki/Lamarckism#mediaviewer/File:Giraffe23.jpg

http://en.wikipedia.org/wiki/Natural_selection#mediaviewer/File:Antibiotic_resistance.svg

http://en.wikipedia.org/wiki/Sexual_selection#mediaviewer/File:Paradesia_decora_Keulemans.jpg

http://en.wikipedia.org/wiki/Mutation#mediaviewer/File:Chromosomes_mutations-en.svg.

http://en.wikipedia.org/wiki/Genetic_drift#mediaviewer/File:Random_genetic_drift_chart.png

http://en.wikipedia.org/wiki/Genetic_drift#mediaviewer/File:Founder_effect_with_drift.jpg

http://en.wikipedia.org/wiki/Nucleic_acid_sequence#mediaviewer/File:Kooditabel.png( genetic code).

http://en.wikipedia.org/wiki/Neutral_network_%28evolution%29#mediaviewer/File:Neutral_network.png

http://en.wikipedia.org/wiki/Phylogenetic_comparative_methods#mediaviewer/
File:Primate_Testes_Allometry_Sexual_Selection.jpg

http://evolution.berkeley.edu/evolibrary/misconceptions_teacherfaq.php

http://evoled.dbs.umt.edu/lessons/evidence.htm

https://bealbio.wikispaces.com/

http://en.wikipedia.org/wiki/Hybrid_%28biology%29#mediaviewer/File:Zeedonk_800.jpg

http://en.wikipedia.org/wiki/Speciation#mediaviewer/File:Punctuated-equilibrium.svg

http://en.wikipedia.org/wiki/Speciation#mediaviewer/File:Speciation_modes.svg

http://en.wikipedia.org/wiki/Extinction#mediaviewer/File:Bufo_periglenes2.jpg

http://commons.wikimedia.org/wiki/File:Stegosaurus_Senckenberg.jpg

http://commons.wikimedia.org/wiki/File:Berlin_Diplodocus.jpg

http://en.wikipedia.org/wiki/Extinction#mediaviewer/File:Edwards%27_Dodo.jpg

http://en.wikipedia.org/wiki/Extinction#mediaviewer/File:Ectopistes_migratoriusMCN2P28CA.jpg

http://en.wikipedia.org/wiki/File:Extinction_intensity.svg

http://en.wikipedia.org/wiki/Bacterial_taxonomy#mediaviewer/File:Haeckel_arbol_bn.png

http://en.wikipedia.org/wiki/Gene#mediaviewer/File:Human_genome_by_functions.svg

http://en.wikipedia.org/wiki/Gene#mediaviewer/File:Gene.png

http://en.wikipedia.org/wiki/DNA#mediaviewer/File:DNA_Structure%2BKey%2 BLabelled.pn_
NoBB.png

http://en.wikipedia.org/wiki/DNA#mediaviewer/File:Maclyn_McCarty_with_Francis_Crick_and_
James_D_Watson_-_10.1371_journal.pbio.0030341.g001-O.jpg

http://en.wikipedia.org/wiki/Genetics#mediaviewer/File:Genetic_code.svg

http://en.wikipedia.org/wiki/Genetics#mediaviewer/File:Eukaryote_tree.svg http://en.wikipedia.org/
wiki/Genetics#mediaviewer/File:Ecoli_colonies.png

http://en.wikipedia.org/wiki/Reproductive_isolation#mediaviewer/File:Reef0484.jpgIn

http://en.wikipedia.org/wiki/Sexual_conflict#mediaviewer/File:Bio_166_revised_wiki_se

http://en.wikipedia.org/wiki/Reproductive_isolation#mediaviewer/File:Drosophila_speciation.svg

http://en.wikipedia.org/wiki/Evolution#mediaviewer/File:Speciation_modes_edit.svg

http://en.wikipedia.org/wiki/Ring_species#mediaviewer/File:Ring_species_seagull.svg

http://en.wikipedia.org/wiki/On_the_Origin_of_Species#mediaviewer/File:Darwin_divergence.jpg

http://en.wikipedia.org/wiki/Outline_of_physical_science#mediaviewer/File:Partial_ordering
_of_the_sciences_Balaban_Klein_Scientometrics2006_615-637.svg http://en.wikipedia.org/wiki/
Rock_%28geology%29#mediaviewer/File:DirkvdM_rocks.jpg

http://en.wikipedia.org/wiki/Fossils#mediaviewer/File:Index_fossils.gifExamples of index fossils

http://en.wikipedia.org/wiki/Fossils#mediaviewer/File:CambrianRusophycus.jpg

http://en.wikipedia.org/wiki/Coacervate#mediaviewer/File:Coacervats.JPG

http://en.wikipedia.org/wiki/Origin_of_life#mediaviewer/File:Blacksmoker_in_Atlantic_Ocean.jpg

http://en.wikipedia.org/wiki/Co-evolution#mediaviewer/File:Yuccaharrimaniae.jpg moth

http://en.wikipedia.org/wiki/Ecological_speciation#mediaviewer

http://en.wikipedia.org/wiki/Fossil#mediaviewer/File:Microfossils.JPG

http://en.wikipedia.org/wiki/Animal#mediaviewer/File:AnimalsRelativeNumbers.png

http://en.wikipedia.org/wiki/Trace_fossil#mediaviewer/File:Cheirotherium_prints_possibly_Ticinosuchus.JPG

http://en.wikipedia.org/wiki/Lamarckism#mediaviewer/File:Giraffe23.jpg

http://en.wikipedia.org/wiki/Natural_selection#mediaviewer/File:Antibiotic_resistance.svg

http://en.wikipedia.org/wiki/Sexual_selection#mediaviewer/File:Paradesia_decora_Keulemans.jpg

http://en.wikipedia.org/wiki/Mutation#mediaviewer/File:Chromosomes_mutations-en.svg.

http://en.wikipedia.org/wiki/Genetic_drift#mediaviewer/File:Random_genetic_drift_chart.png

http://en.wikipedia.org/wiki/Genetic_drift#mediaviewer/File:Population_bottleneck.jpg

http://en.wikipedia.org/wiki/Genetic_drift#mediaviewer/File:Founder_effect_with_drift.jpg.

http://en.wikipedia.org/wiki/Nucleic_acid_sequence#mediaviewer/File:Kooditabel.png

http://en.wikipedia.org/wiki/Neutral_network_%28evolution%29#mediaviewer/

http://en.wikipedia.org/wiki/Phenotypic_plasticity#mediaviewer/File:Phenotypic_plasticity_diagram.svg

http://en.wikiversity.org/wiki/Dominant_group/Evolution#mediaviewer/File:Primate_skull_series_with_legend.png

**F. B. Mandal**

# Contents

# CHAPTER 1

# Introduction

Speciation is the source of diversity in life on earth. It is the mystery in evolutionary biology (Figure 1.1). Darwin wrote not much about species in the "Origin of Species" but suggested natural selection. Genes, cells, living matters, and their interaction with environment are manifested as life. Integration of various branches of biology and interdisciplinary study involving biology, physics, chemistry, and mathematics now present the biological sciences in a new way. Evolution generates new species with various adaptations in varied environments. Knowledge about gene, cell, and species helps understand the evolution. Living organisms are organized structures. They grow, reproduce, adapt, regulate internal behavior, gain energy, and respond to stimuli. Cells, the membrane -bound bags , consist of proteins, nucleic acids , and other biomolecules. Robert Hooke, the father of microscopy called the chambers of cork as cells for their look like the monk's rooms. Cell controls some chemicals pass in, or out. It serves as individual in acellular organism. Many cells make multicellular organism. Prokaryotes and eukaryotes are the 2 of 3 domains of the living world. Eukaryotic cells (Figure 1.2), are the complex, larger than prokaryotes (Figures 1.3 and 1.4), contain nucleus but prokaryotes do not. There is a cell membrane, but not always cell wall. Instead flagella and pili, cilia help in cell movement. In eukaryotes nucleus contains most DNA, ribosome makes proteins and mitochondrion produces ATP. DNA carries out the properties of life. Some viruses use RNA instead of DNA. In RNA, uracil replaces thymine.

Leeuwenhoek, using quality glass as lens, first observed bacteria, yeast, and animalcules now called protests. The best known protist is the *Amoeba*. Many amoebae are shelled. *Amoeba* was previously named *Proteus animalcules*. *Amoeba proteus*, the largest protest, is rounded, membrane- bound, irregular blob with protrusions and mixture of granules which flow as thick jelly. *Amoeba* moves and ingests food. Its nucleus reproduces and split one *Amoeba* into 2 which grow and divide again.

Evolution is a pathway of arising present-day life forms from common ancestors. It helps explain similarities and differences in

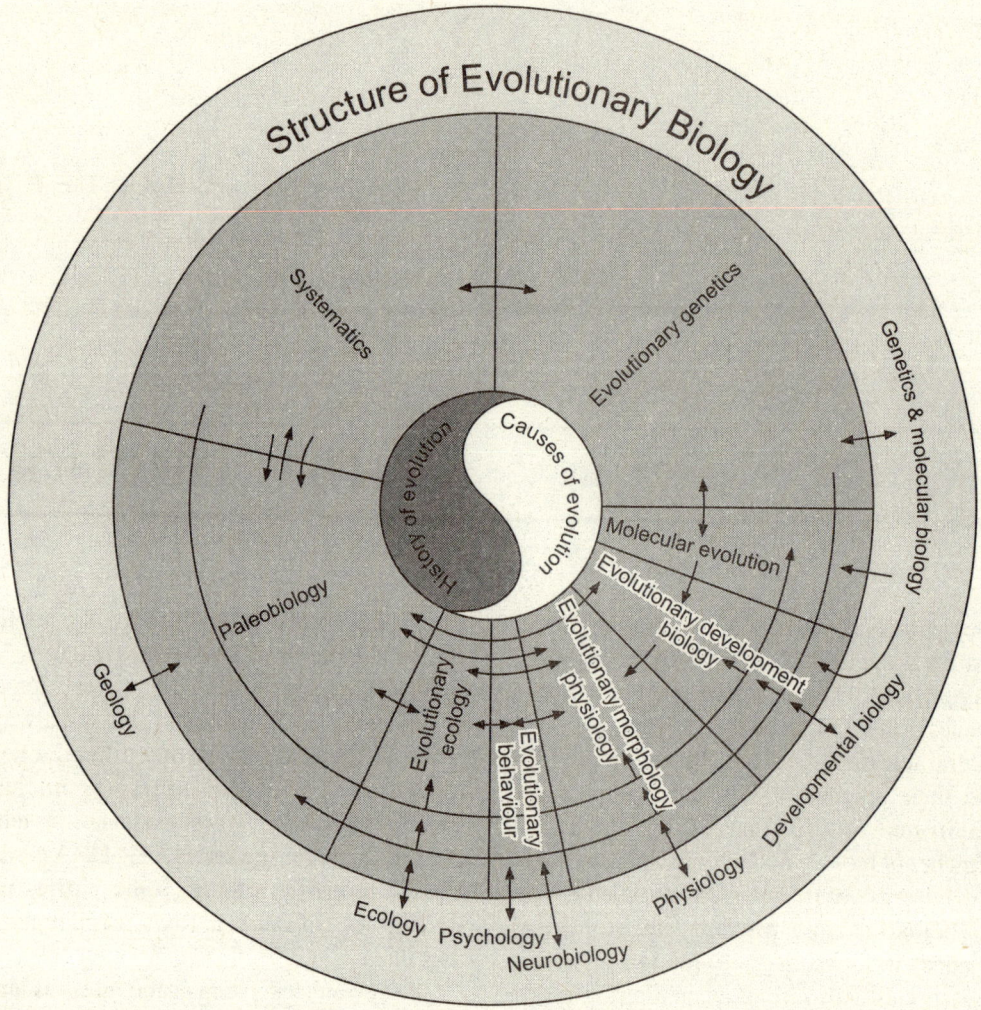

**Figure 1.1**   Structure of Evolutionary Biology

organisms as well as small-scale and large-scale changes within populations. Evolution occurs in population during transfer of heritable change from one generation to the next. Genetic variation initiates evolutionary change which is governed by natural selection. Organisms with advantageous traits enable to survive better and pass their genes into next generation through the natural selection. All life forms changes over time and various species share common ancestor. Evolutionary relationships can be represented on family tree called **phylogeny**. Small-scale evolution is the change in gene frequency from one generation to the next. Large scale evolution is the descent of different species from the common ancestor. The evolution is passing copies of genes into the next generation through reproduction, not always about progress or survival.

Evolution happens as genes mutate and individuals are selected. It may be microevolution, or macroevolution. Large

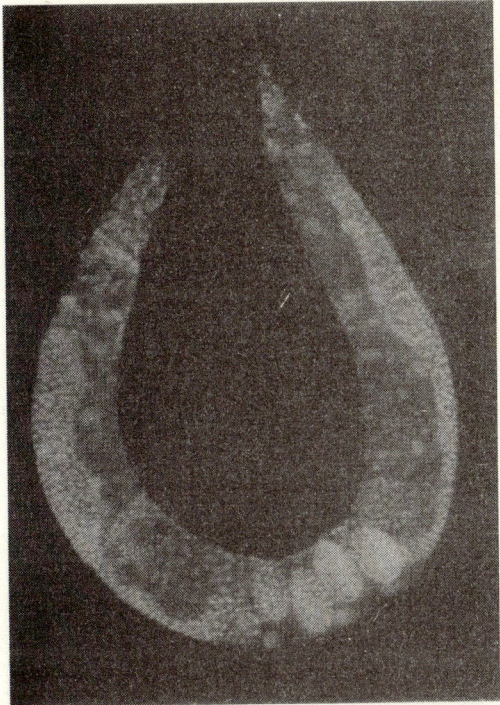

**Figure 1.2**   *C.* elegans

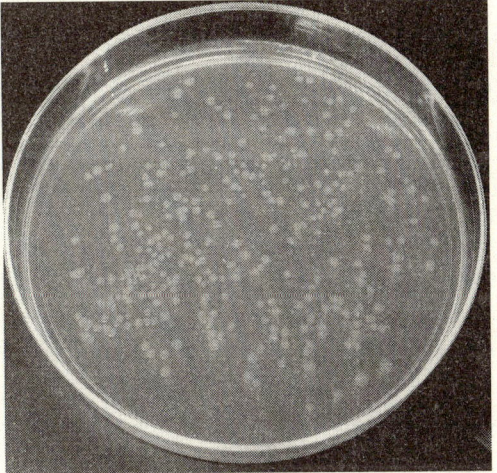

**Figure 1.4**   *E.* coli colony

is cumulative microevolution. Link between organisms via descent to common ancestor is often called evolution. Evolution unifies various disciplines in biology and explains life history strategies. Micro evolutionary theory states that natural selection optimizes the existing genetic variation in population to maximize reproductive success. This helps to interpret biological traits and their importance. Macro evolutionary theory explains functioning of living things. Organisms modify over time by cumulative natural selection. Distribution of gene based traits across group is explained by splitting of lineages and production of new traits by mutation. Evolution, the result of interactions of the potential of a species, increases species number and genetic variability in their offspring. It explains the fossil records and molecular similarities in diverse species. Different plant, animal, and microorganisms are related by descent from the common ancestor. Evolution, the change in hereditary make up of groups of organisms over generations links the present and past biodiversity. **Evolution** produces new species which Darwin called descent with modification. Existing biodiversity is the result of billion years of evolution.

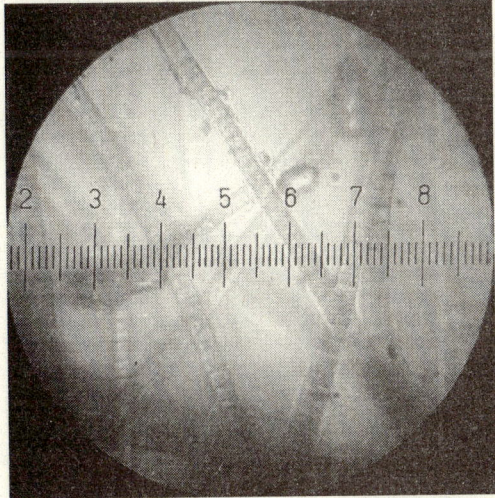

**Figure 1.3**   Cyanobacteria

change like formation of new species is the macroevolution .This is viewed as different from the micro evolution by some. Others view such difference as arbitrary - macroevolution

Evolution is a change in gene pool (total genes in a population). The English moth, an example of observed evolution had 2 color, light and dark. Dark moths were less than 2% of population before 1848.Frequency of dark moth increased later. The 95% moths in Manchester and highly industrialized areas became dark by 1898. Their frequency was less in rural area. The light colored moths became dark colored moths. A single gene primarily determines moth's colour. High incidence of dark colored moths was due to change in gene pool. The increase abundance of dark moth was due to natural selection. During England's industrial revolution in late 1800, soot released from factories darkened the birch trees inhabited by moths (Figure 1.5). In that sooty background, birds observed better the light colored moths and ate them. Dark moths survived till reproductive age and left offspring which caused their high incidence, an example of natural selection. Frequency of black moths increased due to microevolution. Population, a collection of individual is

evolving. Each individual harbors different set of traits. Single organism is not typical of entire population when variation exists within the population. Individual organism keeps the same genes throughout their life and do not evolve. Change in ratio of different genetic types causes evolution of population.

Big bang theory states the origin of universe before 10 to 20 billion years. About 4.5 billion years ago, the sun, earth, and rest of the solar system formed from a nebular cloud of dust and gas. The early earth was different from the present day earth. Evidence for the bacteria extends back to above 3.5 billion years. Evolution of life caused dramatic changes in composition of original ($O_2$ less) earth's atmosphere. Geologic time is estimated from rock sequences by estimating the age of rocks. Interaction among the solid earth, oceans, atmosphere, and organisms results in evolution of earth system. Changes through earthquakes and volcanic eruptions are noted, but processes like mountain building and plate movements take place over millions of years.

**Figure 1.5** *Biston betularia*

Fossils found in rocks of increasing age show the lineage of living things, from the single-celled organism to *Homo sapiens*. All vertebrates have a common body plan with segmented body and hollow main nerve cord along the back. This shows that all vertebrates are descended from the common ancestor and they have diverged through evolution. Comparison of difference in DNA sequences among organisms provides evidence for evolutionary events that cannot be found in the fossil record. The evolution by the random genetic variation and natural selection makes sense.

**Species** is a group of interbreeding organisms .It produces fertile offspring and cannot breed with members of other such groups. Variation is the genetically determined difference in characters of members of same species. **Natural selection** is the greater reproductive success in individuals of a species arising from genetic characters that confer advantage in an environment. Different species although look dissimilar exhibit similarity in their internal structures and common ancestry. Species acquire features through adaptation. Naturally occurring variations in populations are sometimes selected. **Adaptations** are the changes in structures, behavior, or physiology that enhance survival and reproductive success of organism. Species extinction occurs when adaptive characters of a species do not allow its survival. Evolution is the progressive change. Humans have become more upright, intelligent, and complex. Many phylogenetic lines have resulted in larger organisms and complex behaviors. Such trends show that progress is not necessary for evolution. Reduction in complexity and other signs of progress are found in about all phylogenetic lines. Parasites are the smaller, less complex, but more specialized in comparison to their non-parasitic ancestors.

The molecular clock deciphers the relationships between living organisms by examining the difference between the DNA sequences of organisms. The comparisons can be done using the proteins. Every living cell uses cytochrome c (a protein) in energy metabolism, which is identical in human and chimpanzee. An 86 % overlap in molecules is found in humans and rattlesnakes, and 58 % overlap between us and brewer's yeast. Fossils show that many organisms that lived before are now extinct.

## 1.1 Evolution: A Contemporary and Unifying Process

Evolution by natural selection is continuing. Evolution of human pathogens poses serious public health problems. **Natural selection** has amplified resistant strains caused by genetic variation. Causative agents for malaria, tuberculosis have acquired resistance to antibiotics. Use and overuse of antibiotics have effect on selection for the resistant populations. Rapid evolution is occurring in many organisms. Rats have acquired resistance to the poison. Many insects and agricultural pests have evolved resistance to pesticides. Some plants have evolved tolerance to toxic metals and reduced their interbreeding with nearby non tolerant plants. New plant species arises through cross breeding between the native and exotic plants. Many closely related species have splitted from the common ancestor recently. Fossil records shed light on speciation. Whales and dolphins have evolved from a group called mesonychids.The shape of teeth of some mammal show that they crushed and ate turtles. This mammal called *Ambulocetus* moved between sea and land with front forelimbs and powerful hind legs with large feet adapted for paddling. Later fossil in the series, *Rodhocetus* from Pakistan, show small function in hind limbs and more back flexibility. This species perhaps did not venture onto land often. *Brachiosaurus* fossils from Egypt and the US present a whale, with front flippers for steering and flexible backbone. This animal still has the hind limbs. Modern whales evolved from a group of hoofed mammals into species that were more adapted to aquatic life which are reduced in modern whales. *Homo sapiens* are the hominoid that includes orangutans, gorillas, and chimpanzees. The first members of *Homo* had evolved before 1.5 million years. Based on sequencing of DNA found in mitochondria, it is proposed that a small group of modern human evolved in Africa about 0.15 million years ago.

Evolution, a unifying concept of interdisciplinary importance is in operation. The mechanisms of evolution have changed over time. Anaximander before 2500 years thought that a gradual evolution had created the world's organic coherence, from a formless condition with transformation of aquatic species into the terrestrial ones. Darwin and Wallace simultaneously proposed mechanism for evolution in 1858. They realized that though most species produce sufficient offspring than can survive, size of overall population remains about the same. Organisms reproduce with the constraints of food, space, and other resources. The offspring often differ with one another in way that is heritable. After generations, nature selects those individuals best suited to an environment, a process called natural selection. Natural selection acts variously on individuals in a population which explains the variety of organisms and species.

In the mid-19th century, naturalists recognized several thousands plant and animal species. In the mid 20th century, biologists estimated about 2 million species. Investigations in tropical rain forests have multiplied the prior estimates at least 10 fold. Similarities among closely related species are common. Distantly related species also share anatomical and functional characters. Bones in a whale's front flippers are arranged in same way as bones in our arms. Similar developmental stages are found between the egg and embryo. Extinct species are related in various ways to organisms living today. Living organisms use the same biochemical process to carry out the basic life processes. Proteins which make up cells and catalyze chemical reactions are identical in various species. Some human genes that code for proteins differ slightly from the corresponding genes in fruit fly. Living forms use the same system to pass genetic information from one generation to another. Different organisms share many characters of structure and function as they are related to one another. Million of existing different species are related by descent from the common ancestors. Fossils of microorganisms show that life emerged on earth before 3.8 billion years and reveals profound changes in kinds of organisms that inhabited over its long history. Trilobites that populated the seas millions of years ago no longer crawl about. More than 99% species that have ever lived on earth are now extinct, due to species extinction, or species evolution.

Mendel, from the crossbreeding of varieties of peas, shows that organism acquire traits through discrete units of heredity, which later called genes. **Variation** produced through the traits is the raw material on which natural selection acts. Slight and dramatic variation occurs through mutations in genes. When a mutation enabled an organism to survive, or reproduce effectively, it tends to be preserved and spread in population through natural selection. Evolution thus depends on mutations and natural selection. **Mutations** cause genetic variation. Natural selection selects the useful change from the deleterious one. Selection through natural processes of favored variants explains many observations. Individuals of the same bird species might be larger and darker in northern part, but smaller and paler in southern part due to advantages of large size and dark coloration in forested and cold regions. If the species occupies the entire range constantly, genes for light color and small size flow into northern population and vice versa, prohibiting their separation into distinct species.

The important source of new species is the **geographical isolation** which serves as barriers to gene flow. After a time, they become so different that they cannot interbreed. In the 1950s, the molecular structure of proteins was determined. Certain proteins serve the same function in different species, with similar amino acid sequences. This evidence was consistent with idea of common evolutionary history. Discovery of structure of DNA extended the study of evolution. Sequence of chemical bases in DNA specifies the order of amino acids in proteins and determines the synthesis of specific proteins. Thus, DNA is the source of change and continuity in evolution. Modification of DNA through occasional changes, or rearrangements in base sequences emerge new traits and new species. All organisms use the same molecular codes to translate DNA base sequences into protein amino acid sequences. This suggests that all organisms alive now share the common ancestor. Selection could work on existing genetic variation of new generation, which occurs not in response but randomly over time for evolution (Figure 1.6)

## 1.2 Discovery of Missing Link

Before 1967, no specimen of wasps, or other Hymenopterous was known from the fossil record for interpreting the ancestry of ants. Many fossils of ants dating back 50 million years are known. These were different species from those existing now, but their bodies still possessed the basic body form of modern ants.

**Missing link** of ant evolution was often cited as evidence against evolution. Evolution does not design new organisms. New organisms emerge from inherent genetic variation in organisms. Genetic variation is the random, but natural selection is not. Natural selection tests the combinations of genes in members of a species and proliferates those that confer more ability to survive and reproduce. Evolution is not the simple product of random chance.

The discovery of a fossil fish Tiktaalik has shown the existence of a transitional form between fish and amphibian. The fossil is claimed to be 375 million years old. With its bony forelimbs, eyes and nostrils on a flattened head, it perhaps lifted itself with its fins while wading through streams. Several problems appear when thinking this transitional form. The bones in the forelimbs are inserted in the muscle and do not join the axial skeleton, the same bone structure as found now in living fish. Tail and pelvic fins are absent in the fossils. The presence of bony fins does not establish that the fish is developing into amphibian. The coelacanth and lungfish believed to be the ancestors to modern amphibians in the past have not been established.

## 1.3 Evolution and Ecology

Animals and plants do not live and evolve in isolation. Physical factors of environment and species together make pressure for diversification. An animal is a host for parasite, or prey. A plant competes with other plants for space and light, can shelter parasite, and provides food for herbivores. Interactions within complex community create evolutionary forces due to constraints and opportunities. Competitors among same species create constraint. An example of evolution involves 13 species of finches now called **Darwin's**

**finches** on the Galapagos Islands. Single year of drought on the islands could drive evolutionary changes in the finches. Drought reduces supply of easily cracked nuts, but allows the survival of plants producing tough nut and favors birds with strong and wide beaks that can break tougher seeds, producing bird populations with such traits. A new species of finch might arise in about 200 years. MacArthur studied 3 North American warblers of the same genus that feed on insects in coniferous trees in the same areas, often in the same trees. Macarthur's observed that the 3 were actually specialists: one fed on insects on major branches near trunk; another occupied the mid regions of branches and ate from various parts of foliage; and the third fed on insects occupying the finest needles near periphery of tree. The 3 warblers occurred together but they were not competitors for same food source. Often species that evolve together in the same ecosystem do so by highly interactive process. Natural selection favors organisms with defense against predation. Predators face selection for traits that overcome those defenses. Such co evolutionary competitions are common. Some plants produce and store chemicals that deter insects. Some insects combat with the deterrent to eat about free of competitors. Another **co evolution** is the introduction of rabbits and myxomatosis virus into Australia. After rabbits were introduced to Australia, they multiplied and threatened the wool industry as they grazed on the same plants as sheep. To control the rabbit, the virulent myxomatosis virus, was introduced into Australia. Within a decade, rabbits became more resistant to virus, and

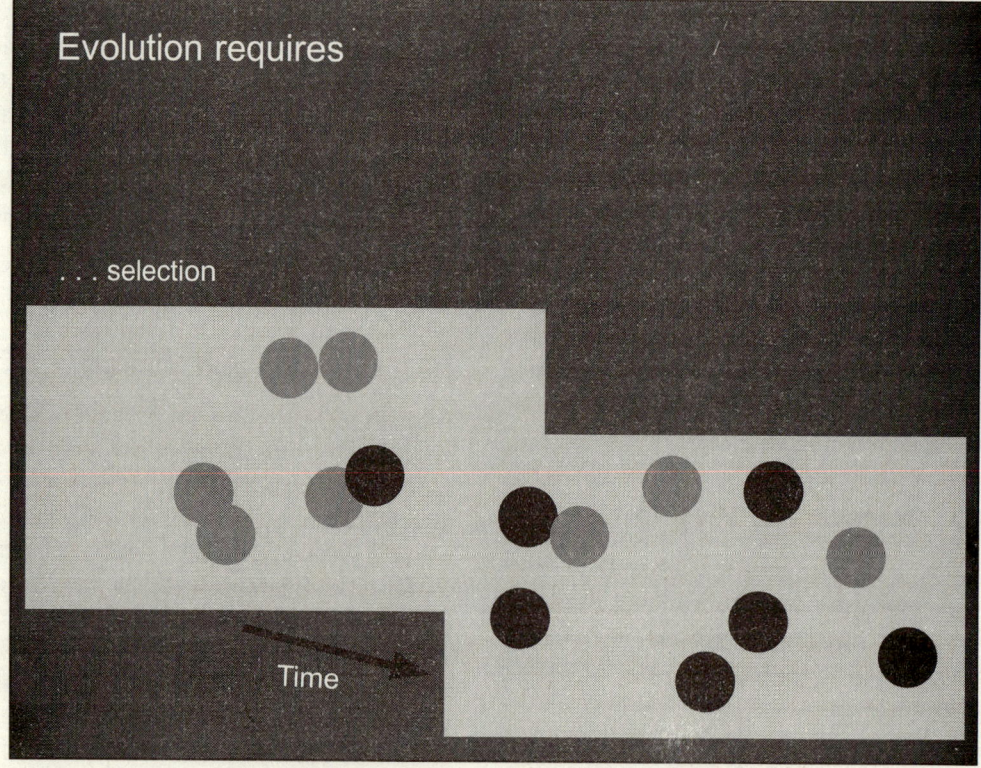

**Figure 1.6** Evolution requires selection and time.

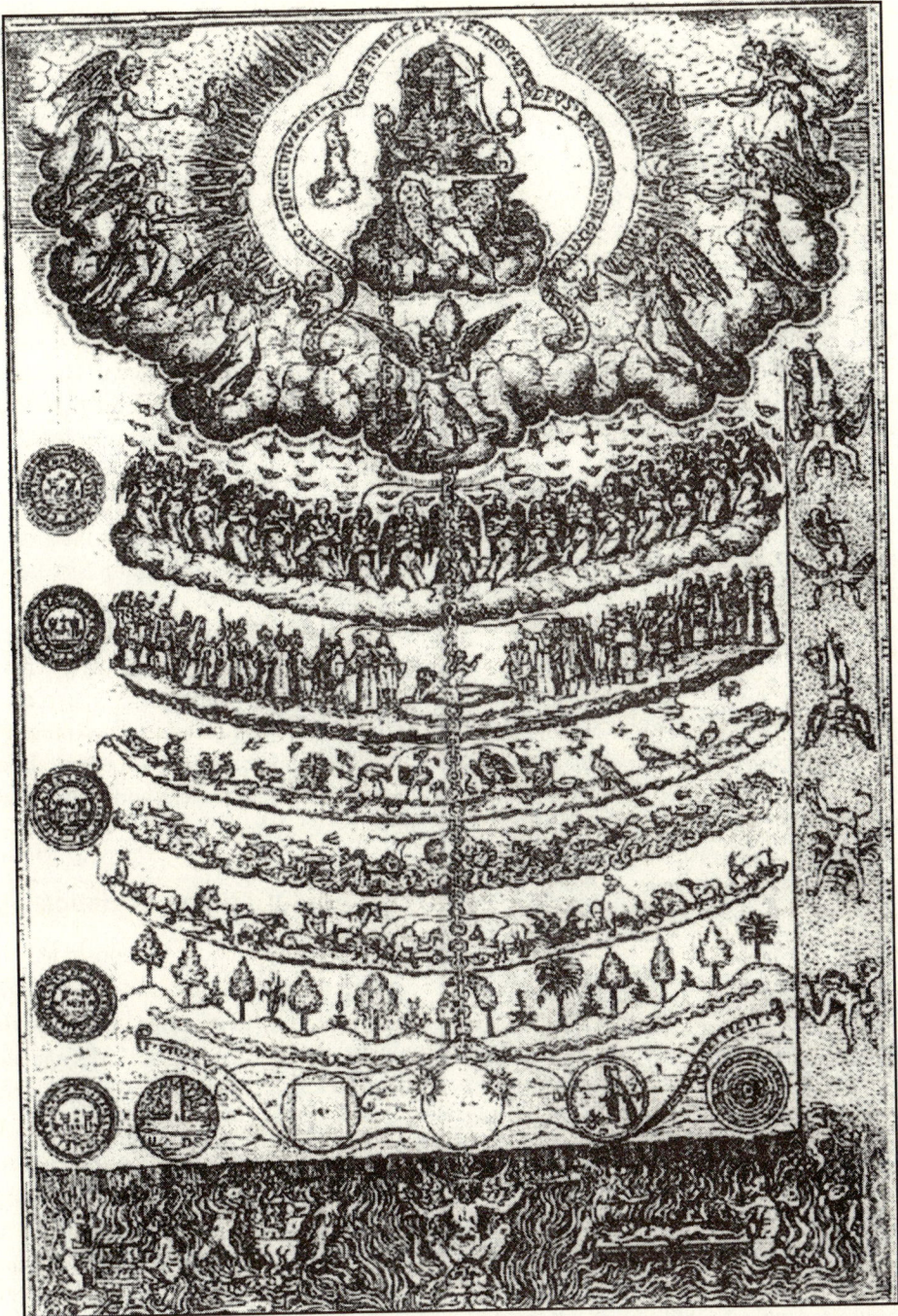

**Figure 1.7** *Scala naturae* : The great chain of being

virus evolved into a less virulent form, allowing both host and pathogen to coexist.

## 1.4.  Evolution and Taxonomy

Aristotle's *Scala naturae* (Figure 1.7) was adopted as the product of perfect creation by the early Christians. Individual can be grouped as similar kinds, or species. Many species are closely related to one another. **Classification** of species and their arrangement into groups began with publication of *Systema Naturae* by Linnaeus. Linnaeus knew 7 dog-like species, gave each a double name and identified about 9,000 species. Systematics now recognizes about 1.5 million species. Categorizing and describing the species is yet to be completed. Smaller invertebrates, many bacteria and microorganisms remain to be discovered. Number of species of plants and animals in world may be yet to be estimated.

Evolutionary relationship resembles the branches of trees. Closely related species are grouped in genera, families, and so on. The result is hierarchical diagram showing evolution of different species from the common ancestor. Under some circumstances, remains of organisms undergo chemical change in which original material is replaced by molecules that form stone. Thus, the organic remains of living things are fossilized. The orders in which the sediments are deposited, the most recent layer of rocks remain on top and the oldest layer on bottom. Fossils in each layer consist of organisms that lived at the time the layer was formed. Fossils in the lower layer shows species that lived earlier than those found in upper layers. Relative position of fossils tells which are older and which are younger. Sedimentary rocks are formed when solid materials carried by wind, or water accumulate in layers and are compressed by overlying deposits. Sedimentary rocks contain **fossils** formed from the parts of organisms deposited along with

other solid materials estimate the difference in ages of 2 fossils by noting the thickness of rock that separates them. Many soft-bodied invertebrates like worms and jellyfish lack hard parts like bones and shells, and decay without becoming fossilized. For mammals, birds, reptiles, and amphibians, death is followed by the skeleton being dismembered and the bones scattered. Tiny fossils first reveal the presence of bacteria 3.8 billion years ago, and animals consisting of more than a single cell are known from about 670 mya. The organisms that lived between these 2 dates lacked hard parts and were rarely preserved as fossils. Then, about 570 mya, a dramatic change took place at the beginning of the Cambrian, with animals having calcified shells and other body coverings got better chance of becoming fossilized. The Cambrian seas were populated with various invertebrates. The earliest vertebrate fossils are known from about 500 mya. Thereafter early amphibians and reptiles appeared. Birds and mammals appeared in fossil record about 200 mya, and dinosaurs appeared about 225 mya and disappeared suddenly about 160 million years later.

## 1.5  Evolution and Genetics

Mendel's work failed to demonstrate heritable genetic change as it could not explain the appearance of novel traits. He only worked the operation of inheritance. Morgan showed the creation and inheritance of genetic change. Morgan bombarded the fruit fly population with X-rays and discovered a white eye mutation in a red eyed population. Emergence of white eyes shows the development of novel traits. The mutation occurred in a male fly. Morgan bred the white eyed fly with a red eyed fly and found none in 2nd generation with white eyes. He then cross bread the 2nd generation flies and found some males with white eyes, showing both inheritance of genetic changes and sex linked inheritance. From this study, Morgan ended the location of gene on specific chromosomes. It was unknown that DNA is the carrier of genetic information until 1943. Avery discovered genetic material, and then work began to figure out the exact structure of material. In 1948, it was determined by using X-ray diffraction that DNA had some helical structure. In 1953, Watson and Crick described the structure of DNA.

## 1.6 Evolutionary Relationships in 3 Domains of Life

In 1977, Woese identified a 3rd domain of life, named Archaea, based on ribosomal RNA (rRNA) sequence comparisons. The other 2 domains are Eubacteria and Eukarya. Archaea shares some physical features with Eubacteria but tend to live in environments which were supposed to be prevailed on early earth. Archaea was the ancestor to both Eubacteria and Eukarya. The Archaea is genetically and biochemically different from Eubacteria, and are not considered as ancestor of Eubacteria. Archaea is genetically more similar to eukaryotes than Eubacteria and is often treated as a sister to Eukarya. Eukarya genes appear to evolve from both Archaea and Eubacteria, and a genome fusion is proposed. Archaea gave their informational genes to Eukarya, and Eubacteria have given Eukarya their operational genes. Rather than an evolutionary tree of life, a ring of life is suggested. The Archaea and Eubacteria fused using the processes of endosymbiosis and lateral gene transfer to cause Eukarya.

## 1.7 Analogous Structure

Structures that serve a common function in various species, with different ancestor are called analogous structures and the phenomenon is called **analogy**. Wings have same function in insects and birds, but evolutionary similarity between them is absent. Wings of birds and bats look similar in structure though they evolved separately, but are related as they are adapted to similar environment. This process is called **convergent evolution**. Analogous traits are adopted due to this evolution. The tusks of elephants and the gnawing front teeth of the beaver, both are the incisors teeth. They are inherited from common ancestor, but modified in course of evolution according to the use. Both new and old world vultures look similar. Both have featherless necks and heads and feed on carrion. Old vultures belong to eagle's family. New world vultures belong to stroke's family. Old vultures have eyes to find food. New vultures use olfactory senses and sight to search food.

Analogy is different from homology where the structures are similar because they have common embryonic origin. Limbs of tetrapods and arthropods are analogous. The limbs of arthropods appeared after the Cambrian explosion. Tetrapods evolved from fish about 370 mya. The limbs of arthropods and tetrapods were evolved separately. The legs of vertebrates and insects serve the same purpose with different structure and evolutionary history. They have 2 different origins. Two insects of the same species look similar because of the same color, or spots. Bat and bird appear similar for their wings. Human, whales, and lizards have similar skeleton structure with different habitat and lifestyle. Humans throw a ball using hand. Whales use fins for swimming. Lizards use limbs to climb. Each one of them is similar in structure, but different in details. Analogy states that the structures are similar not due to origin, but due to functional similarities. Analogy appears due to similar challenges and problems faced by species. Evolution then works for development of structures.

## 1.8 Homologous Characteristics

Homologous characteristics are used to build the phylogeny. A homologous characteristic is one that may appear different in 2 dissimilar species, but is fundamentally same and indicates a common ancestry. An example would be the homologous characters of 4 limbs (tetra pods). Birds, bats, mice and alligators have 4 limbs. Both the ancestor of tetra pods and their descendents have inherited the feature of 4 limbs. So, the presence of 4 limbs is a **homology**. Birds and bats have wings, but mice and alligators have not. The wing of a bird shows a major difference to that of a bat. Bat wings have skin stretched between their bones. Birds have feathers extending along their arm. This suggests that they didn't inherit wings from the common winged ancestor. They are similar in function, but different in structure. Bird and bat wings are analogous, having separate evolutionary origins, but they serve the same function. Sometimes this is called convergent evolution.

Adaptations do not always provide clear selective advantage. Adaptations from molecular to organismal may not be perfect, but adaptations providing a selective advantage must be good for survival and reproductive fitness. The punctuated equilibrium provides another tempo of speciation in the fossil record of many lineages. It does not refute or overturn the evolutionary theory, but rather increase scientific richness.

## 1.9 Evolution of Feather

The origin of birds has been a major problem for Darwinism with issues like evolution of feathers. Feathers are the complex designed structures used in flight and are today found only on birds. Even though feathers are found

in the Cretaceous Period in amber, the fossil record does not provide strong evidence for feather evolution. The feathers have allowed birds to spread to every corner of the globe. Their songs, coloration, and flight make them a marvel. Each feather consists of main shaft, barbs, and barbules, which possess small hook lets and grooves to hold the feather together. The nervous system controlling flight is complex. The muscles, lungs, tendons, feathers and brain must have evolved at once. Each part may not provide a survival advantage. The appearance of feathers suddenly in the fossil record supports the special creation theory. The origin of flight in evolutionary hypotheses ranges from tree-down gliders to ground-up flyers. Feathers and bird bones are abundant in the fossil record, with 79% of the 329 living families known from fossils. Transitional structures between scales, skin, and feathers are not known, suggesting the sudden appearance of birds.

## 1.10 Dinosaur's Evolution into Birds

Disappearance of dinosaurs in a short time is a major puzzle. Their wide distribution and adaptation to diverse climates deepens the mystery. Up to 9 major extinction events are supposedly preserved in the fossil record. The extinction that occurred at the end of the Cretaceous was the most extensive. Different theories on extinction consider temperature, nutrition, and food supply. The meteorite extinction theory is the most prevalent. A large meteorite struck the earth, forming debris in the air that cooled the atmosphere. Iridium concentration around the globe, presence of shocked quartz, and other evidence provide support to the theory. The cooling from the debris perhaps caused the **extinction** of the dinosaurs. Massive volcanic activity caused the cooling effect, and released toxic gases for killing the dinosaurs. Dinosaurs simply evolved into the birds. The theropods, including *T. rex* and *Deinonychus* are the probable ancestors. Reptiles are cold-blooded, and birds are warm-blooded. Birds have highly complex mechanisms to regulate body temperature. Some believed that dinosaurs were warm-blooded without convincing evidence. Dinosaurs are divided into saurischian and ornithischian based on structure of the pelvis. The lizard-hipped theropods are supposed to evolve into the birds.

## 1.11 Raw Materials of Evolution

Darwin's theory of natural selection was not widely accepted, even within scientific community. The rise of Mendelian genetics, beginning in 1900 with rediscovery of Mendel's experiments with garden peas, convinced many scientists that mutations were the real engine of evolution. This is **discontinuous evolution**, whereas evolution by natural selection is **continuous evolution**. Mutations are the basis for evolutionary change. Hugo De Vries proposed that macro mutations were the real engine of evolution. Some geneticists began studying variation in crop plants, such as corn and wheat. Unlike Mendelian variation, in which certain traits show either dominance or recessive ness, most traits were the result of many genes coding for same traits. This produced continuous variation, necessary to create the "raw material" for natural selection. Evolution, generally the progressive change over a long time by which later organism differs considerably from the earlier ones, leads to formation of new species. Various fossil remains are found in different strata of the earth, which are dated by various techniques. Earlier strata contain fossil forms that later strata lack, and vice-versa; and generally, earlier forms are structurally simple than later ones.

Species appear and disappear. Human and many other species now exist on earth are relatively late comers. Life begets life, and no life emerges from non-life. Since we have not monitored species causing new species in our lifetimes, this inference is based on indirect evidence. Biologists rely on fossil discoveries to support the idea of species change and also point to metabolic, genetic, morphological, and other uniformities, to strengthen this first hypothesis.

Probably, similar organisms must have descended from common parents, but completely independent parallel trees of life occurred under uniform natural laws. Direct body of evidence for changes possible in life is experience of plant and animal breeding by men, creating varieties with little similarity to the originals. Examples are found in change of thin wild grasses to rich domestic wheat or barley, or wolves into dogs of all shapes, sizes and colours. Nowadays genetic engineering provides evidence of possible change.

## 1.12 Evolutionary Trends

Evolution spans nearly 4 billion years associated with the origin and extinctions of millions of species and the struggles of many individual organisms. Evolutionary trends are identifiable patterns with the overall evolution of a trait in a given direction in a group for a long period. The trend is the single most important component in evolution (McKinney, 1990). Some view it as the only consistent, directional change within lineages. A long-term directional change in a summary statistic for a clade, such as the mean (McShea, 2005) and a directional character gradient through time in a well-defined monophyletic clade (Gould, 1990) are also called traits. **Trends** are the primary phenomena of evolution at higher levels and longer time scales (Gould,

2002). Its nature, generality, underlying causes, and significance should not be overlooked, or overstated. McShea (1998) listed 8 potential large-scale trends, including overall directional changes in "entropy, energy intensiveness, evolutionary versatility, developmental depth, structural depth, adaptedness, size, and complexity." The patterns of change involving increases (McNamara, 2006) in body size and morphological complexity are the well known and form the basis of most of the examples (Erwin, 2014). Organisms today are larger and more complex than they were in the past. In the start, all lives were almost small and simple, but the largest and most complex species are still alive today. An increase in the average value of a given trait counts basically as a trend, and biological systems are often characterized by extensive variation. Comparisons of fossils and inferences drawn from phylogenetic analyses are necessary to establish the presence of a trend. The Cope's Rule is often taken as a global trend that applies to many lineages. More detailed analyses of particular groups have shown it not to apply sometimes (Jablonski, 1997), or to be local rather than global. A large-scale evolutionary trend is a pattern of directional change occurring over long periods. The detection of such a pattern does not provide an explanation for it. The question may arise about the patterns of change that have occurred within component lineages over short timescale. In particular, whether the internal dynamics of the larger trend have involved consistent change in all lineages, or whether the trend represents the net outcome of a more complex internal dynamic (McShea, 1994) is a question. One may investigate the causes which must have a basis behind the internal dynamics that add up to a trend. Evolutionary biologists have developed a series of analytical methods for understanding the dynamics underlying trends, the causes that generate them, and both adaptive and non-adaptive bases behind the causes.

## SUMMARY

1. Knowledge about gene, cell, and species helps understand the evolution. Interaction among genes; cells, living matters, and environment are manifested as life.

2. Evolution generates new species with various adaptations in varied environments. Living organisms are organized structures. Many cells make multicellular organism.

3. Evolution is a pathway of existing biodiversity from common ancestors. It helps explain similarities and differences in organisms as well as small-scale and large-scale changes within populations. Evolution occurs in population during transfer of heritable change from one generation to the next. Genetic variation initiates evolutionary change which is governed by natural selection. Organisms with advantageous traits enable to survive better and pass their genes into next generation through the natural selection. All life forms changes over time and various species share common ancestor.

4. Small-scale evolution is the change in gene frequency from one generation to the next. Evolution happens as genes mutate and individuals are selected. Macro evolutionary theory explains functioning of living things. Organisms modify over time by cumulative natural selection .Evolution, the result of interactions of the potential of a species increase species numbers, genetic variability in their offspring. It explains the fossil records and molecular similarities in diverse species.

5. Comparison of difference in DNA sequences among organisms provides evidence for evolutionary events that cannot be found in the fossil records. The evolution by the random genetic variation and natural selection makes sense. Natural selection is the greater reproductive success in individuals of a species arising from genetic characters that confer advantage in an environment.

6. Different species look dissimilar exhibit similarity in their internal structures, chemical processes, and common ancestry. Evolution by natural selection is continuing. Evolution of human pathogens poses serious public health problems.

7. More than 99% species that have ever lived on earth are now extinct, due to species extinction, or species evolution. Distantly related species share anatomical and functional characters. Modification of DNA through occasional changes, or rearrangements in base sequences emerge new traits and new species.

8. Natural selection explains change in organisms and adaptation explains why they change. Structure and function of an organism are directly related to its environment. Numerous environmental mechanisms influence adaptive evolution.

9. Evolution spans nearly 4 billion years associated with the origin and extinctions of millions of species and the struggles of many individual organisms. A large-scale evolutionary trend is a pattern of directional change occurring over long period.

# REVIEW QUESTIONS

## Short Answer Questions

1. What is microevolution?
2. What is macroevolution?
3. Define Natural selection.
4. Define species.
5. Comment on the missing link in evolution

## Long Answer Questions

1. Why evolution is a unifying concept?
2. Discuss the relationships between evolution and ecology
3. Describe the relationships between evolution and taxonomy.
4. Comment on the homologous structure.
5. Comment on the analogous structure.
6. What are the raw materials of evolution ?

# REFERENCES

Erwin, D. H. 2014. Evolutionary Trends. Oxford Bibliographies.

Gould, S.J. 1990. Speciation and sorting as the source of evolutionary trends, or 'things are seldom what they seem'. In: McNamara KJ, editor. Evolutionary trends. Tucson, AZ: University of Arizona Press. p. 3-27.

Gould, S. J. 2002. The structure of evolutionary theory. Cambridge, MA: Harvard University Press

Jablonski, D. 1997. Body size evolution in Cretaceous molluscs and the status of Cope's rule. Nature; 385:250-2.

McKinney, M.L.1990. Classifying and analysing evolutionary trends. In: McNamara KJ, editor. Evolutionary trends. Tucson, AZ: University of Arizona Press. p. 28-58.

McNamara, K.J. 2006. Evolutionary Trends. Encyclopedia of Life Sciences. John Wiley & Sons, Ltd.

McShea, D.W. 1994. Mechanisms of large-scale evolutionary trends. Evolution; 48:1747-63.

McShea, D.W. 1998. Possible largest-scale trends in organismal evolution: eight "live hypotheses". Ann Rev Ecolog. Syst; 29:293-318.

McShea, D.W. 2005.The evolution of complexity without natural selection, a possible large-scale trend of the fourth kind. Paleobiology; 31(Suppl.):146-56.

# CHAPTER 2

# Evolutionary Biology and Its Applications

Thales (640-546 B.C.) believed that all life came from water. Xenophanes (576-480 B.C.) and Herodotus (484-425 BC) recognized the remains of living forms as fossils .Democritus (500-404 B.C.) differentiated between organic and inorganic systems. Empedocles (495-435 B.C.) detailed explanation of origin of life with fire, air, earth and water, and 2 forces- love and hate. Proper mixture of materials, forces and body parts was believed to create a functional living organism .An improper mixture resulted in reduced survival.

Hippocrates (460-370 B.C.) stated the inheritance of acquired characters. He advocated the principle of use and disuse that states a part in use become elaborate and a part not in use is lost. Variation was due to differences in environments and habits .Plato introduced the idea of animate or living cosmos where all physical and living systems were believed to be linked into a harmonious whole - a supernatural harmony that could be understood through pure thought. Aristotle (384-322 B.C.) rejected metaphysical

speculation, but accepted the idea of fixity of species. He attempted to find principles that agree with observed facts and developed the beginning of biology. He first used the gradations in form among organisms to arrange them into a ladder like scale with man at the top of living world - a teleological system.

Early Christians adopted Aristotle's *Scala naturae* as the product of perfect creation and unchanging. During the middle ages, the Arabs carried on the Greek tradition of natural explanations for biological phenomena. The Christian church in Paris in 1209 banned Arabic literature dealing with natural explanations. Archbishop James Ussher (1581-1656) calculated that the earth was created 4004 BC on Sunday, October 23. Real science emerges with explanations of physical processes in the Age of Enlightenment (1650-1800). Science then slowly spread into the study of life. In 1600s, Francis Bacon (England) and Renee Descartes (France) suggested that natural laws surpassed Christian dogma in explaining natural phenomena. They suggested that species change through the actions of physical processes.

Voltaire (1694-1778) questioned the idea that species were fixed quantities and stressed the variation within species. He questioned the idea of great chain of being with humans being the ultimate living things. He saw living things as all part of a continuum of equal. Comte de Buffon (1707-1788) studied fossils, noted changes in fossil record and involved both improvement and degeneration of parts. Calculated from rates of sedimentation and thickness of sedimentary deposits, the earth was estimated to be more than 160, 000 years old. *Histoire Naturelle* (1749) shows that all species have an internal mold which organizes the body and arose by spontaneous generation. He states that similar species share a common ancestor and modified by climate. James Hutton (1726 - 1797) using physical and geologic processes concluded that the earth was inconceivably old. Geological observations should be explained by uniformitarianism. Charles Lyell (1797-1875) championed uniformitarianism. In the 1700s, geologists studying fossils noticed many forms that were no longer living. Similarly, many forms in deep geological layers persisted through many layers and then disappeared. Geology established the idea of an old earth that had changed through time.

**Catastrophism**, the view before uniformitarianism, suggests that major catastrophes and perhaps unknown forces accounts for major geological features and abrupt changes in the fossil record. It was consistent with Christian beliefs about creation and destruction. Some speculated that there had been multiple creations followed by destruction.

Erasmus Darwin (1731-1802), the grandfather of Charles Darwin recognized organisms' struggle for existence, speculated about change in species through time, suggested

human's origin from primitive species, and inheritance of acquired characteristics. Carolus Linnaeus (1707-1778), in *Systema Naturae* (1735) (Figure 2.1) provided a classification of plants and animals with species organized into genera, and into families, orders, etc. He did not believe in change or common descent and saw all species as fixed entities. Linnaeus's purpose was to discover the pattern in God's creation.

**Figure 2.1**   Carolus Linnaeus

Jean Baptiste Lamarck (1744-1829) (Figure 2.2) was interested in the apparent close fit between organisms and their environment. He suggested species could change in response to their environment. He rejected Aristotle's great Chain of Being as due to a single change of progress and accounted for the apparent change by transformation -successive rounds of spontaneous generation. Lamarck's model did not allow for extinction of forms His ideas were published in Philosophie Zoologique

(1809). Lamarck proposed a natural and testable hypothesis about the evolution. He noted that organisms interacts with the environment and adaptations could account for complex adaptations. He considered diversity from the lowest forms of life to the advanced and included humans in his scheme of evolution although did not attempt to test his ideas. Lamarck's ideas were highly criticized by many.

**Figure 2.3**   Georges Cuvier

**Figure 2.2**   Jean Baptiste Lamarck

Georges Cuvier (1769-1832), (Figure 2.3) is well known for his studies of vertebrate anatomy. He argued about the lack of evidence for change of species, and promoted the idea that species were fixed entities. Because of the apparent age of earth, fossil diversity, and geographic variation within species, many began to revise their views of a static and unchanging living world. Darwin (Figure 2.4) had an interest in natural history from childhood. At 16, he entered Edinburgh University to study medicine, but could not stomach surgery. His second option was to study for the clergy at Christ's College, University College, Cambridge University. His interests in natural history continued and he

**Figure 2.4**   Charles Darwin

became friends with Henslow and Sedgwick. Upon graduation (1831 - age 22) he received an invitation to become the naturalist on the H.M.S. Beagle. The Beagle expedition aimed to map the coastline of South America and then continued through the Pacific, around the world. His job was to collect plants, animals, and rocks to document the natural history. The trip took 5 years. Many natural theologians were investigating the lives of animals and plants then. Geologists were documenting fossil change and stirring up doubt about age of the earth and literal truth of genesis. Physical scientists were favoring testable natural explanations. Natural theologians was discussing about the change in living things through time.

Ideas about biological change and interactions between organisms prior to Darwin influenced his thinking, but origin of modern evolutionary biology can be traced back to Charles Darwin. Darwin spent as much time as possible on land. He had many inland trips in South America, some lasting weeks. He collected many specimens and many were sent to England from South America before the Beagle went to the Pacific. Darwin's collections were well known before he returned to England in late 1836. Each continent had its own unique plants and animals. Even, when they have similar climates the organisms are different. Australia has marsupial mammals. South America has many placental mammals. Armadillos are found only in the new world. The only mammals in New Zealand were bats.

Fossils found in an area are most similar to living forms in that area. e.g. *Glyptodon* and Armadillo. Deeper stratigraphic layers had plants and animals that were different from living species of shallower stratigraphic layers. Oceanic islands had their own unique plants and animals. Plants and animals on oceanic islands were similar to species found on nearest continent. Species found in each area seemed to be well suited to the environment they inhabit. His observations made him question the fixity of species. He knew the application of **uniformitarian** principle in geology and

wondered if there were physical explanations for his biological observations.

Gould examined birds from the Galapagos, and told Darwin about the variety of mockingbirds. Darwin knew that each of the Galapagos Islands had its own unique form of tortoise. These facts compelled Darwin to consider the possibility that each of mockingbirds was descended from a single ancestor. Uniformitarian thinking required the mechanism for past events in terms of processes that could be demonstrated in the present. Darwin hypothesized that evolution had occurred. Species had changed through time. Darwin mapped coral reefs in the Indian Ocean, observed the forest inhabiting species in Brazil, collected birds and tortoises from the Galápagos Islands. After his return to England, Darwin sketched his theory of evolution by means of natural selection which shapes life. Darwin observed the variations in barnacles, crustaceans and published monograph (two-volume) and showed the variation in barnacles in size and shape. Darwin found striking similarities in embryos of animals like fish and birds and astonished by similarities in behavior of humans and apes. Armadillos were the descendants of glyptodonts. Species evolved on each land mass independent of other land masses and to be well suited for conditions in which they lived. Two pieces of evidence lead Darwin to an explanation of how evolution had occurred.

1. Selective breeding of plants and animals produce new varieties.

2. The extreme potential for animals and plants to multiply results in struggle for existence. Thomas Malthus wrote "Essay on the Principle of Population" (1798) and noticed that human and animal populations have the potential to grow geometrically: $2 \rightarrow 4 \rightarrow 8 \rightarrow 16 \rightarrow$ etc. And in doing so, they outstrip their food

supply. Populations remain relatively stable as the birth rate increases so does the death rate. Increased population growth results in suffering in members of the population. Darwin concluded that there is a **struggle** for **existence** and mutationism emerged from the Mendelian genetics, which showed the inheritance of discrete characteristics. As characters and their inheritance appeared discrete, the gradual evolution that Darwin had described was rejected by mutationists.

Mutations and their causes were studied by Hugo De Vries (1848-1935) (Figure 2.5) and Thomas Hunt Morgan (1866-1945) (Figure 2.6). Each rejected natural selection and argued that new species arise as mutations. Richard Goldschmidt (1878-1958), an advocate of mutationism is known for his idea of hopeful monsters. Major mutations mostly lethal, result in large changes in body form, but some survive

**Figure 2.6**  Thomas Hunt Morgan

to produce new and very different type of organisms. Before Mendel's laws were rediscovered, many biologists were studying continuous variation because many traits showed **continuous variation**. They invented mathematical tools to describe and study continuous variation. This was the beginning of statistics. These biologists were biometricians who were on right path, but were sidetracked after rediscovery of Mendel's laws, as Mendel's laws seemed to apply only to discrete variation.

One of the first biometricians Francis Galton (1822-1911), a cousin of Darwin, was the supporter of idea of evolution by natural selection. He pioneered studies of heritability of behavioral characteristics and coined the phrase "**nature versus nurture**" to explain the importance of inheritance and environment in explaining human behavioral traits. The biometrics and Mendelian view points were rectified when quantitative traits with

**Figure 2.5**  Hugo De Vries

continuous variation was showed to account for polygenic inheritance. Many genes, each influencing the phenotype and each inherited in a Mendelian fashion produced continuous phenotypic variation. Mendelian genetics had grown to encompass polygenic systems and population genetics was developed.

## 2.1   Trends in Evolution

Horses since the late 1800s provide an unambiguous demonstration of evolutionary trends over their 55 million year history, including an increase in body size, reduction in number of toes and adaptations of teeth for grinding (MacFadden, 1992). Analysis of horse fossils over the past 20 years has shown that the first 35 million years of their evolution involved no significant change in body size. The trend toward larger size resulted from an increase in size as diversity of the group expanded. Reductions in size also occurred in some genera (Gould and MacFadden, 2004). Rather than a linear progression toward larger body size, **fossil horse** macroevolution is characterized (https://laelaps.wordpress.com/xmlrpc.php) by 2 distinctly different phases. From 55 to 20 mya , primitive horses had body sizes between 10 and 50 kg. From 20 mya until the present, fossil horses were diverse in their body sizes. Some clades became larger those that gave rise to *Equus*, others remained relatively    static    in    body    (http://creationsafaris.com/crev200503.htm) size, and others became small over time. Comparing modern genera (*Equus*) with the earliest members (*Hyracotherium*) may reveal an average increase in size, but this provides an oversimplified view.

### *Dynamics in Trend*

McShea (1994) defined 2 kinds of trend dynamics- driven and passive. **Driven trends** are those underlying dynamic which is homogeneous, whereas passive trends result from dynamics that are at least locally heterogeneous. The dynamics of driven trends occur primarily in one direction and apply to most component lineages. The **passive trend** is the result of complex dynamics operating in different directions in different lineages, or at different times (McShea, 1994). The characteristics of changes in component lineages, or at smaller time scales underlie a large-scale trend. If the dynamics occur consistently in one direction, then the trend is driven; if the dynamics vary, then the trend is passive. A driven trend in which descendant species consistently have larger bodies on average than their ancestors, or a passive trend in which lineages begin at small size such that a bounded increase in variance results in an increase in average size.

A driven trend caused by natural selection operates among individuals. A passive trend resulting from a developmental constraint limits change in one direction. Natural selection for larger body size on larger individuals help them to be more effective predators or acquiring better territories relative to smaller individuals, or developmental constraints related to limitations on organ function at very small sizes. Passive evolutionary trends may represent increases in overall diversity in component lineages over time, but one in which expansion is limited to a single direction-for example, if ancestral species exhibit small bodies near the minimum possible size then diversity can only expand in direction of larger maximum size (Stanley, 1973). Identifying a trend as driven may not automatically imply that it results from adaptive change, but it does highlight the need to investigate this possibility further. Discovering more complex dynamics while evaluating whether a trend is driven or passive may help to direct further investigations, for example, by showing which lineages follow the trend and which do not, with differences allowing hypotheses to be formulated and tested regarding the causes of dynamics. Perhaps most importantly, the driven versus passive classification can have implications for how the evolutionary process is interpreted in broad sense. McShea (1994) noted that driven trends permit extrapolations from small scale to large-scale and vice versa, whereas passive trends do not. For example, a trend caused by consistent( Gregory, 2008) patterns of change occurring within populations (microevolution) implies that large-scale evolutionary patterns (macroevolution) have been the result of small-scale processes amplified through deep time, whereas a trend resulting from higher-level processes above the species level requires an expanded, hierarchical view of macroevolution (Gould, 2002). This has significant implications for the degree to which future patterns of evolution can be predicted. Raising the example of a presumed trend toward increased intelligence in primate lineage, which is often assumed to imply that further increases in intellectual prowess can be expected in human lineage over time, McShea (1994) noted that:

*If such a trend in primates exists and it is driven, that is, if the trend is a direct result of concerted forces acting on most lineages across the intelligence spectrum, then the inference is justified. However, if it is passive, that is, forces act only on lineages at the low-intelligence end, and then most lineages will have no increasing tendency. In that case, most primates species-especially those out on the right tail of the distribution like ours-would be just as likely to lose intelligence as to gain it in subsequent evolution .Determining whether trends are driven or passive is an important aspect of their study. To this end, evolutionary biologists have developed several tests that can be applied to fossil data to address this question. Often these are used together, in part because no single test provides a conclusive designation on its own.*

## Causes of Trends

Whether the dynamics underlying a particular trend are driven, passive, or some combination of both, they in turn call out for an explanation based on identification of their underlying causes. There are numerous processes capable of causing either driven or passive trends, which by large are not mutually exclusive and may interact in interesting ways (Gould, 2002). Some of these relate to process operating within populations, or what can be considered standard neo-Darwinian evolution, and may involve either external factors or internal ones. Still others exert their influence only at higher levels, such as through sorting among species, and are, therefore, part of a broader, macro-evolutionary view of evolution (Gould, 2002). There is disagreement among evolutionary (Gregory, 2008) biologists as to whether population- or species-level processes predominate in creation of most large-scale trends (Hallam, 1998), but it is worth considering the various possible causes that have been proposed.

## New Trends

A single glass beaker can contain billions of microbes with vast variation on which natural selection can work. A thousand generations of bacteria may span a few weeks. The experiments on microbes shed light on molecular changes of organisms during adaptation to new challenge. They reveal that natural selection can alter behavior and the social relationships among microbes over a week or months. Scientists observed the origin of new species. Richard Lenski set up one experiment in 1988 with a single *Escherichia coli*. *E. coli* had emerged as the best-understood microbe. It grows quickly on a sugar diet. Lenski challenged them with cycles of feasts and famines to see if the bacteria would be altered by natural selection. In each generation, some **bacteria** would mutate. Few mutations might help them grow and reproduce faster which would out compete the other bacteria. In time, natural selection might transform the bacteria. Lenski separated the bacteria into 12 identical lines, each keeping in seperate flask. He preserved the history of evolution. Microbes can be frozen without killing for years even decades. Once the bacteria thaw out, they resume life. So Lenski and others began freezing some *E. coli* from all 12 lines in experiment every 500 generations. This frozen bacterium helps Lenski to measure evolutionary change precisely than was possible in early study. He and his students can thaw out the bacteria from early in history of a line and then put them in Petri dish with their descendants. The scientists then observe how fast the 2 populations of bacteria grow under similar conditions. They also sequence the bacterial DNA at each stage of experiment to pinpoint the mutations that were favored by natural selection.

By the early 1990s, Lenski had clear evidence about the evolution of bacteria. They were growing faster than their ancestors and it wasn't just one line that grown faster, but all 12. The longer experiment run, they evolved more. So he moved the bacteria to new flasks every day, kept building up his frozen fossil record. After 50,000 generations, the bacteria grow more than 75% faster than they did at the beginning of the experiment. The rate of improvement has slowed down, but they are getting better. The bacteria in 12 lines have become about twice as big as their ancestors, perhaps due to the side effect of mutations that made them better able to survive their daily famines. Lenski discovered more **parallel evolution** when he began to zero in on some mutations that had arisen in bacteria. He and his colleagues found few key mutated genes in about all of the lines.

However, the similarity was not exact. Each line had various mutations of the same genes. The evolved *E. coli* then faced a new challenge. Scientists changed the bacteria's diet from glucose to maltose. Some lines of *E. coli* could not feed on maltose. Other lines thrived on new food. Bernhard Palsson and others fed *E. coli* glycerol. After 44 days of Palsson's experiment, the bacteria were growing twice as fast as their ancestors. Palsson sequenced the genomes of some bacteria, starting with ancestor and end up in final generation. He and his colleagues pinpointed all the mutated genes in each line, inserted copies of mutated versions of genes, one at a time, into ancestral bacteria. Sometimes, these evolved genes immediately allowed the bacteria to grow faster on glycerol. But the order of insertion of genes made a big difference on effect of each gene. Some genes could speed up the growth of *E. coli* only if Palsson had already inserted some of the other evolved genes. On their own, some of those genes were harmful, slowing the bacteria down.

Genes work together in an organism. The effect of a mutation on any one gene depends on makeup of other genes, called epistasis. This explains the thrive of some bacteria on maltose and failure of some. The pattern of random mutations that arose in each line was different. Once a specific beneficial mutation arose and spread through one lines of bacteria, it changed the effects of future mutations on bacteria. Each line accumulated some different mutations as the result, although they all adapted to a glucose diet. In some lines, the combination of accumulated mutations allowed them to thrive on maltose. In other lines their glucose-adapted genes made them unable to feed on different sugar.

In the early 1990s, Julian Adams used a single microbe to produce a colony with a low supply of glucose. Unlike Lenski, he replenished their sugar. The bacteria began to evolve. Natural selection did not favor a single strategy. When the bacteria were kept in Petri dishes, they grew into 2 types of colonies. Some formed big splotches, while others formed small ones. While repeating the experiment, once again, he discovered that some bacteria made big splotches and others made small splotches. It took about 200 generations for 2 types of microbes to emerge. Adams realized that a single clone was evolving in time and again into 2 distinct *E. coli. E. coli* grows more slowly on acetate than on glucose. Adams noted that some *E. coli* evolved into acetate-feeders, become more efficient at feeding on acetate than their ancestors. In fact, they turned the waste of glucose-feeders into their own food. The acetate-feeders grew slowly, but they weren't driven to extinction because they were advantageous of a food that the faster-growing bacteria weren't eating.

Michael Doebeli and others supplied *E. coli* both with glucose and acetate. After 1,000 generations, the bacteria had evolved into big and small colonies. The colonies were different from the big and small colonies that Adams had produced. The big colonies fed on glucose until it ran out, where upon they started to feed on acetate. The small colonies switched over to acetate soon, before the glucose ran out. When *E. coli* feeds on glucose, it keeps the genes for digesting acetate highly repressed. If it made both sets of enzymes at the same time, they get snared in a metabolic traffic jam. When the time comes to switch to acetate, the bacteria first destroy the enzymes for glucose. In the small colonies, the bacteria evolved to no longer repress their acetate genes. They made enzymes for both molecules. As their enzymes interfered with each other, the bacteria grew slowly on glucose, and thus produced small colonies. The big colonies contained bacteria that continued to feed only on glucose at first, and then slowly switched over to acetate after the glucose finished. The small colonies were able to feed on acetate. The big colonies slowly retooled their metabolism.

## 2.2 Gene-neuron-hormone-environment Interaction

A good example of gene- environment interaction is the schizophrenia. About 1% of the population suffers from the schizophrenia. Children of normal parents adopted by a schizophrenic parent had 3% chance of schizophrenia. If their biological parents were schizophrenic, but their adoptive parents didn't, they had a 9% chance. Schizophrenia is more genetic than environmental. A child raised by schizophrenic parents whose biological parents had the disease, had a 17% chance of getting the Schizophrenia. Thus, the effects can be more than the additive when genes and environment work together. In some environments, a gene does the opposite thing in changed environment.

When testosterone is removed, aggression levels drop. Without testosterone the males could be more aggressive. Aggression isn't contained in hormone; it is contained in interactions between hormones. Combining 2 hormones together not result in a combination of 2 behaviors that they modulate. The brain monitors hormone levels and the rate at which hormone levels are changing. A **fixed action pattern**, a set of behaviors is hard-wired. Human born knowing the smile, and laugh. We aren't born being afraid of spiders. We learn that spiders are scary. Located at the center of the brain, the limbic system controls emotion. It sits above the hypothalamus, which controls the body's physical state via hormones (http://trendingsideways.com/index.php/the-biology-of-human-behavior-robert-sapolskys-key-insights/).

It controls the hypothalamus. Each part of limbic system controls different behavioral aspects. They all fight over control of the hypothalamus. They inhibit one another. The parts of limbic system are connected to each other by frontal cortex, located in forehead. The frontal cortex is larger in humans than any other animal. It inhibits parts of limbic system. Because of our frontal cortex, we choose to do something hard instead of something easy. Morality, personality and social reasoning, have a lot to do with the frontal cortex. Behind the frontal cortex is the anterior cingulate cortex which lights up when we feel pain, and when somebody we care about feel pain. A part of brain devoted to empathy. The amygdala, a part of limbic system, controls fear and aggression which are closely related to one another. The amygdala controls male sexual motivation, explaining sexual aggressiveness in males. .

The nucleus accumbens, another part of limbic system, revolves a great deal around dopamine and plays important role in motivation and addiction. It doesn't just light up when we receive reward, it lights up when the reward is anticipated (http:// trendingsideways.com/index.php/the-biology-of-human-behavior-robert-sapolskys-key-insights/). The part of brain lights up when we are morally disgusted and when we are disgusted by rotten food. There is an association between literal and moral disgust in every language. We associate things like the warmth of a person's personality with literal warmth of temperature. If we find the opportunity to wash our hands, we feel morally clean. On some level, the brain takes these metaphors seriously.

There is no gay gene .A single gene can not cause anything. No single strand of DNA is found in all gay individuals. Homosexuality may have more to do with prenatal development and fetal environment than genetics. If genes play a part, the question

arises why evolution would select for a trait that prevents or reduces the chances of reproduction. Siblings of gay individuals are more attentive parents on average and gay individuals are very helpful toward their sibling's children. Humans are not the only species that commits murder, rape, or war. In some species, males resort to kill other male's children to improve their own **reproductive success.** Among chimps, even females commit infanticide against other female chimps. The role of **aggression** appears to have something to do with status. Much of the biology of aggression is hard wired in amygdala. Context is something that we learn from culture. We learn when to be aggressive, and when not to. This is the role of frontal cortex.

## 2.3    Issues of Evolutionary Biology

Evolutionary biology discusses

- The history of life from the common ancestor, their phylogeny; times of their origin and extinction, change in their characters, origin of hereditary variations; various processes affecting the variations; co-acting processes of change; process of rapid change, natural selection, genetic drift, mutation, anatomical, behavioral, molecular and other features of organisms; and origin of new species.

- Relations between distribution of species and current, or historical geographical features, observed natural selection in laboratory and real world, mathematical studies of effect of selection principles on changes in complex systems.

Genetic changes are perhaps random. Chemicals and cosmic rays cause mutations and genetic changes. Selection introduces a degree of preference. Some mutations improve the survival chances, some decrease them, and most are the neutral. Natural selection is like the gravitational force. Mutations are like random excursion of water molecule. Environment is like the landscape. The evidence for evolution is extensive, the main sources being the pliability of the form and behaviour of organisms. Similarities between existing creatures, occurrence of same biochemical components and systems in various organisms suggest a common origin.

The fossil record reveals coherent sequences of changes over time. The geological record dates the fossil species. Genetic features of organisms, especially DNA sequences, confirm lines of descent and time of changes. **Fossils** are relics of living creatures from the past. Because rocks are often datable, many fossils can be reliably datable. Fossils provide a sparse record of life forms of past. Above 250,000 fossil species are discovered. Number of known fossils is increasing rapidly with discovery of new sites. Classic example is the **evolution of horse** where sequence begins with *Eohippus* (0.4 metres long), but is renamed *Hyracotherium* following the taxonomic rules. Sequence continues with *Mesohippus* (35 million years old and 0.6 metres long), *Merychippus* (15 million years old and 1 metre long), *Pliohippus* (8 million years old and 1.3 metres long) and finally *Equus*, the same as modern horse (1 million years old and 1.6 metres long) (Figure 2.7).

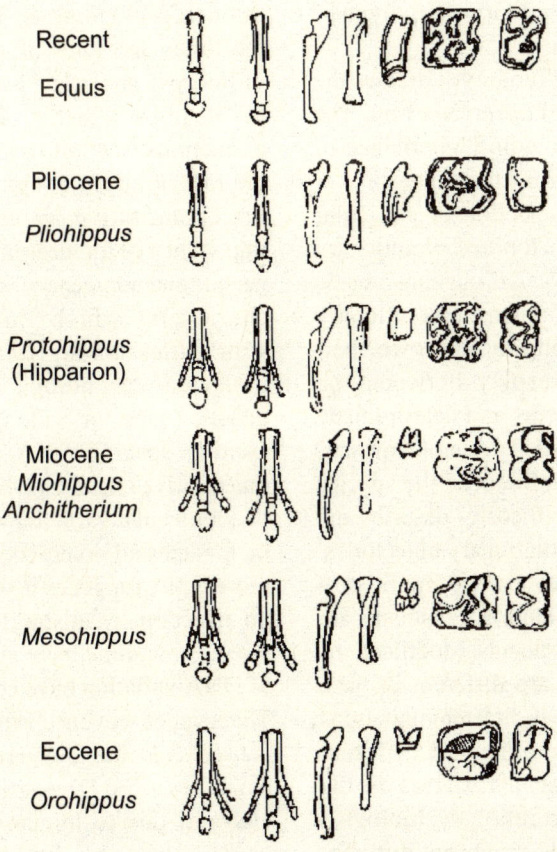

Recent
Equus

Pliocene
*Pliohippus*

*Protohippus*
(Hipparion)

Miocene
*Miohippus*
*Anchitherium*

*Mesohippus*

Eocene
*Orohippus*

**Figure 2.7**  Genealogy of the Horse.

Sequence of changes in this lineage of horse ancestors can be tracked. Every species is unique and are linked to another by common descent. All species share some biological properties due to inheritance of features from their common ancestor. Studying one species help understands other species. Studies in mouse have led to major insights into human biology. Degree of relatedness of 2 organisms determined by amount of time that has elapsed since their common ancestor shows the extent of sharing their biology. siRNA, regulates protein levels in cells of plants and human, shows promise for drug development. Distantly related organisms share common genes and pathways. Mutation in *Escherichia coli* and *Saccharomyces cerevisiae* helped identify the defective genes in hereditary nonpolyposis colon cancer in human.

Most disciplines of biology addresses the **proximate causation** of observed event. The **ultimate causation** including inheritance of feature by a species from the ancestors, or preference of natural selection of particular feature over others is the topic of evolutionary biology. Evolution is an active, ongoing process that affects living organisms. Evolutionary science states that living system owe their properties due to interplay between the stochastic (random) events and deterministic processes. Random mutations, asteroid impacts, and other events have influenced the species evolution. Probabilistic theories describe the likelihood of various evolutionary trajectories. Some adaptations to environment are predictable; other characters of organisms are the result of historical accidents. Modifications of forelimbs for flight are different in bats, birds, and pterodactyls, as different mutations presented natural selection with different options in such lineages. Variation is the important object to evolutionary biologist. About every character is somewhat different among the individuals of a population.

Evolutionary perspective on variation has implications for normality and abnormality, and differences in human characteristics.

**Phylogenetic methods** are used to estimate relationships among species as the species share derived features stem from recent common ancestor. Apes, rat, whales, share recent common ancestor with one another. Chimpanzee is closely related to humans than to gorillas. Long sequences of DNA reveal more similarities and differences among species than found in their anatomy. DNA sequences are used to estimate how recently variants of a gene carried by different people arose from the single ancestral gene. Fossils are used for 2 major kinds of study. One traces the evolutionary changes in characters of lineages through geologic time. The other determines the times and rates of origin and extinction of lineages and related such changes to other events in earth history. Each of 5 great mass extinctions was followed by great increase in the rate of origin of species and higher taxa.

Quantitative genetic analysis is an important tool for measuring and distinguishing genetic and non-genetic variation in phenotypic characters which involves measuring similarities among relatives. DNA based molecular technologies construct detailed genetic maps for wide range of species and identify specific DNA regions that control quantitative characters. **Neutral theory** holds that molecular variation within species should be greater and divergence among species are more rapid, for genes in which mutations have no effect on organisms' fitness than for those where mutations have strong effcct. Studies of DNA variation have confirmed this model. The level of sequence variation among species is used as a clue to determine the function of newly described DNA sequence. **Evolutionary changes** due to human activities, or other causes like environment changes, or introduction of new species are recorded.

Changes in food supply due to drought in the Galapagos Islands caused substantial evolutionary change in beak size of finch within few years.

## 2.4 Sub- disciplines of Evolutionary Biology

Evolutionary biology includes many sub disciplines that differ in their subject matter and methods. Some of the major sub disciplines are:

**Behavioral Evolution:** This deal with evolution of adaptations like courtship, mating systems, and cooperation. Changes in neural, hormonal, and developmental mechanisms are also objects of evolutionary study. Adaptive differences among species in memory, learning patterns, and other cognitive processes are reflected in differences in structure of brain. Structure, physiology, and life history patterns often evolve in the concerted way.

**Evolutionary Developmental Biology:** This deals with evolution in processes that translate the genetic information contained in an organism's DNA into its anatomical and other characters. It aims to describe variation at genetic level that affects survival and reproduction. Comparative data on genetic and mechanical bases of development in vertebrates and other organisms shed light on mechanisms of human development and help understand the bases of hereditary and other congenital defects in humans, and may be useful in gene therapies. Rapid evolution of antibiotic resistance in previously susceptible pathogens presents a critical need for evolutionary study.

**Evolutionary Ecology:** This field discusses the evolution of life histories, diets, and other ecological features; their affect on composition and properties of community and ecosystem, and evolution of species in response to each other. It considers the evolution of short or long life spans, as well as broad and narrow distribution of species. Evolution and evolutionary history that affect the species number in community is also dealt with. New and resurgent diseases have emerged as major threats to public health which can be countered in several ways. Screening for phylogeny of organisms related to known pathogens may help to identify pathogens with potential to enter the human population. Studies of new and emergent pathogens can elucidate their origins, their rates and modes of transmission, and ecological circumstances leading to outbreaks, or to evolution of greater virulence. Studies of model systems, including organisms related to known pathogens can identify mechanisms of virulence and genetic and environmental factors that influence drug resistance.

**Evolutionary Genetics:** Molecular and classic genetic methods help understand the origin of variation by mutation and recombination, describe genetic variation pattern within and among populations and species, and attempt to discover the role of genetic drift, gene flow, and natural selection on variation. Mathematical theory of evolutionary genetics interprets genetic variation and predicts evolutionary changes. It provides a strong foundation for understanding evolution of genome structure and life histories.

**Palaeobiology:** This field deals with massive evolutionary patterns in fossil record, discusses the origins and fates of lineages, evolutionary changes in anatomy through time, and geographic and temporal variations through geologic past. It addresses the physical and biological processes and historical events that shape the evolution. It deals with problems ranging from the change in distribution of species to evolutionary responses of major groups to catastrophic and gradual environmental changes. These data calibrates the rate for phenomena as mutations in nucleotide sequences.

**Evolutionary Physiology and Morphology:** This field looks at the biochemical features of organisms that provide adaptation to their environments and ways of life. It begins to define the limits to adaptation. Such limits may restrict a species' distribution, or lead to its extinction. It deals with the form and function of change in feature in relation to each other during evolution.

**Human Evolution:** Scientists dealing with human evolution use concepts, methods, principles, and information from evolutionary systematics, genetics, ecology, paleontology, and animal behavior. Researchers study genetic variation and processes that affect evolution in contemporary human populations along with human behavior and psychology.

**Molecular Evolution:** This field investigates the history and causes of evolutionary changes in nucleotide sequences of genes (DNA), their structure, number, organization on chromosomes, and other molecular phenomena and searches for evolution ranging from phylogenetic relationships among species to mating pattern in population.

**Systematics and Phylogeny:** Documenting the diversity of potentially useful organisms is the foundation for further work. This recognition leads to funding by pharmaceutical companies for biodiversity inventories in Costa Rica and elsewhere. Phylogenetic aspect of systematics is important for pointing researchers toward species that are related to those in which potentially useful compounds, or metabolic pathways have been found, since related species may have similar, perhaps even more efficacious, properties. Systematics of bacteria, protist, fungi, and other organisms are very poorly known. Systematists distinguish species, infer phylogenetic relationships in species, and classify species based on evolutionary relationship. It helps understand variation and nature of species. Such knowledge is indispensable for inferring the history of evolution and for understanding the mechanisms of evolutionary processes.

## 2.5 Achievements of Evolutionary Biology

The genes governing the first steps in embryonic development through specifying the axes and major body regions of embryo are similar in organization, sequence and function in vertebrates and insects. Some mouse genes, implanted in a fly's genome, instruct the fly genes to act their normal developmental functions. Evidence of **common ancestry** is given by nonfunctional DNA sequences called pseudo genes- dead genes without function, but is shared by many species. The rudimentary wings of flightless insects descended from flying ancestors attest evolution. Fossil evidence of evolutionary transitions support the common ancestry based comparisons among living species.

Evolution of terrestrial amphibians from fishes and similar evolutions can be traced in fossil record. Phylogenetic inference methods provide evidence on relationships among organisms. History of evolution of particular characters can be inferred from their distribution on a phylogenetic tree. Social behavior has evolved at least 15 times among insects, since each of the 15 groups of social species is related to different nonsocial group. Comparisons of closely related social insect species have shown that sociality has evolved in steps. Phylogenetic studies have revealed some events in history of life. Mitochondria and chloroplasts descended from free-living bacteria became intracellular symbionts.

Phylogenetic inference methods produce gene trees within species and in species. Gene trees reveal about the history of populations, like their age, size, and subdivision.

Different characteristic evolve at different rates within a lineage. Every organism is the patchwork of characters that have changed largely in recent past and others that have changed little over many years. This is true for DNA sequences and phenotypic features. Individual anatomical features and clusters of features evolve rapidly sometimes in a lineage, and seldom at other times. In fossil record, this pattern is known as stasis, interrupted occasionally by short periods of rapid change- a pattern called **punctuated equilibrium**. Evolutionary radiation produces distinct lineages from the common ancestor within short time in association with evolution of new adaptation to provide access to new resources or a new way of life, or with the extinction of taxa that had previously dominated the ecosystem.

Patterns of diversification and extinction are known from the fossil record. Early marine organisms increased rapidly in diversity, and remained in a roughly stable level for much of the Paleozoic era. Their diversity then dropped to perhaps 4% of their previous diversity during the greatest mass extinction. Diversity then rebounded rapidly and has increased ever since. The separation of continents causes diversification and increases the global diversity. Throughout the fossil record, there is turnover-extinction and origin of taxa. Extinction pattern in coastal marine invertebrates over geologic time suggests that tropical reef dwellers are most vulnerable.

Mathematical theory of **population genetics** describes the interplay and relative importance of mutation, recombination, genetic drift, and natural selection. Natural selection exerts a stronger force than mutation on many

phenotypic characters as direction and rate of evolution is driven by selection though mutation is necessary for evolution. Certain forms of natural selection and population structure maintain genetic variation instead of eroding it.

Neutral theory predicts that the greatest variation should occur in the functionally less critical parts of gene. Mutations induced in various parts of a gene of the *Escherichia coli* those differ little in different bacteria species impaired enzyme function, whereas mutations in widely varied regions in species had little effect. Populations are highly variable genetically. Extensive genetic variation exists within and among populations. No 2 humans except for identical twins are genetically identical. High level of genetic variation may allow populations to evolve rapidly in changing environments.

The reservoir of genetic variation plays the role in the success of artificial selection of desirable traits in crops and domestic animals. Rapid evolution is found in the development of insecticide resistance in many insects. The evolution can be studied directly. Studies of bacteria show that adaptive evolution can be based on new mutations, not just on preexisting variation. Observed adaptive changes often have deleterious side effects, which, if sufficiently great, limit further adaptation. Subsequent genetic changes sometimes occur that compensate these side effects. Blowfly attacking sheep evolved resistance to diazinon. The resistant populations initially showed retarded development and physical abnormalities. These traits later diminished due to selection of other genes that ameliorated the deleterious effects. Much has been learned about genetic changes that underlie speciation. In some cases, **speciation** occurs rapidly, while in others, interactions among many genes cause speciation slowly and gradually. Certain speciation like the polyploidy is more prevalent in plants than in animals. Some wild plants evolved by polyploidy are re-created in the laboratory.

Many forms of natural selection are known. Selection operates through differences in survival and female reproduction, and in male mating success which creates competition among males, or preference by females for males with certain traits. **Sexual selection** causes elaborate male behaviors and anatomical features like huge antlers in deer and bright plumes and elaborate displays of many male birds. Traditionally, natural selection was differences in survival, or reproduction among phenotypically different individuals within populations. We now know that selection can reside in differences in survival, or reproduction among genes among whole groups of individuals, and among species or higher taxa. **Genic selection** can be potent. Selfish genes are genes that spread more copies of themselves through population than other genes. **Transposable elements** are DNA sequences that replicate and spread throughout the genome which may not benefit, but even harm the organism, or the species. Theories based on natural selection explain the evolution of many puzzling characteristics. From a long list, 2 examples (cooperative behavior and senescence) are described below.

**Cooperative Behavior: Altruistic behavior**, causes helping other individuals to rear offspring, seems difficult to explain. Such altruistists divert energy that they could otherwise use for their own reproduction, or survival. Kin selection states that an individual who aids others may bequeath fewer of its own genes to next generations, but it may more than compensate for this by enhancing the survival and reproduction of its relatives, which possess many same genes. Most cooperative behavior is directed toward relatives.

**Senescence:** Mathematical theory of life histories shows that offspring born late in a parent's life contribute less to future population numbers than offspring born earlier. Reproducing late in life contributes fewer genes to the population than early reproduction. Genetic advantage of surviving to reproduce may be selected because of their effect on early reproduction, but cause senescence as the side effect as supported by studies. The genes that enhance survival, or reproduction early in life may have deleterious side effects in later life. Antagonism between hosts and parasites and between prey and predators, lead to **evolutionary arms races** in which each change occurs in response to changes in other. Plants have evolved diverse chemical defenses against herbivores and pathogens including nicotine, caffeine, and salicylic acid that human have used for diverse purposes. Each such defense has been overcome by some insects.

In salamanders, evolution in genes that affect the production of hormones influences the rate and time of development, results in species with juvenile characters throughout their adult life. Some salamanders that grow only to miniature size fail to develop certain bones and have altered skulls. In *Drosophila*, master switch genes regulate the action of other genes lower in command hierarchy that determine the identity and features of segments of insect's body. Homologue of these genes exists in mammals and other animals. These master genes regulate lower-level genes that differ from one taxa to another and produce different characteristics. **Master genes** that regulate the development of flowers have some similarities in DNA sequence to animals' master genes. DNA sequences show that human are closely related to the chimpanzees. The more than 98% DNA sequence similarity of humans and chimpanzees implies that they diverged from the common ancestor about 6 to 8 mya.

Early hominids with apelike features are

made yearly in eastern Africa. The oldest hominid fossils are about 4.4 million years old, approaching the time of common ancestry as suggested by DNA data. That all contemporary human populations are descended from a single African population that spread into Eurasia about 200,000 years ago by replacing *Homo sapiens* that previously occupied this region is controversial. Genetic differences among modern human populations in various parts of globe have had little time to develop. There are some regional genetic differences in characters like facial features and frequencies of blood groups, but all human populations are genetically similar. Human genetic variation is found within, rather than in populations. Thus, if all humans go extinct except for a single tribe, at least 85% of the genetic variation that exists today would still be present in future population.

## 2.6 Evolutionary Biology in Other Fields

Role of evolutionary biology in other fields is briefly described below :

### Genetic, Systemic and Infectious Diseases

Variant genes or chromosomes in combination with environmental factors and individual's genetic constitution at other loci cause genetic disease. Conditions associated with old age, learning disabilities, and behavioral disorders cause human suffering. Alleles at one, or more genetic loci, which range in frequency, cause genetic disorder. Allele frequencies determine the reason for the frequency of **deleterious allele**, and estimate the inheritance of allele. The high frequency of alleles for sickle-cell and other defective hemoglobin in some geographic locations signaled the maintenance of these alleles in populations perhaps by some

agent of natural selection. Their geographic distribution showed an association with malaria.

These alleles prevalent as heterozygous carriers confer great resistance to malaria. Thus, a heterozygous fitness advantage maintains deleterious alleles in population. Children inherit genetic diseases, especially, if they have family history of such diseases. Genetic counseling relies on pedigree analysis and frequency of particular allele in population at large to estimate the likelihood of inheritance of genetic defect. Health consequences of marriage among related individuals can be explained on theories of population genetics. Carriers of deleterious alleles can be identified and normal alleles can be substituted for defective ones in genetic therapy.

Human death is associated with chronic systemic diseases, like coronary artery disease which result from interactions between gene and environment. To study the linkage between genes and systemic disease is difficult. Some genes with their known biochemical, or physiological functions are called the candidate genes. Molecular genetic variation at such candidate loci in general human population exists. Evolutionary phylogenetic techniques estimate a gene tree from this genetic variation which shows the history of genetic variants of candidate gene. If any mutation altered risk for systemic disease, the entire branch of gene tree bearing mutation shows similar disease association. Gene tree analyses are used to discover genetic markers that are predictive of risk for Alzheimer's disease, risk for coronary artery disease, and response of cholesterol levels to diet.

Viruses, bacteria, protists, fungi, and helminths cause **infectious diseases**. New species of malaria-causing protozoan (*Plasmodium*) evolves. Progress in controlling malaria in the Mediterranean region was slow until the discovery of 6 almost identical species of *Anopheles*, of which only 2 transmit malaria. Population genetics are indispensable to know the mode of reproduction of pathogens and their carriers and their population structure. Multiple genetic markers in *Salmonella* and *Neisseria meningitidis*, reveals that bacteria reproduce mostly asexually, but rarely transfer genes by recombination among distantly related strains.

Immunological variations traditionally used to classify strains of these bacteria are not well correlated with the genetic lineages neither by multiple genetic markers or with variations in pathogenicity. Thus, predicting these traits in public health require multiple genetic markers. Population genetic methods can estimate rates and distances of movement of disease-carrying organisms, which affect disease transmissions and potential for control. Molecular analysis of a gene in a mosquito species showed that the gene had recently spread in 3 continents, which prove the insect's efficient dispersal capability. The rapidity of evolution in natural microorganism populations, many with short generation times and huge populations has important implications. The pathogen was expected to adapt to consistent, strong selection, like that produced by widespread use of drugs.

HIV and malarial protozoan have become resistant to antimicrobial drugs. Many organisms are resistant to drugs, because antibiotic resistance genes are often transferred between species of bacteria. Evolution of **drug resistance** has greatly increased the cost of therapy. Increased morbidity and mortality has raised fears about many infectious diseases. Such a grim future may be averted by reducing selection for antibiotic resistance. WHO has recommended the judicious use of antibiotics. **Virulence of pathogens** evolves rapidly. Theory of parasite/host co evolution predicts that greater virulence may evolve when transmission among hosts increase. Major outbreaks of influenza and other pandemics are believed to be caused by evolution that transpired in crowded cities and among mass movements of refugees. HIV has evolved higher virulence due to high rates of transmission by sexual contact and sharing of needles by intravenous drug users.

## Physiological Functions

Genes in the major histocompatibility complex (MHC) play critical role in cell mediated immune responses. MHC contributes to rejection of tissue transplants. Some MHC alleles are linked with autoimmune diseases like juvenile diabetes and crippling arthritis. Genetic variation in the MHC is too great. MHC genes must be under some kind of balancing selection that maintains variation. Some human MHC alleles are genealogically closer to some chimpanzee alleles than other human alleles providing clear evidence that natural selection has maintained variation for at least 5 million years by roles of different alleles in combating different pathogens, but its exact role need more study.

## Plant and Animal Breeding

Concepts such as heritability, components of genetic variance, genetic correlation, and experimental elucidation of phenomena like inbreeding depression, hybrid vigor, and basics of polygenic (quantitative) variation, play central roles in agriculture. Recent example of development and application of techniques using molecular markers to locate the multiple genes for continuously varying traits, like fruit size and sugar content and to identify the metabolic function of these genes is mentionable. In the past, a few model organisms like *Drosophila* was well known genetically to provide such information. It is now possible to map genes of interest in any organism.

A widely planted, genetically uniform crop is the suitable for plant pathogens, or other pests, which adapt to it and spread rapidly. Potato blight causing widespread famine in Ireland in 1840's is one example. The epidemic of southern corn leaf blight in the United States in 1970 caused an economic loss of about $1 billion (1970dollars). More than 85% of nation's acreage of seed corn had been planted with strains carrying a genetic factor (Tcms) that prevents development of male flowers, which was used for producing uniform hybrid varieties. The Tcms factor made the corn susceptible to a mutant fungus *Phytopthora* infestans, which rapidly spread through the Corn Belt.

Genetically uniform crops are now used widely for economic efficiency. Building up germplasm banks of various crop strains such as drought tolerance and pest resistance is essential. Important source of potential useful genes is wild species related to the crop. The cultivated tomato is self-fertilizing and harbors little genetic variation among the available varieties. It originated in Andean South America and reached North America via domestication in Europe. Studying evolution of tomato led to the fact that it has many relatives native in Chile and Peru, and these species carry wealth of genetic variation. More than 40 disease resistant genes are found in these native species, and 20 of them are transferred into commercial tomato stock. Fruit quality traits have been improved and resistance to drought, salinity, and insect-pests is expected, for about 4- to 5-fold increase in agricultural production. Systematics of tomatoes, together with ecological genetics and an understanding of plant's breeding system, formed the foundation for successful application that is being tested in many other crops. Genetic engineering, which transfers genes from any species into any other, makes available, for agricultural and other purposes. The vast genetic library of organisms carry a tremendous variety of genes for traits such as heat tolerance,

disease- and insect resistance and many other potential features. To use this library in future, the library should be preserved.

## Conservation of Biodiversity

Alteration of habitats, intentional and unintentional harvesting of natural populations and other human activities constitute a grave threat to persistence of many species. Evolutionary biology and ecology work hand-in hand in addressing these issues .Evolutionary systematics, biogeography, and ecological genetics provide information needed to develop guidelines for conserving genetic diversity. Previous crises in **biodiversity** can be seen in fossil record and evolutionary paleontologists can use these records as natural experiments on consequences of biodiversity loss, characteristics of species most at risk, and nature and time scale of biotic recovery. Much extinction in geologic past was followed by outbreaks of weedy disaster species. Much more needs to be learned about this process. Similarly, past biodiversity crises are associated with marked declines in primary productivity. This fact is relevant to future human welfare; in that human now consume an estimated 25% of global primary productivity.

The minimal population sizes necessary for species to retain sufficient genetic variation, to avoid inbreeding depression and to adapt to diseases, climate change, and other perturbations. Factors causing extinction, role of multiple populations in long-term genetic and ecological dynamics of species, role of interactions among species in maintaining viable populations, and effects of co evolution among interacting species on dynamic processes in ecosystems are important to understand. Some conservation efforts rely on germplasm banks (for plants) and captive propagation. Population genetic theory plays a crucial role in these efforts. Inbreeding depression in small captive populations can be avoided by applying population genetic principles. The winds of change in evolutionary biology are coming from developmental biology, microbiology, ecology, behavior, and from the cultural studies. The construction of a new evolutionary paradigm is continuing.

Evolutionary principles are applicable to conservation of rare and endangered species and ecosystems, and to environmental management issues that bear directly on human health and welfare. New methods of environmental remediation and restoration of degraded land have developed. Some grasses and other plants have become adapted to soil highly polluted with nickel and other toxic heavy metals. Studies of systematics, genetics, and physiology of these plants have laid the foundation for techniques for re-vegetating and stabilizing soils which was made barren by mining activities and even for detoxifying metal contaminated soil and water. Some bacteria have the capacity to metabolize mercury to a less toxic form, and their genes for this capacity have been transferred into plants in laboratory experiments.

Plants that have evolved the capacity to hyper accumulate heavy metals and thus withstand toxic soils are now being used commercially as a clean-up technology. Studies of evolutionary ecology of seed dispersal and germination are playing a role in reforestation of overgrazed land in tropical America and in the revegetation of landfill sites. Evidence from populations evolving at different temperatures may also help us to predict the diversity of responses to climate change and speed with which various populations can adjust to it.

Evolutionary biology is playing a major role in addressing biodiversity crisis. An important consideration is which species, ecological communities, or geographic regions merit the most urgent conservation efforts, since there are economic, political, and informational limits on the number of species we can save. Among the conservation roles of evolutionary biology are:

1. Using phylogenetic information to determine which regions contain the greatest variety of biologically different, unique species;
2. using the data and methods of evolutionary biogeography to identify hot spots-regions with high numbers of geographically localized species;
3. using genetic and other methods to distinguish species and genetically unique populations;
4. using population genetic theory to determine the minimal population size needed to prevent inbreeding depression and to design corridors between preserves to allow gene flow, both of which maintain the ability of populations to adapt to diseases and other threats;
5. using the theory of life histories and other characteristics to predict which species are most vulnerable to extinction;
6. using genetic markers to control traffic in endangered species.

### Pest Management

Plant pests, chiefly insects and fungi, take heavy economic toll in crop losses and control measures. Excessive use of chemical pesticides has resulted in more than 500 species of insecticide resistant insects in the last 40 years. Some are resistant to all known insecticides. In the United States, pesticide resistance has added $1.4 billion to annual cost of crop and forest product protection. Rotational use of different control measures and judicious combination of chemical with non-chemical controls gives some help.

Two non-chemical-methods, use of natural enemies and resistance breeding have profited greatly. Natural enemies which are specialized predators, or parasites of pest species are often sought in the pest's region of origin, once potential enemy such as parasite is found. If an enemy is approved for introduction, large numbers must be bred for release. In this stage, application of evolutionary genetics is crucial to prevent the parasite stock from becoming inbred, or unconsciously selected for characters that could impair its effectiveness. The strategy aims to select for resistance in crop plants by screening for genes that provide resistance in laboratory, or in field plots, and then crossing those genes into crop strains with other desirable characters. A **pest** may adapt to a resistant crop strain as readily as it adapts to chemical insecticides. At least 6 major genes for resistance to the Hessian fly have been bred into wheat. In each case, within a few years of planting of new strain, the fly overcame the resistance. For every resistance mutation in plant, a corresponding mutation in the fly nullified its effect. Despite new alternative methods of pest management, judicious use of pesticides will undoubtedly remain indispensable. Evolution of pesticide resistance in insects, nematodes, fungi, and weeds is a serious economic problem that should receive major attention. This will require studies of genetics and physiological mechanisms of resistance, population dynamic studies, and modeling of methods to limit or delay evolution of resistance. Evolutionary considerations will be important in evaluating many alternative methods of pest management, such as mixing different crops or crop varieties, or developing transgenic crops that carry resistance factors to protect them against insects or other pests. Experiments have shown, for example, that tobacco pests can adapt to transgenic tobacco carrying a bacterial toxin, highlighting the need for studies of genetic variation in insect responses to transgenic crops.

There is enormous potential for **transgenic** use of the innumerable secondary compounds and other properties of wild plants that protect them against insects and pathogens. Experimental and phylogenetic screening of these natural resistance factors should prove rewarding. The evolutionary ecology concerned with secondary plant compounds and the interactions between plants and their insect and fungal enemies is relevant to this effort. It will be important to analyze the physiological effects of natural resistance factors on pest organisms, the mechanisms by which some insects and fungi overcome their effects, and genetic variation in responses of target species to natural resistance factors. Evolutionary studies of parasite/host interactions, using both model systems and human parasites and pathogens, are only beginning to determine the conditions that lead parasites to become virulent or more benign. Evolutionary study need to develop a general predictive theory of the evolution and population dynamics of pathogens and their hosts, especially for rapidly evolving organisms such as HIV and for rapidly migrating host species like modern humans. Analyses of genetic variation in resistance to pathogens in humans and other hosts are also needed.

### Genetic Engineering

Proposals abound for introducing various traits into crop plants with questions about their potential risks. It needs tests to be sure that a foreign gene does not unpredictably interact with a crop plants own genes to generate harmful effects. The likely risk is that such genes could spread by cross pollination into wild plants related to crops and cause them to become vigorous weeds. Because genes are often transferred between species of bacteria, natural bacterial populations could acquire features from engineered bacteria that render them vigorous and harmful. Methods for determining the fitness effects of genes and measuring rates of gene exchange among populations and species would have valuable applications.

## Forestry and Fisheries

Knowing the genetic structure of populations and species by statistical analyses of genetic markers distinguishes stocks of fish species that migrate from different spawning grounds. Such distinctions have important management and political implications in cases like salmon industry. Spawning locations and those where the fish are harvested have an economic interest in stocks. In forestry, nurseries where commercial stocks of conifers are developed and grown are subject to genetic contamination by air-borne pollen from wild trees. Developed methods are useful for determining the distance that pollen travels and for measuring levels of contamination, which affect the seed's market value.

Analyzing the genetic basis of desirable traits, like growth rate and insect resistance, in conifers would contributes to hybrid breeding and genetic engineering programs. Several evolutionary studies have been important in managing commercial and sport fisheries. Molecular genetic markers will aid researcher's in distinguishing breeding populations and migration routes of species like cod and salmon. Studying the growth rate and age at maturity will enable to evaluate the genetic and demographic effects of harvesting on fish populations. For certain fish species that are widely stocked, genetic and physiological studies of adaptation to and fitness in different environments will be useful. Stocking plans will also include the use of transgenic fish, which are in the early stages of development.

## Useful Natural Products

Almost all pharmaceutical products use living organisms, or products originated from biological processes in organisms. The search for fossil fuels is based largely on age correlations among sedimentary deposits which are based on fossilized protozoan, molluscs, and other organisms. Many living species may prove useful as future crops or in medical, energy, industrial, or research applications. Indeed, organisms can be considered living capital. Over 20,000 plants are listed by the WHO for **medicinal purposes**, a substantial fraction of these really are effective. Malaria was treated until very recently by quinine from the cinchona tree. Recent discoveries of other medicinally useful plant compounds abound. Taxol, a compound found in the Pacific yew, has shown promise in treating breast cancer. The rosy periwinkle of Madagascar contains 2 chemicals that are useful for fighting leukemia and which have increased childhood leukemia survival rates from 10% to 95%. Natural plant products are used as scents, emulsifiers, and food additives. An extract from horseshoe crabs is the basis for the lysate test, widely used in pharmaceutical industry to test for presence of bacteria. **Microorganisms** provide not only products, but also biochemical processes useful in biosyntheses, biodegradations, and biotransformation. Pharmaceutical and other industries have started programs for screening natural products in expectation of more such discoveries. Neurobiologists seeking inhibitors of neurotransmitters led successfully to venoms of certain snakes and spiders, organisms that have evolved such inhibitors to overcome their prey. Fungi release antibiotics to control bacterial competitors. Plants harbor many thousands of compounds to ward off their natural enemies. Evolutionary ecological study of such adaptations has only begun to reveal compounds that merit further attention. Pharmaceutical and other industries are searching for novel products and processes by screening plants, animals, and microorganisms. The search for development of novel products and processes raises serious issues in patent law, international law, and the publication of scientific data that are beyond the scope of this book, but which will affect the engagement and activities of research scientists.

**Genetic diversity of human and economically important organisms** Research on human genetic diversity will complement the Human Genome Project, which would sequence the entire human genome. Such research will provide data at the molecular level on the immense genetic diversity that exists within human populations. The techniques of population genetics and phylogenetic analysis will be applied to explore information on human genetic variation to determine the history of populations, and will continue to provide tools for identifying genetic lesions associated with inherited diseases and defects. Evolutionary comparisons of human DNA sequences with those of other species will provide insight into gene functions. Population geneticists will analyze the genetic bases of interesting variable traits, such as reactions to allergens.

**Genes** that provide adaptations to environmental factors such as pathogens and diet will be identified by studying genetic differences among/within populations. The methods used by evolutionary geneticists will be applied to human diversity to elucidate cases of complex inheritance of disease and to study genotype/environment interactions-the differential expression of traits like disease resistance under different environmental conditions. **Population genetics** methodology also uses linked genetic markers to determine the likelihood that an individual carries genes of particular interest. As evolutionary geneticists improve these methods and apply

them to data on human genetic diversity, it will be possible to use molecular markers confidently and accurately for such purposes as counseling individuals on the likelihood that they or their children will carry a genetic disease , determining paternity, and forensic analysis.

Production of food, fiber and forest products has historically been improved by exploiting genetic variation. Scientists use QTL (quantitative trait loci) mapping and other methods to locate the genes for, and elucidate the mechanistic bases of important plant traits, like resistance to pathogens and to environmental stresses. Such studies help understands the adaptations of plants to environmental factors. Similar studies on wild plants will locate genes for useful traits that can be genetically engineered into crops. Research programs of this kind will use principles and information from studies of plant phylogeny and adaptation. The critically important task of developing and maintaining germ plasm banks will continue to depend on studies of variation within/among populations.

### Adaptation

Antibiotics, resistance factors for use in transgenic crops, and other useful natural products are likely to be found by studying the chemical mechanisms of competition among fungi and microorganisms, the defenses of plants against their natural enemies, and waxes, steroids, terpenes, hormones, and innumerable other compounds that organism use for diverse adaptive ends. **Genetic and physiological studies:** Bacteria, yeasts, and other microorganisms have exceedingly diverse metabolic capacities. They have been the source of penicillin, polymerase enzyme used in DNA sequencing, and important industrial processes of fermentation, biosynthesis, and biodegradation. Industry anticipates that "great advances in bio-processing can be expected from future exploration of the yet unexplored biodiversity of the land and sea".

### Bioremediation

Bioremediation refers primarily to use of organisms in cleaning up spills and toxins, treating sludge, and restoring degraded soils. Evolutionary biology contributes to **bioremediation** by identifying species, or genetic strains with desirable properties, by understanding the agents of natural selection that cause such properties, and conditions that favor the persistence of useful organisms. Bacteria that can degrade polychlorinated biphenyls (PCBs) and other persistent contaminants are known, but it is unknown whether this capability is characteristic of certain species, or evolves in situ due to selection of new mutations. The community of bacteria involved in waste-water treatment undergoes a change in composition during the process, but the roles of turnover of species versus genetic change in metabolism of persistent species are not known. Evolutionary genetics and systematics, together with microbial ecology and physiology, should continue to make important contributions to these and other questions in bioremediation.

### Unplanned introductions

Many serious pests, including weeds, insects, red-tide dinoflagellates, and zebra mussels, cause damage in which they are non-native. **Quarantine procedures** are intended to prevent such introductions. Genetic engineering has caused concern about the escape of vigorous, genetically novel microorganisms, plants, fishes, or other organisms and about the possibility that gene for novel capacities could spread by hybridization from transgenic organisms into wild ones, transforming benign species into novel pests. Evolutionary biologists

have been active in assessing such risks. Studies of gene flow between and within species and evaluations of fitness effects of genes must complement ecological studies of relevant organisms, if we are to predict the possible unintended effects of transgenic releases. Traditional role of systematics in identifying introduced organisms will continue to be important.

Of the anthropogenic activities, the most universal possible effect is global warming. Desertification, salinization of fresh water and acid rain are more local, but still with profound effects on biological resources. Predicting and possibly forestalling the effects of such changes are important. We need to understand for better the conditions under which populations adapt to environmental changes versus migrating, or becoming extinct, what kinds of species will follow these courses and conditions favoring "breakouts," in which new species adapt to and disperse rapidly into novel environments. Agriculture and urbanization have produced many novel environments, and such break-out species may not be benign.

Many species have adapted rapidly, and many have not, but a full theory of vulnerability versus potential for rapid adaptation is needed. **Paleobiological studies** can complement genetic and ecological studies by providing detailed histories of changes in composition of communities and distributions of species under past environmental changes and can also help us to develop generalizations about kinds of species and communities that are most vulnerable.

## 2.7   Landmark Discovery in Evolutionary Biology

1. Charles Darwin publishes "On the origin of species by means of natural selection" and establishes the theory of evolution in 1859.

2. Gregor Mendel paper "Experiments in plant hybridization" became formalized as Mendelian laws of inheritance in 1866.

3. Johann Friedrich Miescher extracts DNA from the nuclei of WBCs in 1869.

4. Erich von Tschermak-Seysenegg, Carl Correns and Hugo de Vries, rediscover independently Mendel's basic laws of inheritance in 1900.

5. Walter Sutton and Theodor Boveri propose that chromosomes bear heritary factors following Mendelian laws in 1902.

6. Thomas H. Morgan establishes the chromosomal theory of inheritance, discovers the recombination of homologous chromosomes at meiosis in 1910.

7. Oswald T. Avery, Maclyn McCarty and Colin MacLeod identify DNA as the transforming principle in the year 1944.

8. Erwin Chargaff discovers regularity in proportions of DNA bases. In organisms, the amount of adenine (A) equaled that of thymine (T), and guanine (G) equaled cytosine (C) in the year 1950.

9. J. Watson and F. Crick (Figures 2.8, 2.9) discover the structure of the DNA which meets the unique requirements for a substance that encodes genetic information in the year 1953.

10. Discovery of mRNA by Sydney Brenner, Francis Crick, Francois Jacob and Jacques Monod in the year 1960.

11. Werner Arber, Hamilton O. Smith and Daniel Nathans discover restriction endonucleases in 1968. Frederick Sanger and Walter Gilbert develop techniques for DNA sequencing in 1977.

12. Kary B. Mullis (Figure 2.21) invents and helps to develop the polymerase chain reaction (PCR) in 1983.

13. With the 1,830,137 bp of *Hamophilus* *influenzae* sequenced, the first genome of a free living organisms determined in the year 1995

**Figure 2.8** James Watson

**Figure 2.9** Francis Crick

**Figure 2.11** Edward Wilson

**Figure 2.12**　Robert L. Trivers

**Figure 2.15**　William D. Hamilton

**Figure 2.13**　Alfred R. Wallace

**Figure 2.16**　Stephen J. Gould

**Figure 2.14**　Ronald Aylmer Fisher

**Figure 2.17**　George G. Simpson

**Figure 2.18**   Alan Wolfe

**Figure 2.19**   Plato

**Figure 2.20**   Lynn Margulis

**Figure 2.21**   Kary B. Mullis

14. The nematode, *Caenorhabditis elegans* was sequenced in 1998
15. The fruit fly, *Drosophila melanogaster* was sequenced in 2000
16. Homo sapiens was sequenced in 2001 (Figure 2.11-2.20)

## SUMMARY

1. Erasmus Darwin recognized organisms' struggle for existence, speculated about change

in species through time, suggested that humans were derived from more primitive species, and told about inheritance of acquired characteristics. Lamarck proposed a hypothesis about the evolution. Curvier argued about the lack of evidence for change of species and promoted the idea that species were fixed entities. Darwin hypothesized that evolution had occurred. Species had changed through time.

2. Species evolved on each land mass independent of other land masses.

3. A trend caused by consistent( Gregory, 2008) patterns of change occurring within populations implies that large-scale evolutionary patterns have been the result of small-scale processes amplified through deep time, whereas a trend resulting from higher-level processes above the species level requires an expanded, hierarchical view of macroevolution (Gould, 2002).

4. Evolutionary biology discusses:

   (a) history of life from the common ancestor, their phylogeny; times of their origin and extinction, change in their characters, origin of hereditary variations; various processes affecting variations; co-acting processes of change; process of rapid change,

   (b) natural selection, and genetic drift ,mutation, anatomical, behavioral, molecular, and other features of organisms; and origin of new species,

   (c) relations between distribution of species and current or historical geographical features, observed natural selection in laboratory and real world, mathematical studies of effect of selection principles on changes in complex systems are the topics of evolutionary biology.

   (d) Genetic changes are perhaps random. Chemicals and cosmic rays cause mutations and genetic changes. Selection introduces a degree of preference. Some mutations improve the survival chances, some decrease them, and most are the neutral.

   (e) Natural selection is like the gravitational force. Mutations are like random excursion of water molecule. Environment is like the landscape. The evidence for evolution is extensive, the main sources being the pliability of form and behaviour of organisms. Similarities between existing creatures, occurrence of same biochemical components and systems in various organisms suggest a common origin.

   (f) Fossil record reveals coherent sequences of changes over time. Geological record dates the fossil species. Genetic features of organisms, especially DNA sequences, confirm lines of descent and time of changes. Fossils are relics of living creatures from the past. Because rocks are often datable, many fossils can be reliably datable. Fossils provide a sparse record of life forms of past. Above 250,000 fossil species are known. The ultimate causation including inheritance of feature by a species from the ancestors, or preference of natural selection of particular feature over others is the topic of evolutionary biology.

   (g) Evolution is an active, ongoing process that affects living organisms. Evolutionary science states that living system owe their properties due to interplay between stochastic (random) events and deterministic processes. Random mutations, asteroid impacts, and other events have influenced the evolution. Long sequences of DNA reveal more similarities and differences among species than found in their anatomy. DNA sequences are used to estimate how recently variants of a gene carried by different people arose from the single ancestral gene. Fossils are used for 2 major kinds of study. One traces the evolutionary changes in characters of lineages through geologic time.

(h) Some mouse genes, implanted in a fly's genome, instruct the fly genes to act for their normal developmental functions. Evidence of common ancestry is given by nonfunctional DNA sequences called pseudo genes-- dead genes without function, but is shared by many species.

(i) The rudimentary wings of flightless insects descended from flying ancestors attest evolution. Fossil evidence of evolutionary transitions supports the common ancestry based comparisons among living species.

(j) The neutral theory of molecular evolution predicts that the greatest variation should occur in functionally less critical parts of gene. Mutations were induced in various parts of a gene of *Escherichia coli*. Mutations in those regions that differ little in different bacteria species impaired enzyme function, whereas mutations in widely varied regions in species had little effect. Populations are highly variable genetically.

(k) Extensive genetic variation exists within and among populations. No 2 humans except for identical twins are genetically identical. High level of genetic variation has implications. They may allow populations to evolve rapidly in changing environments.

(l) The reservoir of genetic variation plays the role in success of artificial selection of desirable traits in crops and domestic animals. Rapid evolution is found in development of insecticide resistance in many insects. Studies of bacteria show that adaptive evolution can be based on new mutations, not just on pre-existing variation. To study the linkage between genes and systemic disease is difficult. Some genes with their known biochemical, or physiological functions are called the candidate genes.

(m) Molecular genetic variation at candidate loci in general human population exists. Evolutionary phylogenetic techniques estimate a gene tree from this genetic variation which shows the history of genetic variants of candidate gene. If any mutation altered risk for systemic disease during evolution, the entire branch of gene tree bearing the mutation shows similar disease association.

(n) Population genetic methods can estimate rates and distances of movement of disease-carrying organisms, which affect disease transmissions and potential for control. Molecular analysis of a gene in a mosquito species showed that gene had recently spread among 3 continents proving the insect's efficient dispersal capability. The rapidity of evolution in natural microorganism, many with short generation times and huge populations have important implications.

(o) Evolutionary geneticists and evolutionary ecologists need to develop a general, predictive theory of evolution and population dynamics of pathogens and their hosts, especially for rapidly evolving organisms like HIV and for rapidly migrating hosts like modern humans. Analyses of genetic variation in resistance to pathogens in humans and other hosts are also needed. Clearly animal and human traditions based on learning cannot be explained by the gene alone.

# REVIEW QUESTIONS

## Short Answer Questions

1. State the name of contributors in evolutionary biology.
2. State the theory of Lamarck.
3. Who wrote, "On the origin of species by means of natural selection".
4. Define evolutionary ecology.
5. Define evolutionary genetics.

## Long Answer Questions

1. Discuss the modern synthesis.
2. Discuss the evolution of evolutionary concept.
3. Discuss the gene-hormone-environment interaction in light of recent study.
4. Write an essay on the application of evolutionary biology.

# REFERENCES

Gould, S.J. 2002. The structure of evolutionary theory. Cambridge, MA:Harvard University Press.

Gould, G.C, McFadden, B.J. 2004.Gigantism, dwarfism, and Cope's rule: nothing in evolution makes sense without a phylogeny. Bull. Am. Mus. Nat. Hist, 285:219-37.

Gregory, T.R. 2008. Understanding evolutionary trees. Evo. Edu. Outreach, 1:121-37.

Hallam, A.1998. Speciation patterns and trends in the fossil record. Geobios, 30:921-30.

http://creationsafaris.com/crev200503.htm

http://trendingsideways.com/index.php/the-biology-of-human-behavior-robert-sapolskys-key-insights/

https://laelaps.wordpress.com/xmlrpc.php

Linné, Carl von, 1707-1778.Systema naturae 1735 : facsimile of the first edition / Carolus Linnaeus ; with an introduction and a first English translation of the "Observationes" by M. S. J. Engel-Ledeboer and H. Engel. Soulsby no.: Post-Soulsby no. 3928

MacFadden ,B.J. 1992.Fossil horses. Cambridge, UK: Cambridge University Press.

Malthus, T. 1798 .An Essay on the Principle of Population. An Essay on the Principle of Population, as it Affects the Future Improvement of Society with Remarks on the Speculations of Mr. Godwin, M. Condorcet, and Other Writers. London. Printed for J. Johnson, in St. Paul's Church-Yard ,1798. Electronic Scholarly Publishing Project

# CHAPTER 3

# History of Life and Modern View of Evolution

Life is a self-sustaining system capable of Darwinian evolution. Living systems contain linear strings of information based on DNA. The best example of this is the leucine zipper in which leucine residues interdigitate with one another at regular positions. Leucine zipper was self-replicable but incapable of mutation in the same way like DNA (Lee et al., 1997). When isoleucine or valine were introduced near the leucine-leucine interdigitation, the isoleucine could template not only itself, but also the valine, and vice versa. In **DNA**, however, when a guanosine residue mutates to adenosine, the adenosine no longer efficiently pairs with cytosine. The mutant offspring of DNA of necessity compete with their parents for resources; but the mutant offspring of peptide replicators do not suggest that the basis of evolution is not only survival, but competition itself (Ellington, 2012).Mythologies say that god created life and on the 3rd day of creation, God first brought out plants. Fish, birds, mammals, and finally man appeared. Hindu mythology views life as the creation of Brahma. Various theories regarding the origin of life are as follows:

## 3.1 Cosmozoic Theory

Helmholtz believed that microbe came to earth along with meteorites and comets from space and transformed into higher organisms in water. NASA's record of bacteria-like organisms as fossils on martin rock from Antarctica suggests that life came to earth from the outer space. Cosmozoic theory state that some planet with primitive life collided the earth and seeded it with life. Proponents of this theory are Richter, Helmholtz and Arrhenius.

## 3.2 Spontaneous Generation Theory (Abiogenesis)

Thales and others believed that life regularly originated from the non-living matter. The worms and frogs can appear in mud, maggot in meat and mice in refuse. The theory was accepted for about 2000 years. Francisco Redi first shown, the maggot does not appear, if cooked fish and meat are kept in covered containers. Pasteur used flasks to trap dust and micro-organisms. The boiled broth in flasks did not rot, while broth kept in open flasks rotted, proving that spontaneous generation theory was wrong.

## 3.3 Biogenesis

Redi proposed that life begets life and living animal found in the mud, water, or meat originates from the spores, eggs, or hibernating animals.

## 3.4 Biochemical Origin of Life

The theory simultaneously proposed by Oparin and Haldane states that life originated by chemical reactions, after the cooling of primitive earth. Carbohydrates, fats and proteins consist of C, $O_2$, $H_2$ and $N_2$ which were present in the primitive atmosphere. As the earth's temperature dropped to about 1000°C, the elements combined to form $CO_2$, CO, $H_2O$, $NH_4$, $CH_4$, $H_2$, and $C_2H_2$. In absence of free oxygen, the atmosphere was highly reducing. Water vapors along with other gases were formed in turbulent atmosphere. Further drop in temperature condensed vapors and gases, brought down heavy rains to create oceans. Hot metal carbides on the earth's surface reacted with rain water producing $C_2H_2$ gas that filled the atmosphere. $NH_4$ donated $NH_2$ to acetic acid to produce the amino acid like glycine. All amino acids formed in primordial broth have the same structure. Carbohydrates and fatty acids, the chains of C, $H_2$ and $O_2$ could form chemically. Amino acids formed polypeptide chains by amide bonding to form proteins in high energy and high temperature oceans. Such protein globules formed **protobiont** or protein aggregate or poteinoid microsphere or **coacervate** which could grow in size by absorbing protein from substrate

and then divide by binary fission and probably had enzymatic activity.

Lipids formed in water, coated the surface of coacervates, changed them into the Eubionts which maintained their shape, grown and multiplied by fission. Eubionts with primitive cell membrane perhaps had the primitive enzymes and genetic material to store information. Various protein chains inside the eubiont made nucleoproteins and finally the simple RNA in the energy filled atmosphere. As per Cairns & Smith, the earliest genes were self-replicating minerals associated with proteins, and coded for necessaey proteins. The coding material, nucleic acids appeared later and selected out against minerals. RNA, the first genetic material acted as enzymes and catalyzed their own replication. DNA was formed by RNA with the help of reverse transcriptase. Such organisms appeared about 4000 mya and were like today's PPLOs (Pleuropneumonia-like organisms). These are called progenotes, the ancestor of all life, which were slow growing, slow replicating heterotrophs with few genes, few proteins and a plasma membrane. They fed on organic molecules present in primordial broth. Synthesis of organic substances using enzymes by the cell evolved later. Progenotes then caused complex **prokaryotes**, like bacteria containing free DNA and cell-wall. Fossils of cyanobacteria from 3500 million year old rocks have recovered. Prokaryotes appeared to evolve into eukaryotes, when DNA got covered in nuclear membrane about 1600 mya. Thus, the single-celled organisms caused all life.

### *Evidences of Biochemical Origin of Life*

Miller (Figure 3.1, 3.2) artificially created a primitive earth like condition in a flask in which water boiled to produce steam in a big chamber.

**Figure 3.1** Harold Urey (Left), Stanley Miller (Right)

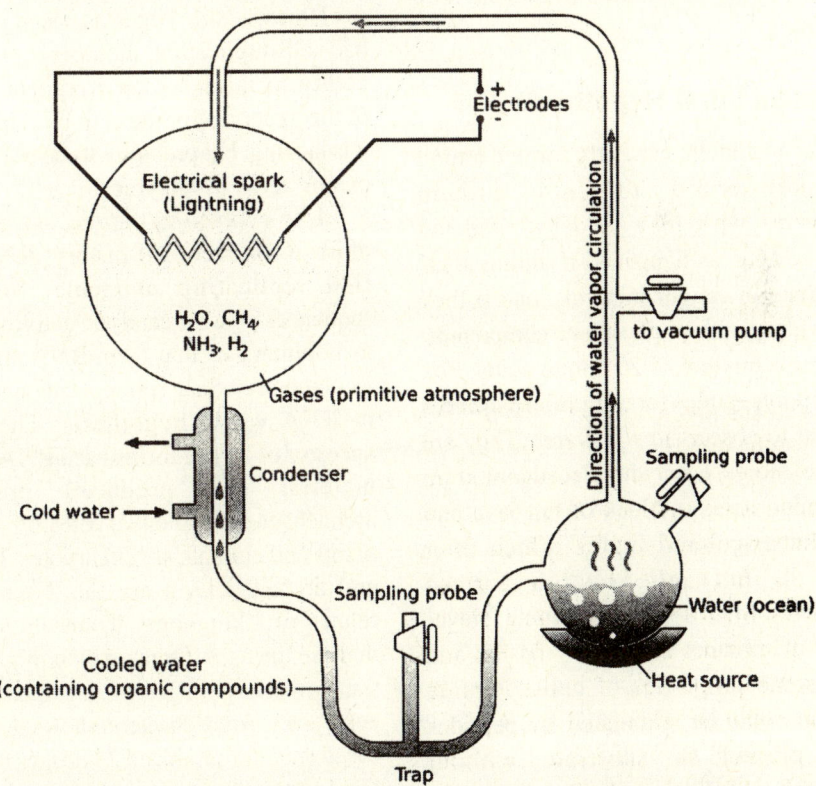

**Figure 3.2** Experiment on the origin of life by Miller and Urey.

$CH_4$, $NH_4$, $H_2$ and water vapors were exposed to electric sparks and then cooled to obtain liquid. After about week, Miller obtained glycine, alanine and aspartic acid. Bahadur obtained almost all amino acids by exposing paraformaldehyde, ammonia and ferric chloride to sunlight. Pavlovskaya and Pasynskii exposed a gas mixture to UV radiation and found glycine, alanine, aspartic- and glutamic- acid. Lowe and others prepared many organic compounds by heating liquid hydrogen cyanide and aqueous ammonia to 90°C for 18 hours. Abelson found that a mixture of CO, $N_2$, and $H_2$ would form hydrogen cyanide and when subjected to UV radiation formed serine, aspartic acid, glycine, alanine and glutamic acid. Schopf supported the natural formation of amino acid in primordial oceans and detected 22 amino acids in 3000 million years old rock formation.

## 3.5 Protein-first Hypothesis

According to Calvin, naturally formed amino acids polymerized to form protein microspheres, when they are dehydrated and exposed to heat, as happened if amino acids are absorbed in clay or minerals, and if they reacted with cyanide or phosphate compounds. By heating a mixture of 20 amino acids, Fox obtained polypeptides forming **microspheres** equivalent to coacervates in water. They are called protenoids. Later she placed a mixture of dry amino acids in block of hot lava and got an amber-colored liquid, which after dilution in hot salt solution formed microspheres. Microsphere containing 2 layer non-fatty membranes divided by fission and showed some properties of cells. Peptide production could be stimulated by peptides already present in substrate without involvement of nucleic acids. Some of them probably acted as enzymes.

## 3.6 Gene-first Hypothesis

Formation of nucleotides is the unlikely event in primitive earth as it needs catalytic action of enzymes to bind nucleoproteins, sugars and phosphates together. It is suggested to be happened naturally on young earth. Ponnamperuma synthesized ribose and deoxyribose sugars by irradiating a mixture of $CH_4$, $NH_4$ and $H_2O$ and showed that formaldehyde in aqueous solution under suitable condition condensed into simple sugars. Schramm showed that polynucleotide was formed when phosphorus-containing compounds were kept under dehydration. Oro and Ponnamperuma showed the formation of adenine from the hydrogen cyanide in special conditions, and purines and pyrimidines were formed under influence of UV radiation. As per Horowitz, life began as naked genes which had self-duplication, mutagenic property, and ability to influence environment for supply of the materials for the cell. Evidence to prove it is lacking, but probably some polynucleotide chains acted as enzymes.

Life evolved in the sea, stayed there for most of the recorded history. **RNA** was the first replicating molecule. Some RNA sequences possess catalytic activity. Some act as polymerases that form RNA strands from its monomers. Such self replication is explained in RNA world hypothesis. The common ancestor of all life forms used RNA as genetic material and produced prokaryotes, archaebacteria and eukaryotes. Protist, fungi, plants and animals are eukaryotes. **Eukaryote** and archaebacteria are the 2 most closely related of 3 kingdoms. Translation is similar in these lineages. Organization of genome and transcription is different in prokaryotes than eukaryotes. Archaebacteria shows that the RNA based ancestor produced 2 lineages that formed DNA genome and evolved mechanisms to transcribe DNA into RNA. The first cell was probably thermophilic, fermentative and anaerobic in absence of oxygen. Rocks as old as 3.5 billion years old contain prokaryotic fossils. Warrawoona, the rocks from Australia provide evidence of organized bacterial communities into structures, the stromatolites found throughout the world.

Bacterium is the only life form found in the rocks for long period. Eukaryote appeared about 1.5 billion years ago. Fungi-like organisms appeared about 900 mya .In photosynthesis, the organism trapped sunlight to produce sugar about 3.4 billion years ago. The first photosystem PSI uses light to convert $CO_2$ and $H_2S$ to glucose with sulfur as bye product. After about a billion years, the 2nd photosystem evolved. Organisms with PSII use both photosystem to convert $CO_2$ and $H_2O$ into glucose with $O_2$ as bye product. Oxygenic (or $H_2S$) **photosynthesis**, using PSI, is found in purple and green bacteria. Oxygenic (or $H_2O$) photosynthesis, using PSI and PSII, occur in cyanobacteria. Since oxygenic bacterium is a lineage within cluster of an oxygenic lineage, PSI perhaps evolved first which corroborates with geological evidence.

Green plants and algae use both photosystems using chloroplasts which originated as free living bacteria that were engulfed by ur-eukaryotes and finally entered into endosymbiotic relationship. **Endosymbiotic theory** of L. Margulis of eukaryotic organelles is now accepted. When the DNA inside chloroplasts was sequenced, the gene sequences were similar to free-living **cyanobacteria** sequences than to sequences from plants the chloroplasts resided in. After the evolution of photosystem II, $O_2$ levels increased. Dissolved $O_2$ in oceans increased like atmospheric $O_2$ which is referred as holocaust. $O_2$, a very good electron acceptor, can damage living organisms. Many anaerobic bacteria die almost immediately in presence of $O_2$. Animals can avoid cellular damage due

to $O_2$.When $O_2$ began building up in environment, materials already present neutralized it. High iron concentrations in sea was oxidized and precipitated and iron layers deposited on sea floor. It attained high concentrations to be dangerous to living organisms and many species became extinct. Some continued to thrive in anaerobic microenvironments and several lineages freely evolved respiration.

The purple bacteria used aerobic respiration reversing the flow of molecules through their C fixing pathways and modify their electron transport chains. Purple bacteria enabled the eukaryotic lineage to become aerobic. Mitochondria, the endosymbionts like chloroplasts of eukaryotes play role in cellular respiration and formed symbiotic relationship in early eukaryotic history. Few groups of eukaryotic cells had mitochondria. Few lineages picked up chloroplasts later. Red algae picked up ur-chloroplasts from cyanobacterial lineage. Green algae picked up different urchloroplasts from prochlorophyte, a group related to cyanobacteria. Animals appeared before the Cambrian, about 600 mya. The first animals dating before the **Cambrian** found in rocks near Adelaide, Australia are the Ediacarian fauna. It is unclear if these forms have any surviving descendants. Some look likes cnidarians; others like annelids. The Cambrian explosion, perhaps the result of higher $O_2$ concentration enables large organisms with high metabolism to evolve due to spreading of shallow seas producing variety of niches to produce variety of animals. Cambrian animal fossils show various organisms from the Canada.

Animal life before the Cambrian was not modern .Photosynthetic protests and algae formed the base of food chain. Following the Cambrian, the number of marine families reached about 200.The Ordovician explosion larger than the Cambrian around 500 mya followed this explosion, produced many Paleozoic fauna. The Cambrian fauna declined slowly during that time. By the end of the Ordovician, the Cambrian fauna had mostly given way to Paleozoic fauna and about 400 marine families stayed till the end of the Permian. Plants evolved from ancient green algae about 400 mya. Both groups use chlorophyll a and b. Plants and green algae store starch in their chloroplasts. Plants and fungi invaded the land about 400 mya. The first moss-like plants required moist environments to survive. Developments of waxy cuticle allowed some plants to exploit inland environments. True vascular tissue is absent in the mosses to transport fluids and nutrients which limits their size. Vascular plants evolved from the mosses. The first vascular land plant is *Cooksonia*.

Arthropods followed plants onto the land. The first land animals are myriapods. Vertebrates moved to land by the Devonian. *Ichthyostega*, an amphibian, the first known land vertebrates evolved from lobe-finned fish (Rhipidistians). Amphibians evolved into the reptiles with scales to decrease water loss and shelled egg permitting young for hatching on land. The earliest well preserved reptile is the *Hylonomus* from the rocks in Nova Scotia. The Permian extinction was the largest extinction. Cambrian fauna went extinct. Paleozoic fauna decreased from about 300 families to about 50 and about 96% of all species became extinct. The modern fauna including fish, bivalves, gastropods and crabs expanded since the **Ordovician** and now includes above 600 marine families. Paleozoic fauna held steady about 100 families. A 2nd extinction shortly followed the Permian and kept animal diversity low for a while.

During the Carboniferous and Permian, ferns and their relatives dominated the land. After the Permian extinction, gymnosperms evolved seeds from the seedless fern ancestors,

to disperse effectively and became abundant. Gymnosperms evolved pollen, encased sperm for out crossing. Dinosaurs evolved from the archosaur with an upright stance. Amphibians and reptiles walked in undulating pattern as their limbs were modified fins. Their gait was modified from of swimming of fish. Dinosaurs evolved an upright stance similar to upright stance freely evolved in mammals for continuous locomotion. Dinosaurs were warm-blooded. Birds originated from the saurischian dinosaur.

**Angiosperms** evolved from the gymnosperms. Origin of fruits and flowers helped gymnosperms to become dominant. Fruits helped in animal-based seed dispersal and deposition of fertilizer. Flowers facilitate mostly insect based pollen dispersal. Over 3/4ths of all living plants are the angiosperms. Insects evolved from the primitive arthropods. Their mouth parts are modified legs. Over half of all named species are the insects, of which one third are the beetles. At the end of the Cretaceous, a minor mass extinction results in demise of all lineages of dinosaurs saving the birds. Up to this time, mammals were confined to nocturnal, insectivorous niches. Dinosaurs appeared and diversified. *Morgonucudon* exemplifies the first mammals. Mammals originated from therapsid reptiles exemplified by *Diametrodon*. Most successful lineages of mammals are humans. Earth was in a state of flux for 4 billion years during which abundance of different groups varies. New lineages evolve and radiate causing extinction of older lineages. Organisms modify their environments causing further evolutionary change. Diversity of organisms increased although interrupted many times by mass extinctions. Diversity reached a peak prior the arrival of humans. Increased human population decreased biodiversity at an ever-increasing pace.

Earth formed about 4,570 mya after a collision that formed the moon about 40 million years later, cooled quickly to have oceans and an atmosphere. The oldest report of life on earth dates to 3,000 mya, although fossil bacteria date back to 3,400 mya. Geochemical evidence for presence of life goes to 3,800 mya. Colonies of different bacteria were the dominant life on earth. Oxygenic photosynthesis enabled them in oxygenation of atmosphere about 2,400 mya. This atmospheric change increased their effectiveness as nurseries of evolution. In eukaryotes, cells were complex in internal structures and their evolution became rapid due to the ability to transform oxygen from a poison to a potential source of energy. This innovation may have arisen from the primitive eukaryotes capturing $O_2$-powered bacteria as endosymbionts and transforming them into mitochondria. The earliest report of complex eukaryotes with mitochondria dates from 1,850 mya.

Multicellular life is composed only of eukaryotic cells. The earliest report for it is the Francevillian group fossils from 2,100 mya. Specialization of cells for different functions first appears between 1,430 mya and 1,200 mya. Sexual reproduction may be a prerequisite for cell specialization. The earliest known animals are **cnidarians** from about 580 mya. Early animal fossils are rare as they had no mineralized hard parts. Spread of life from water to land required protection against drying out and supporting against gravity. The earliest report of land plants and land invertebrates date back to 476 and 490 mya respectively. The lineage producing land vertebrates evolved later, but rapidly between 370 mya and 360 mya. Successful land plants caused an ecological crisis at the end of the Devonian, until the origin and spread of fungi that could digest dead wood. In the Permian, synapsids dominated land, but **Permian-Triassic extinction** event came close to wipe out complex life. The extinctions were sudden at least to vertebrates. During the slow recovery from this catastrophe, archosaurs became the most abundant.

One archosaur group, the dinosaurs, was the dominant land vertebrates for rest of the Mesozoic. The birds originated from one group of dinosaurs. During this time, mammals' ancestors were small, nocturnal insectivores. After the **Cretaceous-Paleogene extinction** event (65 mya) the non-avian dinosaurs were killed. The birds were the only surviving animals. The mammals increased rapidly in size and diversity. Rapid rise to dominance of terrestrial ecosystems was due to the co evolution of pollinating insects. Social insects originated at the same time. They constituted only small parts of the insect "family tree", now form over 50% of the total mass of all insects.

**Remark:** Acceptable theory on the origin of life is still a major task in contemporary evolutionary biology with philosophical (Griesemer, 2008) and education challenge (Lazcano and Pereto, 2010). Some evolutionary studies had incorporated spontaneous generation to explain the origin of life without direct divine causation (Farley, 1977). Lamarck (1986) suggested the starting point of the on-going transformative process of living beings as an event of spontaneous generation. Lamarck's ideas on chemistry hindered an open acceptance of the transformation of mineral substances into living ones( Tirard, 2006).Whether the existing biodiversity have descended from a single stem, or diverse stems, or whether life started once or at different times left unanswered by Haeckel ( Fry, 2000, Richards, 2008). The confrontation between Louis Pasteur's experiments to discard spontaneous generation and the assumption of a natural origin of life within a Darwinian context was considered as a dead end (Fry, 2000). The origin of life from inorganic matter through chemical transformations on the primitive earth was speculated by several scientists (Kamminga, 1988)

## 3.7  Viruses and Sub cellular Replicators

**Viruses** are the sub cellular replicating entities. Sub cellular entities include viruses, viroids, transposons, prions satellites, plasmids. They require cells for replication and follow the Darwinian model of descent with modification and natural selection. Most viruses are unrelated with old ancestry. Multiple independent origins of viruses are suggested by different means. Some viruses are DNA based, others are RNA based. More fundamental variations among viruses exist. Viruses may be originated by spontaneous creation by cells when mutation creates a piece of DNA or RNA those codes for endless replication. Evolutionary process could be additives and reductionist. Selective pressure reduces the size of genetic code and eliminates unnecessary elements so that copies of the code can be made faster and with few resources. Thus, the rate of replication for smaller pieces of code will be faster which will overpopulate larger ones. This applies to both cellular and sub cellular evolution. Under some conditions, reduction of genetic code is the primarily selected characteristic. Viruses and other sub cellular replicators become less complex as a large part of their life cycle is completed in their host cells through generations. As the host cells for sub cellular replicators become complex, the sub cellular replicators shed genetic code that performs redundant functionality to code that exists in host cell. Thus, the existing viruses are less complex than original ancestral viruses.

**Viroids** are much smaller than viruses. They have no protein coat. Viroids can be replicated in test tubes. Satellites, or virusoids, are viroids which depend on virus for their replication and spread. Satellites, the small pieces of RNA replicate and get encapsulated into protein capsule of helper virus. Sometimes, the virus cannot be replicated without **satellite**. Sometimes the satellite is like a parasite of virus. Plasmids are autonomously replicating segments of DNA that integrate into host DNA. Sometimes plasmids affect the behavior of their host, causing the host bacteria to join with other bacteria and inject copies of plasmid. Some plasmid kill cells, while others are beneficial. **Transposons** replicate and move about the DNA of a cell. They are thought of as segments of DNA that only operate within a cell, though they can spread from one organism to another. Transposons are called selfish DNA which are replicated and move about the genetic code with no apparent function for the host. They have an indirect effect of mixing up the genetic code and increasing the rate of genetic change.

**Prions** are the first known replicators without nucleic acids. They are made only of protein. Prions discovered in 1980s, cause several infectious diseases, including mad-cow disease. Prions are the products of proteins found in brain tissue. The prions cause the brain related diseases. Major problem with the diseases caused by prion is that protein is difficult to destroy than nucleic acids. DNA and RNA are easy to breakup, but proteins are resistant and long-lasting. With few exceptions, these entire sub cellular replicators are inert. Metabolic activity of some viruses last for short periods. Some viruses do nothing and just take in by cells. Some actively attach to cells, transfer their genetic code. Replication of viruses is determinative. Cells have instructions to take genetic code (DNA, or RNA) and act based on the structure of code. When a piece of DNA, or RNA is in a cell, the cell take that piece of code and integrate it back into the nucleus, replicate it, or transcribe/reverse transcribe it. When the viral code enters a cell, the cell takes the code and starts working the instructions in code. The cell treats viral DNA, or RNA the same way that it treats other DNA, or RNA in cell. Sometimes the virus codes for enzymes that are absent in cell, which causes different behavior. This preceded the way any chemical reaction proceeds. Sub cellular replicators are passive agents. The cells actually reproduce them. Viruses have no active role in their own construction. At the time of virus reproduction, the original virus doesn't even exist.

## 3.8 Parallel and Convergent Evolution

Sometimes evolutionary change follows common pathway in 2 or more unrelated, or distantly-related organisms due to similar pressures from the environment. It culminates in unrelated organisms with similar external characters even though they did not have the same ancestor. This is **parallel evolution** as found in distantly-related plant families that have evolved to parasitic mode of existence. Some plants have evolved separately into mycotrophic mode of existence to obtain nutrients from the mycorrhizal fungi, which are parasitic on nearby forest trees and shrubs. Photosynthetic pathways, such as CAM (crassulacean acid metabolism) and C-4 photosynthesis, have evolved separately in distantly-related plant families.

There are different hypotheses for monophyletic origin of flowering plants (angiosperms). Vessels appear in 2 distantly related group viz., the gnetopytes and angiosperms. This illustrates parallel evolution (Homoplasy), which under similar environment in distantly-related organisms results in plants and animals that are morphologically similar in overall appearance. This is convergent evolution. Some authors use these 2 terms interchangeably. North American cactuses and

South African euphorbias belong to different plant families. They are distant relatives in phylogeny of flowering plants and have succulent, thick stems for retaining water. They have spines for protection, and are adapted in arid desert regions with low and erratic rainfall. Some African euphorbias without flowers are inseparable from their North American counterparts.

Biologists often distinguish the convergent evolution from the parallel evolution assuming that when a phenotype evolves, the genetic mechanisms are different in distantly related species (convergent), but similar in closely related species (parallel). The same phenotype might evolve among populations by changes in different genes. Conversely, by changeing the same gene, similar phenotypes would evolve in distantly related species. Perhaps, the distinction between the convergent and parallel evolution is a false dichotomy. All

**Figure 3.3** Pill bugs and Pill millipede converged for defenses.

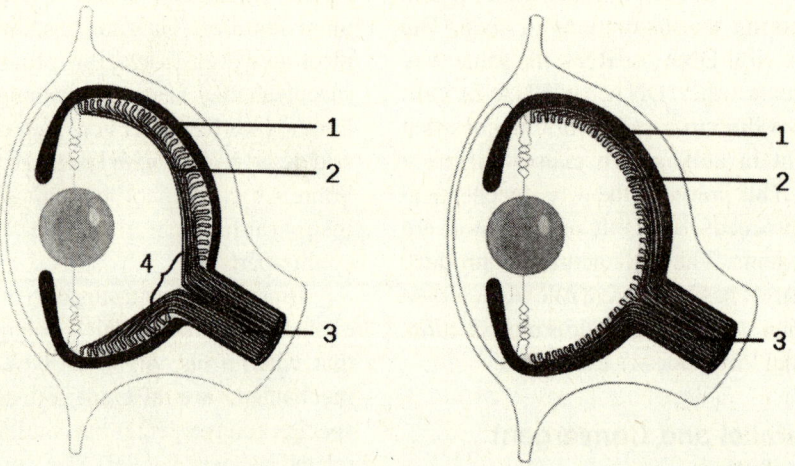

**Figure 3.4**    In the camera eye development, the nerve fibers pass in front of the retina. There is a blind spot(4), where the nerve pass through the retina. In the Octopus, the eye is constructed with the nerves attached to the retina and do not have a blind spot.

instances of the independent evolution of a given phenotype are convergent and their phenotypic similarity can be called convergent evolution (Figures 3.3, 3.4).

In Australia, many marsupials resemble North American placental mammals. Australia's flying phalanger is similar to the North American flying squirrel. Both glide through air with their parachute-like fold of skin between the front and hind legs. The North American preying mantis and mantispid shows the convergent evolution. Although they differ greatly in size, these 2 insects are similar in appearance. Their triangular heads have large eyes. They have a pair of raptorial front legs; other 2 pairs of legs are used for walking. They belong to 2 different insect orders. Mantids belong to the Orthoptera. Mantispids belong to the Neuroptera. The mantispid is smaller than preying mantis with short antennae. Mantispids have the 2 pairs of membranous wings with a network of veins (nerves) typical of the Neuroptera. The name Neuroptera is derived from Greek and means

nerve wing. The wings are tent like on the body, unlike that of mantid's wing.

Leathery forewings of mantids lie flat on the abdomen. A pair of membranous hind wings is folded beneath the forewings. Mantispids completes life-cycle through egg, larva, pupa and adult. Mantids undergo incomplete metamorphosis with egg, nymph and adult. A female mantispid lay stalked eggs on leaves and wooden structures. The freshly hatched larvae search for spiders. They enter the spider's egg sac by direct penetration, or they climb onto the female spider for entering the egg sac. While the matispid is waiting for female spider to build an egg sac, it enters the spider's book lungs and feed on the blood of spider. The mantispid enters the egg sac before completion of spinning of protective silken case by female spider. Once inside the egg sac, the mantispid dine on spider eggs and grow. Mature larva then spin a cocoon and metamorphose into a pupa within the spider's egg sac. It emerges as adult a few weeks later.

Convergent evolution is the development of similar body and adaptation patterns in 2 unrelated species. Every species has its own characteristic traits. To survive in constantly changing environment, species alters their structures. Often 2 different species undergo similar modifications to adapt to the same environment. Thus,these species acquire similar structures through independent changes. Differences among convergent-, parallel - evolution and evolutionary relay need understanding.

Evolutionary relay refers to similar types of changes that occur in 2 species, but at different time. The dorsal fins of an extinct ichthyosaur and a shark are the same type of adaptive alteration in 2 distinct species that evolves at different times. A **parallel evolution** refers to changes occurring in related species at the same period. The prickly extrusions are found on new world porcupines and old world porcupines. The pattern is adapted by species of the same family. Convergent evolution is different as it occurs in the same time involving species of different ancestry. It occurs due to ecological living patterns, like the food habits of 2 animals of different ancestry as found in marsupials and placentals. Both have certain similarities arising out of their type of living. Saber-tooth formation in the cat family of both types and the flying squirrels exemplify it. Characters that emerge due to this evolution are analogous structures. The likeness that evolves in different species following this evolution is termed homoplasy (Figure 3.5). The influence of convergent evolution in development and adaptation is a matter of ongoing debate. **Convergent evolution** is a real phenomenon which cannot be credited to a given set of guidelines as of today.

The independent evolution of similar features in different evolutionary lineages is common, perhaps ubiquitous (Conway Morris, 2010), but without clear-cut interpretation.

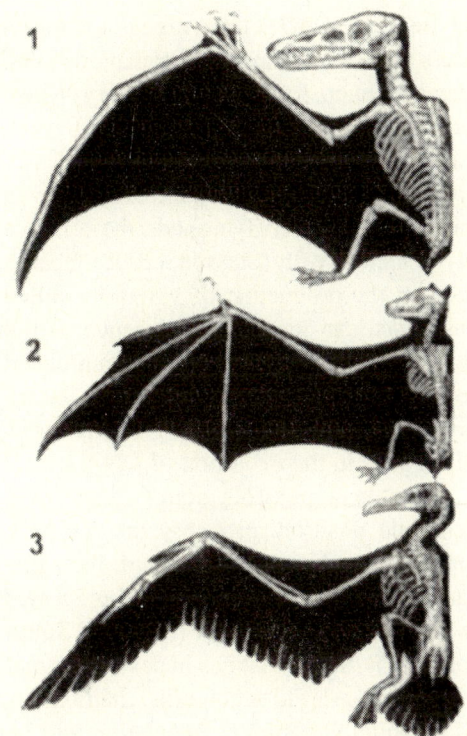

**Figure 3.5** Wings of Pterosaur (1) bat (2) and birds (3) showing hormoplasy.

Convergent evolution is considered as evidence of adaptation. Phenotypic similarity of distantly related taxa that occur in similar environments is the strong evidence that natural selection has produced evolutionary adaptation (Conway Morris, 2003). The rationale for this assumption is that process other than natural selection could lead fast-swimming aquatic animals like sharks, tunas to evolve a streamlined body form. Convergent evolution is phenotypic similarity that is independently derived in 2 or more lineages, instead of similarity from inheritance from the common ancestor. Recognizing such derived similarity needs a phylogenetic perspective (Ridley, 1983) which led to development of statistical methods that test whether taxa occupying similar environmental conditions tend to evolve in similar ways (Huey and Bennett, 1987). Convergent evolution occurs for reasons unrelated to adaptation and natural selection. Random evolutionary change cause species to become more similar to each other than were their ancestors. The rates of convergence can be high in circumstances in which clades are diversifying under the influence only of genetic drift (Stayton, (2008). Shared biases in the production of variation, called constraints, can lead to convergence. If the possible variants are limited, then unrelated species can produce the same variations, which may be fixed in population by genetic drift. This is seen in the evolution of DNA.

As only 4 possible states exist for a nucleotide position, probably the distantly related taxa independently acquire the same change by chance. Developmental processes are the source of constraint (Maynard Smith et al., 1985). Some changes in developmental process are easier to accomplish than others, and potential variation in population will be biased toward those changes. As a result, evolution is likely to proceed in that direction, due to random drift or selection. If taxa share the same developmental system, they evolve in same way, producing convergent evolution. In salamanders, the number of digits on limbs is determined in part by cell number in limb bud. Evolutionary changes that reduced the number of cells resulted in convergent reduction in digit number from 5 to 4 in many species (Wake, 1991). Jaekel and Wake (2007) provide many examples from salamanders. Shared patterns of genetic correlation similarly bias the production of phenotypic variation and channel evolution in certain directions. This idea states that evolution may proceed most readily along lines of least genetic resistance (Schluter, 1996); species with similar genetic correlations evolve in similar way of shared epistatic or **pleiotropic effects** because convergence (Leroi et al., 1994). Convergent evolution occurs in taxa that face similar

selective environments that is taken as evidence of operation of natural selection and is not indicative of adaptation. If traits evolve convergently for reasons unrelated to natural selection, then we should not expect a correlation between possession of trait and occurrence in a particular selective environment (Pagel, 1994). If a trait evolves convergently in multiple taxa for reasons unrelated to natural selection by chance all the taxa may occupy the same selective environment. The possibility that convergent evolution of similar traits in similar environments could occur coincidentally is enough to cast serious doubt on comparative approach to identifying adaptation. Workers reject the application of probabilistic thinking to historical evolutionary analyses (Kluge, 2005). Statistical approaches are used in other sciences. However, these approaches are inapplicable to historical evolutionary events (Felsenstein, 2004).

Independent acquisition of similarities in characters in organisms is called homoplasy. But if this similarity is found in closely related animals that have descended from common ancestor, it is called parallelism and is caused by parallel evolution. As animals are closely related, they respond to selection by modifying organs in a similar way. If **homoplasy** occurs in systematically distant types, similarities will fall under convergence, as in dolphin and fish. Ungulates inhabiting forests and grasslands have descended from a common ancestor *Condylartha* and have evolved parallel cursorial adaptation by having hoofed unguligrade locomotion, grazing or browsing habit and identical general body organization. Similarly, creodonts, the archaic carnivores gave rise to modern carnivores belonging to Felidae and Canidae families, which show striking similarities in dentition, claws, body structure and way of hunting. There are 3 different sources of **phylogenetic parallelism**: Due to similar genetic factors such as parallel mutations and gene recombination .Parallel natural selection acting on homologous organs, leading to similar adaptations. Parallel selection act on analogous organs due to identical environment leading to superficial similarity of characters. Chromosomes of *Drosophila pseudoobscura* and *D. miranda* show homologous genes in inverted and translocated arrangements on the chromosomes. Eye pigmentation in *D. melanogaster* and *D. pseudoobscura* is controlled by homologous genes that show parallel evolution. **Parallel mutations** are seen in genes controlling colour of European snails, *Cepea hortensis*, *C. nemoralis* and *C. vindobonensis*. In snails belonging to family Helicidae, namely, *Murella, Tyrheniberus, Rossmaessleria, Iberus, Levantina* and *Eremina*, the shells showing identical shape and structure of shells exhibit parallel evolution. If environmental conditions are identical, parallel natural selection often results in parallel evolution. Protective coloration and resemblance in animals results

from such a selection. Thus narrow and elongated body and legs in insects have evolved freely in stick insects, *Limnotrochus* and *Neides* and pond skaters.Broad leaf-like wings have developed in *Phyllium crurifolium*, grasshoppers, plant bugs and praying mantis. If the wing colours show parallelism in sympatric species and it gives protection to species, it may give rise to mimicry as found in Papilionidae, Danaidae, Heliconidae, Pieridae, and Satyridae.Relatedness and common ancestry is the key factor to distinguish parallelism from convergence. Similarity in evolution of marsupials in Australia and their counterpart placentals in other parts of world can be called parallel evolution, since as mammals they have evolved from common ancestor but relatedness ends there, and they have diverged far apart genetically. In appearance and habits, marsupial mole (*Notoryctes*) is similar to placental mole (*Talpa, Thylacinus* to *Canis, Dasyurus* to placental *Felis, Petaurus* to *Glaucomys* and

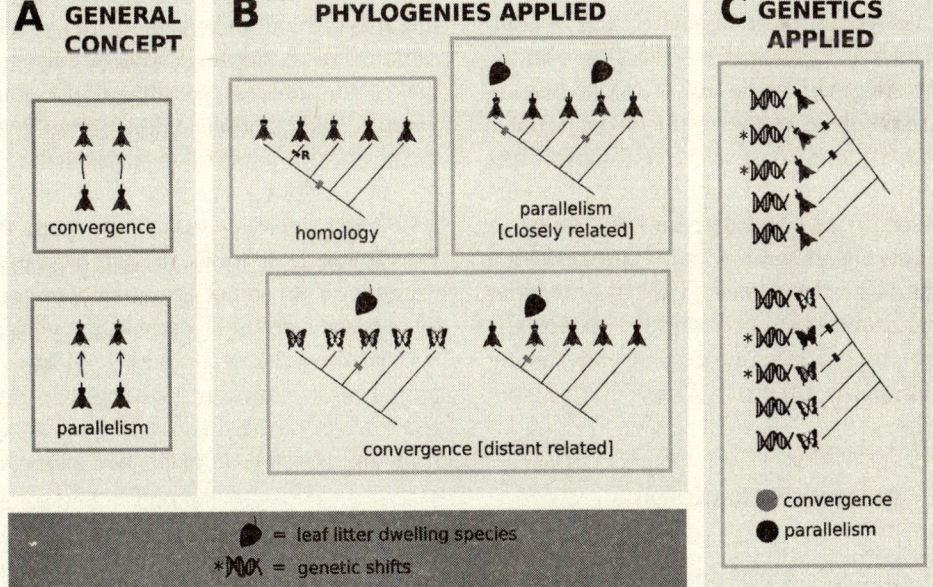

**Figure 3.6**   Changing definitions of convergence and parallelism over times.
(A) General Concept (B) Phylogenies Applied (C) Genetics Applied

*Dasycercus* to *Mus*. These animals are definitely **ecological equivalents** but have diverged far apart to be categorized in parallel evolution (Figure 3.6).

Difference between convergent and parallel evolution is contentious and the terms partly predate modern concept of genetics and development (Wake, 1999). Parallel evolution sometimes refers convergent evolution achieved through same genetic and developmental pathways. Nonparallel convergence refers to similar phenotypic result achieved in other way. Occurrence of parallel evolution suggests the possibility that range of produced variation is limited or some variants may originate often than others. Natural selection may be driving convergence, but internal constraints favour the available options. Possible better solutions might not attain due to limitations (Scheck and Wagner, 2004). Nonparallel convergence suggests that natural selection has gained optimal solution for environment induced problem. That a phenotype is never produced within or among species like an even number of legs in adult geophilomorph centipedes (Arthur and Farrow, 1999) suggests that the lack of species bearing that phenotype is the result of an intrinsic constraint, instead of selective disadvantage. Relatively little variation for a trait occurs intraspecifically and derived traits found as variants in related species suggest that evolutionary change is favoured by limitations on available variation and a role for the variation-generating process in constraining evolutionary direction (Wake and Larson, 1987).

## 3.9   Divergent Evolution

**Divergent evolution** is a cumulative process of natural selection leading to formation of new species, from the same ancestral stock. This is a gradual process of evolution.

Organisms of same species, with slight different variations in genetic code, are subjected to different environmental conditions. Over a period of time, every group and its succeeding generations face different environmental conditions and challenges. Rules for survival are varied for these groups now, which once belonged to the same ancestral stock. Natural selection promoted different characters in these groups. Over a long period, small and gradual changes in the features and constitutions of their generations cumulatively give rise to a new species in both groups. They now look different from their common ancestral stock. Thus, they diverge from ancestral line and form their own new species.

The diverged groups classified as separate species are subjected to varied environmental conditions. They differ more in their appearance. The need of adaptation modifies in sub sequent generations through selecting conducive traits from the varied gene pool. After millions of years, they diverge so much in their features, appearance, and functions that it is difficult to believe that they had the same ancestors. The underlying machinery that makes this process possible is the genetic variance and inheritance. It is type of genetic memory that is passed along through the DNA to every offspring, one each from the mother and father, creating variance in gene pool. Development of limbs in vertebrates is an example of divergent evolution. The arm in human, the side fin of a whale, the wing of a bat and the paws of a cat evolved from one primal stock of ancestors. Divergent evolution is a force of nature that creates the variety of form and structure in plants and animals. .

## 3.10  Fossils of Doushantuo Formation

The 580 million year old Doushantuo formation harbours microscopic fossils that may represent ancient bilaterians. Some were described as animal embryos and eggs. Some may represent the remains of large bacteria. The fossil, *Vernanimalcula* is interpreted as a coelomate bilaterian, but may simply be an in filled bubble. These fossils form the earliest hard-and-fast evidence as opposed to other predators.

**Cambrian substrate revolution:** Traces of organisms, moving on and directly underneath the microbial mats, that covered the Ediacaran sea floor are preserved from the **Ediacaran period** about 565 mya. They were probably made by organisms resembling earthworms in shape and size. The burrow-makers have never been found preserved. The burrowers probably had bilateral symmetry. They fed above the sediment surface and forced to burrow to avoid predators. Around the start

of the Cambrian, new types of traces first appear, including well -known vertical burrows like *Diplocraterion*, and traces attributed to arthropods like *Cruziana* .The vertical burrows show that worm-like animal acquired new behaviors and new physical capabilities. Some Cambrian trace fossils show that their makers possessed hard but not mineralized exoskeletons. An Ediacaran **trace fossil** was formed when an organism burrowed below a microbial mat. Burrows provide firm evidence of complex organisms. They are more readily preserved than body fossils, to extent that absence of trace fossils is used to imply the absence of large, motile bottom-dwellers. Cambrian explosion presents real diversification and is not preservational artifact as burrowing became established. Burrowers disturbed on the sea floor aerated it, mixing

(a) *Opabinia* made the largest single contribution to modern interest in the Cambrian explosion.

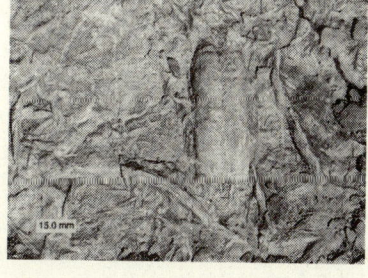

(b) *Rusophycus* and other trace fossil from the Gog formation, Middle Cambrian, Lake Louise, Alberta, Canada.

(c) *Dickinsonia costata,* Ediacaran organism of unknown affinity, with a quilted appearance.

(d) Fossil of *Kimberella*, a triploblastic bilaterian, and possibly a mollusc.

**Figure 3.7** Organisms created scope for higher diversity.

$O_2$ into the toxic muds. This created the bottom sediments suitable for inhabiting a range of organisms - creating new niches and scope for higher diversity (Figure 3.7).

**Ediacaran organisms:** At the start of Ediacaran period, most acritarch fauna which had remained relatively unchanged for millions of years became extinct, and replaced with range of new, larger species. This radiation, the first in fossil record, is followed soon after by unfamiliar, large, fossils dubbed the Ediacara biota, which spread for 40 million years until *Dickinsonia costata*, an Ediacaran appeared at the start of the Cambrian. Most Ediacara biota were at least few cms long, larger than any earlier fossils. The organisms belong to 3 distinct assemblages and increase in size and complexity in time. Many organisms were unlike anything that appeared before or since, resembling discs, mudfilled bags, or quilted mattresses. One paleontologist proposed to classify the strange organisms as separate kingdom, Vendozoa. Some early forms of the phyla at the heart of the Cambrian explosion were interpreted as early molluscs, echinoderms and arthropods. *Kimberella* was at least a triploblastic bilaterian animal. These organisms are central to debate about the abruptness of the Cambrian explosion

## 3.11 Skeletonization in Ediacaran-Early Cambrian

The first Ediacaran and lowest Cambrian skeletal fossils represent tubes and problematic sponge spicules. Monaxon siliceous spicule aged around 580 mya known from Doushantou are the oldest sponge spicules. Formation in China and deposits of same age in Mongolia caused change in interpretation of these fossils as spicules . In the late Ediacaran-lowest Cambrian, many tube dwelling organisms appeared. It was organic-walled tubes and chitinous tubes of sabeliditids that prospered

up to beginning of Tommotian. Mineralized tubes of *Cloudina*, Sinotubulites and other organisms from carbonate rocks formed near the end of Ediacaran period from 549 to 542 mya and triradially symmetrical mineralized tubes of anabaritids from uppermost Ediacaran and lower Cambrian. Ediacaran mineralized tubes are often found in carbonates of stromatolite reefs and thrombolites. They could live in an adverse environment.

Although they are hard to classify as other Ediacaran organisms, they are important in 2 other ways. First, they are the earliest known calcifying organisms Secondly, these tubes are device to rise over substrate and competitors for effective feeding and to lesser degree, they serve as armor for protection against predators and adverse conditions of environment. Some *Cloudina* fossils bear small holes in shells. It is possible that holes are evidence of boring by predators sufficiently advanced to penetrate shells. A possible evolutionary arm race between the predators and prey attempts to explain the Cambrian explosion. In the lowest Cambrian, the stromatolites were decimated, beginning colonization of warm-water pools with carbonate sedimentation. Anabaritids and Protohertzina fossils were the first. Such mineral skeletons as shell, sclerites, thorns and plates appeared in uppermost Nemakit-Daldynian. They were the earliest gastropods, halkierids, hyoliths and other organisms.

The beginning of Tommotian has been marked an explosive increase of number and variety of molluscs, hyoliths and sponges, along with complex of skeletal elements of unnamed animals, the first brachiopods, archaeocyathids, tommotids and others. This sudden increase is partially an artefact of missing strata at Tommotian section, and most fauna began to diversify in a series of pulses through Nemakit-Daldynian and into Tommotian. Some animals had sclerites, thorns and plates in Ediacaran, but thin carbonate skeletons cannot be fossilized in siliciclastic deposits. Older fossils indicate that mineralization long preceded the Cambrian, probably defending small photosynthetic algae from single-celled eukaryotic predators.

## 3.12 Small Shelly Fauna

Fossils known as small shell fauna are found in many parts of world and date from just before the Cambrian to about 10 million years after start of the Cambrian. These are a mixed collection of spines, sclerites, tubes, archeocyathids and small shell like those of brachiopods and snail-like molluscs -but all tiny, mostly 1 to 2 mm long. These fossils are common than complete fossils that produced them. They cover the period of time that otherwise lack fossils and supplement the conventional fossil record and allow fossil ranges of many groups to be extended.

## 3.13 Echinoderms and Early Cambrian Trilobites

The earliest trilobite fossils are about 530 million years old, but the class was already diverse and worldwide, showing their presence around for some time. Fossil record of trilobites begins their appearance with mineral exoskeletons - not from the time of their origin. The earliest accepted echinoderm fossils appeared a little bit later, in Late Atdabanian; unlike modern echinoderms. All these ancient Cambrian echinoderms were not radially symmetrical which provide data points for end of explosion or at least indications that the crown groups of modern phyla were represented.

## 3.14 Fauna from the Burgess Shale

A fossilized trilobite from the Burgess shale preserves soft parts - antennae and legs. The Burgess shale and similar lagerstatten preserve soft parts of organisms, provide a wealth of data to classify enigmatic fossils. It often preserved complete specimens only otherwise known from fragmented parts like loose scales, or mouthparts. Majority of organisms and taxa in these horizons are entirely soft bodied and absent from the rest of the fossil record. Since a large part of ecosystem is preserved, the ecology of community can be reconstructed. Assemblages may represent a museum: a deep water ecosystem that is evolutionarily behind the rapidly diversifying faunas of shallower water. As the lagerstatten provides mode and quality of preservation that is virtually absent outside the Cambrian, lots of organisms appear completely different from the conventional fossil record. This led workers to attempt to classify these organisms into extant phyla. Preservation mode was the rare in preceding Ediacaran period, but known assemblages show no trace of animal perhaps indicating absence of macroscopic metazoans.

## 3.15 Precambrian Life

Changes in abundance and diversity of some fossil have been interpreted as evidence for attacks by animals. Stromatolites, the stubby pillars built by microorganisms, are a major constituent of fossil record from about 2,700 mya, but their abundance and diversity declined steeply after about 1, 250 mya due to grazing and burrowing animals. **Precambrian** marine diversity was dominated by small fossils called acritarchs. This term describes almost any small organic walled fossil, Stromatolites near Helen Lake, Banff National Park, Canada, from the egg cases of small metazoans to resting cysts of different kinds of green algae. Acritarchs appeared around 2,000mya underwent bloom around 1,000 mya, increased in abundance, diversity, number of spines and size, and complexity. Their increasingly complex spiny forms in the last 1 billion years are perhaps due to need for defence against predation. Small organisms from the Neoproterozoic era show signs of anti-predator defenses. Consideration of taxon longevity supports an increase in predation pressure around this time. The rate of evolution in Precambrian was too slow with persisting unchanged cyanobacterial species. Cambrian animals did not appear from nowhere at the base of the Cambrian. Their predecessors had existed for hundreds of millions of years.

## 3.16 Early Cambrian Crustaceans

Crustaceans, one of the 4 modern groups of arthropods, were rare throughout the Cambrian. Convincing crustaceans were once thought to be common in Burgess shale-type biotas, but none was crown group of true crustaceans. **Microfossils** are the record of crown group crustaceans in the Cambrian. The Swedish Orsten horizons was found to contain later Cambrian crustacea, but only organism's smaller than 2 mm are preserved. This limits the data set to juveniles and miniaturized adults. More information is found in organic microfossils of Mount Cap formation, Mackenzie Mountains, Canada. This Cambrian assemblage consists of microscopic fragments of arthropod's cuticle, which is left behind when rock is dissolved with hydrofluoric acid. Diversity of this assemblage is like that of modern crustacean faunas. Analysis of feeding machinery found in formation shows that it was adapted to feed in a precise fashion. This contradicts with the other early Cambrian arthropods, which fed by shoveling anything they could get their feeding appendages into their mouths. This feeding machinery belonged to large organism and would have provided potential for diversification. Specialized feeding apparatus allows several approaches to feed

and develop, and creates several different approaches to avoid being eaten.

Major metazoan groups appeared in few million years of early to mid-Cambrian. Evidence of Precambrian metazoa is gradually accumulating. If Ediacaran *Kimberella* was a mollusc-like protostome, protostome and deuterostome lineages must have split before 550 mya. Even if it is not protostome, it is the bilaterian. Fossils of modern-looking cnidarians are found in Doushantuo lagerstätte. The bilaterian and cnidarian lineages must have diverged 580 mya. Trace fossils and predatory borings in *Cloudina* shells are signatures of Ediacaran animals. Few fossils from the Doushantuo formation are believed as embryos and one as bilaterian coelomate, although such views are not universally accepted. Predatory pressure has acted on stromatolites and acritarchs for around 1,250 mya( Figure 3.8).

Appearance of animal was gradual and their evolutionary radiation was also not very rapid. The **Cambrian explosion** was no faster than any other radiations in animals' history. Some innovations linked to explosion like resistant armour only evolved once in animal lineage. This makes a lengthy Precambrian animal lineage harder to defend. While the phyla may have diversified in this period, representatives of the crown- groups of many phyla do not appear until later in Phanerozoic. Mineralized phyla that form the basis of fossil record may not be representative of other phyla, as most mineralized phyla originated in benthic setting. Fossil record is consistent with the Cambrian explosion that was limited to benthos. Range of different organism designs, or ways of life rose sharply in the early Cambrian. The idea that disparity was high throughout the Cambrian before subsequent decrease is discarded. Disparity remains relatively low throughout the Cambrian, with modern levels of disparity only attained after

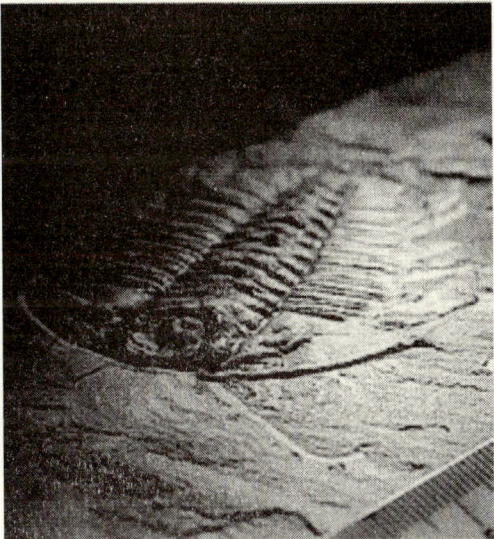

(a)  A fossilized trilobite, an ancient type of atthropod. This specimen, from the Burgess shale preserves soft parts– the antennae and legs.

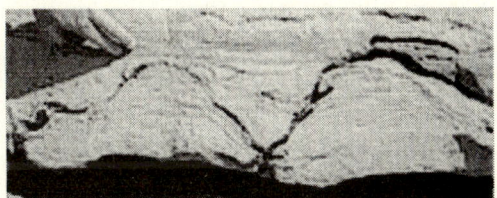

(b)  Stromatolites (Pika Formation, Middle Cambrian) Near Helen Lake, Banff National Park, Canada.

(c)  Modern stromatolites in Hamelin Pool Marine Nature Reserve, Western Australia.

**Figure 3.8**  Stromatolites and trilobite fossil

the early Ordovician radiation. Diversity of many Cambrian assemblages is similar to today, and high level diversity is thought by some to have risen smoothly through the Cambrian, stabilizing somewhat in the Ordovician. Despite the evidence that moderately complex animals existed before and possibly long before the start of the Cambrian, the pace of evolution was fast in the early Cambrian. Possible explanations for this may be developmental and ecological changes. Explanation must explain the timing and magnitude of the explosion.

## 3.17 The Cambrian Explosion

It is often noted that the number and diversity of fossils of multicellular plants and animals increases greatly in the fossil record from about 550 mya. Such sudden appearance of complex life forms, many sharing their most fundamental characteristics with modern species, is called the Cambrian explosion which caused rapid appearance around 530 mya, of major animal phyla accompanied by major diversification of animals, phytoplankton and calcimicrobes. Before 580 mya, most organisms composed of individual cells were simple, rarely organized into colonies. Over the next 70 or 80 million years, rate of evolution increased. Rapid appearance of fossils in Primordial Strata is known. Long-running puzzle about appearance of the Cambrian fauna abruptly and from nowhere focuses on 3 points: whether there was a real mass diversification of complex organisms over a relatively short time during the early Cambrian; what might have caused such rapid change; and what it would imply about the origin and evolution of animals.

William Buckland realized that dramatic step-change in fossil record occurred around the base of the Cambrian. The lowest Silurian stratum showed the origin of life on earth, though others including Charles Lyel differed. Darwin considered this sudden appearance of many animal groups with no known antecedents to be the grave single objection to his theory of evolution. He reasoned that earlier seas had swarmed with living creatures, but their fossils are not found due to imperfections in fossil record. Charles Walcott proposed that an interval of time, the Lipalian, was not represented in fossil record or did not preserve fossils and that ancestors of the Cambrian animals evolved during this time.

History of life on earth goes back to at least 3,850 million years. Rocks Warrawoona in Australia contain fossils of stromatolites at that time. Fossils of more complex eukaryotic cell, from which all animals, plants and fungi are built, have been found in rocks from 1,400mya in China and Montana. Rocks dating from 580 to 543 mya contain fossils of the Ediacara biota. Organisms were so large that they were perhaps multi-celled, but very unlike to any modern organism. P. E. Cloud argued for a period of eruptive evolution in early Cambrian, but in 1970s there was no idea of arrival of modern-looking organisms of Middle and Late Cambrian.

Harry B. Whittington and colleagues in the 1970s concluded that several fossils from the Burgess Shale were complex and different from living animal. The common organism, *Marrella* was an arthropod, but not a member of any known arthropod class. Organisms such as 5-eyed *Opabinia* and spiny slug-like *Waptia* were so different from any organism that they represent different phyla. Whittington and Gould proposed that *Opabinia* made the largest single contribution to modern interest in the Cambrian explosion. Other analyses argue that complex animals similar to modern types evolved well before the start of the Cambrian.

Radiometric dates for much of the Cambrian, gathered by detailed analysis of radioactive elements contained in rocks, have only recently become available and for only a few regions. Relative dating is often sufficient for studying processes of evolution, but this too is difficult because of the problems involved in matching up rocks of the same age across different

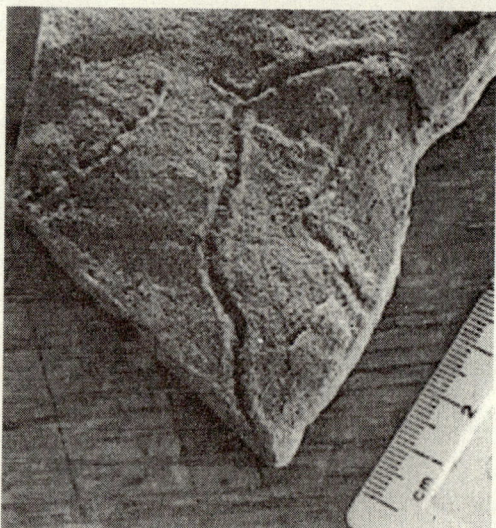

**Figure 3.9** An Ediacaran trace fossil, made when and organism burrowed below the microbiol mat

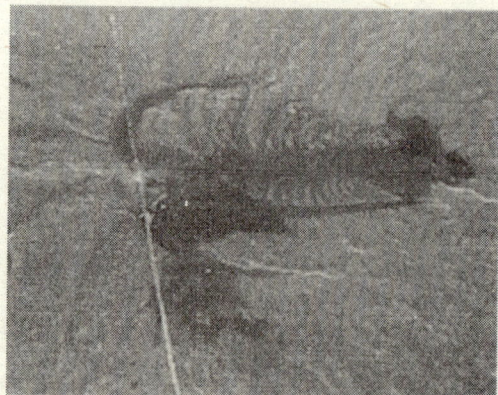

**Figure 3.10** Marrella specimen illustrating how clear and detailed the fossils from the Burgess Shale lagerstätte are

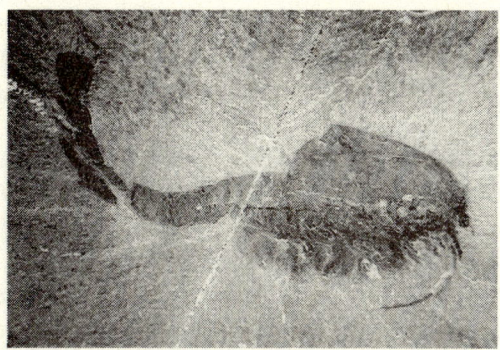

**Figure 3.11** *Waptia*

continents. Therefore, dates or descriptions of sequences of events should be regarded with some caution until better data becomes available (Figures 3.9, 3.10,3.11).

**Uniqueness of explosion:** Cambrian explosion can be viewed as 2 waves of metazoan expansion. First, the rise in diversity, as animals inhabited on Ediacaran sea floor, following the second expansion in the early Cambrian when they established in water column. Diversification rate seen in the Cambrian explosion is unparallel among marine animals. It affected all metazoan clades. Later radiations, like those of fish in Silurian and Devonian periods, involved few taxa with similar body plans. Although the recovery from Permian-Triassic extinction started with about few animal species as Cambrian explosion, recovery produced far few new types of animals. The early Cambrian diversification opened up a range of previously unavailable ecological niches. When all these occupied space limited space existed for wide-ranging diversifications to occur again, because strong competition in all niches and among incumbents usually had advantage. If a wide range of empty niches had continued, clades would continue diversifying and become disparate for us to recognize them as different phyla. When niches were filled up, lineages continued to resemble one another long after they diverge, as limited scope existed for them to change their life-styles and forms. There were 2 similar explosions in evolution of land plants. After a cryptic history beginning about 450 mya, land plants underwent a rapid adaptive radiation during the Devonian period about 400 mya. Angiosperms originated and rapidly diversified during the Cretaceous period.

## 3.18 Changes in Environment

**Increase in Oxygen Levels:** Earth's earliest atmosphere contained no free oxygen. The oxygen is the product of billions of years of evolution. The concentration of $O_2$ in the atmosphere has raised for about the last 2.5 billion years. The shortage of $O_2$ might well have prevented the rise of large, complex animals. Amount of $O_2$ an animal can absorb is mostly determined by area of its oxygen absorbing surfaces; but amount needed is determined by its volume, which grows faster than oxygen absorbing area, if an animal's size increases equally in al directions. An increase in concentration of $O_2$ in air, or water would increase the size to which an organism could grow without its tissues becoming starved of $O_2$. Members of Ediacara biota reached metres in length; clearly oxygen did not limit growth. Other metabolic activities would have been inhibited for lack of $O_2$. Convincing correlation between oxygen levels and evolution is absent. Thus, oxygen was perhaps not a prerequisite to complex life than water or primary productivity.

**Snowball Earths:** In the late Neoproterozoic, the earth suffered massive glaciations in which most of its surface was covered by ice, which caused a mass extinction producing genetic bottleneck. The diversification may produce Ediacara biota, which appears after the last Snowball Earth episode that occurred long before the start of the Cambrian.

**Calcium Concentration of the Cambrian seawater:** Perhaps the volcanically active midocean ridges caused a massive and sudden surge of calcium concentration in oceans, making it possible for marine organisms to build skeletons and hard body parts. A high influx of ions could have provided by widespread erosion that produced Powel's Great unconformity.

**Size and Diversity of Planktonic Animals:** Total mass of plankton has been similar to modern levels since early in Proterozoic. Before start of the Cambrian, their corpses and droppings were too small to fall quickly on the sea bed, as their drag was about equal as their weight. They were destroyed by scavengers, or by chemical processes before they reached the sea floor. **Mesozooplankton** is the plankton of larger size. Early Cambrian specimens filtered microscopic plankton from the seawater. These larger organisms would have produced droppings and corpses that were large enough to fall fairly quickly. This provided new supply of energy and nutrients to mid-level and bottom of the sea, which provide new possible ways of life. If any of these remains sank uneaten to sea floor, they could be buried which would have taken some carbon out of circulation, resulting in an increase in concentration of breathable $O_2$ in the seas. Initial herbivorous mesozooplankton were probably larvae of benthos. A larval stage was probably an evolutionary innovation driven by increasing level of predation at sea floor during Ediacaran period. Metazoans have an amazing ability to increase diversity through co evolution. Evolution of **predation** may cause one organism to develop defence, while other developed motion to flee. This caused the predator lineage to split into 2 species: one that was good at chasing prey, and another that was good at breaking the defences. Co evolution is more subtle, but in this way, great diversity can arise. Three quarters of living species are animals and most of the rest have formed by co evolution with animals.

## 3.19 Ecosystem Engineering

Evolving organisms inevitably change the environment. Devonian colonization of land had planet-wide consequences for sediment cycling and ocean nutrients, and was likely linked to the Devonian mass extinction. Similar process could occur on small scales in oceans, for example, the sponges filtering particles from water and depositing them in the mud in a digestible form; or burrowing organisms making previously-unavailable resources available for other organisms.

**Discredited Hypotheses:** As our understanding of the events of the Cambrian becomes clear, proposed causes are now include the evolution of herbivory, vast changes in speed of tectonic plate movement, or of cyclic changes in earth's orbital motion, or operation of different evolutionary mechanisms from those that are seen in rest of Phanerozoic eon. Explosion may not have been a significant evolutionary event. It may show a threshold being crossed: for example a threshold in genetic complexity that allowed a vast range of morphological forms to be employed.

**Early Ordovician Radiation:** After extinction at the Cambrian-Ordovician boundary, other radiation occurred to establish the taxa that would dominate Palaeozoic. This radiation doubled the total number of orders, tripled families, increasing marine diversity as typical of Palaeozoic, and disparity to levels about equal to now.

## 3.20 Geological Time Scale

When specific events occurred in the history of life on earth is often difficult to defect. Biologists depend on dating the fossils containing rocks and on molecular clocks in the DNA of living organisms. Scientists now can measure the variation between different species at molecular level and estimate the time that has passed since a single lineage

**Table 31**  Geological Time Scale

| Age in billion years | |
|---|---|
| 3.8 | Beginning of life, the first life developed in undersea alkaline vents and was perhaps based on RNA rather than DNA. The common ancestor gave rise to 2 main groups of life- bacteria and archaea. |
| 3.5 | The oldest single-celled fossil date from this time. |
| 3.46 | Some single-celled organisms feed methane |
| 3.4 | Rock formations in Western Australia |
| 3 | Viruses were present. |
| 2.4 | Oxygen builds up in the atmosphere. Dissolved oxygen in the oceans makes the iron "rust" and sinks to the seafloor, producing striking banded iron formations. |
| 2.3 | Earth freezes possibly as a result of a lack of volcanic activity. Melting ice indirectly leads to more oxygen being released into the atmosphere. |
| 2.15 | First fossil evidence of cyanobacteria. |
| 2 | Eukaryotic cells with nucleus come into being. Eukaryotic cells evolved when one simple cell engulfed another, and the 2 lived together. The engulfed bacteria become mitochondria, to provide eukaryotic cells with energy. Later, eukaryotic cells engulfed photosynthetic bacteria . The engulfed bacteria evolved into chloroplasts that give green plants their colour and allow them to extract energy from sunlight. |
| 1.5 | The eukaryotes divide into 3 groups: the fungi and animals diverge into separate lineages to evolve separately. |
| 900 mya | The first multicellular life develops .Choanoflagellates are the most closely related to multicellular animals |
| 800mya | The early multicellular animals first divide into the sponges . The latter being known as the Eumetazoa. |
| 780 mya | A small group, the placozoa breaks away from the rest of the Eumetazoa. Placozoa are thin plate-like creatures consisting of only 3 layers of cells. They may be the last common ancestor of all the animals. |
| 770mya | The earth freezes over in another "snowball Earth" again. |

| | |
|---|---|
| 730 mya | The ctenophores split from other multicellular animals. They rely on water flowing through their body cavities to take oxygen and food |
| 680 mya | The ancestor of cnidarians breaks away from other animals. |
| 630 mya | Some animals for the first time evolve bilateral symmetry. Small worms called Acoela may be the closest surviving relatives of the first bilateral animal. The first bilateral animal was perhaps a type of worm. *Vernanimalcula guizhouena,* dates from around 600 million years ago and supposed to be the earliest bilateral animal. |
| 590 mya | The Bilateria undergo a evolutionary split into the protostomes and deuterostomes.The deuterostomes finally include all the vertebrates, plus the Ambulacraria. The protostomes become the arthropods, various types of worm, and rotifers. |
| 580 mya | The earliest known fossils of cnidarians date to about this time |
| 575 mya | The Ediacarans appeared around this time and persisted for about 33 million years. |
| 570 mya | A small group Ambulacraria breaks away from the deuterostomes. This group becomes the echinoderms, hemichordates and Xenoturbellida. The sea lily, the "missing link" between vertebrates and invertebrates, occurred around this time. |
| 565 mya | Fossilized animal trails suggest movements of some animals. |
| 540 mya | As the first chordates emerge among the deuterostomes. The sea squirts as tadpole-like chordates metamorphose their lives into bottom-dwelling filter feeders like a bag of seawater attached to rock. Their larvae now resemble tadpoles. |
| 535 mya | The Cambrian explosion begins with the rapid appearance of new life forms |
| 530 mya | The first true vertebrate appears from jawless fish that has a notochord which is probably like a lamprey, hagfish. The first clear fossils of trilobites look like oversized woodlice. These invertebrates grow to 70 cms in length, proliferate in the oceans. |
| 520 mya | Conodonts looking like eels appear. |
| 500 mya | Animals explored the land at this time. The first animals to do so were euthycarcinoids - believed to be the missing link between insects and crustaceans. *Nectocaris pteryx* lives around this time. |
| 489 mya | The Great Ordovician biodiversification event starts, leading to increase in diversity. Many new varieties appear. |
| 465 mya | Plants begin colonizing the land. |

| | |
|---|---|
| 460 mya | Fish split into the bony - and cartilaginous fish. The cartilaginous fish have skeletons made of cartilage rather than the harder bone. They finally include sharks, rays and skates. |
| 425 mya | The coelacanth splits from the rest of the lobe-finned fish. |
| 417 mya | Lungfish splitted from the other lobe-finned fish |
| 400 mya | The oldest insect lives about this time. Some plants grow woody stems. |
| 397 mya | The first tetrapods evolve from intermediate species like *Tiktaalik*, probably in shallow freshwater. |
| 385 mya | The oldest fossilized tree dates from this period. |
| 375mya | *Tiktaalik* lives about this time. The fleshy fins of its lungfish ancestors evolve into limbs. |
| 340 mya | The first major split occurs in tetrapods. The amphibians branch off from the others. |
| 310mya | Within the rest tetrapods, the synapsids and sauropsids split from one another. The first synapsids had distinctive jaws. They are sometimes called "mammal-like reptiles". |
| 320 to 250 mya | The pelycosaurs dominate the land. The famous example is *Dimetrodon*, with a sail on its back |
| 275 to 100 mya | The therapsids evolve and survive until the early Cretaceous. A group of them called the cynodonts develops dog-like teeth and evolves into the first mammals. |
| 250 mya | The Permian period ends with great mass extinction wiping out vast number of species, including the last trilobite .In the oceans, the ammonites evolve about this time. Several groups of reptiles colonize the seas, |
| 210 mya | The fossil called Protoavis suggest that some early dinosaurs are evolving into birds at this time |
| 200 mya | Another mass extinction strikes, causing the dinosaurs extinct. Proto-mammals evolve warm-bloodedness . |
| 180 mya | The monotremes break apart from others. Few monotremes including duck-billed platypus and echidnas survive. |
| 168 mya | A flightless dinosaur called *Epidexipteryx* lives in China. |
| 150 mya | *Archaeopteryx*, the first bird, lives in Europe. |
| 140 mya | Placental mammals split from their cousins, the marsupials. The majority of modern marsupials live in Australia. |

| | |
|---|---|
| 131 mya | *Eoconfuciusornis*, a bird lives in China. |
| 130 mya | Emergence of the first flowering plants . |
| 105-85 mya | The placental mammals split into 4 major groups- laurasiatheres, euarchontoglires , Xenarthra and afrotheres. |
| 100 mya | The Cretaceous dinosaurs reach their peak in size. The giant sauropod *Argentinosaurus* lives about this time. |
| 93 mya | The oceans become starved of oxygen. Twenty-seven per cent of marine invertebrates are wiped out. |
| 75 mya | Splitting between the ancestors of modern primates and ancestors of modern rodents and lagomorphs |
| 70 mya | Grasses - though it will be several million years before the vast open grasslands appear. |
| 65 mya | The K/T extinction wipes out vast number of species. The ammonites are also wiped out. |
| 63 mya | The primates split into 2 groups, haplorrhines (dry-nosed primates) and strepsirrhines (wet-nosed primates). |
| 58 mya | The tarsier splits from rest of haplorrhines. |
| 55 mya | The Palaeocene/Eocene extinction occurs. Sudden rise in greenhouse gases causes temperatures soaring and drastically changes the planet. |
| 50 mya | Artiodactyls evolves into whales. |
| 48mya | Indohyus lives in India. |
| 47 mya | The famous fossilized primate, "Ida" lives in northern Europe. |
| 40 mya | New first simians diverge from the rest of the group, colonizing South America. |
| 25 mya | Splitting occurred between apes split and Old World monkeys. |
| 18 mya | Gibbons become the first ape to split from others. |
| 14 mya | Orangutans diverge from other great apes. |
| 7 mya | Gorillas diverge from other great apes. |
| 6 mya | Humans evolve from their closest relatives; chimpanzees and bonobos. |
| 2 mya | A 700-kilogram rodent, *Josephoartigasia monesi* lives in South America. It is the largest known. rodent |

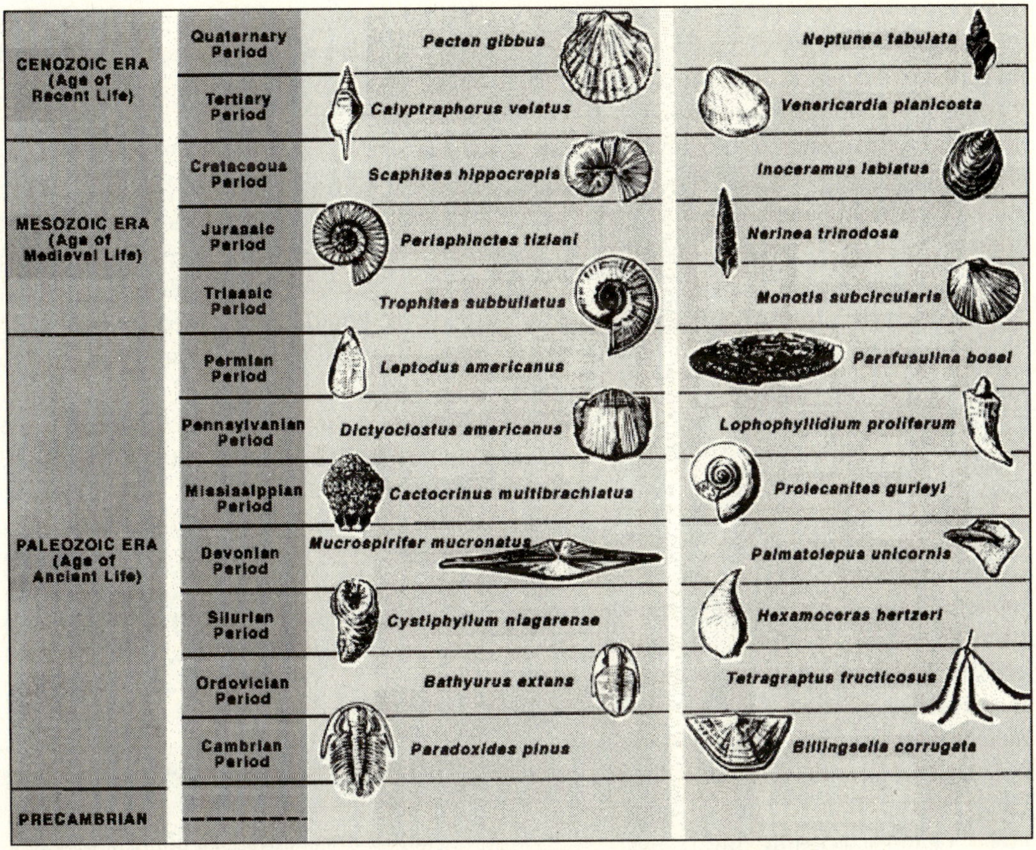

CENOZOIC ERA (Age of Recent Life)
Quaternary Period
Pecten gibbus
Neptunea tabulata
Tertiary Period
Calyptraphorus velatus
Venericardia planicosta
MESOZOIC ERA (Age of Medieval Life)
Cretaceous Period
Scaphites hippocrepis
Inoceramus labiatus
Jurassic Period
Perisphinctes tiziani
Nerinea trinodosa
Triassic Period
Trophites subbullatus
Monotis subcircularis
PALEOZOIC ERA (Age of Ancient Life)
Permian Period
Leptodus americanus
Parafusulina bosei
Pennsylvanian Period
Dictyoclostus americanus
Lophophyllidium proliferum
Mississippian Period
Cactocrinus multibrachiatus
Prolecanites gurleyi
Devonian Period
Mucrospirifer mucronatus
Palmatolepus unicornis
Silurian Period
Cystiphyllum niagarense
Hexamoceras hertzeri
Ordovician Period
Bathyurus extans
Tetragraptus fructicosus
Cambrian Period
Paradoxides pinus
Billingsella corrugata
PRECAMBRIAN

**Figure 3.12** Geological time scale showing index fossils

900
600
300
0
Families of Organisms
Five Great Mass Extinctions
Mass Extinctions
Modern-Type Species
Paleozoic-Type Species
Cambrian-Type Species
Paleozoic
Mesozoic
Ce
V C O S D C P Tr J K T
600
400
200
Geologic Time (10⁶ years)

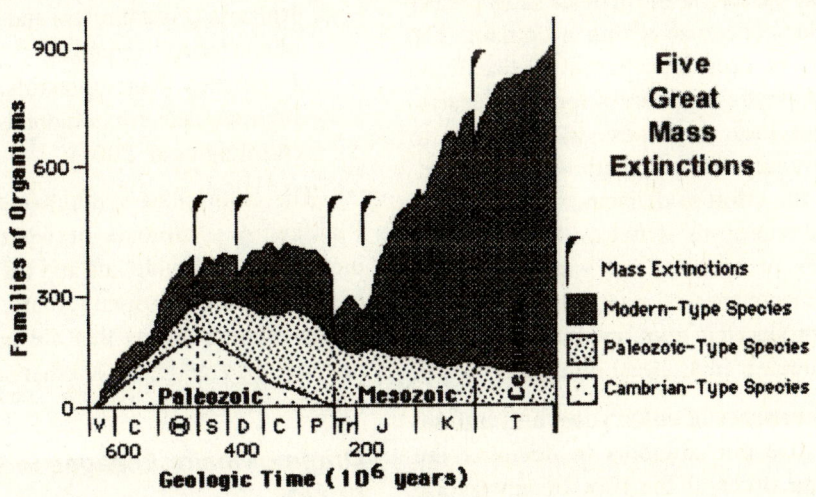

**Figure 3.13** Geological time scale showing mass extinctions

split into different species. However, the geological time scale that provides the picture of evolution beginning from the primitive life forms to *Homo sapiens sapiens* is very important in understanding evolutionary biology. The following account is based on the article," The Evolution of Life by Michael Marshall (2009) (See also figures 3.12 & 3.13.)

## 3.21 Modern View about Evolution

**Origin of protein folds, viruses, and cells:** There seem to exist ~1,000 or few thousand structural protein folds, the relationships between which are unclear (Coulson et al, 2002).For several classes of viruses, especially for positive strand RNA viruses and nucleo-cytoplasmic large DNA viruses, evidence of monophyletic origin is found. Evidence of common ancestry for all viruses is absent. The cells of archaea and bacteria have distinct membranes, non-homologous enzymes for membrane biogenesis, and non-homologous core DNA replication enzymes which makes the reconstruction of cellular ancestor of archaea and bacteria difficult.

**Major branches (phyla) of bacteria and archaea:** Bacteria and archaea show greater molecular coherence within a domain. The topology of deep branches in archaeal and bacterial phylogenetic trees remains elusive. The trees lack robustness with respect to gene(s) analyzed and methods employed. Despite the effort to delineate higher taxa of bacteria, consensus is not on horizon. Two branches of archaea, euryarchaeota and crenarchaeota are well established. This split is not reproduced in trees, and further divisions remain unclear in archaeal domain.

**Major branches of eukaryotes and animal phyla:** Despite attempts to decipher the branching order at the root of eukaryotic phylogenetic, there is little progress. The

situations seem to be a star phylogeny, with 5 or 6 established super groups, but relationship between them remain unresolved (Adl et al, 2005). During the Cambrian explosion all the diverse body plans of animals appear to have emerged almost instantly (Kerr, 2002). Molecular clock suggests that the Cambrian explosion is an artifact of fossil record. Actual divergence occurred earlier (Blair and Hedges. 2005). In popularized pattern, relationship between animal phyla remains controversial. Appearance of bushes in tree of life (TOL) is attributed to cladogenesis that appears to be characters of transitional epochs in evolution. Erosion of phylogenetic signal results in poor resolution of phylogenetic trees for ancient divergence events. TOL exists and is resolvable in practice; full resolution might never be attained (Rokas et al, 2005). The 5 eukaryotic super groups are:

1. Plantae (green plants, green algae, red algae)
2. Chromalveolates (alveolates, including Apicomplexa, ciliates, dinoflagellates, and stramenophiles including oomycetes, diatoms, and others)
3. Unikonts (Animals, fungi, Amoebozoa)
4. Rhizaria (Foraminifera and other poorly characterized groups)
5. Excavates (kinetoplastids, euglenids, diplomonads, trichomonads, and other) (Keeling et al, 2005).

The chloroplast symbiosis gave rise to Plantae. A symbiosis between primitive unicellular eukaryote and red alga led to the emergence of Chromalveolata. The other endosymbiotic events that are postulated to underlie the emergence of other super groups (Figures 3.14, 3.15, 3.16).

### Origin of major lineages in super groups

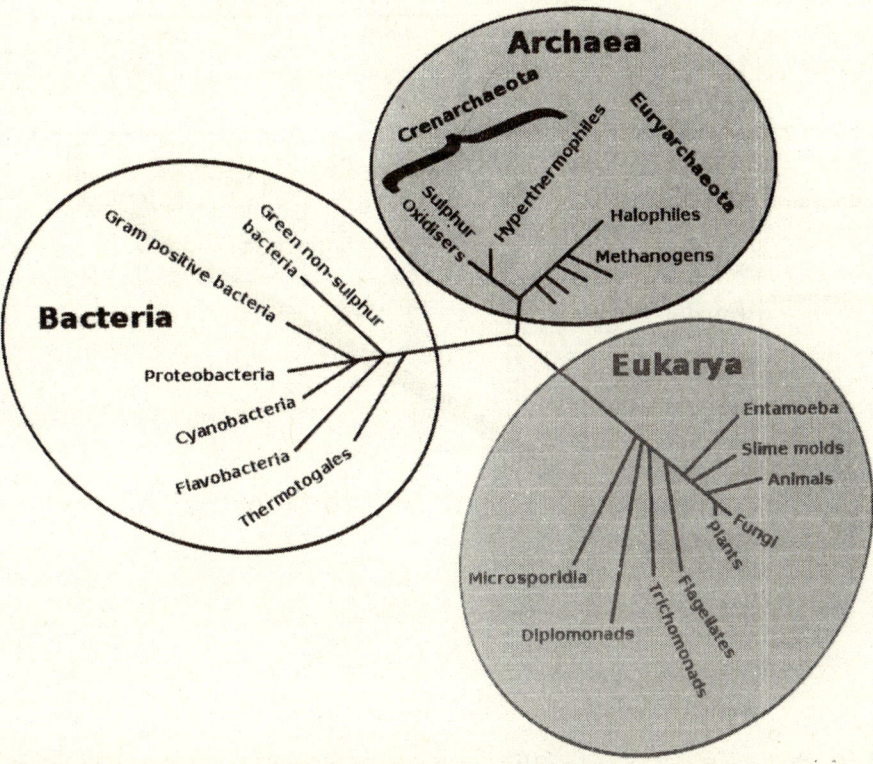

**Figure 3.14** Three domains of life

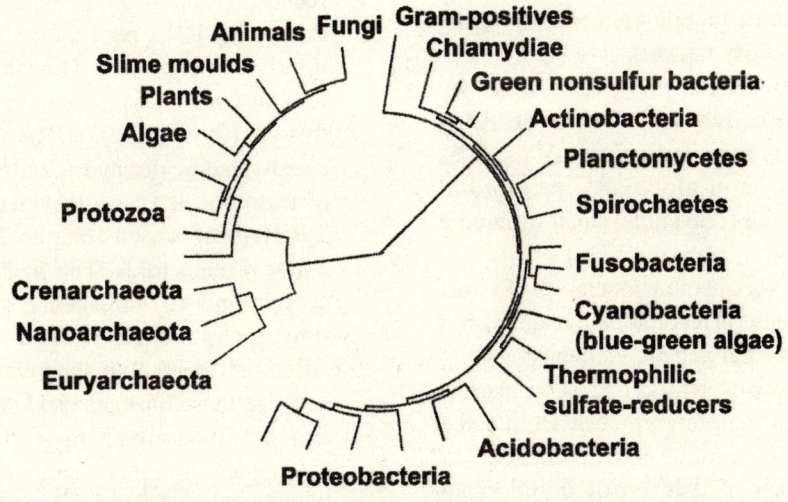

**Figure 3.15** Tree of life based on sequenced genomes

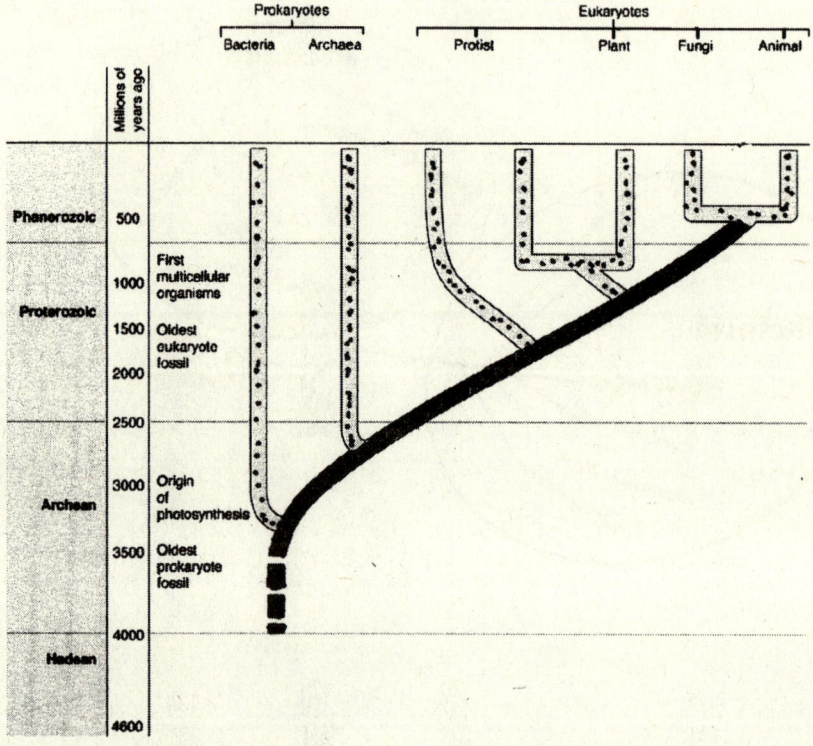

**Figure 3.16** Phylogenetic tree representing approximate time of appearance and relationships of major groups

Such origins are due to:

1. Invasion of mobile elements; rewiring of regulatory networks;
2. The emergence of protein folds by recombination and fusion of RNA segments encoding primordial peptides and/or non-globular primordial polypeptides containing small structured units;
3. The emergence of major classes of virus-like agents via recombination and fusion of primordial genetic elements;
4. Emergence of archaeal and bacterial cells, via gene sampling process, from same primordial pool of genetic elements at later stage. Existence of initial rapid evolution with extensive mixing and matching of genetic elements is proposed for each transition in early studies. Primordial RNA pool might have been highly concentrated (Baaske, 2007) creating conditions for recombination and fusion (Koonin, 2007). The first trees ever to emerge during life's history were gene trees which consolidated at the first 2 BBBs that caused complex RNA and major protein folds. The next 2 BBBs correspond to emergence of major lineages of bacteria and archaea. Attempts to resolve the relationships between these lineages yield conflicting and hardly compelling results. The emergence of archaea and bacteria via the respective BBBs perhaps followed by new inflationary phases in which the

archaeal and bacterial cells formed distinct communities of cells that exchanged genetic material. Membranes of these earliest archaea and bacteria were leakier than those of modern prokaryotes (Mulkidjanian et al, in press), making cells susceptible to DNA uptake. Genetic exchange was enough to preclude the origin of individual lineages. During this stage, archaeal and bacterial cells existed as physical entities. Selection affected genes and gene complexes. Evolution was rapid with rampant gene flow between cells. The proto-bacteria and proto-archaea might have thrived as colonies at **hydrothermal vents** (Figure 3.17) from which first cells have emerged (Koonin and Martin, 2005).

**Figure 3.17**  Deep sea hydrothermal vents

The proto-bacteria and proto-archaea would produce distinct communities with intensity of gene flow between these being lowered than that within each community. Individual bacterial and archaeal cellular lineages would emerge, when gene sampling produced advantageous combinations with tighter membranes curtailing HGT. The origin of eukaryotic cell and emergence of eukaryotic super groups can be perceived as a single BBB event. Symbiosis between an archaean and an proteobacterium is the triggering event of eukaryogenesis which instigated a new wave of inflation during which the archaeal host DNA was bombarded by genes and selfish elements from symbiont, leading to rearrangements within the hybrid cell, starting with formation of endomembranes and nucleus. Existence of stem phase of eukaryotic evolution is shown by existence of many pan-eukaryotic sets of paralogous genes at this stage. During this phase, emerging eukaryotic cell developed phagocytic capacity and engulfed additional bacteria resulting in multiple, transient symbioses and transfer of varying number of bacterial symbiont's genes to **eukaryotic genome**. Thus, mitochondrial endosymbiosisis triggered all aspects of eukaryogenesis, specially the evolution of phagocytic capacity. Stable symbiosis would result in termination of inflation for host subpopulation and emergence of new cellular lineage, one of eukaryotic super groups.

Now the **evolutionary tree** of eukaryotes is best illustrated as a bush (polytomy) of 5 or 6 super groups (Adl, 2005). Consensus on relationship between super groups is absent.

The Plantae super group including glaucophytes, red algae, green algae, and green land plants derived following the endosymbiosis of a cyanobacterium within ancestral eukaryotic cell. In chromalveolate hypothesis (Cavalier-Smith, 1999), a secondary endosymbiosis, the engulfment of a red alga by non-photosynthetic eukaryotic cell, gave rise to Chromalveolata (Hackett et al, 2007). Additional secondary symbioses occurred at base of several major lineages belonging to super groups like *Diatom*, euglenid and dinoflagellate .Variety of transient and stable bacterial symbionts has been detected in numerous individual unicellular eukaryotes, showing the establishment of symbiosis, although evolution of symbionts into bonafide organelles appears to be rare.

Chromalveolate hypothesis state that secondary loss of endosymbionts is not uncommon. It occurred independently in multiple branches of Chromalveolata like ciliates and several lineages of heterkonts, dinoflagellates and apicomplexans (Bhattacharya et al, 2004). Losses of morphologically distinguishable endosymbionts in other super groups do not appear implausible. The chromalveolate hypothesis may oversimplify the history of secondary endosymbiotic events (Bhattacharya et al, 2004). This would not bear on generality of endosymbiosis coinciding with BBB in this context. The deep analogies between the BBB model and the cosmological model of eternal inflation are also suggested.

First, the eternal inflation model and they way it reinterprets the Big Bang need to be summarized. The 20th century concept of origin and evolution of our universe involved expansion of space, time, and energy-matter from an initial state. Primordial explosion of initial state has dubbed the Big Bang. The Big Bang model is proved effectively by discovery of low-energy cosmic microwave background radiation (Weinberg, 1977).

## 3.22 Bush of Life: Typical Tree with Unresolved Branches

The tree was generated from simulated data using Tree View program (Page, 1996). The notion of communal ancestor is developed into a specific scenario (Koonin and Martin, 2005) that derived information from the comparative-genomic study that the enzymes of membrane biogenesis and core DNA replication machineries in archaea and bacteria were non-homologous. In this picture, LUCA a diverse population of genetic entities inhabited networks of inorganic compartments at hydrothermal vents. Exchange of genetic materials between compartments, primordial analog of HGT, is an inherent feature of this model. Transition from selection for individual genetic entities to selection for selfish cooperatives occurred followed by independent escape of the first membrane-bounded cells (Koonin and Martin, 2005). Numerous attempted escapes might occur, but only 2 successful ones leading to archaea and bacteria. Evolution of non-membrane-bounded and compartmentalized populations of diverse genetic elements include virus and develops in virus world concept (Koonin , 2006). Virus world stems from fact that a set of genes encoding essential proteins involved in viral genome replication, packaging, and virion formation, are shared by many dissimilar virus groups. The early stage of evolution is envisaged as virus-like with indistinct progenitors of viral and cellular genomes but the segregation of parasites and cooperatives occurred. Major classes of extant prokaryotic viruses emerge directly from primordial genetic pool. **Eukaryogenesis** was the 2nd melting pot of virus evolution, where major eukaryotic viruses emerged through recombination between bacteriophage and cellular genomes. Modern protein folds evolved by recombination of ancient peptide modules as folds have independent and polyphyletic origins. They all derive from the same recombining pool of genetic elements encoding primordial peptides (Lupas et al, 2001). High concentration of RNA attainable in networks of inorganic compartments at hydrothermal vents (Baaske, et al, 2007) has added credibility to the idea that recombination and fusion of RNA played major role in pre-biological evolution (Koonin, 2007). The HGT is widespread among prokaryotes (Lawrence and Bartel, 2003), the principal early discovery of genomic revolution (Koonin et al. 2000) led to a reappraisal of the TOL concept. Since HGT is substantial and involves all categories of genes, molecular phylogeneticists would have failed to find true tree, as the history of life cannot be represented as a tree (Doolittle, 1999). The TOL still could be saved by redefining the TOL as the consensus tree of relatively stable core of highly conserved genes (Wolf et al, 2002). The plurality in biological evolution including vertical, tree-like inheritance and exchange of genetic material between life forms traditionally viewed as HGT was emphasized. Replacing the tree-centered monism by a plurality of pattern in evolutionary models (Doolittle WF, Bapteste, 2004) was called for. The early evolution outlined that pre-cellular evolution of life followed qualitatively distinct pattern from pattern of next evolution. The distinct early phase of evolution perhaps involved extensive exchange of genetic material that assumes different forms resulting in rapid evolutionary innovation and precludes origin of distinct lineages.

Emergence of complex rRNA occurs by fusion/recombination of smaller RNA segments. The first of 3 original great BBBs that might have shared a physical substrate, the primordial gene pool, were abiogenically compartmentalized. This BBB would cause the tree pattern of evolution for 1st time (Grishin, 2001). Emergence of major classes of viruses in primordial gene pool of genetic elements encoding hallmark viral genes. The 2nd of the 3 great BBBs occurring in primordial gene pool (Koonin et al 2006 ). Emergence of 2 prokaryotic cell types, archaea and bacteria occurred through recombination, fusion, and sorting of diverse genetic elements in primordial gene pool. The third and last of 3 great BBBs occurred in primordial gene pool. Crucial processes involve formation of selfish cooperatives, extensive transfer of genetic material between compartments, and sampling of genes into emerging protocells. Probably of numerous trials on cell formation, only 2 types were fixed (Woese, 2002). Extensive gene exchange between protoarchaeal and **protobacterial** cells with leaky membranes within primordial microbial mats occurred near hydrothermal vents. More constrained process of gene sampling, with numerous trials on more robust cells capable of departing primordial mats. Extensive gene flows from endosymbionts to host chromosome(s) accompanied with massive invasion of introns and pervasive genome rearrangement. Distinct symbiotic events caused the 5 super groups of eukaryotes. Woese proposed that the early stages of evolution including that of the Last Universal Cellular Ancestor (LUCA), involved rampant **horizontal exchange** of genetic material between primordial life forms such that individual lineages could not form (Woese, 2002). This communal ancestor has a physical history but not a genealogical one (Woese, 1998).

## 3.23  Life and Randomness

Finding order among the diversity and complexity of the living organisms is the major task in biology. Linnaeus brought order out of chaos by classifying the animals and plants. Mendel's law of inheritance provided the basis for the advances in biology. The natural selection was a step forward in explaining evolution through generations. In the continual intraspecific competition, the winners successfully produce offspring, passing on the advantageous traits. Randomness in evolution is less documented. It shuffles things backwards and forwards and provides the material for natural selection. All the evolutionary change is built on the foundation of randomness which in some cases directly produces order. The role that randomness plays in evolution differs with the size of the organism.

**Size and Randomness:**   Evolution leads from small to big, from simple to complex. There is important trend in control of effect of randomness. In microorganisms, random events are common, but with increase in size and complexity a decrease in role of chance occurs. Thus the increase in size, increase in complexity, and decrease in part played by randomness are interrelated and go together in evolution. In comparison to small organisms, large organisms are protected from the vagaries of chance. They cling to essential randomness for their existence. All novelty is founded on generation of new genes, which results from the directionless and random appearance of new mutations. So, the large animals and plants by a variety of mechanisms limit chance through sexual system which is ruled by natural selection. The progression over years does not mean the elimination of simple, small organisms. They are maintained by selection or lack of selection. So the selection causes the progressive changes in one group of organisms taking cognizance of interdependence of organisms. Animals could not exist without plants. Role of any organism in size-complexity randomness spectrum exists as it fits in and is part of community.

Prokaryotes, protozoa, numerous invertebrates, fungi, and lower plants still exist now. They are evolving. Lamarck believed in innate and inevitable tendency towards perfection of all organisms and supposed that new and simple forms were continually produced by spontaneous generation. The lowest forms like infusoria and rhizopods have remained for a large period in about their present state. Looking on randomness and its progressive decreasing role reveals an unnoticed phenomenon. Eukaryotic microorganisms might have more morphological variation as they are relatively untouched by natural selection. For any character for which selective advantage of organism may not be apparent, the safest assumption is that it is an adaptation. That stable morphological traits could originate during evolution by chance is often dismissed .Stephen Jay Gould and Richard Lewontin argue that the justifications of calling something an adaptation is totally absent. As the upper size limit increases over time, a corresponding increase in period of development produces mature morphology. This involves an increase in complexity and effectiveness of mechanisms of control, and one result is a progressive stifling of influence of randomness. Role of randomness is different in micro and macro organisms. In large organisms, many steps occur in their extended development under genetic control. If an unfavorable mutation occurs in one of those steps, it blocks development, and embryo dies. This is called **internal selection**. The chances that any such mutation could be beneficial are unlikely as all the subsequent steps are dependent on early step, so any change is likely to be deleterious. Development eliminates undesirable random mutations. The larger the organism, the longer the sequence of developmental steps with more possibilities of internal selection. In small organisms, with few developmental steps, one random change affects the morphology and often the whole organism. Thus, generating the masses of different organism is possible, many of which might be unaffected by natural selection.

**Randomness** is the backbone of Darwinian evolution particularly in the form of heritable variation which must be controlled. Sexual mechanism provides right amount of variation that makes evolution by natural selection. The right amount of variation leads to greater reproductive success. Sexuality, an important element in natural selection is essentially ubiquitous. **Sexual reproduction** yields fewer offspring per parent and is more costly than asexual reproduction. But, if that cost were not paid, evolution will not occur. In simple forms, asexual cycles are often interspersed with sexual ones. The former are clones without variants and are found in environment where they can multiply rapidly. The latter appear in changing environment where variation might produce some individuals that cope with change. When the larger and more complex animals and plants appear on earth, asexual phase disappears about completely. With few exceptions, large animals and higher plants do not reproduce asexually which is the apex of control in organic evolution. Randomness is an underlying foundation of all of Darwinian evolution for organisms. Mutations are random and without mutation there could be no change. So randomness of mutation is retained at all levels, or stages of evolutionary progress.

## 3.24 Chance in Evolution

The role of chance in evolution was recognized early as mutations were random. For years, the randomness was understood at the level where one base in DNA chain is substituted for other. Attempts to show that in some cases mutation might force change in particular direction stands invalid. That mutations are

**Figure 3.18**   Sewall Wright

and controlled. Genetically identical twins show differences in their life span and other features. Many events during development are random leaving their imprint on adult. C. H. Waddington called this developmental noise. Michael Lynch states that natural selection accounts for everything and random events play role in evolutionary change. Mainly concerned on the evolution of the genome he emphasizes the randomness of mutation as the key in shuffling the genome recombination. He argues that random molecular changes result in directional push over time without natural selection. That chance plays role in evolution has been promoted by a number of workers.

random has long been accepted. Many aspects of evolution besides mutations involve chance. The genetic events involved in sexual reproduction are peppered with chance. The egg and sperm with diploid number of chromosomes arise with chromosome reshuffling during meiosis. The genes any one gamete contains vary and the nature of such variation is a subject of chance. Random events involved in producing the genetic variation in individuals are the fodder for natural selection. Sewall Wright (Figure 3.18) in the 1930s first pointed out that random event changes the genetic makeup of a population. He called this stochastic evolutionary change drift which is important in small population, for the variant genes it possessed would be the ones that remained during the population expansion. Such bottleneck called the founder effect (as it might lead to the origin of new species) allows the chance event of drift to produce change. It is called the **Adam and Eve effect**. The changes in the gene pool of a population can be determined by chance. C.E. Finch and T.B.L. Kirkwood noted that the life span of any animal is entirely random, the result of accidents and other factors is consistent

## 3.25 Genotypes and Phenotypes

The similarities between living organisms are not confined to structural features, but extend to microscopic scale. Except viruses, all living organisms are composed of single or many cells. In eukaryotes along with multicellular species and some unicellular species, cells are bounded by a membrane and contain cytoplasm and nucleus, which carries the genetic material. The rest of unicellular organisms, the prokaryotes, bacteria and archaea are simple cells in which genetic material floats in cytoplasm with no sub cellular structures and nucleus. Viruses reproduce inside cells of other organisms and consist of a chemical coat on genetic material. Cells are mini factories that produce the chemicals for organism, generate energy from food, and provide body structures. Most chemical products of cell are proteins, some of which are enzymes. Some proteins function in storage, or transport like binding to iron and storing it in liver or to carry oxygen via blood. Communication proteins like hormones circulate in blood to control many functions. Signaling **proteins** found on cell surface communicates with other cells. Structural proteins form skin and bones.

Proteins are composed of amino acid chain. Only 20 different **amino acids** found in living world show similarity between organisms and support evolutionary hypothesis of common ancestor.

Genetic material of all species is organized in chromosomes, composed of a DNA molecule and a protein coat. Different chromosomes controlling different functions are characterized by different size and structure of their DNA. Number of chromosomes is strongly species-specific. DNA consists of 2 helices of alternating sugar and phosphate with each sugar binds to one base: adenine (A), cytosine (C), guanine (G), and thymine (T), called DNA bases. Each base in one helix binds to base at corresponding position of other helix, but A binds only with T and C with the G .The portion of DNA corresponding to a locus is the gene and may take one of several forms, called alleles. Chromosomes of same type can contain different alleles at the same loci. Most genes function to code the structure of protein. Each triplet of genetic letters detects an amino acid and sequence of triplets detects the amino acid chain of protein. Since there are 64 different triplets and only 20 amino acids, some triplets code the same amino acid.

DNA portions control the activity of cell; determine the protein to be produced in the cell diversification. In asexual species, all cells are haploid. An individual produces an offspring by replicating its genetic material, where chromosomes are first duplicated and then the cell divides into 2 identical daughter cells. In sexual species, some polyploid cells form a group of homologous chromosomes. There are 2 mechanisms of sexual reproduction. In diploid species, all cells are diploid except for eggs and sperms which are haploid. In production of gametes, each pair of homologous chromosomes exchanges alleles at some loci, then all new chromosome pairs are duplicated, and cell divides into 4 gametes,

each with one chromosome for each type. Union of egg and sperm restores a diploid cell, which develops into new individual by cell divisions.

The second mechanism of sexual reproduction characterizes sexual haploid species, where fusion of haploid cells produces polyploid cells in which recombination of homologous chromosomes occur before cell division gives rise to new haploid cells. Asexual reproduction may occur in some sexual haploid species as occasional, or predominant reproduction. The haploid cells are temporary in first mechanism of sexual reproduction, but **polyploid** cells are temporary in second. In both cases, homologous chromosomes in polyploid cells are inherited one from each parent and the genetic material of the progeny is a mix of that of the parents.

The genome is the set of all possible chromosomes characterizing an individual of species and is identified by all allelic forms of all genes of species. **Genotype** is a specific genome realization given by chromosomes. In asexual and in sexual haploid species, the genotype is identified by one chromosome for each type and is found in all the cells. In diploid species, genotype is given by pairs of homologous chromosomes and is found in all cells except gametes.

## 3.26 Ancient Population Genetics and Phylography

While authentic ancient DNA has been recovered from warmer climates, consistently cold places are the rich source of material for ancient DNA analysis. The first analyses of changes in genetic diversity within populations through time took benefit of young, museum-preserved skins. One population-level analyses amplified mtDNA from skins of 3 geographically isolated populations of Panamint kangaroo rat in California. These data of skins were compared with data collected from the same localities. In 1988 the populations had remained genetically isolated from each other throughout the period spanning the sample ages. Pocket gophers from Yellowstone National Park are genetically isolated from nearby populations for at least 2400 years. **Climate change** is the driver of most of these demographic changes. The first large, ancient DNA population data set contained more than 600 sequences from North American bison, ranging from only 100 to more than 55,000 years old. These data showed evidence for a peak in bison diversity around 35,000 years ago followed by a rapid decline toward extinction.

The timing of beginning of this decline was surprising as it predated the peak of the last ice age and first appearance of large numbers of human's in North America, 2 competing hypotheses about the cause of mass extinction. More sophisticated demographic models further resolved the bison demographic history, revealing that around 13,000 years ago, bison narrowly escaped extinction. This bottleneck was followed by rapid recovery of genetic diversity that persists today. Horses peak in genetic diversity slightly after bison peak in North America, probably because they were better able to survive once the steppe grasslands began to disappear at the onset of the last ice age. It is clear that climate change played a major role.

### Ancient DNA and Genomics

Ancient DNA uses DNA sequence data recovered from poorly preserved organisms, usually deceased for years. Ancient DNA data can provide unique picture on evolution of populations and species. The field started in the early 1980s with recovery of first ancient DNA sequences from preserved muscle of a quagga, a relative of zebra. In 1984, 2 short fragments of mitochondrial DNA (mtDNA)

from dried muscle of quagga (*Equus quagga*) were cloned which first described DNA in dead tissues. After 3 years, sequences cloned from a mummified human liver and recovery of DNA from Egyptian mummies was reported. Svante Paabo, the father of ancient DNA, aimed in genetically characterizing evolutionary history of Egyptian mummies. In 1985, he found 2 members of the Alu family of human repetitive DNA sequences from an Egyptian mummy. When the same fragment of *quagga* DNA was amplified using the PCR, 2 differences were found between the quagga and its closest relative, the plains zebra. Paabo used the PCR to assess DNA survival in differently aged remains and showed that ancient DNA sequences undergo modifications including strand breaks, DNA crosslink, and modified bases. DNA damage and contamination are the 2 biggest problems. Degradation occurs through the action of endogenous nucleases. In some cases including rapid desiccation or deposition in very cold, dry or salty environments, these enzymes degrade before they destroy the DNA. In ideal circumstances, exposure to oxygen and water slowly break down the surviving DNA. The continuous breakdown of DNA results in few surviving, non fragmented molecules per sample. The common form of hydrolytic damage in ancient DNA is deamination, in particular the conversion of cytosine to uracil. This results in template DNA being read as thymine, rather than cytosine, and in erroneous incorporation of an adenine in the complementary strand. This form of DNA damage account for nearly all mis-incorporated bases from geographic regions where preservation conditions favor long-term DNA survival.

Amplification and sequencing of 2 mitochondrial fragments of extinct Tasmanian wolf confirmed that Tasmanian wolf was closely related to Australian marsupial carnivores than to more similar-looking marsupial carnivores, showing that their shared morphological features have evolved independently. **Ancient DNA** is used to place extinct species in molecular phylogenies. One result reveals that mammoths are closely related to elephants. Complete mitochondrial genomes of mammoths and mastodons later revealed that mammoths are more closely related to Asian elephants than to African elephants. Ancient DNA from the Oxford dodo showed that this was a type of pigeon. Ancient nuclear DNA from the remains of extinct New Zealand moa revealed that the 3 described species were only 2, and that vast size difference used to distinguish the species from each other was due to pronounced sexual dimorphism. Ancient DNA can be recovered from any element commonly bone, teeth, hair, seeds, muscle, or eggshells, from mixed materials like coprolites and soil. DNA extracted from coprolites provides genetic information about defecator and about individual's last few meals. DNA present in soil is generally naked, not bound to anything and not protected from bombardment by damage-inducing environmental events. Sedimentary DNA does make it possible to identify when and where species were present.

With publication of Neanderthal genome, we now have more information about precisely which mutations distinguish us from the chimpanzee. Prior to 2010, Neanderthals and humans shared a derived allele at the FOXP2 locus, which is involved in speech and language. The draft Neanderthal genome revealed large genomic regions that have been under positive selection since our divergence from Neanderthals. These regions include genes associated with human-specific maladies including autism spectrum disorder and type 2 diabetes. Methods to target and capture specific regions of DNA provide a promising route to refine these observations and improve

our understanding of what makes species look and act the way they do. Shotgun sequencing hundreds of individuals for population genomic analyses is still too expensive. The approaches are in development to capture specific fragments of DNA from **DNA libraries** which can then be bar-coded, pooled, and sequenced together on a next-generation platform. This approach allows hundreds or thousands of loci to be sequenced simultaneously from hundreds of individuals. It provides a solution to matrilineal bias of using only mitochondrial DNA, and more power to detect changes in genetic diversity associated with either particular environmental events or episodes of natural selection. The results of analyses incorporating ancient DNA data have ranged from obvious to surprising. It is clear that perspective gained from these data has benefited many aspects of evolutionary research.

Recent technological advances collectively known as next-generation sequencing have been embraced by the ancient **DNA community** which allow millions of sequencing reactions to happen in parallel by creating micro reactors and/or attaching DNA molecules to solid surfaces or beads prior to sequencing. These technologies explore more fully the amount and quality of DNA preserved in ancient specimens and makes it feasible to obtain larger amounts of ancient data in much less time-consuming and often less expensive way. The first complete ancient genomes were published in 2001, long before next-generation sequencing was the state of the art. Two teams published mitochondrial genomes from two species of moa. Each 17,000-bp mitochondrial genome was pieced together from overlapping 350-600 bp fragments amplified via PCR. These genomes were the proof that ancient genomics is feasible, and could provide useful evolutionary information. The moa genomes were used to estimate the timing of the divergence between ratite birds and provide a temporal framework for the break-up of Gondwana into smaller continental fragments. Five years later, 2 complete mitochondrial genomes of mammoth were published using different techniques. One group pieced together the mammoth mitochondrial genome by targeting only longer, intact fragments, between 1200 and 1600 base pairs in length. A second developed a multiplex PCR approach to co amplify non overlapping fragments of mammoth mitochondrial DNA in a single PCR, greatly speeding up the process of data generation and significantly reducing the amount of sample required to perform the experiment. In the same year, a third group took mammoth mitochondrial-genome sequencing into next generation. They used the Roche 454 technology to shotgun sequence a permafrost-preserved mammoth bone. Of the 13 million base pairs of mammoth DNA they recovered, 222 reads, each around 89 bp long, mapped to mammoth mitochondrial genome. In a shot gun-sequencing approach, the entire DNA extracted from a particular specimen is made into a library and that DNA library is then sequenced.

As a result, sequences are generated not only from the target specimen but also from any bacteria or other organisms that may have colonized the sample during its preservation history, and any DNA that may have contaminated sample during processing. The sample used in this first study was remarkably well preserved: 45.4 percent of the sequences from the genomic library were identified as mammoth DNA, remainder likely coming from organisms colonizing the sample after its deposition. Libraries that were later used to sequence the complete nuclear genome of the Neanderthal contained only 1-5 percent Neanderthal DNA. In this case, enzymes targeting specific sequences present in bacterial DNA were used to chop up bacterial DNA in the DNA libraries, thereby increasing the ratio of Neanderthal to contaminating DNA. Draft ancient genomes have been published for a mammoth, a 4000-year-old Pale Eskimo from Greenland, a Neanderthal, and a previously unknown hominin from Denisova Cave in Siberia.

## 3.27 Adaptive Radiation

Many millions species exhibit diversity in color, size and behavior. **Adaptive radiations** occur in groups of distinct but closely related species that have evolved from the common ancestor in a short time. Such radiations help reveal the causes of their evolution. Due to natural selection during and after speciation, descendant species varies morphologically or physiologically for exploiting different environments. Absence of constraints from competitor species governs adaptive radiation. The same evolutionary path as often taken by various organisms in same environment shows the guiding force of environment. Taxonomic groups vary in their intrinsic potential to diversify as they possess traits that are key evolutionary innovations or as they readily exchange genes through hybridization. Invasion of an underexploited environment allows species to initially multiply at a high rate, and diversify morphologically and ecologically. The fossil record and reconstructions from molecular phylogenies show that both speciation and diversification rates later decline. Experiments in bacteria replicate the diversification pattern in observable time. Bacteria respond to ecological opportunity by diversifying into a maximum number of ecologically differentiated types.

Some problems involved in inferring history can be explained with studies of living organisms. Direct tests of radiation theory can be done with microorganisms. Bacteria have the enormous advantage of short generations and rapid evolution. Introduced to a plate of agar, an inoculum of asexual bacterium *Pseudomonas fluorescens* diversifies into 3 main morphological types. Diversification in a microcosm is an analogue of radiations of sexually reproducing organisms into many species over millennia. In the longest-running experiment performed so far, genetic changes affecting metabolism of *E. coli* have been investigated for more than 50, 000 generations. Microbes can be stored in freezer, then taken out years later to compare their performance with that of descendants of parent population to test the hypothesis that competitive ability evolves. The *P. fluorescens* experiment shows a repeated pattern of evolutionary diversification into 3 main ecotypes recognized by their distinctive morphology. Colonizing type gives rise to 2 more: a biofilm-producing wrinkly spreader and a fuzzy spreader. They are specialized on different parts of environment. Through repeated mutation they give rise to clones that are morphological and metabolic variations of 3 main themes. Variants within a spatial niche compete for resources and replace one another. Number of morphologically distinct clones reaches a peak through time and then declines, thus overshooting the long-term carrying capacity of environment. Range of morphological variation-the disparity-shows a different pattern. It rises to a maximum and remains there. Bacteria in absence of predation and competitor species respond to ecological opportunity by diversifying into a maximum of ecologically differentiated types. What happens then depends on environment. Either it remains fixed, as imposed by investigators, or changes with concomitant changes in community. Little is known about how gradually changing environment affects adaptive radiation, or how adaptive radiation of one group facilitates

further radiation through positive feedback from other organisms with which it interacts .An increase in aridity in last couple of million years altered the speciation-extinction balance in the Caribbean birds. It opened up new niches in the Galapagos and Darwin's finches responded by evolving seed-eating specialists.

Adaptive radiations are becoming increasingly quantitative, experimental, and comparative to understand general properties and differences, according to time of occurrence, taxonomic group, and particular environment. Vertebrates and some plant groups have dominated investigations of extant groups so far, although the recent exploitation of microcosms for experimental investigation has revealed an enormous potential residing in microorganisms. Additional experimental potential at level of ecological communities has scarcely been tapped. A second source is genetics-gene expression of ecologically important traits during development-for an understanding of comparative evolvability in different lineages. A third is speciation, introgressive hybridization, and interrelationship of two. Experimental investigations have a larger role to play in both revealing and testing the causal factors that observations imply. Inferences about how radiations unfold will improve as analytical methods are refined.

The term adaptive radiation was coined in 1902 by Osborn. Simpson viewed the evolutionary radiation of marsupial mammals, as various lines of descent from a common ancestor arising simultaneously and diverging in different directions. The products of an adaptive radiation are known to be adapted to exploit the environment in different ways. In the last 30 years, the range of well studied extant organisms has increased. An adaptive radiation is the product of differentiation of an ancestral species into an array of descendant species that differ in the way they exploit the environment. When differentiation has proceeded rapidly, evolutionary transitions from one state to another can be interpreted. Angiosperm plants, dinosaurs, and marsupial mammals are typical examples at high taxonomic levels. Typical examples at lower levels are Darwin's finches on the Galapagos Islands, honeycreeper finches, *Drosophila*, spiders, the silvers word alliance of plants in the Hawaiian archipelago, cichlid fish in the Great Lakes of Africa, and *Anolis* lizards in the Caribbean. They comprise several to many species, the species vary morphologically in conspicuous ways and relatedly, they occupy diverse ecological niches. Most species were derived from a single ancestor in their present environment, and most diverged relatively rapidly. Cichlid fish in Lake Malawi are an outstanding example. Hundreds of species were derived from one or few common ancestors in the last 2 million years, and diversified into trophic forms, like algae-, snail-, insect-, and fish eating animals. Their mouth and teeth morphologies reflect their diets, and for this reason the variation is inferred to be adaptive, the product of diverse natural selection. The rock-dwelling Mbuna of the genus *Tropheops* alone comprises 230 species. Ole Seehausen has calculated that one new species arose every 46 years.

## 3.28 Ecological Theory

A radiation starts with rapid multiplication of species and diversification of morphological types. As it proceeds, both species proliferation and morphological evolution slowed down. Species multiply at a high rate, diversify phenotypically and ecologically through invading an underexploited environment. Speciation and diversification decline as competition increase for diminishing variety of unexploited and underexploited resources.

**Phenotypic differentiation** caused by natural selection arising from differences in environments, competition for resources, and speciation governed by both processes. Quantitative analysis of hundreds of the cichlid fish species shows that some groups diversify without radiating. *Plethodon* salamanders diversified rapidly early in their history, and slowly later, but speciation was largely allopatric, resulting in slight morphological diversity and relatively little adaptation. The 46 extant species exist in few niches. Speciation may occur repeatedly through diversification of mating signals by sexual selection like the color of cichlid fish, vocalizations of birds, pheromones of moths, and stridulation of crickets. In such examples of diversification through random drift or founder effects, the role of ecological factors in speciation can not be ruled out, and the distinction between adaptive and non-adaptive causes of species proliferation is sometimes blurred. Non-adaptive processes perhaps contribute to an adaptive radiation. A particular radiation may have heterogeneous causes like adaptive and non-adaptive processes, natural selection and sexual selection, and competition and predation.

## 3.29 Evolutionary Limits and Constraints

Evolution leads to rapid changes in characteristics of organisms, limits evolution arise from a lack of genetic variation, loss of well-adapted genotypes in populations due to gene flow, trait interactions leading to trade-offs, and/ or the difficulty of evolving simultaneous changes in several traits. Signatures of genetic constraints at molecular level include a loss of functional genes due to mutational decay. Evolution rapidly changes morphology, physiology, and behavior of species, but it can be constrained. Fossil record shows long periods of morphological stasis.

Some lineages show little change in appearance across millions of years. Cockroaches, tuatara, cycads, and horseshoe crabs provide well-known examples. Such lineages often reached an evolutionary dead end and constrained to narrow **ecological niche**. Effects of selection can be powerful, as exemplified in artificial selection in generating crop plants and domesticated animals. Although pesticide resistance is widespread in insect pests, weeds, and fungi that cause plant disease, many agricultural chemicals have remained effective against pests for several decades. In such situations, the pest populations do not evolve resistance, evidently as they are constrained in some way. A population or species might have the potential to evolve, but other factors like movement of genes among populations and/or trait interactions make evolutionary change difficult despite ongoing selection. Both fundamental limits and other constraints prevent populations and species from adapting to new environments. Constraints restrict species to live in a particular environment. Thus, limits and other constraints drive biodiversity; without them, there might be few common species adapted to a wide range of conditions, majority restricted to a narrow range of ecological conditions. Explanations for evolutionary constraints can be divided into 2 categories. Those that reflect the nature of genetic variation required for evolution and adaptation, and those have their origin in ecological processes to which populations and species are exposed.

Natural selection can not change a trait if the trait lacks genetic variation. In absence of genetic variation, any increase or decrease in mean value of a trait after selection can not be passed on to next generation. Take a population of individuals derived from a single clone. Barring mutation, all individuals in population could be genetically identical. Because of differences in environmental conditions experienced by individuals, they could differ in appearance and performance. Even if these differences affect fitness of individuals, differences can not be passed on to next generations, because all individuals are genetically identical. Although natural populations can be genetically diverse, they may nevertheless have traits lacking genetic variation. One possibility is that populations and species lack specific genes that are required to adapt to new environment. Antarctic marine fish have lost the genes coding for proteins and regulatory mechanisms needed to live in warm environments, preventing colonization of warmer waters. Many *Drosophila* species lack copies of genes coding for heat shock proteins (HSPs) essential for surviving in hot conditions, where HSPs protect other proteins from degradation. In such situations, absence of appropriate functional genes represents an evolutionary limit. Constraints are associated with development that influence body plans and morphological options of species .At molecular level, this limit can be overcome by a change in genome, such as a duplication of another gene, or set of genes that evolves a new function and restores the proteins required for living in warm water or surviving hot conditions. Such changes might then produce an evolutionary lineage consisting of species with ability to colonize a warmer environment. Genes are also subjected to DNA decay as mutations accumulate to become functionally inactive. **DNA decay** occurs when there is no selection on a gene to remove mutations that lead to inactivation. The genomes of species contain many genes in a state of decay and on way to becoming nonfunctional pseudo genes. Once DNA decay and eventual gene loss have occurred, function may not be restored unless there is a further duplication event and evolutionary changes to produce a new function.

As traits are typically affected by several

genes, genetic variation in a trait may be lost when there is a cumulative effect of molecular changes at multiple loci, or when a key regulatory gene in a developmental pathway is inactivated. Absence of genetic variation in a trait can be detected through the inability of selection to change the distribution of the trait when artificial or natural selection is imposed in particular direction. It can be detected through a loss of heritable variation in a trait, which is often estimated from family studies. Heritability reflects the extent to which variation in a trait is determined by genetic rather than environmental factors. Heritability estimates for morphological traits are high in out breeding species, while they tend to be lower for behavioral and physiological traits. Heritability estimates for natural populations of animals and plants can be variable, for traits that are important in determining the ecological niche.

Distribution of various *Drosophila* species coincides with their levels of resistance to desiccation and cold stresses. Species sensitive to cold and dry conditions tend to be confined to moist and warm tropical rain forests, others that are widespread have a high resistance. Comparisons of heritability for these resistance traits across species show that sensitive tropical species tend toward a low level of heritable variation (Kellermann et al., 2009). This explains why such species are restricted in their distribution based on a lack of genetic variation preventing them moving out of their warm and moist habitats. One mechanism likely to drive a loss of genetic variation is DNA decay. When a species becomes restricted to an area for its dependence on a host plant with a narrow distribution for food, or for breeding or because it becomes confined by physical barriers, there can be a loss of purifying selection for specific characteristics and their underlying genes .In absence of purifying selection, genes start decaying as

they accumulate mutations that may become fixed by genetic drift. The decaying decreases the evolutionary potential of a species. It is not clear how often DNA decay limits further evolutionary change. Strong selection contributes to a loss of genetic variation in traits. If directional selection for increased expression of a trait persists for many generations, the allele's favored by selection is expected to increase and go to fixation. Once this occurs across all the loci affecting a trait, genetic variation in trait is expected to decrease toward zero, preventing any further selection response until the strong directional selection is alleviated, and new mutations are accumulated. Evolutionary constraints are the key determinant of biodiversity. If all species could successfully adapt to changing conditions, far fewer species would exist, whereas constraints promote species-level biodiversity. The tropical areas have high number of species as tropical species suffer from evolutionary constraints and lack adaptive potential. The tropical *Drosophila* has a low evolutionary potential to adapt to colder and drier conditions when compared to widespread species. Many ecologically specialized species lack adaptive potential as low level of genetic variation, or presence of strong pleiotropic interactions of both prevent them from adapting. This may prevent insect herbivores and parasites from expanding their diets to include new hosts, and prevent the spread of species from humid tropics into drier and cooler climate zones. DNA decay and mutation accumulation are more likely in some regions-perhaps species become more easily confined to a narrow-range host plant or because of a geographic barrier.

Recognizing adaptive constraints is important for conservation of species and ecological communities. It helps to identify groups of species at potential risk because of an inability to adapt, or even entire communities if common patterns exist across species groups. Unless they are sufficiently phenotypically plastic, they may be prone to extinction due to disease, climate change, or other stressors. If evolutionary constraints arise due to gene flow, alter levels of gene flow for adaptive changes in marginal populations might be possible. Evolutionary constraints are important in applied biology. When pests and weeds are unable to evolve resistance to pesticides, there is the potential to continue using the pesticides. Understanding the action of pesticides and genetic basis of resistance, help to predict the likelihood of resistance developing in the evolutionary lineage of pests. Where evolution is likely, it possibly slow the rate of evolution by ensuring ongoing gene flow between susceptible populations and those under selection. This practice is adopted in management of resistance to toxins introduced into crop plants, where resistant crops are interspersed with susceptible cultivars to slow adaptation by pests.

## 3.30 Trade-Offs

The evolutionary constraints due to DNA decay and loss of genetic variation arise due to absence of genes, nonfunctional gene, lack of genetic variation, and the pleiotropic effects. Genes have potential for pleiotropic effects because proteins encoded by genes are embedded in networks of biochemical processes .These networks influence expression of multiple traits. Genes regulating the expression of other genes can have pleiotropic effects by influencing multiple networks. Because of complex and indirect ways in which genes influence phenotypes, selection for a decrease or increase in a trait favor a set of underlying allelic changes that impact other traits. Genetically based trade-offs occur through pleiotropy when these effects influence traits closely related to fitness but in opposing

directions. In insects a genetic trade off between development time and reproduction exists, as alleles promoting fast development lead to early emergence and early reproduction by adults, but a cost is paid in that adults emerge at a small body size with reduced reproductive output. In plants a genetic trade-off between flowering time and seed production exists, with early-flowering plants tend to produce small flowers. These trade-offs may result in different traits being favored in different environments. Insect populations living in cool conditions are under strong selection to complete development within short growing season, resulting in small body size and reduced overall reproductive output. Populations of an invasive wetland plant in cold latitudes are genetically adapted to flower early at the cost of a small flower size, compared with populations of the same species in warmer areas (Colautti et al., 2010). In such cases, a genetically based reduction in reproduction can become an evolutionary constraint, if reproductive output is no longer sufficient to sustain a population. An allele that increases insecticide resistance by boosting a detoxification mechanism will be favored when chemical is present, but is selected against when it is absent, perhaps because detoxification mechanism is energetically costly. Such information point to a mechanistic basis of a trade-off, but might not provide information about an evolutionary constraint, because other mechanisms of insecticide resistance like decreased sensitivity of target site of chemical might be selected instead, and these other mechanisms might not be associated with a **trade-off**. For this reason, trade-offs are often characterized by looking at genetic correlations that reflect the effects of many underlying genes. These correlations can be measured through family studies by considering inheritance of multiple traits across generations.

Negative genetic correlations for traits

affecting fitness in opposing directions, like association of rapid development with reduced reproductive output, are then taken as evidence of trade-offs that contribute to evolutionary constraints. A negative genetic correlation does not reflect an evolutionary constraint; unless negative genetic correlation is strong. The possibility still exists that selection can proceed at least partly independently on 2 traits. Some alleles may affect one trait but have no pleiotropic effects on other traits, and might then still be selected. Trade-offs can be investigated by carrying out selection experiments, designed to test whether selection on one trait is invariably associated with fitness costs involving a different trait. Many selection experiments in insects, mice, and worms explore whether an increase in early reproduction is invariably associated with a decrease in longevity as a result of allocation of resources to reproduction rather than to maintenance. This trade-off was proposed by G. Williams to explain constraints on evolution of increasing life span. When testing for constraints due to pleiotropy, it is important to note genetic interactions among traits due to chromosomal linkage from those due to pleiotropy. If genes are closely linked along a chromosome, alleles affecting traits end up in **linkage disequilibrium**-an allele that increases development time might by chance end up linked to an allele affecting reproductive output but from a different, closely linked gene. Although pleiotropic effects can be difficult to distinguish from linkage in practice, linkage associations between traits are expected to be lost as recombination occurs among the linked loci, and so need not impose long lasting constraints.

Lenski and others placed *E. coli* bacteria at 20°C, at which they maintained and monitored adaptation after 2000 generations. Bacterial populations improved in their competitive performance at 20°C by 8 percent,

but at cost to performance at 40°C of 20 percent. These were interpreted as evidence of trade-offs associated with temperature adaptation. *E. coli* adapts to low temperatures, but cost that constrains evolution in an environment where temperatures fluctuate. Evolutionary constraints due to pleiotropy can be distinguished from those due to lack of genetic variation and by presence of genetic variation in each of correlated traits. In former case, each trait is expected to show genetic variation. This situation applied to most traits under selection in prairie legume. In selection experiments, the persistence of genetic variation in traits even after a selection limit is reached can point to pleiotropy, rather than lack of genetic variation acting as an evolutionary constraint.

## 3.31 Gene Flow in Marginal Populations

Gene flow occurs when individuals or propagules move from one population to another and contributes to genetic constitution of other population. This enhances or retards evolutionary adaptation. The former occurs when gene flow increases genetic variation by introducing new genetic variants into a population that can be selected to increase fitness. When too much **gene flow** occurs, the effects of selection can be overwhelmed by an influx of nonadapted genotypes. Gene flow can than act as an evolutionary constraint. In some populations, decreased density of individuals occurs compared with centrally located populations. This results in directional gene flow into marginal populations in some situations which may be sufficient to retard adaptation in marginal population. This process might then be sufficient to prevent expansion of the species. Directional gene flow acts as constraint, but the empirical data for this hypothesis are limited. Marginal populations are often as dense as centrally located populations, making unidirectional gene flow from high density central populations to low-density marginal populations unlikely. When gene flow is measured across multiple populations with molecular markers, it often seems to occur in both directions rather than only from central to marginal populations. The strongest evidence for this evolutionary constraint comes from transplant experiments suggesting that populations can survive outside their normal range.

## 3.32 Knowledge from Chemistry and Physics

Of the 3 major events in evolution of Universe, the first is the Big Bang, the 2nd is the formation of earth, and the 3rd is the origin of life. Species are believed to be arisen in a set of material bodies. The observed similarities developed the evolutionary tree which can be detailed to individual species of organisms. Changes of habitat of organisms are not given in their description in branches of trees. Maynard-Smith & Szathmàry (1995) end that organism's increase in complexity is neither universal nor inevitable. All biological materials are made from atoms in and energy from the environment. Let's start from an analysis of the energy inputs to environment and organisms considering them as one system.

**Energy from the Environment:** Energy drives geological and biological evolution. Energy moved from the high-temperature interior to lower surface temperature of 300 K. It continued for all the period of earth's existence. The oxidic rocks gave earth a hard heat-insulating mantle with water covering it around 4.0 Gyr ago. **Radioactive decay** of potassium, thorium, and uranium in interior

of earth partly maintains temperature above 5000 K. Decomposition of unstable minerals on or near the surface particularly in ocean vents, like black smokers caused energy transfer. This surface heating was compensated by energy absorbed at surface of earth from the sun and in atmosphere. Together, the 2 raise the surface temperature from 250 K to about 300 K. Loss of heat from earth has fallen over time due to loss of some green house gases. The sun has gained radiative strength. Surface temperature of earth has remained about constant for all the period from ca 4.0 Gyr ago to date with variations of ±25 K. Material evolution perhaps has occurred in a thermostat at 300 K (Cavalier-Smith et al., 2006). Thermostat kept water liquid on surface permitting the material flow. Chemicals are inorganic in nature like minerals, including water. Energy flow caused the formation and loss of unstable organic chemicals in synthesis/degradative cycles in cells. The organic systems are based on H, C, N, O, S and P, and contain several mineral elements. Energy trapping in such system from high-energy chemicals or sun can create unstable organic chemicals of high-energy content or as physical gradients of concentration of charge or particles across boundaries. Certain elements can be energized chemically as their chemical bonds have sufficient kinetic stability and cause formation of specific organic compounds of long life. Total system increased ability to absorb energy, during evolution of environment/organism system. Evolution is driven by energy flow. Material cycles increases rate of energy degradation. If the cycle was in steady-state flow, only change would be the entropy production and there would have been no evolution.

**Basic Material Flow on Earth:** Energy flow occurs from the hot source to cold sink. It can go through intermediate stages in which material absorbed energy physically, or chemically. Physical changes arise from increase in random motion (temperature) and/ or increase in directed flow. The first appears as a change in equilibrated or steady-state structures, solid→Liquid→gas, associated with increased random kinetic energy. The second related to creation of gradients appears as directed kinetic energy in steady-state organized patterns of flow. The two is readily distinguished. On earth, there are phases of ice and water in containers of fixed structures and water vapour with random thermal motion in atmosphere. Both the first 2 phases have ordered physical boundaries. Solids have ordered internal structure causing static shapes. Water in atmosphere is also observed in clouds, the examples of material in organized flow. Clouds come in closely, reproducible dynamic shapes. Cumulus clouds are formed from low-down turbulent flow of water droplets falling under gravity but dispersing again at higher temperature as vapour that rises to reform droplets at lower temperature again. In high stratus clouds, the droplets and vapour circulate in streaming motion.

When we replace the energization flow process of physical condensation and vaporization by chemical flows in which hundreds of chemicals take part in cyclic energized transformations from the environment to cells and back, it is possible to group large numbers of species and individuals into types. Species and individuals within types are characterized by small differences in shapes, behaviour, or molecular sequences. The circulation of water can be traced as clouds fall as rain, forming streams and rivers. As the river flows, it erodes the minerals off its banks and flow them towards the sea. Water energized into clouds. As river flow slackens, the eroded material settles forming new shapes of riverbanks and deltas. This is the non-cycling waste product of water

flow. The deposited material is called sediment or soil. Its evolution with continuous water flows allowed much organic life on land to evolve. Material flows evolve. Clouds, rivers and land show that organism is far from the only systems with characteristics of developing organized flow. The simplest chemical system, illustrating organization and order, is the formation of ozone layer by sunlight on oxygen. Ozone layer evolved in last 1 Gyr. Humans have introduced chemicals interfering with flow. These systems are in general cases predictable and they follow physical/chemical principles. General behaviour causes steady state of flow, which die but again recur due to fluctuation of environment. This steady state represents the maximum flow and maximum energy degradation (Morowitz, 2002). The essence of life is organic-a mixture of energized flows of inorganic and organic molecules. This energized chemical system, with physical boundaries, the cells, is very complicated.

**Organic Systems:** Sometime after the Big Bang, the giant stars formed and then chemical elements arose, 92 of them fall in the Periodic Table of atoms. They were formed in a predictable pattern of abundances (Fraústo da Silva & Williams, 2001). Later their dispersal at high temperature and then cooling and condensation, caused the planetary system including earth. We need to consider only the elements on surface which through abundance were made available and so were irradiated. These were mainly in atmosphere, and somewhat dissolved in sea, C ($CO$, $CO_2$, $CH_4$), H ($H_2$, $H_2O$, $H_2S$), N ($NH_3$, $N_2$), S ($H_2S$) and in minerals, which are soluble to some very different( Williams, 2007) degrees , giving the following ions also in the sea: $Na^+$; $K^+$; $Mg^{2+}$; $Ca^{2+}$; $Mn^{2+}$; $Fe^{2+}$; $CO^{2+}$; $Ni^{2+}$; $Cu^+$; $Zn^{2+}$; MO or W ($MoO_2^{-4}$ or $WO_2^{-4}$); Se ($H_2Se$); P ($HPO_2^{-4}$); V($VO^{+2}$); and $Cl^-$. These elements, with perhaps one or 2 others, are important in possible initial organic systems. Of about 20 elements listed, there are representatives of some 14 and possible 17 groups of elements with strong chemical differences in Periodic Table. These 20 elements are essential to life and evolution. We call the system in an organization, organic. Environment has not been constant over time (Williams & Fraústo da Silva, 2006). Evolution has occurred in environment and organisms. Both have developed cooperatively through the nature of energization of system of molecules. Evolution was directional once the flow became to and from compartment cells and in all probability prior to coding by the *energization* of organic material in kinetically stable reduced states of carbon.

**Kinetic Stability and Organic Energization:** A system of irradiated or energized chemicals has a considerable lifetime if it is to enter a flow system of any complexity and it must not disperse. Molecules made from N, H, C, and O, and to a less extent including P and S can be persistent. Chemicals made from these basic building blocks are thermodynamically unstable but kinetically stable at 300 K and are reduced relative to very simple environmental compounds, such as $CO_2$ and $H_2O$. Probably the original constructs of a living cell were uniquely linked with larger organic molecules of this kind, which we see now, i.e. lipids, proteins, saccharides and nucleic acids. A characteristic of this unique set of organic molecules is that their elements in compounds are reduced relative to CO ($CO_2$) and $N_2$ by incorporation of hydrogen. They are mostly negatively charged to keep many of them soluble in water. Lipids are insoluble in water and readily form vesicle membranes preventing dispersion of internal contents. They could generate thus the precursors of cells. These vesicular protocells are believed to cause the origin of

life (Russell & Hall, 2006).

The reductive chemistry, which later produced kinetically stable cells, generated oxidized waste and rejected some materials into environment.Light, is not the only source of energy. The cycles in complicated form are oxidized material like sulphur from $H_2S$ or oxygen from $H_2O$, where the amount and availability of water made itself the preferred source of hydrogen in evolution. Energy applied rejects some available elements that could be poisonous to system. The selective rejection is linked to the origin of protocells and life. Adaptation to waste drives evolution allowing greater use of energy. The kinetics of compounds made from inorganic ions in water, here the sea, are different from those of organic compounds in that generally they equilibrate quickly between free ionic forms, various oxidation states and bound conditions between organic molecules (Fraústo da Silva & Williams, 2001). Their compounds of high thermodynamic stability have low kinetic stability. We shall treat all the ions $Na^+$, $K^+$, $Mg^{2+}$, $Ca^{2+}$, $Mn^{2+}$, $Fe^{2+}$, $Co^{2+}$, $Ni^{2+}$, $Cu^+$ ($Cu^{2+}$), $Zn^{2+}$ and $Cl^-$ as in effective but different equilibria with their compounds in different oxidation states in both protocells and the environment, even though distribution to binding sites may require carriers. All these ions exist in sea and cells in controlled amounts and certainly differently in sea and cells at various times during evolution (Fraústo da Silva & Williams 2001).

In the protocell, they had to be, energized in their concentrations by pumps as they cannot cross membranes. As one group is much energized, same kind of synthesized organic molecules is found in central parts of all cells and certain free concentrations of ions remain fixed there. In the steady state protocells and cells, the outward/inward pumping to/from the cytoplasm maintains similar kinetic gradients of free ions including $H^+$ across all outer cell membranes. Protocells like all cells had to reject $Na^+$, $Ca_2^+$ and $Cl^-$ to the sea so as to reduce osmotic pressure ($Na^+$, $Cl^-$) and prevent conglomeration and precipitation of organic molecules ($Ca_2^+$). $K^+$ and $Mg_2^+$ are then allowed into the cells to balance mainly the negatively charged soluble organic molecules. There are restrictions in large outward pumping, depending on $M_2^+$ ion concentrations .Although these ions in bound states operate catalytic or structural functions but a switch in evolutionary conditions later has made it necessary to pump these ions selectively in new ways either in or out of cells effectively. There is a grave problem in maintenance of iron levels in cells, since in oxidizing environment, free iron ions are in very low concentrations, $10^{-17}$ M Fe (III). Condensation and removal of water requiring energy leads to most metabolites from H, N, C and O and to all biopolymers, proteins, saccharides, nucleotides, and lipids. The removal of water is due to pyrophosphate hydrolysis as nucleotide triphosphates. In Pyrophosphate hydrolysis, the phosphate is incorporated into all nucleotides and many lipids. The energy is required to form all these phosphorylations again, causing prior formation of pyrophosphate. This chemical was required before any complex biomolecule could be formed. Pyrophosphate is synthesized from 2 phosphates using the energy of a proton gradient in first condensation reaction. The proton gradient is made from light (Williams, 1961), with an electron transfer circuit causing reduction. Underlying condensation and the formation of many cellular chemicals are then the need for phosphate.

In the sea, phosphate cannot exceed about $10^{-6}$ M due to the presence of $Ca^{2+}$ ions. The cellular need for phosphate can be met by pumping it into cells, where free calcium reduces in concentration. Here, energy is again required. Selecting each of 20 chemical elements is one of the necessities. Many elements are required in cells, especially their functions in reduction and condensation and in physical stability. In the first of these, many organic elements are incorporated in compounds by formation of kinetically stable bond and/or by binding at equilibrium after pumping of ions. Incorporation of oxygen was and is needed in all life. For this, organisms need molybdenum, or tungsten as a catalyst. The incorporation of oxygen is solved through the formation of carboxylates from aldehydes. These elements exchange oxygen atoms with water readily at low oxidation/reduction potential which existed before there was oxygen (Williams & Fraústo da Silva, 2002). Life depends from its beginnings upon C, H, N, O, S, P and common metal ions, molybdenum and tungsten.Lipids when agitated, cause vesicles and postulate that these vesicles can concentrate material. A code will become essential only when we come to reproductive life, which we shall put on one side as an accident.

**Markers of System Evolution:** Importantly, the long-period changes of metal elements (ions) and isotopes of non-metals (Holland, 2006) can be followed in sedimentary mineral deposits like red bands and baryte, $BaSO_4$. Changes in use of inorganic elements in cells, mostly bound with successive introduction of organic proteins. Certain proteins are strong markers of long-period evolution, which can be directly related to newly expressed parts of coded DNA. Changes in availability in environment and use of $Ca^{2+}$, $Zn^{2+}$, $Mn^{2+}$, $Fe^{2+}$, $Co^{2+}$, $Ni^{2+}$ and $Cu^+$ ($Cu^{2+}$) and their proteins in cells increases and then cellular content changes. We can trace them against the known background of $O_2$ increase and appearance of $SO_2^{-4}$ and $NO^{-3}$ and some non-metal compounds. For these reasons, changes in 20 available elements and a broad view of their compounds in cells are the best markers of the evolution. Inorganic ions had the connections with environment.

Cells use a required series of proteins, including internal carriers inward pumps, transcription factors; protein partners themselves, outward pumps, ion channels and cellular buffers. Chemistry of selective equilibrium binding of all the ions is now well understood (Fraústo da Silva & Williams 2001) and we can therefore find and rationalize the most suitable 'signature' amino acid sequences in proteins for different metal ions. The set of individual time patterns of ion-binding proteins and enzymes can be found in genome. It is also possible to give a development pattern for different ions in relation to introduction of chemotypes. The cells evolve with different elements, and their protein partners are differently organized (Dupont et al., 2006). Molybdenum shows connection to genome. The non-metals H, C, N and P, are much involved in cell chemistry, but their compounds did not change greatly in cytoplasm due to conservative nature of all life in its production of proteins, saccharides, nucleotides, and lipids. There was a relatively small change in carbon chemistry after oxygen increased in the atmosphere, which we find in some extra cellular proteins, in higher organisms. A larger change of environmental nitrogen shows role of essential primitive metal elements in life. The main metal ions should be followed as early markers of evolution together with changes of O, N and S. Later new markers appear, e.g. Cu and Zn (Williams, 2007).

**Effect of Oxygen on the Environment:** Waste oxygen from biological activity did not accumulate quickly in atmosphere. The large amount of reduced inorganic materials in environment rapidly equilibrated with it to create oxidized states of elements. The $O_2$ pressure was buffered for a long period. The major materials which are powerful fast

reductants were the free $H_2S$ and $Fe^{2+}$ from sulphides. They were oxidized to $SO_2^4$ and $Fe^{3+}$. Their conversion can be followed by precipitation of $Fe^{3+}$ iron bands in surface geological minerals (Holland, 2006). Some 2 Gyr the free oxygen concentration rose to 1% of present levels (Kasting & Siefert, 2002). The oxidation potential of $O_2$ rose steadily with its partial pressure, reaching present levels say 500 mya ago. In the later periods, oxidation of the sulphides of Ni, Cu, Co, and Zn gathered pace, but vast amount have remained buried (Saito et al., 2003). The vanadium and molybdenum were readily oxidized from sulphides. The sea became rich in these 2 elements in early evolution. Now, non-metals other than sulphur are oxidized and original CO and any NO or $NH_3$ (Miyakawa et al., 2002) are replaced by $CO_2$ and NO–3. The $CH_4$ formed by organic degradation reacts slowly with $O_2$; some of it became buried, causing formation of oils and coal. Most organic debris from cell activity was oxidized by cells. The non- availability of many elements and the gain of many others caused stress on pre-existing chemical flows in organisms.

The early organisms faced a hostile environment, always present as the Na+, Cl– and Ca2+ in the sea, but increasingly from non-availability to C, S, Se, N, and $Fe2^+$ and from the direct poisonous increase of amounts of $O_2$.

The chemotypes are the classes of species, while genotype relates to individual species. The specific genes define individuals. The change of **chemotypes** represents a physical/chemical system explanation. Evolution is a simple directed external chemical change due to biological waste, forcing slow adaptive change on organisms, thus with increasing energy absorption and degradation. Initially, the buffering of $O_2$ was environmental and inorganic by fast $Fe^{2+}$ and $H_2S$ reactions when organisms hardly changed for 2 Gyr . Now the short-term buffering is largely organic (Lovelock, 2000). There is continuing slow oxidation of methane, oil and coal. Bacteria can adapt quickly to long-lasting poisons, organic and inorganic (Hg, Pb, Cd), but it is doubtful, if higher animals can adapt to such changes.

**Chemotype Classification of Organisms:** Changes of concentrations of C, N H, and O in their compounds with time in cells are complicated (Kasting & Ono, 2006). Environment is linked to them through inorganic compounds like $CO_2$, $N_2$, $H_2$ and $H_2O$ which cannot be controlled homeostatically. Changes in availability of these 4 elements and some oxidized modifications of them and of S, I and Se compounds during the chemotype diversification is important. The metal ion concentrations and components, metallomes, have changed with time which form close link between cells and the environment due to fast reactions in both compartments. Both cells and environment compartments linked to their free metal ion concentrations through equilibria, very different in 2, describing evolutionary

changes. The best linkage to C, N, H and O chemistry is via metalloproteins and then of these proteins to DNA sequences. These proteins have a homeostatic link to metal ions they bind. The gene tell us which binding proteins can be expressed .T he metalloprotein concentration in a compartment inform us which are formed quantitatively. Often, only the DNA of a cell type is characterized and we know there are differences between DNA and metal/protein characteristics of the major types of organisms (Williams & Fraústo da Silva, 2006). In the most primitive single cell anaerobes, unavailable dioxygen and copper cannot be (Williams, 2007) used and their protein partners are absent. In the single cell aerobic prokaryotes, there is little use of calcium and few of its proteins, while calcium signalling becomes valuable for single-cell eukaryotes, in which novel calcium-binding proteins (Williams, 2007) appear. Multiellulan organisms make extensive use of Ca, Zn, Cu and Fe proteins and many elements outside cells in controlled fluids. Animals with nerves found a new use of Na+ and Cl in these fluids requiring new pump proteins.

# SUMMARY

1. NASA's record of bacteria-like organisms as fossils on martin rock from Antarctica suggests that life came to earth from the outer space. Cosmozoic theory state that some planet with primitive life collided the earth and seeded it with life. Redi proposed that life begets life and living animal found in mud, water, or meat originates from the spores, eggs, or hibernating animals.

2. Schopf supported the natural formation of amino acid in primordial oceans and detected 22 amino acids in 3000 million years old rock formation. By heating a mixture of 20 amino acids, Fox obtained polypeptides forming microspheres equivalent to coacervates in water.

3. Life evolved in the sea, stayed there for most of recorded history. The common ancestor of all life forms used RNA as genetic material and produced prokaryotes, archaebacteria and eukaryotes. Protist, fungi, plants and animals are eukaryotes.

4. Eukaryote and archaebacteria are most closely related. Translation is similar in these lineages. Organization of genome and transcription is different in prokaryotes than eukaryotes and archaebacteria show that RNA based ancestor produced 2 lineages that formed DNA genome and evolved mechanisms to transcribe DNA into RNA.

5. Fungi-like organisms appeared about 900 mya .In photosynthesis, organism trapped sunlight to produce sugar about 3.4 billion years ago. Animals appeared before the Cambrian about 600 mya. The first animals dating before Cambrian found in rocks near Adelaide, Australia are the Ediacarian fauna .The Cambrian explosion, perhaps the result of higher $O_2$ concentration enables large organisms with high metabolisms, to evolve as spreading shallow seas produced variety of niches to produce variety of animals. Cambrian animal fossils show various organisms from the Canada.

6. Animal life before the Cambrian was not modern .Photosynthetic protests and algae formed the base of food chain. Following the Cambrian, number of marine families reached about 200.The Ordovician explosion larger than the Cambrian around 500 mya followed this explosion, produced many Paleozoic fauna. The Cambrian fauna, declined slowly during that time. New lineages evolve and radiate causing extinction of older lineages.

7. Organisms modify their environments causing further evolutionary change. Diversity of organisms increased although interrupted many times by mass extinctions. Multicellular life is composed only of eukaryotic cells. The earliest report for it is the Francevillian group fossils from 2,100 mya.

8. Viruses are the sub cellular replicating entities. Sub cellular entities include viruses, viroids, transposons, prions satellites, plasmids .As the host cells for sub cellular replicators become complex, the sub cellular replicators shed genetic code that performs redundant functionality to code that exists in host cell. Thus the existing viruses are less complex than original ancestral viruses. When the viral code enters a cell, cell takes the code and starts working the instructions in code. The cell treats viral DNA, or RNA the same way that it treats other DNA, or RNA in cell.

9. Under similar environment distantly-related organisms results in plants and animals that are morphologically similar in overall appearance. This is convergent evolution. Convergent evolution is the development of similar body and adaptation patterns in 2 unrelated species.

10. Divergent evolution is a force of nature that creates the variety of form and structure in plants and animals. Fossils known as small shell fauna are found in many parts of world, and date from just before the Cambrian to about 10 million years after start of Cambrian.

11. Mineralized phyla that form the basis of fossil record may not be representative of other phyla, as most mineralized phyla originated in benthic setting. Fossil record is consistent with the Cambrian Explosion that was limited to benthos, with pelagic phyla evolving much later. Range of different organism designs or ways of life rose sharply in the early Cambrian.

12. That disparity was thought to be high throughout the Cambrian before subsequent decrease is discarded. Disparity remains relatively low throughout the Cambrian, with modern levels of disparity only attained after the early Ordovician radiation. This is often noted

that number and diversity of fossils of multicellular plants and animals increases greatly in fossil record from about 550mya. Such sudden appearance of complex life forms, many sharing their most fundamental characteristics with modern species, is called the Cambrium explosion.

13. Cambrian explosion caused rapid appearance around 530 mya, accompanied by major animal phyla and major diversification of animals, phytoplankton and calcimicrobes. Before 580 mya, most organisms composed of individual cells were simple, rarely organized into colonies. Diversification rate seen in the Cambrian phase of explosion is unparallel led among marine animals. It affected all metazoan clades of which the Cambrian fossils are found.

14. For several classes of viruses, especially for positive strand RNA viruses and nucleo-cytoplasmic large DNA viruses, evidence of monophyletic origin is found.

15. Carl Woese proposed that the early stages of evolution including that of the Last Universal Cellular Ancestor (LUCA), involved rampant horizontal exchange of genetic material between primordial life forms, such that individual lineages could not form (Woese, 2002).

16. Material evolution perhaps has occurred in a thermostat at 300 K (Cavalier-Smith et. al., 2006). Thermostat kept water liquid on surface permitting the material flow. Chemicals are inorganic in nature like minerals, including water. Energy flow caused the formation and loss of unstable organic chemicals in synthesis/degradative cycles in cells

17. Essence of life is organic-a mixture of energized flows of inorganic and organic molecules. This energized chemical system with physical boundaries, the cells, is very complicated.

18. Adaptation drives evolution allowing greater use of energy. The kinetics of compounds made from inorganic ions in water, here the sea, are different from those of organic compounds in that generally they equilibrate quickly between free ionic forms, various oxidation states and bound conditions between organic molecules (Fraústo da Silva & Williams, 2001).

# REVIEW QUESTIONS

## Short Answer Questions

1. What do you understand by the term "abiogenesis"?
2. Name three sub-cellular replicating entities.
3. What are satellites?
4. Define parallel evolution.
5. Define divergent evolution.
6. What are microfossils?
7. Define "geological time scale".
8. Name five eukaryotic super groups.
9. Define genotype.

10. Define phenotype.

11. Define population genetics.

12. Define adaptive radiation.

13. Define biodiversity.

14. What is convergent evolution?

15. Define fossil.

## *Long Answer Questions*

1. Describe the biochemical basis of origin of life.

2. Write a note on divergent evolution in your own words.

3. State the role of the Cambrian explosion in evolution.

4. Discuss the important environmental changes occurred during evolution.

5. Give a tabular presentation of geological time scale in brief.

6. Write a note on the adaptive radiation in light of the modern study.

# REFERENCES

Adl S,M, Simpson, A.G., Farmer M.A, Andersen R.A, Anderson O.R, Barta, J.R., Bowser, S.S.,, Brugerolle, G., Fensome ,R.A., Fredericq, S., James, T.Y.,Karpov, S., Kugrens, P., Krug, J., Lane, C.E., Lewis, L.A., Lodge, J, Lynn, D.H., Mann, D.G., McCourt, R.M., Mendoza, L., Moestrup, O., Mozley-Standridge,S.E., Nerad, T.A., Shearer, C.A., Smirnov, A.V., Spiegel, F.W., Taylor, M.F.2005. The new higher level classification of eukaryotes with emphasison the taxonomy of protists. J Eukaryot. Microbiol ,52(5):399-451.

Arthur, W., and Farrow, M. 1999. The pattern of variation in centipede segment number as an example of developmental constraint in evolution. J. Theor. Biol., 200:183-191

Baaske, P., Weinert, F., Duhr, S., Lemke, K.H., Russell, M.J., Braun, D. 2007. Extreme accumulation of nucleotides in simulated hydrothermalpore systems. Proc. Natl. Acad. Sci. U S A , 104:9346-9351.

Bhattacharya, D., Yoon, H.S., Hackett, J.D. 2004. Photosynthetic eukaryotesunite: endosymbiosis connects the dots. Bioessays, 26(1):50-60.

Blair, J.E., Hedges, S.B. 2005. Molecular clocks do not support the Cambrian explosion. Mol. Biol. Evol., 22(3):387-390.

Cavalier-Smith, T. 1999. Principles of protein and lipid targeting insecondary symbiogenesis: euglenoid, dinoflagellate, andsprozoan plastid origins, and the eukaryote family tree. J.Eukaryot. Microbiol , 46:347-366.

Cavalier-Smith, T., Brasier, M., Embley, T.M.2006. Molecular steps in cell evolution: palaeontological, molecular and cellular evidence of their timing and global effects. Phil. Trans. R. Soc. B. , 361:843-846

Conway Morris, S. 2003. Life's solution: inevitable humans in a lonely universe. Cambridge Univ. Press, Cambridge , UK .

Conway Morris, S. 2010. Darwin at the edge of the visible universe. EMBO Rep. 11:898.

Coulson, A.F., Moult, J. 2002. A unifold, mesofold, and superfold modelof protein fold use. Proteins, 46(1):61-71.

de Duve C. 2002.Oxford University Press; New York, NY.

Doolittle, W.F.1999. Phylogenetic classification and the universaltree. Science, 284(5423):2124-2129.

Doolittle, W.F, Bapteste, E. 2007. Pattern pluralism and the Tree of Life hypothesis. Proc. Natl. Acad. Sci U S A , 104(7):2043-2049.

Dupont, C.L., Young, S., Palenik, B., Bourne, P.E. 2006.Modern proteomes contain imprints of ancient shifts in ocean chemistry. Proc. Natl Acad. Sci. USA., 103:17 822-17 827.

Ellington, A. D. 2012. Origins for everyone. Evo. Edu. outreach, 5: 361-366

Farley, J. 1977. The spontaneous generation controversy from Descartes to Oparin. Baltimore: The Johns Hopkin University Press.

Felsenstein, J. 2004. Inferring phylogenies. Sinauer Associates, Sunderland , MA .

Fraústo da, Silva J.J.R., Williams, R.J.P. 2001. The biological chemistry of the elements-the inorganic chemistry of life. 2nd edn. Oxford University Press; Oxford, UK

Fry, I. 2000. The emergence of life on earth. A historical and scientific overview. New Burnswick: Rutgers University Press.

Griesemer, L. Origins of life studies. In: Ruse, M. editor. The Oxford handbook of philosophy of biology. New York: Oxford University Press.

Grishin, N.V. 2001. Fold change in evolution of protein structures. J.Struct. Biol , 134(2-3):167-185.

Hackett, J.D., Yoon, H.S., Li. S., Reyes-Prieto, A., Rummele, S.E., Bhattacharya, D. 2007. Phylogenomic analysis supports the monophyly of cryptophytesand haptophytes and the association of rhizaria with chromalveolates. Mol. Biol. Evol, 24(8):1702-1713.

Holland, H.D.2006. The oxygenation of the atmosphere and oceans. Phil. Trans. R. Soc. B.; 361:903-916.

Huey, R. B., and Bennett, A. F. 1987. Phylogenetic studies of coadaptation-preferred temperatures versus optimal performance temperatures of lizards. Evolution ,41:1098-1115.

Jaekel, M., and Wake. D. B 2007. Developmental processes underlying the evolution of a derived foot morphology in salamanders. Proc. Natl. Acad. Sci. USA ,104:20437-20442.

Kamminga, H. 1988. Historical perspective: the problem of the origin of life in the context of developments in biology. Orig. Life. Evol. Biosph.,18: 1-11

Kasting, J.F, Ono S.2006. Palaeoclimates: the first two billion years. Phil. Trans. R. Soc. B., 361:917-929.

Kasting J.F, Siefert J.L. 2002.Life and the evolution of Earth's atmosphere. Science., 296:1066-1068.

Keeling, P.J., Burger, G., Durnford, D.G., Lang, B.F., Lee, R.W., Pearlman, R.E.,Roger, A.J., Gray, M.W. 2005. The tree of eukaryotes. Trends Ecol. Evol., 20(12):670-676.

Kerr, R.A., 2002. Evolution. A trigger for the Cambrian explosion? Science, 298(5598):1547.

Kluge, A. G.. 2005. Testing lineage and comparative methods for inferring adaptation. Zool. Scr. 34:653-663.

Koonin, E.V. 2006. The origin of introns and their role in eukaryogenesis: a compromise solution to the introns-early versus introns-late debate? Biol Direct, 1:22.

Koonin, E.V. 2007. An RNA-making reactor for the origin of life. Proc Natl Acad Sci U S A , 104:9105-9106.

Koonin, E.V., Aravind, L, Kondrashov, A.S. 2000. The impact of comparative genomics on our understanding of evolution. Cell ,101:573-576.

Koonin, E.V., Martin, W.2005. On the origin of genomes and cells within inorganic compartments. Trends Genet ,21(12):647-654.

Koonin, E.V., Senkevich, T.G., Dolja, V, V. 2006. The ancient virus world and evolution of cells. Biol Direct , 1(1):29.

Lamarck, J. B. 1802. Recherches sur I'organisations des corps vivants. Paris: Fayard; 1986. Originally published in 1802.

Lawrence, M.S., Bartel, D.P. 2003. Processivity of ribozyme-catalyzed RNA polymerization. Biochemistry, 42(29):8748-8755.

Lazcano, A. and Pereto, J. 2010. Should the teaching of biological evolution include the origin of life? Evo. Edu. Outreach. 3, 661-667.

Lee, D. H. et al. 1997. Emergence of symbiosis in peptide self-replication through a hypercyclic network. Nature, 390(6660), 591-594.

Leroi, A. M., Rose, M. R and Lauder. G. V. 1994. What does the comparative method reveal about adaptation. Am. Nat. 143:381-402.

Lovelock, J.2000.: Homage to Gaia. Oxford University Press; Oxford, UK.

Lupas, A.N., Ponting, C.P., Russell, R.B. 2001. On the evolution of proteinfolds: are similar motifs in different protein folds the result of convergence, insertion, or relics of an ancient peptideworld? J. Struct. Biol, 134(2-3):191-203.

Maynard Smith, J., R. Burian, S. Kauffman, P. Alberch, J. Campbell, B. Goodwin, R. Lande, D. Raup, and L. Wolpert. 1985. Developmental constraints and evolution. Q. Rev. Biol., 6:265-287.

Maynard-Smith, J, Szathmàry, E.. Freeman. W.H. 1995. The major transitions of evolution. New York, NY

Miyakawa S, Yamanashi H, Kobayashi K, Cleaves H.J, Miller S.L., 2002.Prebiotic synthesis from CO atmospheres: implications for the origin of life. Proc. Natl Acad. Sci. USA. ;99:14 628-14 631.

Morowitz, H.J. 2002.. The emergence of everything. Oxford University Press; New York, NY

Mulkidjanian, A.Y., Makarova, K.S., Galperin, M.Y., Koonin, E.V. 2007. Inventing the dynamo machine: On the origin of the F-type and V-typemembrane ATP ases from membrane RNA/protein translocases. Nat. Rev. Microbiol , in press.

Page, R.D. 1996. Tree View: an application to display phylogenetic trees on personal computers. Comput .Appl. Biosci,12(4):357-358.

Pagel, M. 1994. Detecting correlated evolution on phylogenies-a general method for the comparative analysis of discrete characters. Proc. R. Soc. Lond. B ,255:37-45.

Richards, R. J. 2008.The tragic sense of life: Ernst Haeckel and the struggle over evolutionary thought. Chicago. The University of Chicago Press.

Ridley, M. 1983. The explanation of organic diversity: the comparative method and adaptations for mating. Oxford Univ. Press, Oxford , UK .

Rokas, A., Kruger, D., Carroll, S.B. 2005. Animal evolution and the molecular signature of radiations compressed in time. Science , 310(5756):1933-1938.

Russell, M.J., Hall ,A.J. 2006.The onset and early evolution of life. Geol. Soc. Am. Memoir. ; 198:1-32.

Saito, M.A., Sigman, D.M., Morel, F.M.M. 2003.The bio-inorganic chemistry of the ancient oceans. Inorg. Chim. Acta. 356:308-320.

Schluter, D. 1996. Adaptive radiation along genetic lines of least resistance. Evolution, 50:1766-1774.

Schwenk, K., and. Wagner, G. P. 2004. The relativism of constraints on phenotypic evolution. Pp. 390-408 in Pigliucci, M.andPreston, K (eds). Phenotypic integration: studying the ecology and evolution of complex phenotypes. Oxford Univ. Press, Oxford , UK

Stayton, C. T. 2008. Is convergence surprising? An examination of the frequency of convergence in simulated datasets. J. Theor. Biol. 252:1-14.

Tirard, S. 2006. Generations spontanees. In: Lamarck, philosophe de la nature. Paris. Presses Universitatires de Francepp. 65-104

Wake, D. B. 1991. Homoplasy-the result of natural selection, or evidence of design limitations. Am. Nat. 138:543-567.

Wake, D. B.. 1999. Homoplasy, homology and the problem of 'sameness' in biology. Pp. 24-46 in G. R.Bock and G. Cardew, eds. Homology. Novartis Foundation Symposium 222. John Wiley & Sons, New York.

Wake, D. B., and Larson, A. 1987. Multidimensional analysis of an evolving lineage. Science 238:42-48.

Weinberg, S. 1977. The First Three Minutes: A Modern View of the Origin of the Universe. New York, Basic Books.

Williams, R.J.P. 1961. Possible functions of chains of catalysts. J. Theor. Biol. 1:1-17.

Williams, R.J.P.2003. The biochemical chemistry of the brain and its possible evolution. Inorg. Chim. Acta. 356:27-40.

Williams, R.J.P, Fraústo da Silva J.J.R.2002. The involvement of molybdenum in life. Biochem. Biophys. Res. Commun. 292:293-299. doi:10.1006/bbrc.2002.6518

Williams, R.J.P, Fraústo da Silva J.J.R. 2006. The chemistry of evolution: the development of our ecosystem. Elsevier; Amsterdam, The Netherlands.

Woese, C. 1998. The universal ancestor. Proc Natl Acad Sci U S A ,95(12):6854-6859.

Woese, C.R.2002. On the evolution of cells. Proc. Natl Acad. Sci. USA. ; 99:8742-8747.

Wolf, Y I, Rogozin, IB, Grishin NV, Koonin EV. 2002. Genome trees andthe tree of life. Trends Genet, 18(9):472-479.

# CHAPTER 4

# Evidence and Theories of Evolution and Extinctions

There is a variety of evidences for evolution. Evolution is the genetic change in population. Speed and direction of the evolutionary change varies in various species at different times. **Continuous evolution** through generations results in new varieties and species. Failure to evolve in response to environmental change often leads to **extinction**. Evolution is continuing. All life forms evolve from earlier species. Existing species continue to evolve today. Perhaps the primordial life on earth began as result of chance 3.5-4 bya. Darwin concluded that nature could produce changes in organisms through competition for resources. Animals which pass through hard times survive better and live long to produce next generation which would be more suitable to environment. Such changes are more gradual than changes imposed by human breeders artificially. Changing environment over a long period causes some organisms of a species to develop different forms and habits through accumulation of changes. Very slow changes finally become significant. Evolution is the dominant mechanism behind the diversity of today's species.

Evolution could occur without morphological change. Humans are larger now due to proper diet and medicine than in the recent past. Such phenotypic changes, induced by **environmental changes** are not evolution as they are not heritable. Phenotype is the morphological, physiological, biochemical, behavioral properties of organism influenced by genes and environment. Most changes like size differences due to environment are subtle. Large phenotypic changes due to genetic changes are the evolution. Populations adapt to their surroundings by evolution. A trait, or strategy successful at one time may be unsuccessful at another. Scientists maintained and observed a yeast culture for many generations. Sometimes mutation arise that allowed its bearer to reproduce better than others. Such mutant strains crowded out the formerly dominant strains. Most successful strains from the culture were taken at various times. In later experiments, each strain would out- compete the just previously dominant type in the culture. Some earlier isolate could out-compete strains that arose late. Competitive ability of a strain was always better than its previous type, but competitiveness in general sense was not increasing. Any organism's success is related to behavior of its surrounding organisms. Each species modifies its own environment. Organisms take nutrients from and add waste to their surroundings. Waste products benefit other species. Oxygen we breathe is a by- product of plants. Species do not change simply to fit their environment; they modify their environment to suit them as well.

Various types of evidences support the evolution. Some such evidences are described below.

## 4.1  Comparative Anatomy

Comparative anatomy reveals that structure of organisms follows basic expected patterns if organisms are related to each other. Anatomy of closely related organisms and less closely related organisms lead to analogous and homologus structures. **Analogous structures** are similar in function but anatomically different. Fin of fish, fin of whale, and wing of penguin, wing of bird and bat are analogous. **Homologues structures** are similar in structure but functionally different. Structures of organisms show overall anatomical similarity based on relatedness than function. Bat wing is similar anatomically to other mammals than to bird wings, though bird wings and bat wings perform similar functions. Appendages of fish, whales, and penguins serve the same basic function. The overall anatomical structure of whale fins is similar to structure of forelimbs of other mammal than to structure of penguin wings, or fish fins. Structure of penguin wings is similar to structure of bird wings than to

structure of fish fins or whale fins. The expected features that exist in an organism are ones that have precedent in an ancestor. Gill is superior to blow holes and lungs are found in whales. We don't find gills in whales as their recent ancestors didn't have gills; they had lungs and breathed air. Comparative anatomy confirms evolutionary theory by showing that organism have features that do have precedents along phylogenic lines.

Mammals and reptiles differ in skeletal structures, particularly in their skulls. Reptilian jaws have 4 bones. The foremost is called dentary. In mammals, the dentary is the only bone in lower jaw. The other bones are part of middle ear. Reptiles have a weak jaw and a mouthful of undifferentiated teeth. Their jaw is closed by external, posterior and internal adductor muscles. Reptile tooth is single cusped. Powerful jaws in mammals have differentiated teeth. Many teeth, like the molars, are multi-cusped. Temporalis and masseter muscles, derived from external adductor, close the mammalian jaw. In mammals, the secondary palate separates nostril passages and throat, to swallow and breathe simultaneously. Reptiles lack this.

*Procynosuchus* shows an increase in size of dentary and beginnings of a palate. *Thrinaxodon* has reduced the number of incisors. *Cynognathus* shows a further increase in size of dentary. The other 3 bones are located inside back portion of jaw. Some teeth are multicusped and teeth fit together tightly. *Diademodon* shows a more advanced degree of occlusion. *Probelesodon* has developed a double joint in jaw. Jaw could hinge off 2 points with upper skull. Front hinge was probably the actual hinge while rear hinge was an alignment guide. Forward movement of a hinge point allowed for the precursor to modern masseter muscle to anchor further forward in jaw for a more powerful bite. The first true mammal *Morgonucudon* from the late Triassic had all the traits common to modern mammals. These species were not grouped in a single, unbranched lineage. Each represents an example from the group of organisms along the main line of mammalian ancestry. The same set of bones is used to construct all vertebrates. Bones of the human hand grow out of same tissue as bones of a bat's wing, or a whale's flipper. They share features such as muscle insertion points and ridges. The difference is that they are scaled differently which shows that all mammals are modified descendants of a common ancestor. Closely related organisms share similar developmental pathways. Differences in development are evident at the end (Figures 4.1, 4.2).

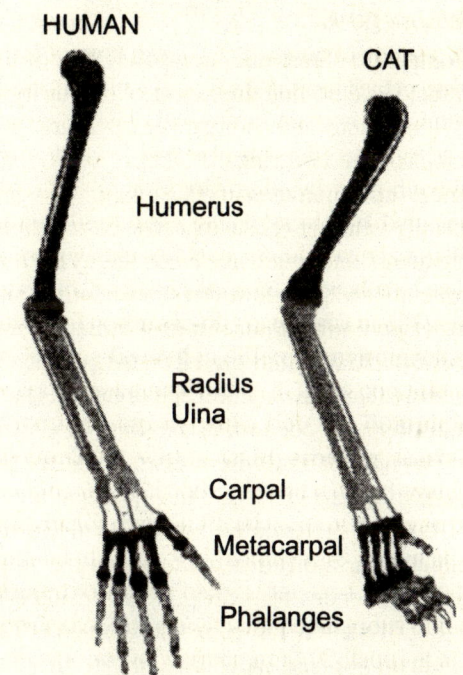

**Figure 4.2** Forelimbs

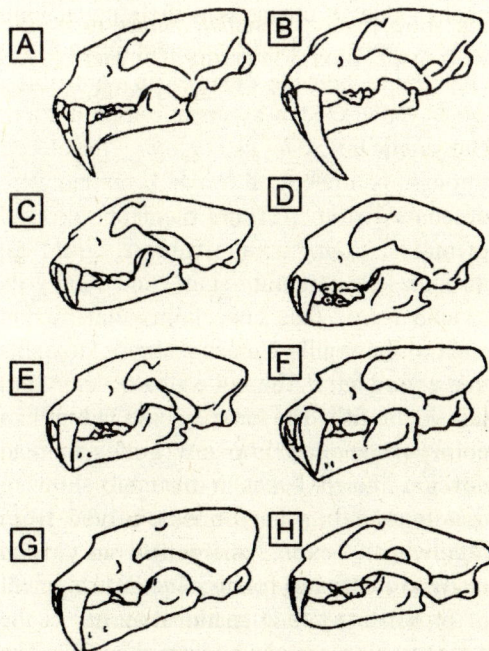

**Figure 4.1** (A) Skull of *Meganteron*, (B) *Similodon*, (C) *Dinofelis*, (D) *Metailurus*, (E) *Machairodus*, (F) *Xenosmilus*, (G) *Homotherium*, and (H) *Panthera*

Plants can be divided into non- vascular and vascular categories. **Vascular plants** can be divided into seedless and seeded. Vascular seeded plants can be divided into gymnosperms and angiosperms. Angiosperms can be divided into monocots and dicots. Each type of plants has several characters that distinguish them from other plants. Traits are not mixed and matched in organism group. Flowers are only seen in plants that carry several other characters to distinguish them as angiosperms. This is the likely pattern of common descent. All the species in a group share traits that they inherit from their common ancestor. Each subgroup would evolve unique traits of its own. Similarities bind groups. Differences show their subdivision. If 2 organisms share a similar anatomy, their gene sequences would be more similar than a morphologically distinct organism.

## 4.2   Comparative Genomics

Comparative genomics compare DNA of life forms to determine the extent of similarity in DNA. Human and chimpanzee DNA are 98% similar. By comparing the DNA of yeast, flies, mice, chimpanzees, and human, one can identify which genes are related to basic cellular processes and genes code for the eye. Gene structure and gene function show relationships in ways similar to anatomical structure and function. Many traits can be coded genetically in more than one way. Related organisms have traits that are coded for in the same way, even in presence of multiple ways to code for the same trait. The DNA of 3 distinct populations of cave fish revealed that in 3 different populations a different mutation of the same gene that codes for skin pigment resulted in loss of function for gene and albinism. The albinism evolved independently in different populations suggesting that mutation arises randomly and is then selected for, or against. When similar traits are evolved independently the structure that produces those traits are not same. For example, the whale fins vs. the fish fin. Cave fish in one cave have evolved their traits independently of cave fish in other caves where cave populations remained separate. Same traits exist in different cave fish populations for different genetic reasons. Organisms retain genes that are no longer expressed. Birds still possess genes that code for production of teeth, despite that bird now does not possess teeth.

## 4.3   Embryology

Complexity and similarity of **embryos** of widely separated vertebrate groups, plus tendency to develop embryos to survive drastic shocks without lasting damage is the evidence of intelligent design. Such similarities are the evidence for descent with modification from common ancestors sharing similar early embryonic stages. This idea was championed by Haeckel, whose theory "**ontogeny recapitulates phylogeny**" is based on Darwin's idea. Darwin pointed to existence of rudimentary and atrophied organs as evidence for descent with modification. Relationships between evolution and development have undergone resurgence during the last decade. The variation between individuals in populations is essential for evolution, especially evolution by natural selection. Researches into **evo-devo** have shown multiple mechanisms by which developmental processes produce extraordinary variation in short time, compared with very long time that Darwin believed for such changes to occur.

Increasing evidence suggests that some evo-devo processes can invert the "classical" sequence of evolutionary adaptation. As per the classical view, characteristics of organisms change due to changes in their underlying genetic code. It is possible for phenotypic changes to precede changes in controlling genes, now known as genetic assimilation. **Biogenetic law** states that organism developed through their ancestral forms. If human were descended from fish and then reptiles, and primates, than human embryo would go through fish -, reptile -stage, and finally its human form. This law claims that human embryos have gills. Human embryos have slits that erased later. The same slits develop into gill slits in fish. The same 2 bones in reptilian embryos develop into jaw bones instead develop into ear bones in mammals showing that maammalian ear bones evolved from reptilian jaw bones. Fossil record has yielded many **transitional forms**. The development of reptilian jaw into mammalian ear is the strongest pieces of evolutionary evidence. The development and later degeneration of gill slits and tail in human, legs in snakes, dolphins, and whales are other evidence for evolution (Figures 4.3, 4.4).

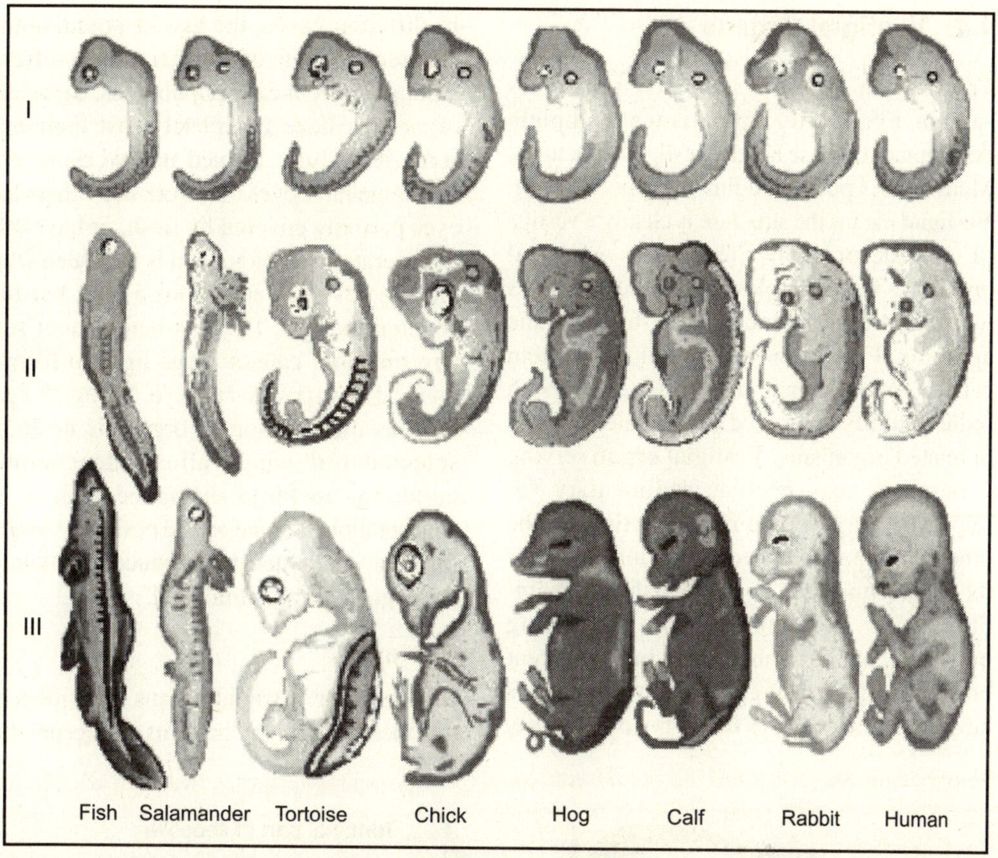

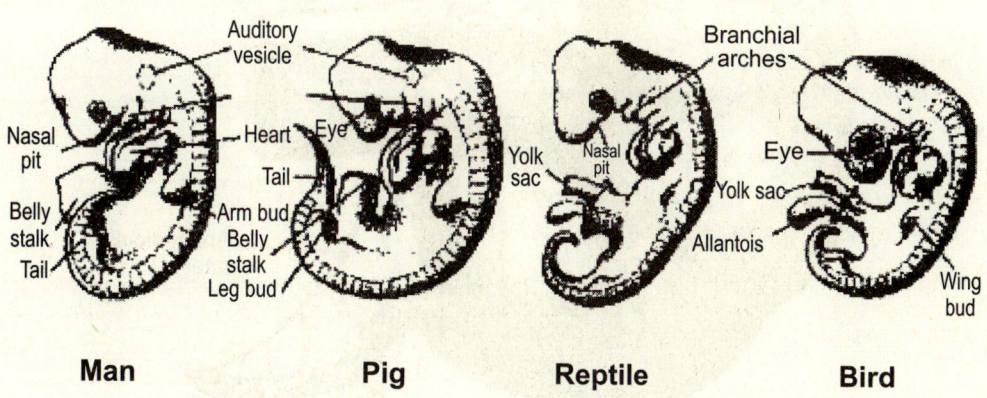

**Figure 4.3** Comparison among embryos of various animals

**Figure 4.4** Homologous similarity in vertebrate embryo

## 4.4    Vestigial Organs

Traces of an organism's ancestry sometimes remain even after organism's complete development. These are the vestigial structures. Many snakes possess rudimentary pelvic bone. **Vestigial** means the structure is clearly a vestige of a structure inherited from ancestral organisms but not useless. Vestigial structures may acquire new functions. In humans, the appendix (Figure 4.5) houses immune system cells. Vestigial organs are less developed with reduced utility compared to equivalent organs in related organisms. **Vestigial organ** serving 2 purposes may become rudimentary for important purpose and remain active for the other. An organism may become rudimentary for its proper purpose. In some fish the swim-bladder is rudimentary for marinating buoyancy, but has transformed into a nascent breathing organ or lung. Many cave organisms have semi-formed eyes that serve no purpose.

In different caves, the age of populations is different and rate of degeneration is different among different cave populations. Some cave organisms have completely lost their eyes, some have fully formed non-working eyes, and some have eyes that work and others have eyes partially covered by flesh, or have other degeneration. Vestigialism is best seen where the eye develops and forms a lens, but flesh covered the eye. The lens here cannot serve any function, cannot focus light to form an image if flesh covered the eye. Vestigial organs arise as the function of organs is no longer selected for, which allows degenerative mutations to build up unchecked, or the degeneration of organ can be positively selected for if the energy used to produce and maintain the organ can be eliminated.

### Remark

Criticism for vestigial organs as evidence is that these organs are the signs of degeneration,

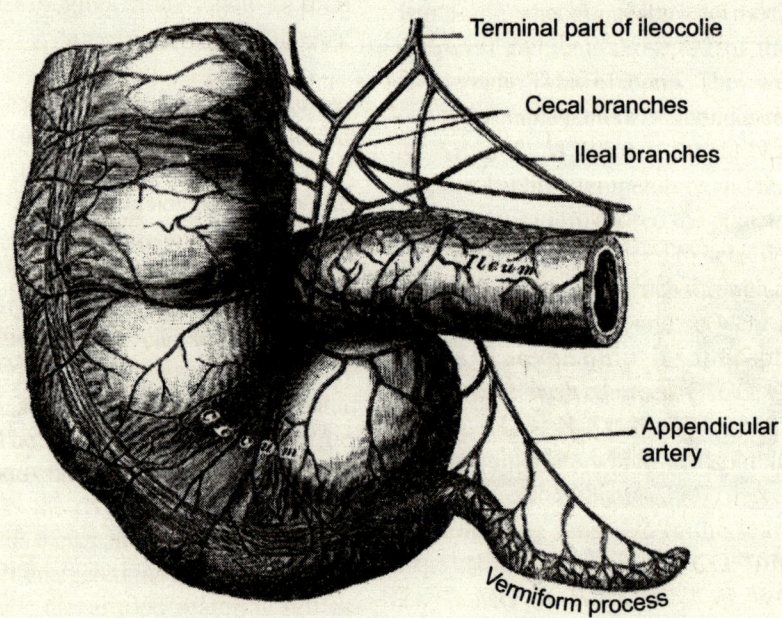

Terminal part of ileocolie

Cecal branches

Ileal branches

*Ileum*

Appendicular artery

Vermiform process

**Figure 4.5**  Vermiform appendix, a vestigial structure in human

not of development of new novel traits. The counter argument is organisms those possess vestigial organs also possess new novel traits. Many cave fish have sensory organs like advanced feelers that their surface dwelling ancestors lack, thus possessing both new traits and degenerated old traits. Vestigial organs alone show descent with modification. Some cave organisms contain vestigial organs and novel traits that was absent in their ancestors. **Cave fish** that are both blind with vestigial eyes and have feelers or other sensory structures for sensing in dark are explained by common descent from an ancestor that had eyes and development of new novel traits that ancestor did not possess. Examples of vestigial organs include the fully formed unusable wings of beetles, vestigial leg bones in whales, snakes, and legless lizards, and wisdom teeth in humans.

## 4.5    Biogeography and Continental Drift

Biogeography focuses the reconstructing the history of biotas and have played an important role in the development of evolution, especially by the works of Wallace and Darwin (Lieberman, 2000; Morrone, 2008; Wiley and Lieberman, 2011). Such research considered species' differentiation across geographic space (Lieberman, 2012) Geographic Information System **(GIS)** has enabled us to analyze the biogeographic pattern preserved in the fossil record. GIS calculates the geographic range of individual species either existing or extinct. For the fossil species, distributions from various localities are plotted on a modern map in GIS. Then, using the program Paleo GIS (Ross and Scotese, 2000), the continents can be rotated back into their original position when the fossils were living. Application of this technique has proven that **invasive species** caused the Late Devonian biodiversity crisis (Stigall, 2010).

Myers and Lieberman (2011) found no statistical evidence for competitive replacement of species. The changes in the physical environment in the Cretaceous Western Interior Seaway appear as the primary factor influencing geographic distribution through time and cause extinction of species. Thus GIS confirms the influence of the environment on macroevolution and the limited role that competition plays during that arena. Remarkable advances in biogeography since Darwin's day include an insight on speciation. New methods for analyzing biogeographic data include phylogenetic approaches and techniques using GIS (Lieberman, 2012).

Past evolutionary pattern are found in natural geographic distribution of related species. Major isolated land areas and island groups often evolve their own plant and animal communities. Before human's arrival, Australia had more than 100 kangaroo species, koalas and other marsupials, but advanced terrestrial placental mammals like dogs, cats, bears, horses were absent. Land mammals were absent from the more isolated islands that make up Hawaii and New Zealand. Each such place had number of plants, insect, and bird species that were found nowhere else. The likely explanation for existence of New Zealand's, Australia's, and Hawaii's unique biotic components is that life forms in such areas have been evolving in isolation from rest of the world for years.

Comparing the animals and plants found at similar latitudes around the world reveals that they are totally unrelated but similar adaptation suggests that they had independently evolved in place, rather than being independently created worldwide. Darwin pointed out that inhabitants of islands lying about the same latitudes worldwide were populated by unrelated animals and plants. They were closely related to organisms located on nearby mainland, providing evidence that they had been derived from those organisms, rather than created in place. These correlations present evidence for descent with modification and were crucial to Darwins in convincing biologists that evolution had happened.

Distribution of several fossil animals and plants at the end of **Permian Period** was puzzling to Darwin. Fossils of extinct animals and plants from Africa, India, South America, and Australia were similar enough to be classified in the same species, yet were so widely separated that required to propose land bridges and transoceanic floating mats of vegetation. Darwin was simply misled by a problem called convergent adaptations. Freshwater mussel is distributed worldwide and appear very closely related. Darwin believed that they were closely related because of their very similar appearance. Recent genetic and taxonomic work on freshwater mussels has showed that they are not closely related. They have evolved very similar appearances due to convergent adaptation to similar environments. The continents were once connected in a giant continent, which later split apart into 5 pieces and drifted to their current locations. Most striking is the relatively sudden and rapid movement of India, which split from Africa about 65 mya and renamed Southeast Asia. This collision, which continues to the present day, raised the Himalaya and explains the striking similarity in species distributions in Africa and India. India is still moving northward into the south of Asia at a rate of about 2 meters per century.

Theory of **continental drift** wasn't recognized or accepted until the mid-20th century. Very strong evidence in support of continental drift across earth's surface over past few billion years explains distribution of closely related but widely separated species that puzzled Darwin. The continents are drifting about the same rate about few millimeters per year. Over many million years, even such a low rate of speed can cause the continents to drift about the surface of earth. Strong evidence suggests that sudden drastic changes can alter course of evolution and geological features of the earth. The evidence suggests that extinction of dinosaurs at the end of **Cretaceous** occurred due to an asteroid collision with the earth. Biogeography deals with study of species distribution. All life is related and organisms should only be present where their ancestors were present. Fossil record confirms that types of organisms are geographically constrained and related to locations of their ancestors. Any given life forms may survive and thrive in a variety of places, but life forms present where they are related to other life forms. A given species could be successful in North America, South America, Australia, Europe but we don't find that species in all those locations. We only find it in a location where fossilized remains of its ancestors exist.

The best example is the ancestral marsupial geographic distribution. Now the only marsupials that exists in North America and Australia, the 2 remotely separated locations. The earliest known fossils of marsupials were known from the Cretaceous. Continental drift suggests the connection of South America with Australia and Antarctica during the Cretaceous period. Marsupial fossils found in South America and Antarctica has not been found anywhere else. Africa was once inhabited by extinct apes closely allied to gorilla and chimpanzee. As these 2 species are now men's nearest allies, our early progenitors perhaps lived on African continent. All the oldest humanoid fossils are found in Africa. Both modern genetics and paleontology support the "**out of Africa**" hypothesis. The human lineage is believed to be originated in Africa (Figure 4.6).

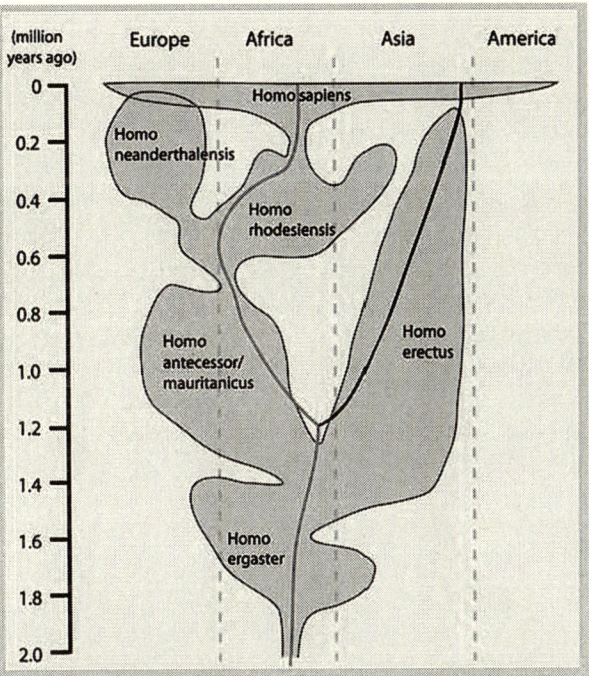

**Figure 4.6** Origin of human in Africa; Temporal and geographical distribution of hominid population (Redrawn from Stringer, 2003).

## 4.6 Homology versus Analogy

Homology is the existence of similar characteristics due to descent from common ancestor as found in the wings of bats and hands of humans. Although these structures appear different, detail examination of bone structure of both shows that the same bones are located in the same relationships with each other and they are homologous. Bats and humans share a common ancestor that had the same bone structure. Genetic analysis of many structures in many animals has showed that what once thought as the result of homology actually is not; often structures that appear to have been derived from common ancestor evolved independently in separate lines of descent called **convergent evolution** (Figure 4.7a,b).

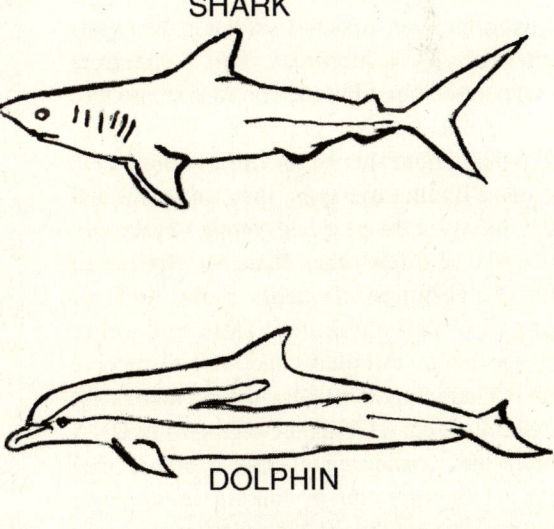

**Figure 4.7(a)**

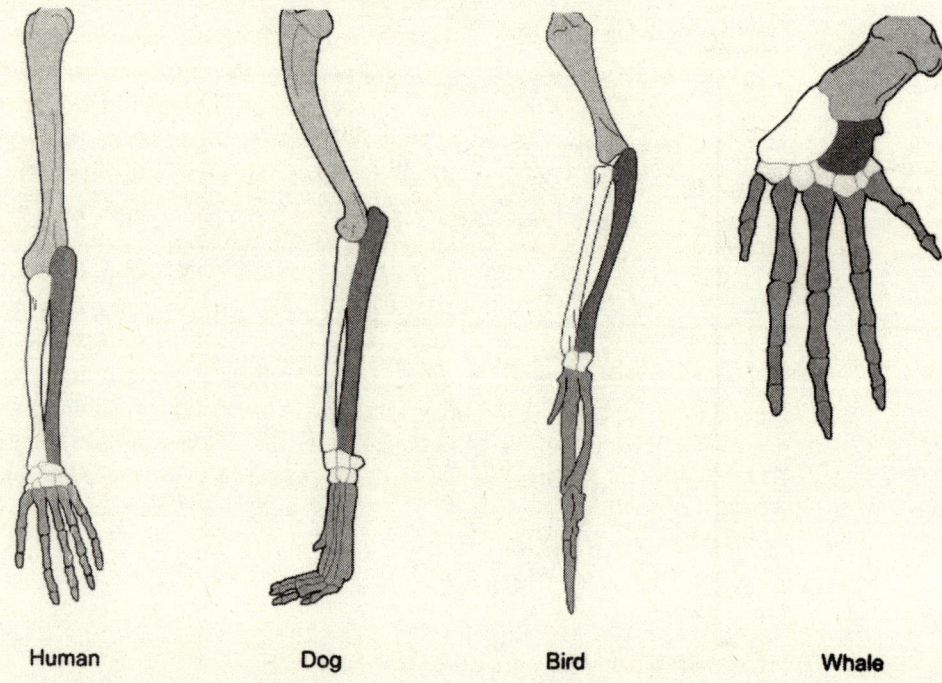

|       |     |      |       |
| Human | Dog | Bird | Whale |

**Figure 4.7(b)**   Homology vs Analogy

## 4.7   Chemical and Anatomical Similarities

Living things are basically similar in their basic anatomical structures and chemical compositions. Single-celled protozoa reproduce by similar division processes. The highly complex organisms begin life as single cell. After a limited life span, they grow old and die. All living things create complex molecules out of C and few other elements. Six out of the 92 common elements make 99% of proteins, carbohydrates, fats, and other biomolecules. All plants and animals receive their characteristics from their parent's by inheriting genes. **Gene**, the segments of DNA molecules contain chemically coded recipes for creating proteins by linking amino acids in specific sequences. All **proteins** are made of only 20 kinds of amino acids. Despite the great diversity of life, the simple language of DNA code is same for organisms, suggesting the fundamental molecular unity of life. Most living things are alike as they require energy for repair, growth, and reproduction from sunlight by photosynthesis, or they obtain it indirectly by consuming green plants and other organisms that eat plants. The arms of humans, forelegs of dogs and cats, wings of birds, and flippers of whales and seals all have same types of bones as they have retained these traits of their shared common ancient vertebrate ancestor. Major chemical and anatomical similarities between living organisms can explain their common ancestry.

## 4.8   Fossil Record

Paleontology deals with prehistoric life as found in the form of the fossils to study the organism's evolution. Establishing in 18th century during study of comparative anatomy by G. Cuvier and developed rapidly in 19th century, Paleontology now uses techniques from biochemistry, mathematics and engineering. Much of the evolutionary history of life is discovered. Paleontology has developed subdivisions. Body fossils and trace fossils are the principal evidence about ancient life. Estimating the dates of the fossils is necessary but difficult. Adjacent rock layers help in radiometric dating, which gives absolute dates that are accurate within 0.5%. Palaeontologists depend on relative dating by solving the jigsaw puzzles of biostratigraphy. Classifying ancient organisms is difficult as many do not fit into the Linnaean taxonomy. Paleontologists often use cladistics to draw up evolutionary family trees. The molecular phylogenetics that deals with relationship between organisms by measuring the DNA similarity in their genomes estimates the dates of species divergence, but there is doubt about reliability of molecular clock. **Paleontology**, the study of ancient life seeks information about several aspects of past organisms "their identity and origin, environment and evolution, and what they can tell us about earth's organic and inorganic past".

Fossils include anything that shows existence of prehistoric organisms. The Latin word *Fossilium* means 'dug out', which include any traces of body of animals and plants buried and preserved by natural causes. Cuvier the 'father of paleontology' studied fossils to develop phylogenies. Remains of animals and plants found in **sedimentary rock** give the idea of past changes through vast periods. Variety of living things exists. Some extinct species were transitional between major groups of organisms. The gaps in fossil record are due to incomplete data collection. One of the first of these gaps between small bipedal dinosaurs and birds, a 150 million year old fossil of *Archaeopteryx* was found in southern

Germany. The animal had jaws with teeth and long bony tail like dinosaurs, broad wings and feathers, and skeletal features which confirmed that bird had reptilian ancestors. Since the 1920's, many intermediate fossils were recorded from Africa. Transitional tetrapod fossil had been an evolutionary gap.

Fossil record states that complex animals and plants were preceded by simple ones and multicelled organisms evolved after the first single-celled ones. Eldredge and Gould developed the concept of **paleospecies** recognizing species from their fossil remains. Rare preservation of soft tissues decreases the diagnostic characters of fossil. Eldredge and Gould state that new species arise by gradual and slow transformation of ancestral population into its modified descendants. This transformation involves large numbers, usually entire ancestral population which occurs over all or, large part of geographic range. These imply several consequences, 2 of which are important. Fossil record consists of intermediate forms linking ancestor and descendant. Any breaks in phyletic sequence are due to imperfections in geological record. Most species evolved as result of **cladogenesis** rather than agenesis which results in multiplication of species and thus diversity of living systems. Species evolved mainly via allopatric speciation of peripheral isolates, or peripatric speciation in which a population of an ancestral species in geographic peripheral part of ancestral range is modified. Even when ancestral and daughter populations come in contact, reproductive isolation exist. Sudden speciation by change in chromosome number in modern populations is rare. **Sympatric speciation** is rare event seen primarily in insect and parasite lineages.

Taphonomy is the study of fossilization. Hard parts of organisms are generally preserved in fossilization and conditions under which hard parts may become preserved are restricted. All this results in fragmented distribution of fossil and parts become fossilized. Living lineage occupies ecological niches adapt to certain geographic features. Dry land species are unlikely to fossilize, whereas marsh or swamp-dwelling species are likely to fossilize. Paleospecies are recognized from the morphology alone. Frequency of **peripatric speciation** as occurs in modern lineages is the rare and a species may produce zero, one, or a few daughter species during its life span. Period of transition between parent species and daughter species is short compared to the period of existence of a species as distinct form. When a small sub-population isolates from the rest of population of species, the particular set of variations in sub-population is smaller than that in rest of the population. Such variation in combination, with suitable features of **geographic locale**, climate, and resources, lead to rapid evolution of reproductive isolation from the ancestral population. Such reduction in variation due to small sub-population size is the **founder effect** (Figures 4.8, 4.9).

Paleontology describes phenomena of the past and reconstructs their causes. Its 3 main elements are description of the phenomena; developing a general theory about causes of change; and applying those theories to specific facts. Palaeontologists often construct a set of hypotheses about the causes and then look for a piece of evidence to show that one hypothesis is a better explanation than others. Discovery by Luis Alvarez and Walter Alvarez of an iridium-rich layer at the Cretaceous-Tertiary boundary made asteroid affect and volcanism, the most favored explanations for the Cretaceous-Paleogene extinction. Paleontology focuses on past life from fossils. Geochemical signatures from rocks help to discover the origin of life, and analyses of carbon isotope ratios may help to identify climate changes and to explain major transitions like the **Permian-Triassic extinction**.

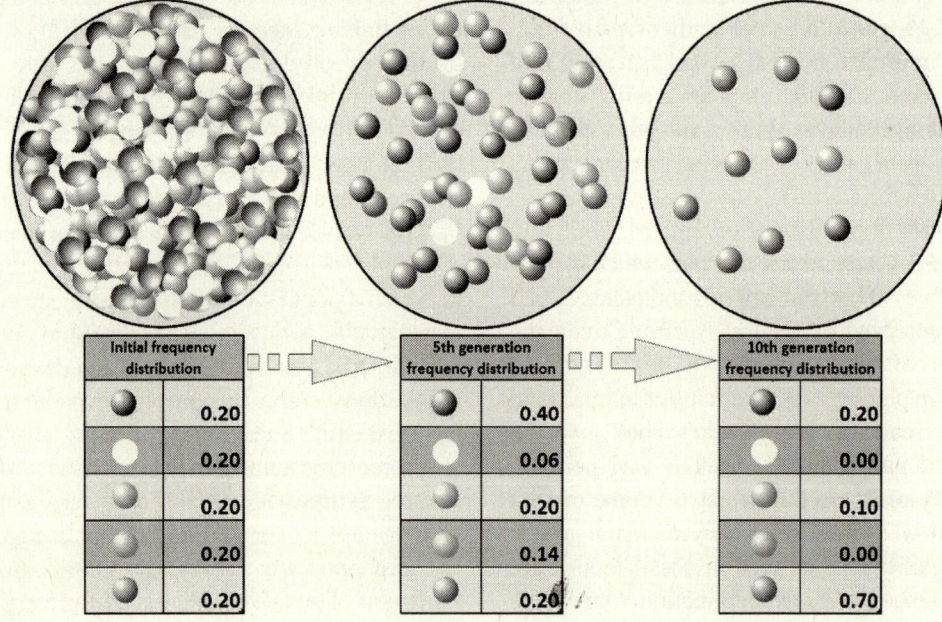

**Figure 4.8** Changes in a population's allece frequency following population bottleneck: rapid and radical decline in population size has reduced population's genetic variation

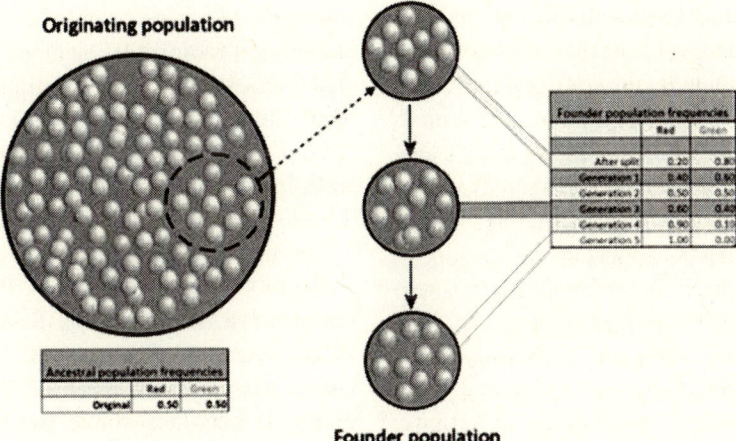

**Figure 4.9**   When few members of a population migrate to form separate new population, founder effect occurs. For a period after the foundation, small population experiences intensive drift. This is fixation of the (red) allele

Molecular phylogenetics often helps by using comparisons of different modern organisms' DNA and RNA to reconstruct evolutionary family trees. It estimate the dates of important evolutionary developments. **Paleoneurology** deals with endocranial casts of species related to humans to know the evolution of human brains.

Vertebrate paleontology deals with fossil vertebrates, from the earliest fish to immediate ancestors of modern mammals. Invertebrate paleontology deals with invertebrate fossils like molluscs, arthropods, annelids and echinoderms. **Paleoecology** deals with the interactions between different organisms like their places in food chains, and the 2-way interaction between organisms and their environment. The evolution of oxygenic photosynthesis by bacteria increased the productivity and diversity of ecosystems, and caused the oxygenation of atmosphere, a prerequisite for evolution of complex eukaryotic cells, which give rise to multicellular organisms. **Paleoclimatology** deals with the history of earth's climate and mechanisms that

have changed it sometimes including evolutionary developments. Expansion of land plants in the Devonian removed more $CO_2$ from atmosphere, reducing the greenhouse effect and aid to cause an ice age in the Carboniferous. Some chemical markers show change in environment at start of the Cambrian in consistency with mass extinction, or massive warming due to release of methane. Such changes cause the Cambrian explosion, although it may result from an increased biological activity. Features of rocks may reveal the time of origin of life on earth and provide evidence of presence of eukaryotes. Analyses of carbon isotope ratios explain major transitions like the Permian-Triassic extinction.

Paleontology reveals the change in living things through time. A substantial hurdle is working out how old fossils are. Beds containing fossils generally lack the radioactive elements required for radiometric dating. This technique is the only means for measuring the age of rocks greater than 50 million years old. We rely on stratigraphy to date fossils. Stratigraphy decipheres the "layer-cake" that is the sedimentary record and is compared to jigsaw puzzle. Rocks form relatively horizontal layers, with each layer younger than one underneath it. If a fossil is found between 2 layers whose ages are known, the fossil's age must lie between the 2 known ages. Because rock sequences are not continuous, rather broken up by faults, or periods of erosion, it is difficult to match up rock beds that are not directly next to one another. Fossils of species that survived for relatively short time links isolated rocks. The conodont *Eoplacognathus pseudoplanus* has a short range in Middle Ordovician. If rocks found to contain any trace of *E. pseudoplanus*, they certainly have a mid-Ordovician age. Such index fossils are distinctive, globally distributed in a short time range. Misleading results are produced if index fossils turn out to have longer fossil ranges than first thought. Stratigraphy and **biostratigraphy** provide only relative dating, often sufficient for studying evolution. Family-tree relationships may help to narrow down the date when lineages first appeared. It is also possible to estimate how long ago 2 living clades diverged - i.e. about how long ago their last common ancestor must have lived if DNA mutation accumulates at constant rate. These molecular clocks provide only approximate timing. They are not precise and reliable for estimating when the groups first evolved that feature in the Cambrian explosion, and estimates inferred by various techniques may vary by a factor of 2. Based on the mode of formation, the fossils can be categorized in several types. Fossilization takes place under specialized conditions.

**Age of Fossils:**   Rate of disintegration of radioactive material is constant and is not affected by environmental factors. Time taken by 50% of radioactive material to disintegrate into stable element is known as its half-life. If half life of an element is known, then the age of fossil can be calculated by finding out the ratio of radioactive element and its stable daughter element using scintillation counter. **Half-life** of Uranium 238 is known as 4.5 billion years. With this rate of decay, one million grams of U238 would yield about 1.7400 grams of Pb206 per year, or in 1 year 1 gram of U238 would produce 1/7,400,000,000 grams of Pb206. The **radioactive isotope** U238 disintegrates and Pb206 by emitting 8 alpha particles and 6 beta emissions through chain of daughter radioactive elements becomes stable isotope. Various radioactive materials have different half-life. Age of recent and very old fossils can be determined by selecting the appropriate radioactive element, if the element is present in that particular rock.

Various methods that use radioisotopes are employed. For example, U235 decays to Pb217 with a half-life of 713,000,000 years. We need to know how much radioactive material was there in the rock to begin with which can be found out by, if ratios in fresh volcanic rock are the same in the past now. Some best methods require volcanic rocks for dating sedimentary layers by dating surrounding volcanic layers. U235 has a very long half-life. So it is useful for very distant dates, but not very short ones as change is too little in isotope ratios over short periods. C14 has a very short half-life (less than 7000 years) and it is useful over periods of tens of thousands of years since over longer periods, there is too little C14 left. .

**Progression in Fossil Record:**  Observed progression in fossil record contributes to development of evolutionary theory. Important evidence for evolution is the existence of so-called **transitional forms** which is like a short lived bridge between 2 different species. All life is a progression of individuals and every individual that has ever existed is a transitional form. Every organism that lives now is a transitional form. Every living thing today has traits that are derived from ancestors. Many transitional forms show development of amphibians from fish, of birds from dinosaurs, of mammals from early tetrapods, of humans from a common ape ancestor. Most life forms in fossil record are probably not ancestors of modern life forms. They are the organisms in lineages that went extinct, but are nevertheless members of populations and can tell us about types and combinations of features that existed among ancestors of modern organisms.

Fossil record is sparse by nature as very small number of organisms that ever lived became fossilized. So we do not have a complete fossil record. Discoveries have shed light on evolution of land dwelling organisms from fish. Transition from aquatic fish to land dwelling tetrapods with legs has been difficult to discern. Since the 1970s many fossils have shed light on this transition. Four legged vertebrates called tetrapods have evolved from fish. Most bony fish that live today are ray-finned fish, fish with delicate rayed fins and a delicate skeletal structure. The tetrapods evolved from the lobe-finned fish with thick heavy bones and thick bony fins. Appendages have evolved from duplications of genome and refining of duplicated segments of DNA through natural selection. Early fish appendages resemble the structures of tails, and appendages evolved from duplications of genes that coded for tails. From these early fins, the limbs were evolved that we now see on land animals. The leg and ability to walk developed among purely aquatic fish before coming out onto land. The first land dwelling vertebrates would have already been able to walk, when they came onto the land.

*Ichthyostega*, an early land dwelling tetra pod whose lineage is extinct, have lived in shallow water with low oxygen content. For this reason, they poke their head above water and breathe air which would have been selected

for. Discovery of *Tiktaalik roseae* fills a gap between earlier aquatic fish and limbed tetrapods. It is unclear how much time *T. roseae* spent on land, but it possesses structures that are transitional between fish and tetra pods. A few remaining lobe-finned fish exist now offer insights into evolution of land animals. Lungfish lives in shallow, oxygen poor environments, and breathe air. This is most closely related to land dwelling tetra pods. Lungfish are the transitional forms between fish and tetrapods, but tetra pods are not descended from lungfish. Lungfish demonstrates the development of lungs and ability to breathe air evolved while fish still lived in water confirming the conclusions. Several species of ray-finned fish, mostly from the order Lophiiformes demonstrate the ability to walk under the water. Their legs and feet have developed from fins. These fish have legs that are composed to ray-fin elements. Development of legs in these fish independently evolved after split between lobe-finned fish and ray-finned fish. Lophiiform fishes demonstrate that walking limb can be selected for among purely aquatic fish. The fossils of *Archaeopteryx* are the famous of transitional fossils. The well known fossils that display transitional bird characteristics include *Archaeopteryx* ( Figure 4.10, 4.11))

**Dinosaur:** Dinosaur is thought not to be an ancestor of modern birds, but a broad spectrum of dinosaurs shows feathers and feather like structures making it likely that bird evolved from dinosaurs. Studies propose that bird evolved from dinosaur, or reptilian ancestor by showing that bird possess dormant genes for teeth production. Different transitional forms establish pathways between ancient reptiles and mammals, but the jaw and ear bones provide an excellent example because jaw and ear bones are commonly preserved. Vibrations from lower jaw are transmitted to single ear bone of reptiles, the stapes

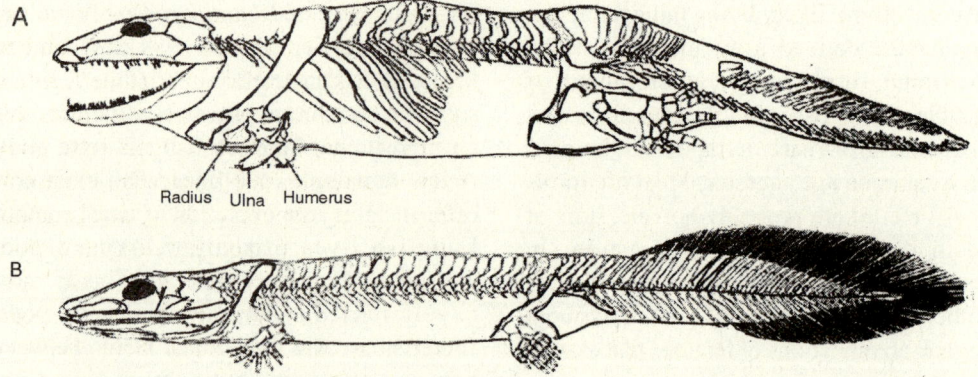

Figure 4.10   (A) *Icthyostega*, (B) *Acanthostaga*.

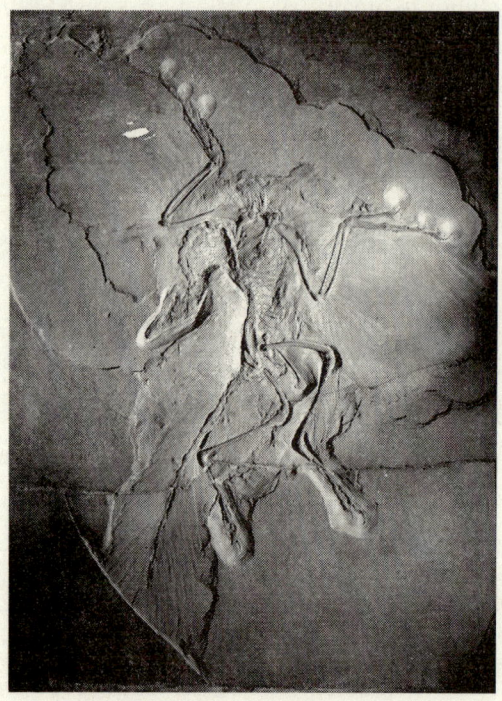

**Figure 4.11**   *Archaeopteryx*

transmitting the vibrations to inner ear. Mammals have 3 ear bones and a single jaw bone.

**Deficiencies in Fossil Record:**   Only resistant body parts fossilize which do not persist in their original form. The rock minerals replace the original remains of organism. **Fossilization** only occurs under circumstances in which the organism is very quickly buried in mud, or sand which doesn't happen in open, dry-land ecosystems. Most fossils are formed in shallow aquatic ecosystems, which exclude most organisms as candidates for frequent fossilization. Geological column had been investigated. Its overall outlines was determined and named before Darwin published the *Origin of Species*. The order of layers in this column is the same, the world over, although the entire column is not known from any single location. There are very large gaps in most layers and in many places whole layers are entirely missing. Darwin noted that the fossil record is everywhere incomplete. We do not have a continuous, unbroken fossil record of the evolution of any living thing, and certainly not for life as a whole. We have bits and pieces of this record, which must be assembled and related to one another by inference. Darwin pointed out that intermediate forms are almost entirely missing from the fossil record. The tremendous volume of fossil material is available. The Smithsonian has analyzed less than 10% of all the fossils in its collection. Millions of fossils in collections around the world, most of which are not analyzed. There are only a few hundred fossils of human ancestors as the human have not existed for very long, and have not left many fossils. The kinds of habitats where the ancestors of humans lived were not conducive to fossil formation.

**Petrifaction:** Petrifaction, literally means "turned into stone" is molecule-by-molecule replacement of organic matter by inorganic compounds like silica, calcium carbonate, and iron pyrites. It takes place in buried conditions, at the lake bottom, ponds and sea, where sediments rich in $CaCO_3$ and silica exist. Over years, inorganic matter replaces the entire bony material, making an exact replica of original. Sediment then transform into sedimentary rocks, in which fossils remain preserved for a long. Most old fossils are petrified, e.g. shells of molluscs, arthropods, and fish skeletons.

**Body Fossils and Trace Fossils:** Fossilization is rare event. Most fossils are destroyed by erosion or metamorphism. Different environments are favorable to preservation of different organism, or parts of organisms. Only the hard parts of organisms are usually preserved, like the molluscan shells. As most animal species are soft-bodied, they decay before fossilization. Of the 30-plus phyla of living animals, two-thirds are absent in fossil records. Unusual environments may preserve soft tissues. Sparseness of fossil record means that organism is expected to exist long before after they are found in fossil record - this is known as **Signor-Lipps effect**.

Trace fossils consist mainly of tracks and burrows, coprolites, and marks left by feeding. They represent the unlimited data source of animals with easily fossilized hard parts reflecting organism's behaviour. Many traces date from earlier than body fossils of animals. While exact assignment of **trace fossils** to their makers is impossible, traces may provide the earliest physical evidence of appearance of

moderately complex animals. Trace fossils are reliable indicator of what life was around and show the diversification of life at the beginning of the Cambrian, with freshwater realm colonized by animals. Cambrian fossil record includes many lagerstätten with soft tissues allowing examination of the internal anatomy of animals, which in other sediments are only found as shells, spines, claws, etc. most significant Cambrian lagerstätten are early Cambrian Maotianshan shale beds of Chengjiang (Yunnan, China) and Sirius Passet (Greenland); the middle Cambrian Burgess Shale (Canada, British Columbia); and late Cambrian Orsten (Sweden) fossil beds. While lagerstätten preserve far more than conventional fossil record, they are not at all complete.

When animals walk on sand and wet soil, they leave trail of footprints of limbless animals and worms leave tracks and trails in mud. If volcanic ash covers these footprints, they can be preserved for long. Mary Leakey discovered footprints of prehistoric man along with the giraffes, elephants, and guinea fowls in Kenya. Annelids, arthropods and molluscs make burrows in soil or bore into rocks or corals and hard wood. As soil hardens, they get preserved to form rock. Sometimes animal excreta are buried and fossilized, which shows the diet of animals. Nodular excreta of salamanders formed in the Carboniferous are found. Huge deposits of guano of sea birds are found in South American seacoast. Impressions of body parts, skin, feathers, and leaves are formed when they are pressed hard against the soft clay, which hardens to form rock. Fossil of *Archaeopteryx* is such an impression. More bird fossils as impressions have been discovered in China, e.g. fossils of *Sinosauropteryx*, *Caudipteryx* and *Confusiusornis*. Moulds are formed by hardening of material surrounding animal body, like volcanic ash, or lava that flows into the sea and traps molluscs and arthropods. Animal body disintegrates leaving only the hollow mould showing contour of animal. If this mould is filled with inorganic material like silica that will harden to form a cast. It will be an exact replica of animal body.

**In Other Materials:** In some coastal areas where there is abundance of petroleum pools and tar-pits, small creature, birds and insects get trapped in them. Asphalt decaying is very slow. If these are buried in soil, they are preserved for long time. Resin is a sticky substance that exudes from coniferous trees and traps small insects like mosquitoes and flies. Resin form **amber** and preserves the insects with all fine details. Most insect fossils are fossilized in this way. Ice is a good preservative even for soft parts. When animals are buried deep in snow that never melts, they are preserved entirely for long. Fossil of woolly mammoth from permafrost in Siberia is such an example, in which even flesh, skin and hair are preserved. **Index fossils** in undisturbed sedimentary rocks and in short geological time period lie in recognizable strata of older rocks below and formed layers above. Based on presence of such fossils in rocks, age of other fossils in same rock can be determined without dating, because such fossils are an index to particular geological period. Ammonites are good index fossils as different species represent specific geological periods in rocks.

## 4.9 Lamarckism

Evolution is the continual random change in the genetic code which enables life forms to survive in an environment with limited resource, creates different life forms naturally and sustains life forms with varied genetic constitution and suitable characters. It eliminates those organisms which cannot adapt. The survived ones reproduce successfully, preserve their characters through inheritance and dominate the particular ecosystem. It is the "Survival of the fittest and smartest".Lamarckism, the first comprehensive theory of evolution is founded on the than existed principle. This is the idea of *scala naturae*, an idea of change in species in varied environments. *Scala naturae* is a hierarchical classification where simple organism remains at the bottom and complex organisms at the peak. The classical *scala naturae* places 4 elements at the bottom, then metals, salts and rocks, mosses and plants, insects, seashells, and so on. Organisms can be classified into simple to complex. Lamarck explained the mechanism for progression and proposed that animal life has inherent ability to become complex. The part of Lamarckism, the **inheritance of acquired characters** is simple. The giraffe lives in savannah, where the trees are taller. This induces a besoin (need) in giraffe to reach the taller branches which caused the neck to grow through the flow of vital fluid. This new long neck passes to offspring. In the disused organ, vital fluid flow out of it, the organ atrophies as found in the reduced eyes of cave animals. The membrane between the digits of swimming frogs, otters, and sea turtles are the result of more swimming. The inter-digital membrane used as paddle caused vital fluid to flow into it. The first principle of Lamarckism is the natural, linear progression in a scale of complexity.

Diversity of life shows that a confounding factor leads to perfection. Organisms adapt to their habitats, leading to diversity of forms. Complex traits arise in individual taxa due to circumstances. Diversity is not a product of continuous biogenesis. Evidence points to the single origin of life and diversity is the result of speciation. Organisms are well-suited to their environment as they go through natural selection. A population of giraffes may possess variable neck sizes. Those with taller necks reach higher tree branches with access to more food for energy and advantage in reproduction.

They would produce more offspring. Assuming a genetic basis of neck length, more offspring with taller necks is likely to be born. The second core of Lamarckism fails as the idea of progression up a scale of complexity fails on a molecular level.

Kimura and Ohta, the founders of dominant neutral and nearly-neutral theories of **molecular evolution** respectively show that mutation are neutral - without effect on fitness of an organism. **Nearly-neutral theory** states that many neutral mutations have small effect. The rest of mutations are deleterious, with only small mutation being beneficial. If there was a preset linear progression to perfection, then all mutations should be advantageous which is not corroborated by evidence. In 1900, both neo-Lamarckists and selectionists were criticized by rediscovery of genetics. **Lamarckism** has the following salient feature:

1. Complex living organism is formed from simple one.
2. Environment does not remain constant rather it changes.
3. Change in environment influences plants and animals.
4. Changes in environment provide new needs for organism.
5. In response to the new needs, organisms develop new structure.
6. Variations in organisms arise through the effects of use and disuse.
7. Continuous use makes a structure developed and disuse makes the structure atrophied.
8. The new structures, developed by organisms in response to environment are acquired characters which transmit generation after generation and finally new species is produced. Lamarckism consists of 4 principles namely internal urge of organism, environment and new needs, use and disuse theory, and inheritance of acquired characters.

1. **Internal urge of organism:** Animals and plants grow; attain the maximum size due to desire. Each part of an organism increases in volume due to an internal urge and inherent ability of animal itself.

2. **Environment and New Needs:** Environment plays important role in evolution. Animals interact with environment for survival; adapt to live comfortably in environment. The animals bring about new structures and new characters in response to environmental change. Plants grow on fertile soils are healthy and luxuriant while the same plants grow on unfertile soil are weak and thin. So the nature of soil and other environmental factors bring about changes in plants.

3. **Use and Disuse Theory:** An organ in constant use develops well. When an organ is not used for a long, it reduces and disappears from the organism. This is called **use and disuse theory**.

**Lamarckism and Neo-Lamarckism:** Lamarckians view that when one generation acquires structure, capacities and habits, through its own exertions, it passes to the next. Cumulative inheritance result in the fixed characters, diversity and gradations of organisms. Through continual use and exercise of functions their capacity increase and modify. Through repetition of voluntary actions, the habitual use originates in them. Fixed habits are correlated with habituation which results in instincts. The non-exercising of function degenerates structure and brings about gradual modification of features of organisms. Temperature, water, moisture in air and other conditions of existence influence the organism simply by playing the role of a sieve, and not themselves molding and transforming, but

## Darwinism vs Neo-Darwinism

| Darwinism | Neo-Darwinism |
|---|---|
| The concept postulated by Darwin to explain evolution | Modified Darwinism according to the recent concept |
| Considers all heritable favourable variation | Considers mutations for evolution |
| Fail to explain the reason for variation | Explains the **variations** |
| Here, the basic unit of evolution is an individual | The basic unit is a population |
| Reproductive isolation is not considered as a major factor in speciation | Here reproductive isolation is a major factor |
| Natural selection refers to the **survival of the fittest** | Natural selection refers to the differential reproduction causing changes in gene frequency |

directly by requiring the production of new developments in living organisms.

**Differences between Darwinism and Neo-Darwinism:** The idea proposed by Charles Darwin explaining the mechanism of evolution is called **Darwinism**. The postulates of this theory are struggle for existence, over production, variations, survival of the fittest, and origin of species. Modified Darwinism according to the recent development in biology is called Neo- Darwinism

## 4.10 Darwinian Evolution

Drwin's teacher, Professor Henslow recommended him as naturalist for the voyage of HMS Beagle. The Beagle's captain, R. Fitzroy judging Darwin as under qualified hired someone else. Fitzroy considered Darwin for gentleman's companion, and amateur naturalist. While voyaging on the Beagle, Darwin read C. Lyell's *Principles of Geology* which inspired him to study the geology of coral reefs. Darwin was not convinced about evolution before the voyage. Such voyage transformed him to a passionate naturalist. He wrote notes and sent many specimens back to museums in England. His notes were the basis for his first book,

*Journal of the Voyages of HMS Beagle*, which he wrote after his return to England.

## 4.10.1 Darwin's Evidence for Descent with Modification

Darwin returned to England after 5-year voyage and started several notebooks on his observation during the voyage. His observations lead him to the idea of evolution by means of natural selection. He observed the patterns of geographic distribution of animals and plants as the Beagle circumnavigated South America. Similar habitat in widely separated locations inhabited by similar species led Darwin to conclude that such species had descended from the common ancestor. He collected animal fossils from South America which were the remains of species no longer present in South America, but were related to species that existed. Darwin suggested that modern species had descended from other similar species which are now extinct. Distribution of species on volcanic islands in the Atlantic and Pacific oceans was recorded and although these islands were located in the same type of habitat, different species inhabited them. Although these species were not at all related to each other, they were related to species located on nearby mainland showing that the island species and nearby mainland species had descended from the common ancestor. He studied the animal and plant breeding very intensively. In about 1837, Darwin was convinced about the occurrence of evolution, but did not stumble on mechanism for evolution until autumn of 1838

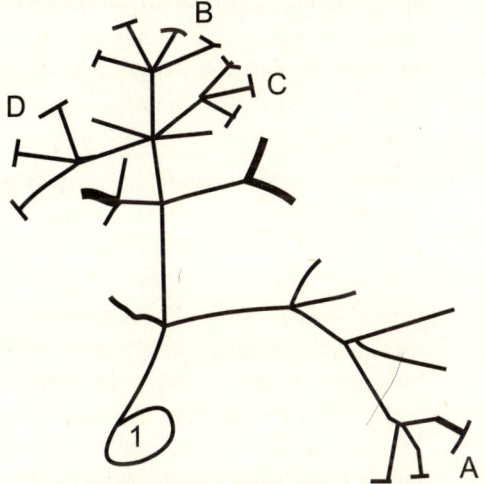

**Figure 4.12**   Figure drawn by Darwin

**Natural selection** may not lead a population with the best set of traits. In any population, certain combination of possible alleles would produce the best set of traits; but there are other sets of alleles that would produce a population almost as adapted. Transition from a local best to the global best may be hindered as the population pass through less adaptive states to make the transition. Natural selection lacks any foresight. It allows organisms to adapt to the existing environment. Structures, or behaviors do not evolve for future utility. An organism adapts to its environment at each stage of evolution. When the environment changes, the new traits get the chance of selected for. Vast changes in population results from the cumulative natural selection. **Mutation** introduces the changes into the population; the minorities of these

changes result in a greater reproductive output of their bearers and are amplified in frequency by selection. **Complex traits** must evolve through viable intermediates. For many traits, it seems unlikely that intermediate would be viable.

Feathers are thought to have evolved as insulation and/or as a way to trap insects. Later, proto-birds perhaps learned to glide when leaping from tree to tree. Thus the feathers that originally served as insulation now became co-opted for use in flight. A trait's present utility is not always indicative of its past utility. It could evolve for only one purpose and may be used later for another purpose. A trait evolves for its present utility is an adaptation; one that evolved for another utility is an exaptation. Speciation is the origin of one to several species from ancestral ones. When species are split into 2 populations due to geographical isolation, they separately accumulate mutations, inversions, translocations, and become reproductively isolated. Isolation is the primary requirement for the formation of species.

### 4.10.2   Pigeons as Model for Natural Selection

Considering several domesticated pigeon breeds as being analogous to natural selection, Darwin noted that all domesticated pigeon breeds are descendants of wild rock dove (*Columba livia*). These breeds were diverse, but they are not separate species. Pigeon breeders believed that various breeds of pigeons were derived from separate types of wild pigeons without any evolution. Darwin concluded that all 700+ breeds of pigeons are derived from the wild rock dove by means of artificial selection which was done unconsciously by animal and plant breeders, for desirable traits for domesticated animals and plants. Greater variation among the domesticated breeds of animals and plants in wild species were found. Darwin noted the tendency of domestic breeds to sprout variations out of their natural environment. Breeders selected new variations that strike their fancy, thus preserving unusual forms that would certainly be lost in nature. Darwin attempted to convince the reality of "**descent with modification**" from common ancestor; and natural selection as the cause of the beautiful adaptation. Darwin used argument from analogy, as he failed to point to any real-world example of natural selection.

### 4.10.3 Examples of Natural Selection

1. **Industrial melanic moth:** *Biston betularia*, (see below) the industrial moth is gray colored moth that perfectly camouflages on tree trunks covered with lichen in England and escapes predation by the birds. With ongoing industrial revolution in England in mid- 19th century, lichens on tree trunks got killed due to smoke. Tree trunks were now bare and dark and made the light gray moth prominent to the predatory birds. Now natural selection favoured dark moths, which camouflage on bare tree trunks. The moth has only one generation per year, and in less than 50 generations, natural selection replaced gray population with black population.

2. **Resistance in mosquitoes and houseflies:** DDT was used extensively over large areas. Initially it killed 99% of mosquito but created a lot of pressure on surviving individuals to mutate. Mutant resistants survived DDT application and became parents. Natural selection preserved the resistant populations and eliminated susceptible ones. This is called an **artificial selection**, due to which now not only mosquito and housefly but also many agricultural pests are resistant to most of available insecticides.

3. **Liederberg's replica plating experiment:** Liederberg conducted experiment on *Escherichia coli* by exposing the susceptible strains to penicillin repeatedly. As the generation time of bacterium is 20-30 minutes only, hundreds of generations were cultured and exposed to penicillin in a short time. Mutations for resistance appeared instantly and quickly replaced susceptible populations by natural selection.

4. **Fluctuation test experiment:** Luria & Delbruck cultured a population of *E. coli* along with bacteriophage viruses, then cultured the mixed samples on agar plates and found similar growth on all agar plates. Instant mutations had appeared for resistance against viruses while in others susceptible strains died out

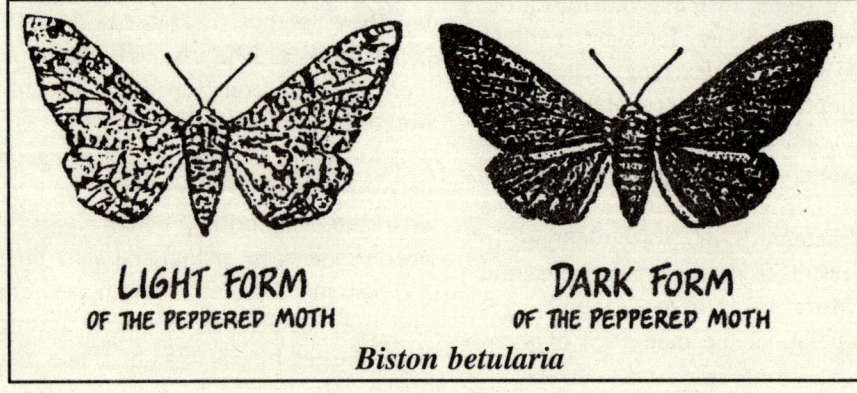

LIGHT FORM
OF THE PEPPERED MOTH

DARK FORM
OF THE PEPPERED MOTH

*Biston betularia*

showing that in populations exposed to environment extremes, the mutants appear, or hidden recessive mutations express and get exposed to natural selection and save the population from possible extinction. Natural selection forces the species to keep improving generation after generation so that they remain in the fittest state to survive in a given environment. Random genetic changes provide raw material for variations and gives natural selection a chance to operate. Natural selection produces systematic heritable changes in population from generation to generation and continue evolution. It is directional phenomenon, producing changes in definite direction to produce new species. Natural selection may be very fast, or very slow depending on the environmental demands and rate of genetic changes.

### 4.10.4   Natural Selection in Microorganisms

Microorganisms reproduce in 3 ways and natural selection may be different accordingly.

1. **In self-reproducing individuals:** When the process of self-reproduction is perfect, there should no room for natural selection. As the process is never complete, mutants appear from time to time. If these mutants reproduce inefficiently, they are removed. But if they reproduce more efficiently, they maintain themselves, causing diversification in population. Protozoa reproduces by binary fission, coelenterates by strobilation and *Hydra* by budding.

2. **In self-fertilizing individuals:** Hermaphrodite or parthenogenic individuals pass the same genotype to their offspring, which are called pure races. Without mutations, **genotype** of each individual will remain unchanged. When such races are exposed to natural selection, some are favoured while others are removed, which alter the ratio. For example, flatworms, lower insects and other invertebrates reproduce by self-fertilization.

3. **In cross-fertilizing individuals:** In cross-fertilizing microorganisms, no 2 individuals are alike genetically and each population has genes in different frequencies. Natural selection increases the frequencies of some genes and decreases the frequency of others. New genotypes are constantly produced by cross-fertilization, and selection exercises its effect. The character becomes heterozygous and dormant and survives for generations, without expression and exposed to natural selection. During this period, environment changes and the character may become advantageous. This also produces heterozygosity in population.

### 4.10.5   Misconceptions about Selection

Selection merely favors beneficial genetic changes when they occur by chance. Selection is not a guided entity; it is simply an effect. Animals are often said to perform some behavior as selection favor it. But the reason is that animals due to their genetic make up perform this behavior tend to be favored by natural selection relative to those who due to their genetic make up don't. The survival of the fittest is used synonymously with natural selection. Survival is one component of selection and perhaps the less important ones in many populations. In **polygynous species**, several males survive to reproductive age, but only few ever mate. Males differ little in their capacity to survive, but in their ability to attract mates - the difference in reproductive success stems from the latter consideration. The word fit is often confused with physically fit. **Fitness** is the average reproductive output of a class of genetic variants in a gene pool.

Selection for one or 2 characters can produce changes in such characters, but selection for many characters lead to extinction. Natural selection cannot cause either background, or mass extinction. It fine tunes organisms to survive and reproduce under existing conditions. A species more finely adapted to its present conditions is more likely to extinct when conditions change. Example is found in extinct cichlid fishes in rocks of New Jersey palisades. Natural selection is the engine of organic change, driving different variants of the same species to diverge until they become new species. Natural selection is the outcome of process with components like:

1. **Variation:** Differences between members of populations with relatively slight differences in physical appearance and behavior.

2. **Inheritance:** Heritable characters pass from parents to offspring. Darwin or any of his contemporary was not aware of coherent theory of heredity. Darwin accepted that many important traits of animals and plants are heritable but he failed to explain their origin. Variations among the individuals are the raw material for natural selection. The huge diversity of living forms and their adaptations have evolved by descent with modification. But the natural selection as the mechanism behind this process was not widely accepted. Those acquired characteristics could be inherited though use and disuse contradicts the blind and purposeless process of natural selection. Naturalists believed that inheritance worked by blending the characteristics of parents. Jenkin pointed out that blending inheritance could eliminate variation within few generations. The variations must have significant effects on survival and reproduction, and they must persist from one generation to next. If all traits are blended, distinctiveness of each variation should lose and the population would remain unchanged. Darwin proposed the occurrence of large numbers of new variations called continuous variations in new generation. His theory of genetic inheritance called **pangenesis** states that all the traits of organisms produce particles of inheritance called pangenes which travel from the anatomical location of trait to sex cells, and from there they pass into the offspring. The amount of variation appears in each generation is insufficient to explain the disappearance of variations. **Pangenes** could not be detected, only inferred, and their effects are not separable from effects predicted by theory of blending inheritance. Most naturalists believed in the occurrence of descent with modification, but they did not believe that it occurred by natural selection.

3. **Fecundity:** Rate of production of offspring by females is called fecundity. Individuals that survive also reproduce, pass onto characteristics into their offspring to survive and reproduce.

4. **Non-Random Survival and Reproduction:** Survival and reproduction are never random. Individuals survive and reproduce as their characters form the basis for evolutionary adaptations. Some species are

polymorphic and exhibit 2 or more distinct morphs. Darwin explains **polymorphisms** as result of selectively neutral variations in population. Modern biologists explain it by referring to natural selection for heterozygous intermediate between 2 morphs. Natural selection act very slowly for long periods, and on few inhabitants of the same region, at the same time by preservation and accumulation of small inherited profitable modifications. Darwin argued that natural selection acts too slowly to be observed and it requires at least 50 years in nature. Darwin provided 2 examples of natural selection in action-- wolves preying on deer and plant-insect co evolution, although these are considered as direct, observable examples of natural selection.

### 4.10.6 Darwin's View on Origin of Species

Darwin believed that species change gradually due to natural selection from his observations in different parts of world during the voyage of Beagle, from which he made 2 deductions and conclusion that natural selection occurs. The diversity of living organisms is explained due to descent with modification. Natural selection certainly can't produce perfect adaptations. All adaptations are compromises.

### Darwin Observations

1. **Overproduction:** Reproductive differential is the ability of species to leave more progeny for the next generation. It is not synonymous with reproductive capacity, which indicates production of more offspring. Leaving individuals for next generation is important for competing species. All beneficial characters, may be behavioral, physiological, physical, or morphological help in reproductive differential. Dice placed same number of white and brown mice in cage that had natural terrain and plenty of food. Their reproductive rate was about same. Dice released a pair of owls in cage and allowed the mice to breed for several generations. In absence of limiting factors in cage except that brown mice could camouflage against rocks whereas white mice were prominent and could be easily spotted by owls. White mice gradually died out, but the brown mice survived and gradually flourished.

Darwin thought that organisms tend to reproduce at higher rate than required. Fishes laid millions of eggs during spawning. Each oyster lay 60-80 million eggs. If a pair of housefly lays all its eggs and if all offspring survive, and reproduce to their full potential, then in one season it will produce $191 \times 10^{18}$ individuals. Elephant is the slowest breeder having reproductive age of 30-100 years, during which it produces only 6 young. This is **overproduction** because if all offspring survive and reproduce, then one pair will produce 19 million offspring in 750 years.

2. **Number is constant:** Earth and its ecosystems have limited space, which cannot support unlimited number of animals. Food supply is also limited. All offspring produced do not reach maturity; only small number attains adulthood and reproduce. Large number dies due to scarcity of space and food, which is the struggle for existence among them. In spite of overproduction organism's number remains constant.

3. **Variations:** Variations are differences among closely related individuals. **Heritable variations** are caused by

aneuploidy, polyploidy, mutations, chromosomal aberrations, hybridization. Somatic variations are short-lived without influence on natural selection. No 2 individuals are alike within a species, race, or cohort. Variations exist in physiology and capacity to starve cold, or heat and other features. All variations are heritable although Darwin did not know the mechanism of heredity.

## Deductions

4. **Struggle for existence:** From the first 2 points that overproduction is going on and number of organisms that can be supported by any ecosystem is constant, Darwin deduced that offsprings struggle among themselves to survive. Many individuals perish in this struggle and few reach adulthood. The struggle among the animals occur in the form of physical combat.

5. **Survival of the fittest:** As the struggle for existence is fierce, only the fittest individuals survive, attain adulthood, reproduce and leave offspring. The fittest individual is physically strong and fights for space, mates and food, and escape predators and survives to reproduce to become parent of next generation.

6. **Natural selection:** When animals overproduce, nature finds vast number of individuals for selection. Nature selects the fittest that become parents of next generation. Unfit and weaker die out. Natural selection is the refining process for improving species, making the species fittest to live in an environment. The species constantly changes and evolution goes on forever. The new characters emerging from random genetic mutations persist and spread, either displacing the old ones, or coexisting with them. If a characteristic is incompatible with surroundings, individuals possess it naturally pass away and no longer reproduce it; if a character is adapted to the environment, it persists and spreads. This is called survival of the fittest. If an individual fail to reproduce, it is irrelevant to evolution.

The Darwin concept of adaptation of species refers to the passive process. Individuals adapt. But species as such do nothing other than live and producing offspring. Some members of a species are sufficiently adapted to the environment to exist. If environment changes and all genetic make up of a species are inadequate, the species fail to exist. If random mutations occur, and species survive in new environment they evolve along same lineage. Some members of species may stray into another geographical area and survive by changing characters due to random mutations, more suitable to new environment. These variations may become pronounced in comparison with original population and a new species evolve, which cannot reproduce with old one. Another group, straying into different geographical area, may evolve differently, and form another distinct species. **Random mutation** coupled with natural selection is the strategy of living organism to maintain life in a changing world. When life almost disappeared on several occasions, the living organisms were forced to take new forms starting from the limited gene pool. Such gains take millions of years. Humans may evolve from bacteria, but the reverse is not true under any conditions. Evolution has a direction, from simple to complex forms of life. Difference in reproductive capability is the natural selection– the only mechanism of **adaptive evolution**; it is differential reproductive success of pre- existing classes of genetic variants in gene pool. Natural selection removes unfit variants as occurs via mutation. Selection removes deleterious alleles and depletes genetic variation. When heterozygotes are fit than homozygote, selection maintains genetic variation called balancing selection as found in sickle-cell alleles in malaria patient. Variation at a single locus changes the rbcs to sickle shape. Presence of 2 alleles for sickle-cell develops anemia. The sickle shaped rbcs (Figure 4.13) precludes them to carry normal levels of oxygen. One copy of sickle-cell allele, coupled with one normal allele provides resistance to malaria. The shape of sickled rbc makes it difficult for *Plasmodium* to enter the cell. Individuals homozygous for normal allele suffer more malaria than the heterozygotes. Heterozygote possess highest fitness of these 3 types and they passes on both sickle-cell and normal alleles to the next generation. Thus, neither allele can be deleted from the gene pool. Sickle-cell allele is found in highest frequency in regions of Africa. Balancing selection is rare in natural populations.

Dark colored moths had a higher reproductive success as light colored moths suffered higher predation. Decline of light colored alleles was caused by decrease in number of light colored individuals. One way alleles can change in frequency is to be housed in organisms with different reproductive rates. Gene is not the unit of selection; neither are groups of organisms a unit of selection. Natural selection favors selfish behavior because true **altruistic act** increases the recipient's reproductive success while lowering that of the donors. Altruists would disappear from the population as the non-altruists could reap the benefits without paying the costs for altruistic acts. Reproductive success has 2 components; direct fitness and indirect fitness. **Direct fitness** is the number of alleles on average a genotype contributes to the next generation's gene pool by reproducing. Indirect fitness is the number of alleles identical to its own. It helps to enter the gene pool. Direct fitness along with indirect fitness is called the inclusive fitness. Natural selection favours traits or behaviors that increase the genotype's inclusive fitness. Closely related organisms share many same alleles. In diploid species, siblings share at least 50% of alleles (Figure 4.14). The proportion is high if the parents are related. Thus helping close relatives to reproduce gets an organism's own alleles to be better represented in gene pool. Natural selection does not causes genetic variation to appear, it only distinguishes between existing variants.

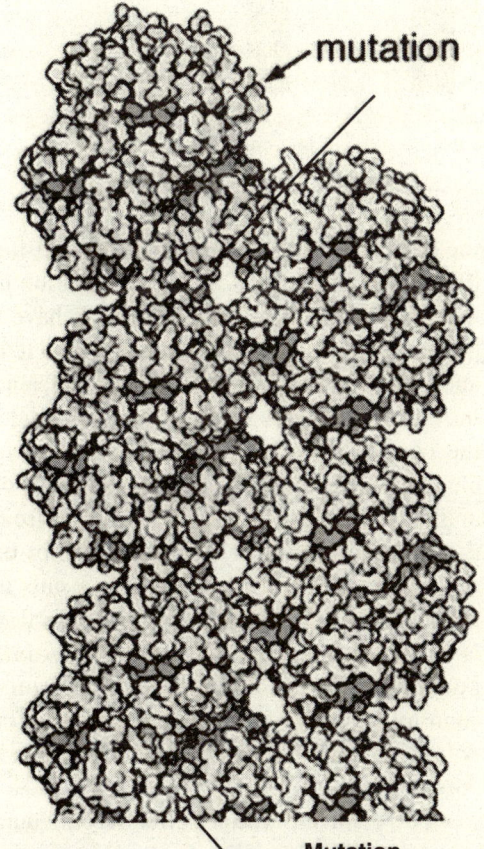

**Figure 4.13** Sickle cell haemoglobin.

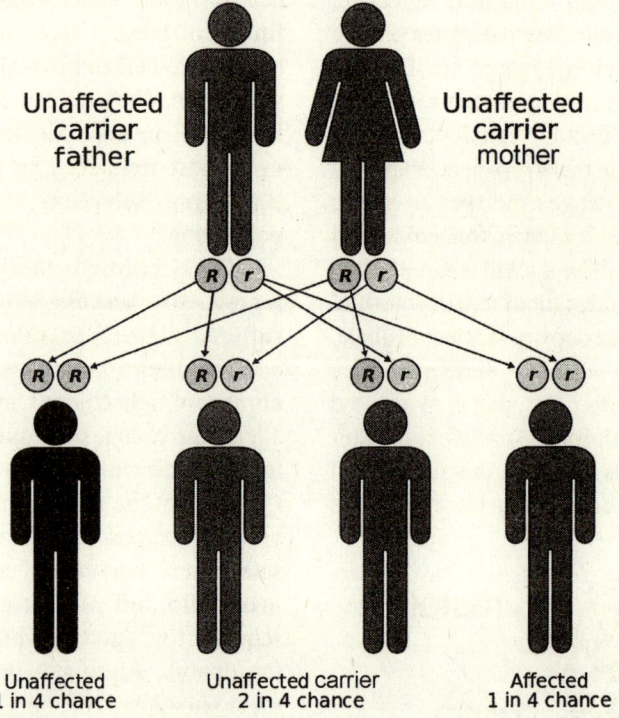

**Figure 4.14**   Sharing the alleles

### 4.10.7  Natural Selection and Evolution

Natural selection is generally described as the mechanism behind molecules-to-man evolution. Natural selection is a process whereby organisms possessing specific characteristics survive better than others in a given environment. Natural selection can change the genetic make-up of populations and involves a loss of genetic information and cannot explain how all of the life-forms on the earth came to be. Evolutionists view the history of life as a single branching tree where all life has come from a common ancestor. Natural selection allows organisms to adapt to their environment. Bacteria can become resistant (Figure 4.15) to antibiotics by losing genetic information, or swapping genetic information with other bacteria that have the resistance. **Antibiotics** bind to a protein in the bacterium to block its function, causing the bacterium to die. If there is a mutation in the DNA which causes the protein to have a different shape, the antibiotic fail to bind to it and the bacterium lives. To become resistant, the bacterium losses the information to make the correct protein. When it is required to compete with other bacteria in absence of antibiotic it is less fit and is eliminated from the population. Natural selection cannot be the driving force of evolution as it results in a decrease of information. Natural selection is the environment's effect on individual phenotype. Individual genetic variations function within the existing genome for polygenic traits and related features to work correctly.

**Polygenic inheritance** offers a continuum of trait expression like hair color, to the polymorphic species with 2, or more dominant

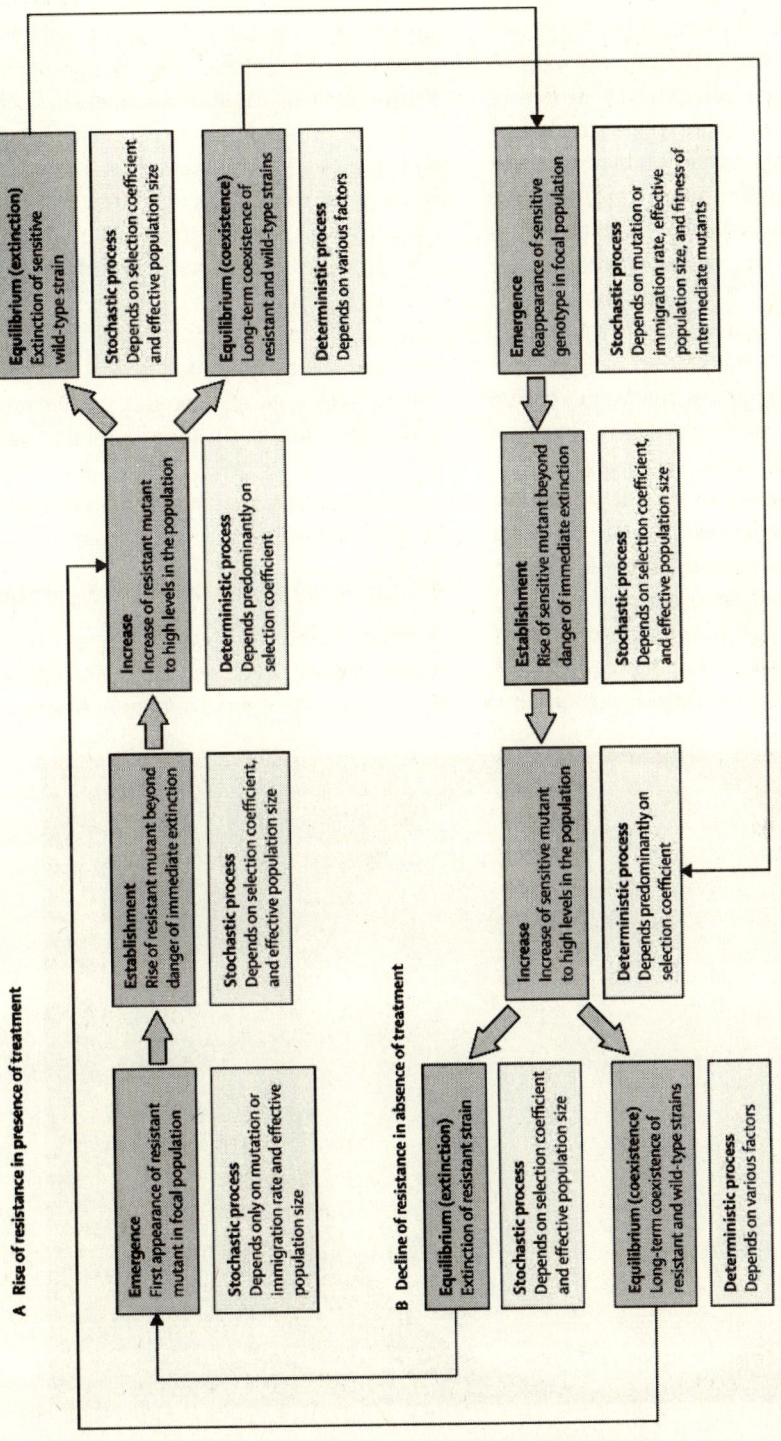

**Figure 4.15** Evolutionary dynamic of resistance in the presence and absence of drug selection. The rise (A) and fall (B) of resistance are characterised by the appearance and competition between resistant and sensitive strains of the pathogen, and the evolutionary process can usefully be subdivided in several dynamical phases

**Source:** www.thelaneet.com/infection. vol. II, March 2011 Evolutionary Bioinformatics 2011 : 7

phenotypes. Polygenic traits and polymorphism produce phenotypic variations that act differently. **Directional selection** operates by selecting one phenotypic extreme in lieu of excluding the other phenotypic extreme and allows the offspring of the favored extreme to reproduce effectively to dominate the population. Consequently the phenotype of next generations moves in a definite direction due to natural selection. It is common when a population colonizes new territory, or when new environmental changes are brought into the system. No environment can teach a chimp to speak. It's all about the variability in genes or environment. Natural selection and genetic drift can occur if there is a genetic difference in population. The diversity of life on earth (Figure 4.16) is the outcome of evolution: an unpredictable and natural process of temporal descent with genetic modification that is affected by natural selection, chance, historical contingencies and changing environments (Figure 4.17). Evolutionary theory is significant in biology, for its unifying properties and predictive features, the clear empirical testability of its integral models and the richness of new scientific research it fosters. The fossil record including abundant transitional forms in diverse taxonomic groups, establishes extensive and comprehensive evidence for organic evolution. Natural selection, the primary mechanism for evolutionary changes, can be demonstrated with numerous, convincing examples, both extant and extinct. Natural selection-a differential, greater survival and reproduction of some genetic variants within a population under an existing environmental state has no specific direction or goal, including survival of a species.

### 4.10.8  Causes of Natural Selection

Natural selection occurs as a result of interactions of organisms with both biotic and abiotic factors of environments. Major agent

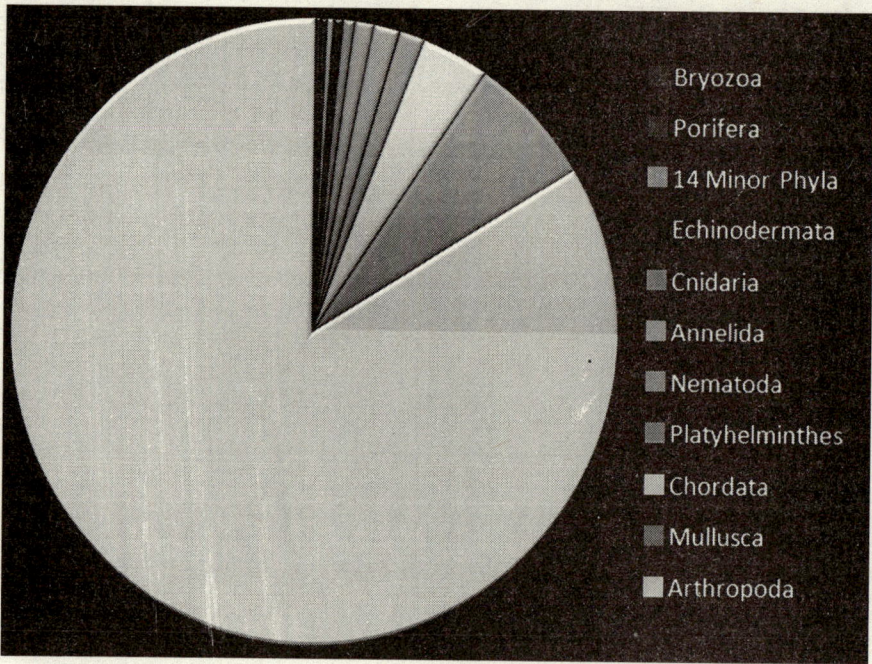

**Figure 4.16**  Diversity of Animals

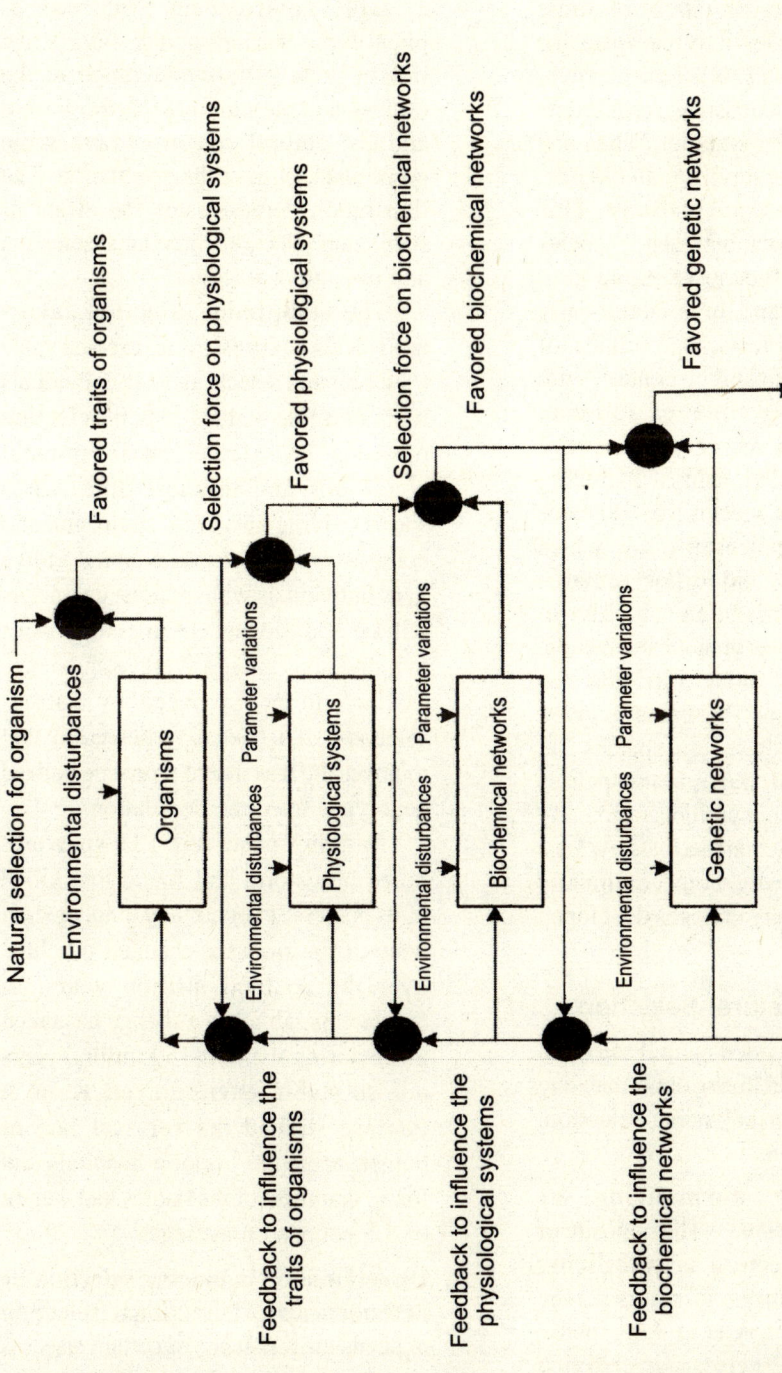

**Figure 4.17** Natural selection process on interplaying of hierarchical biological networks. High-level biological network selection will become selection force on low-level biological network. Natural selection on organisms selects its favored organisms. Once the favored organisms are selected, low-level biological networks have to maintain the favored physiological systems of selected organisms. These favored organisms become the selection force to shape their favored physiological systems. Favored physiological systems will lead to the selector, force on biochemical networks. Favored biochemical networks by natural selection will become the selection force on genetic networks. The lower-level selected networks will feedback to influence the higher-level networks in evolution. Therefore, the natural selection acts on interplaying of the multiple bio-networks

**Source:** Evolutionary Bioinformatics 2011 : 7

of natural selection in biotic environment is competition with other members of same species. Different species may compete for same resources, although to a lesser extent. When 2 species interact extensively, each exerts strong selection pressures on other. When one evolves a new feature or modifies an old one, the other evolves new adaptations. This constant, mutual feedback between 2 species is called **co evolution**. The most familiar form of co evolution is found in predator-prey relationships. In symbiosis, individuals of different species live in direct contact with one another for long time of intimate interaction. **Symbiosis** incorporates several kinds of ecological relationships, including parasitism, in which one species live and feeds on a larger species; commensalisms, in which one species benefits and other remains unharmed; and mutualism, in which both species benefit. Of all types of biotic interaction, symbiosis leads to most intricate co evolutionary adaptations. Whereas a given predator usually preys on several species and may interact with a particular species occasionally, partners in symbiosis typically live together virtually their entire lives. At least one partner and normally both continually adjust to any evolutionary changes developed by other.

### 4.10.9  Types of Natural Selection

Mather and Theoday divided natural selection into 3 categories, viz., stabilizing or normalizing selection, directional or balancing selection, and disruptive selection.

1. **Stabilizing or normalizing or centripetal selection:** The opposite of diversifying selection is **stabilizing selection** because it favors the intermediate phenotype at the expense of the extremes. Whereas a diversifying mechanism signals a changing environment, stabilizing selection shows a stable environment that reduces phenotypic variation and favors the intermediate phenotypic range at the exclusion of the extremes. Figure 4.18(a) and (b) natural selection favors the nonshaded, bulk of the population. The illustration demonstrates the effect of stabilizing selection, directional selection and disruptive selection.

The stabilizing selection favors individuals who possess an average value for a trait and selects against individuals with extreme values. Such selection operates in stable environmental conditions and in short time, when species living in given environmental conditions are perfectly adapted to live in it. Individuals with extreme characters will be at disadvantage in comparisons to individuals having average characters and would be favoured by natural selection. It is negative selection that removes the less fit and more specialized genotypes from the population.

Bumpus observed 133 sparrows killed in a storm and found that killed birds possessed abnormally long, or short wings. *Oppossum* has changed very little over the past 60 million years. In *Sphenodon*, almost no change has taken place during the past 150 million years due to stable environment. Karn & Penrose studied the survival rate of human babies in London hospitals and found that most of the babies had weight of 7.5 pounds on average.

2. **Directional or balancing selection or orthogenesis:** Directional selection explains the resistance of certain diseases to long-standing medicines. The naturally

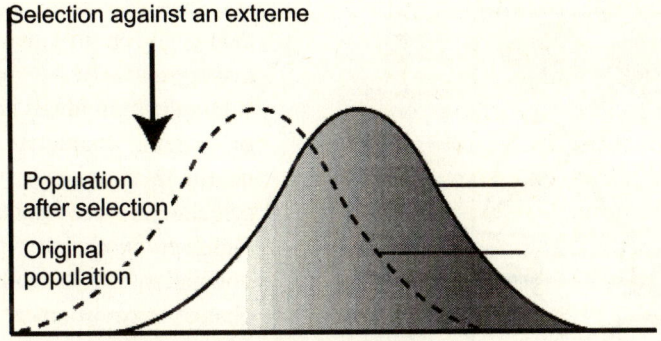

Selection against an extreme
Population after selection
Original population
A graph displaying Directional Selection
Selection against both extremes
Population after selection
Original population
A graph displaying Stabilizing Selection
Selection against the mean
Population after selection
Original population
A graph displaying Disruptive Selection

**Figure 4.18(a)** Various types of selection

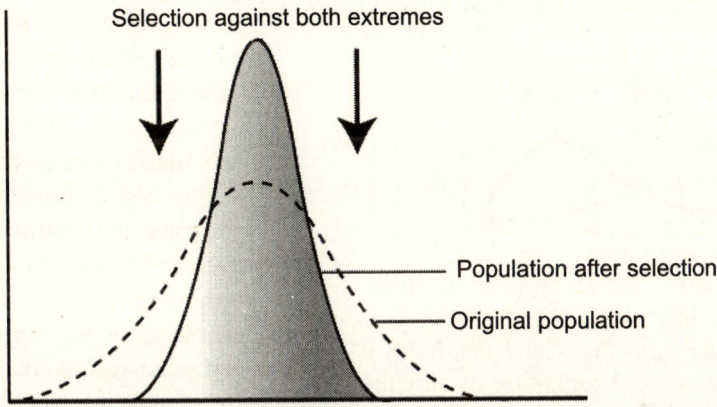

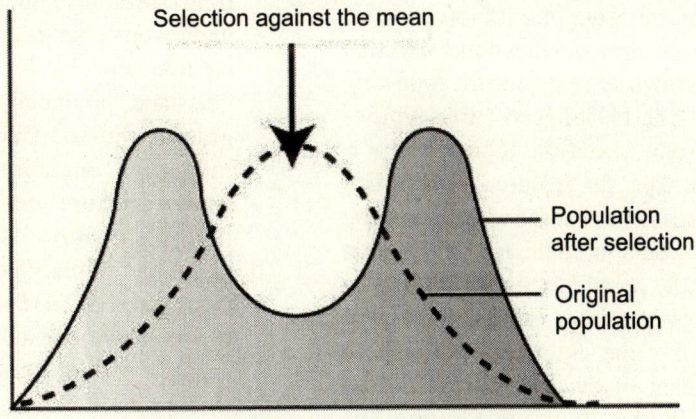

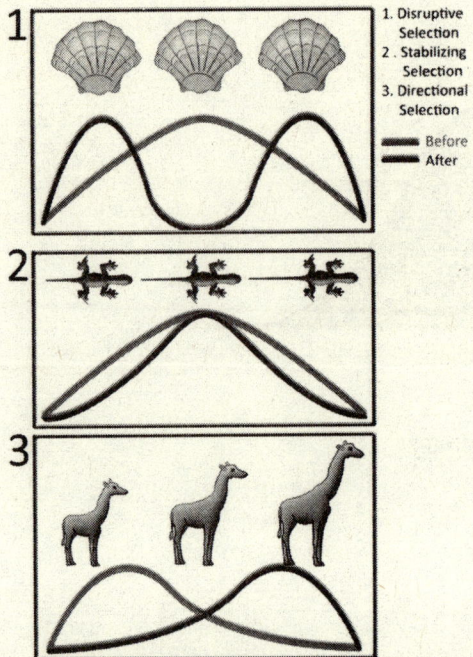

1. Disruptive
   Selection
2. Stabilizing
   Selection
3. Directional
   Selection

▬▬ Before
▬▬ After

**Figure 4.18(b)**    Various types of selection

resistant germ lives and continues to reproduce until formation of an entire population of resistant germs. The phenotype for the germ moves in a direction that was defined by natural selection operating on and favoring an extreme phenotype. Directional selection can be shown by analyzing the following graphs. The bell-shaped curve typifies the normal variation found within a species. The shaded area exemplifies those organisms favored reproduction by natural selection. In the next illustration Graph, the population is the same after many generations. The shaded area shows that the entire population is a descendant of the extreme phenotype found in the previous Graph. The phenotype of the population has changed because of the directional selection forced by natural selection.

Such selection is associated with change in environment, due to which average character becomes non-adaptive. It favours individuals with non-average, or extreme characters, which may be useful in changed environment. This selection is directional and progressive, which produces new types from original population with the ability to survive the change in environment. By keeping the gene from disappearing in heterozygous state, natural selection provides the species with reserves on which it may draw when condition changes. Species with lots of heterozygote and plenty of variations are the fittest ones in changing environments. Such selection favors individuals who possess values for a trait at one end of distribution range of a particular trait and selects against both average individuals and individuals at opposite extreme of distribution. Such directional selection may favor small size and select against both average and large individuals in a population.

Evolution of horse is a good example of directional selection in which small forest dwelling animal, *Hyracotherium*, underwent changes in its body when environment changed from the forest to grassland, producing tall, fast-running, grazing horse. Evolution of black industrial melanic moth (*Biston betularia*) from grey ones in England in 19th century is the directional selection. Replacement of susceptible mosquitoes by DDT resistant strains due to excessive use of insecticide is an example of directional artificial selection (Figure 4.19).

3. **Disruptive or diversifying selection:** **Diversifying selection** favors both phenotypic extremes at the expense of

**Before selection**

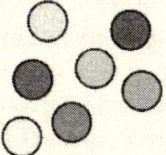

**After selection**

**Final population**

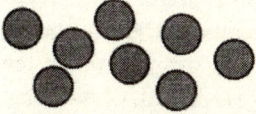

**Resistance level**

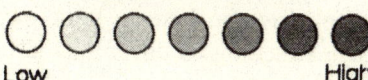

Low　　　　　　　　　　High

**Figure 4.19** Acquiring resistance

**Figure 4.20** Kormodo dragan survive through parthenogeness

intermediate phenotypic ranges. Diversifying selection is the opposite of stabilizing selection. Diversifying selection forms multiple phenotypes that exhibit successful breeding seasons within a population. The result is the prevalence of 2 (or more) morphs within a population. The genes code for proteins which interact with one another to form the body. Majority of the DNA doesn't code for proteins but form a complex network called transcription factors. Natural selection affects the genotype of a population. The environment's affect the individual's phenotype that defines fittest (Figure 4.20).

This selection favors individuals with relatively extreme values for a trait at expense of individuals with average values. **Disruptive selection** favors organisms at both ends of distribution of trait. Such selection pushes some phenotypes away from the population average but at the same time maintains them in population. Polymorphism appears and is maintained in population. Uniformity of characters in young individuals and bimodal or trimodal diversity of characters in adults is considered to be due to disruptive selection. It enhances adaptability of population. **Mega mutations** may produce new types but they are not removed even when they are not advantageous. White tigers and white leopards are common in the populations. The former must have produced to Siberian tiger and the latter to snow leopard by migration. Origin of bird's wing from a reptilian foreleg occurred due to mega mutation, but it may not have been an advantageous character in formative stage but is still maintained in population and became advantageous later. Polymorphism is common in insects, like butterflies, aphids and hoppers, web-spinners and social insects which allow the species to exploit different types of ecological conditions by different forms. Widely distributed species, in mosaic of environmental conditions, show polymorphism.

**Natural Selection and Constraints:** Majority of lineages in a group would evolve in a consistent direction as this change is adaptive. It is the result of directional natural selection which acts on individual organisms. The passive trends involving one-sided expansions in variance are some form of constraint, perhaps internal and nonadaptive in nature. McShea (1994) noted, "the distinction between the passive and driven mechanisms is not necessarily between selection and developmental constraints, nor even between internal and external factors." McShea (2005) pointed out 4 combinations of dynamics and causes related to natural selection and constraints in evolutionary

1. **Driven dynamic caused by natural selection:** The body size data indicate that there is often a tendency for larger individuals to be advantageous relative to smaller members of the population for better defense against predation, improved success as predators, improved success in competition for resources or mates, larger brain size, higher thermal tolerance, and longer lifespan (Hone and Benton, 2005).

2. **Passive dynamic caused by natural selection:** The outcome of natural selection can be conservative and directional. Some forms may prevent changes in certain traits. If such limitation on change occurs primarily in one direction, then selection would prevent decreases in the minimum within a distribution such that any increase in diversity would be in one direction and a passive trend would-be the result. Thus, selection can be a cause of either driven or passive trends, depending on whether it is directional or stabilizing.

3. **Driven dynamic caused by constraints:** The development of organisms consists of a complex and interconnected series of programmed changes that can often be limited in flexibility. Some mutation may be more likely to appear in the population than others, resulting in changes that occur consistently in only one direction-i.e., a driven trend. If this is based on internal constraints on the sorts of changes that are possible, then it would differ from the selective constraints described above. A driven trend may result from a tendency for serially repeated, or modular structures within organisms that begin similar to each other to become more different, simply because there are more ways for such structures to differ than to be the same.

4. **Passive dynamic caused by constraints:** Not all limitations to expanding variance are the result of natural selection. Some simply represent physical limits on the range of morphologies that are possible. The minimum number of cells of which a living organism can be composed is one. If life began as single-celled, then expanding diversity could involve increases in maximum cell number (Valentine et al., 1994).Whether they relate to directional selection, **selective constraints**, or nonadaptive constraints, these causes often are assumed to operate at the level of organisms. Their influence on large-scale trends would involve extending these effects through long periods of time, which is consistent with the neo-Darwinian theory. The factors operating among species can generate

trends at higher levels. Many of these are recognizable as analogs of population-level processes.

Natural selection is conceived as heritable variation in fitness. Fitness is the capacity of individual organisms to survive and reproduce, which determines '**trait fitness**' (Sober, 2006). Variation in fitness is the selection, the driver of ENS which mediates via heredity (Sober, 1984). Heredity involves a correlation between parent character and offspring character. The offspring of an organism should be on average more similar to it than to a randomly chosen member of the population (Godfrey Smith, 2009). Selection will have no evolutionary effect without heredity. Natural selection is the distinct causal process broadly analogous to a force in Newtonian physics (Lewens, 2010). Migration, mutation, and other evolutionary forces are seen primarily as sources of novelty that continually introduce new variety for natural selection to work on, but not as significant drivers of evolutionary change. Sometimes, as with Hull et. al. (2001), the source of novelty is explicitly incorporated into the definition of selection; more usually, it is treated separately or not discussed, with the focus being on variation (Sober, 1984). For Sober evolution is change in gene frequencies but the evolution might in principle occur in non-biological domains. The importance of the phenotype as the unit of selection as opposed to the gene has been a central topic of concern (Kerr and Godfrey-Smith, 2002). It redefines ENS and fitness, re-conceives the role of selection and drift in evolutionary explanations. It defines evolution as change in a population, which in turn is the change in the prevalence of traits in the population, which consists change in the individuals that make up the population.

**Selection and narrow trait fitness:**   For Sober, selection is associated with individual fitness differences. Fitness differences involve differences in rates of survival and reproduction. So Sober defines selection as force of differential rates of survival and reproduction. The expected propensity to survive and reproduce is fitness which is based on ongoing processes of reproduction and survival of individual. Variation in fitness implies variation in rates of reproduction and survival. The rate of reproduction and survival is not treated as an evolutionary force by Sober. If there is variation in fitness, the biasing effect of such variation is referred as selection and treated as a distinct force. If there is no variation in fitness, then force of selection is absent, and any change in population resulting from chance fluctuations in rates of survival and reproduction is swept up into a separate force of drift. The twin processes of survival and reproduction, labelled fitness, only consider as a force, when variation exists in rates of survival and reproduction among subgroups of the population, and then it is called selection. Mutation is considered as a force when it operates regardless of whether there is variation in rates of mutation among the subgroups. Mutation introduces variety into the population, and has a role to play in evolutionary models even when it does not drive evolution in any direction. **Evolutionary forces** bear discrete responsibility for any change in any individual of the population.

### 4.10.10   K-selection and *r*-selection

These 2 categories are made based on population density and reproductive rate at which the natural selection operates. Selection operates in stable environmental conditions in which species live in saturated population densities. Such populations show sigmoid growth curves and live near the carrying capacity. Selection favours individuals with enhanced competitive ability at high population densities near the carrying capacity. Such population has slow growth rate and prevail in non-seasonal tropics. **r-selection** operates on populations having rapid growth rate but low adaptive ness. They rapidly exploit new environment, such as a burnt out forest. Such populations are good pioneers and exploit a new environment with their high biotic potential and no selection pressure. They exhibit J-shaped growth curves, as in seasonal insects.

### 4.10.11   Darwinian Fitness

Darwinian fitness also known as Selective Value or Adaptive Value, a composite of many advantageous and disadvantageous characteristics those confer reproductive advantage to an individual from its genotype. Fittest is not the individual that survives and produces large number of offspring reaching reproductive age. Reproductive success is the key of fitness. **Relative fitness** is symbolized by the letter w, and is inversely related to selection coefficient, $s.w = 1 - s$ or it can be $w = 1 + s$ .frequency of recessive gene a will decrease in next generation owing to lower fitness and that of dominant gene A will increase because of higher fitness. As frequency of recessive gene goes down, selection against recessive gene becomes progressively slower. For instance, if $q = 0.5$ in the beginning, it will take 8 generations to reduce frequency of gene a to 0.1, 100 generations to reduce it to 0.01 and 1000 generations to reduce the value of q to 0.001.

**Different concepts of fitness:** Intuitively, the fitness of a trait is the fitness of the trait, not of the individuals who possess the trait. The fitness of a trait should be a measure of the propensity of the trait to spread. Treating trait fitness as average individual propensity to survive and reproduce gets this wrong in all but a very restricted set of circumstances. The classical account defines fitness in terms of the propensity of individuals to survive and reproduce. If there is heritable variation in propensity of individuals of various types to survive and reproduce, there will be a tendency for the population to evolve such that these 'fitter' types become more prevalent - although this tendency might be counteracted, or overwhelmed by other processes. So it follows though one trait might be 'fitter' in the classical sense that individuals with the trait tend to survive and or reproduce better than the alternative. The overall expectation as to which trait will increase in the population must take into account all biasing effects from all processes causing evolutionary change. This trait is 'fitter' in the sense that we expect it to increase in the population compared to its alternatives as the result of its all-things-considered competitive advantage vis-à-vis said alternatives. Let us call the first sense of fitness, narrow trait fitness and the second sense, broad trait fitness. **Broad trait fitnesses** are most relevant to predict evolution. Variation in broad trait fitness, together with an appropriate source of novel variation, is sufficient to explain the adaptation of organisms to environment, for example the complexity of the eye, and the diversity of organic life. It is for this reason ENS is identified as variation in broad trait fitness. **Narrow trait fitness** is interesting as it is an extremely prevalent contributor to broad trait fitnesses. Because of the central role of survival and reproduction in biological evolution, the focus on narrow trait fitness is perfectly understandable. However, understanding evolution abstractly - especially if we are interested in extending evolutionary ideas to other domains than biology - requires to identify broad trait fitness as the central explanatory concept. Evolution, on this account, is best understood in terms of the interaction of distinct mechanisms of change. In any given population, we can conceive of change as resulting from the operation of one, or more causal mechanisms that act to add or, subtract members from the population, or change the traits of already existing members. Such mechanisms may be treated in a fine-grained way. We may decompose mortality into predation, resource competition, etc., or at the coarse-grained level of individual fitness. Any such mechanisms which produce the systematic increase of one variant over another constitute a competitive advantage for that trait. A trait may be advantaged by one evolutionary mechanism yet disadvantaged by another. The aggregate of these effects will be broad trait fitness, our all-things considered measure of expected evolutionary change.

Heritable variation in fitness is an entrenched, venerable conception of the conditions for evolution by natural selection. It conceives variation as serving as the raw material for selection to operate, with heredity preserving the resultant 'winnowing' or 'sorting', thus gradually building on itself for generation after generation. The classical account of evolution by natural selection must be therefore be abandoned. Heritable variation in fitness is not necessary for evolution by natural selection nor is it sufficient. Heredity itself figures in evolution by natural selection in so far as reproduction figures. The classical account mistakes a specific biological process for a general evolutionary requirement. This means that heredity is not required for selection to occur, and that variation in heredity can lead to evolution by natural selection even if there are no differences in survivability, or fecundity. We cannot try to simply bracket off heredity as another force alongside natural selection. Instead, we can develop a new account of fitness, heredity, and selection that avoids the manifold problems of the classical account and elucidates the relation between selection and evolution in a clear, more elegant way. A more accurate account of natural selection must distinguish between survival

and reproduction, rather than lumping them together as 'fitness', and must treat the role of heredity as specifically related to reproduction, as both a potential source of change in its own right, and as the enabler for variation in fecundity to have an evolutionary effect. It must distinguish between the broad and narrow concepts of trait fitness, and should define ENS in terms of the latter.

Fisher uses a specific definition of fitness consistently. Darwin is fuzzy on sexual selection role in fitness differences. He thought the fitness as the aptitude of an individual organism to succeed in the struggle for existence. Lewontin treats selection as an abstract and conceives evolution in terms of phenotypic change, not primarily in terms of genes. Lewontin defines fitness as the rate of survival and reproduction of a phenotype (Lewontin, 1970). The word '**phenotype**' is ambiguous between denoting a class of individuals, versus denoting a subclass of the traits of a particular individual. One interpretation is that the class of individuals that share a phenotype is variation in the fitness of types of individual which is necessary for selection to occur (Walsh, 2004).Since only individuals have ancestors, descendants, and contemporaneous relatives, Lewontin's 'rate of survival and reproduction of a phenotype' refers to individual fitnesses.

### 4.10.12  Sexual Selection

In many animals, males compete for access to mating with females in the form of contests for dominance among males, or it may involve behaviors, or physical features that females find attractive. Both the conspicuous structures of bizarre courtship behaviours of some males and apparent willingness of females to base their mate choices on these features often seem unlikely with efficient survival and reproduction. Exaggerated ornamentation makes males more attractive to females, and more vulnerable to predators. **Sexual selection** is the special kind of natural selection that acts on traits that help an animal mate. Altruism refers to any behavior that endangers an individual, or reduces its reproductive success but benefits other members. If altruism is encoded in an organism's genes, those genes are placed at risk each time the altruist performs one of its brave behaviours. But natural selection can select for altruistic genes, if the altruistic individual helps relatives who possess the same alleles. This special case of natural selection is called kin selection.

The preference of one mate over another is sexual selection. A **mating dance**, elegant plumage, or overall size and physical/mental characters may influence the decision to mate or not to mate. Some mate and pass on their genes, others will not. The results of this type of selection are closely aligned with directional or diversifying selection, as it affects the overall population structure. Sexual selection is generated by factors within other members of the same species as opposed to an outside event. The favoring of one phenotype over another by the environmental factors creates unequal reproductive rates for individuals in that population. Survival of the fittest means who can reproduce in greater numbers. **Reproductive success** is the essence of natural selection (Figure 4.21).

**Figure 4.21**  Sexual selection in peacock

Darwin states that mating among most animals is based on **female choice** - females are choosy about their mating partner, while males are ready and waiting for females to allow them to mate. These cause females to shape the adaptations of males, by choosing mates based on their characters. Sexual selection can be separated from natural selection, as the 2 processes have separate effects. In many bird species, the grotesquely exaggerated character of males lower the male's chances of survival. Darwin believed that female choice could explain the evolution of the various human races, and the more obvious differences between females and males in species like peacocks.

In many species, males possess distinct secondary sexual characters like the peacock's tail; coloring and patterns in male birds, voice calls in frogs, and flashes in fireflies. Many traits are liability for survival. Natural selection has various components, of which survival is only one. Sexual attractiveness is important component of selection. Sexual selection is natural selection operating on factors that contribute to organism's mating success. Traits that are liability to survival evolve when the sexual attractiveness of a trait outweighs the liability incurred for survival. Male lives a short time producing many offspring is more successful than a long lived one that produces few. The former's genes dominate the gene pool of his species. In many **polygynous species**, where few males monopolize all females, sexual selection causes distinct sexual dimorphism. In such species, males compete with other males for mates which could be direct or mediated by female choice.

In some species females choose, and males compete by displaying phenotypic characters and/or performing elaborate courtship behaviors. The females then mate with males that most interest them. The **good genes model** states that the display shows component of male fitness. A good gene advertises that bright coloring in male birds shows a lack of parasites. Females are cueing on some signal that is correlated with other component of viability. Selection for good genes is found in sticklebacks in which males have red coloration on their sides. Intensity of color is correlated to both parasite load and sexual attractiveness. Females preferred redder males. Redness shows the presence of few parasites. Female may have an innate preference for some male trait before it appears in a population. Females then mate with male carriers when the trait appears. The offspring of these mating have genes for both the trait and preference for trait. Suppose that female birds prefer males with longer than average tail feather. Mutant males with longer average tail feather produce more offspring than short feathered males. In next generation, average tail length increase. As the generations progress, feather length will increase as females do not prefer a specific length tail, but a longer than average tail. Finally tail length will increase to the point with the liability to survival. In many exotic birds, male plumage is often very showy and many species have males with greatly elongated feathers. Sometimes these feathers are shed after breeding season.

### 4.10.13  Convergent Characters

Darwin pointed out that convergent evolution result in traits that appeared similar in unrelated organisms. This lead to false assumptions about common ancestry and phylogenetic lineages. He focused attention on non-adaptive characteristics for accessing phylogenetic relationships. Ever since Darwin, taxonomists have concentrated on non-adaptive characters to classify organisms. This is especially true today; **molecular taxonomy** is based primarily on non-coding nucleotide sequences and satellite DNA. This is because coding sequences tend to be conserved over time,

whereas non-coding sequences tend to diverge randomly at predictable rates.

### 4.10.14  Organs of Extreme Perfection

Darwin's critics pointed out the inconceivability of idea that complex and well-adapted organ like the vertebrate eye, could have arisen without some design, or purpose in mind. Darwin argued that it is possibly not a sudden creation, but from gradual steps, in which structures of increasing usefulness were selected for until something as complex and highly adaptive as the vertebrate eye existed. If any species evolved a trait exclusively for benefit of another species, it could not have done so by natural selection. Darwin points out that often, what seem to be adaptations benefit other species and so natural selection is not undermined.

### 4.10.15  Hybridization

Darwin disagreed that species are created separately, and unbreakable barriers between species prevent real hybridization. There are viable hybrids between many species in nature, undermining the unbreakable barrier argument and natural selection itself could not erect any such barriers. Natural selection cannot select for any form of sterility Darwin asserted that hybrid sterility and/or inviability as the side-effects of divergence of species. Once a species has diverged into 2 separate species and remained separate for long, random changes in hereditary makeup of separate species result in inability to interbreed and form viable offspring. **Hybrid sterility** is a side effect of speciation, not a cause. The sterility of hybrids could not possibly advantageous to them, and could not have been acquired by continued preservation of successive profitable degrees of sterility. The sterility of various species when crossed is so different in degree and fertility of pure species and is so easily affected by various circumstances that it is most difficult to say where perfect fertility ends and sterility begins.

## 4.10.16   Adaptation

Ability to avoid, or recover from predation often makes the difference between life and death, and is one of the strongest components of natural selection. Pressure to adapt is stronger on prey than on predator. If the predator fails to win a contest, it loses a meal; if prey is the loser, it loses its life. Predation was rife long before the start of the Cambrian, in increasingly spiny forms of acritarchs, the holes drilled in *Cloudina* shell, and traces of burrowing to avoid predators. It is unlikely that appearance of predation was the trigger for the Cambrian explosion, although it may have a strong influence on body forms that explosion produced. Intensity of predation does appear to have increased during the Cambrian as new predatory tactics emerged. Natural selection explains change in organisms. Adaptation explains the reasons of change. Structure and function of an organism is directly related to its environment. Numerous environmental mechanisms influence adaptive evolution. We looked at complexity of natural world from the design of above 18,000 species of orchids, to mystery of flight in birds. Understanding the driving force behind the diversity of life is complex. Differences between related animal species living in different parts of world reflected their varied habitat. Lamarck thought that the world's environment is changed, when the animals living off the environment. The animal's characters changed to adjust to their habitat. The adjustments in fit between organisms and environment are the adaptation. The root concept of adaptation grew into Darwin's theory of natural selection, the mechanism that explains the change.

Adaptation is based on the concept that population change over time due to natural selection. Increased survivorship and/or increased reproductive success cause adaptive evolution, when group of individuals in population gain advantage because of sharing the special traits. Such traits may be inconspicuous or elaborate. They may start out as a 2 mm lengthening in nectar-gathering tongue of few orchid consuming moths. The tongue becomes longer in that species as those individuals and their offspring out-reproduce others. The long shape becomes the norm, as the long-tongue allows efficient feeding, and enhances reproductive success. Darwin discovered an orchid with 11 inch long nectar-producing tube in the Madagascar. He predicted a moth that feeds from tube with an 11 inch proboscis. About 50 years later, Darwin's prediction proved true when the moth, *Xanthopan morganii praedicta* with a 12 inch proboscis which fed from, and pollinated Darwin's orchid. The ultimate source of an adaptation is genetic as traits that pass on from one generation to next be influenced by natural selection. Darwin's orchid-and-moth is visible cases of adaptation. A feature of a plant is associated with the corresponding animal feature and both benefit from their interconnected lives in nature. Adaptations work together to make a particular lifestyle work. Many environmental factors are problems and require solutions. Food, climate and predator-prey relationships play important role in selecting for (Figure 4.22).

Most penguins live in extreme environment. These flightless birds provide a wonderful example of multiple evolutionary adaptations. The most difficult tasks the **penguins** face in maintaining their body temperature under the varied land conditions they inhabit and breed, and in the Antarctic water, where they feed. Penguins are homeothermic, with a stable body temperature between 35° and 41°C. Penguins maintain temperature in a climate where sea

**Figure 4.22** Mimicry in butterfly

temperatures stands –2°C and air temperatures stands between 0°C to –60°C. Metabolism and muscle activity generate body heat internally. Blubber under the skin insulate the Penguins, retains heat as in seals, and whales. Densely packed feathers cover the body of the penguin. The base of their feathers are downy, to trap air for better insulation. Penguins keep their feathers in good condition and insulate them from the cold wind and water. By preening, they waterproof themselves by spreading oily secretions from the uropygial gland to other areas of their body.

Their circulatory system helps in retaining and dissipates heat easily. The peripheral arteries and veins are situated close together for exchanging heat, a countercurrent heat exchange system to reflect the to-and-from flow of blood relative to the heart. This raises the temperature of blood flowing from the flippers and legs to body core by drawing it

veins carrying already-warm blood to extremities. Penguins increase blood flow to their flippers to cool down .The Galapagos penguin lives near the equator where it can get hot. They shiver to increase metabolic heat production, and expose their feet to release excess heat. Some seek shelter under rocks to avoid temperature extreme. The Emperor Penguin huddles together to share body heat in the harsher conditions of mainland Antarctica. Penguins are amphibious, feeding at sea and breeding on land.

Three pygoscelid penguins prey mainly on krill and to some extent on fish. They remain under constant threat from their predators like leopard seals, orcas, and fur seals. Penguins are much adept at swimming than walking. They consume one-third less energy at sea than on land. On land, penguins walk, jump, sometimes over long distances for breeding, or to enter the seas. In the water, they are marvelous. Torpedo-shaped bodies and flipper design allow penguins to fly underwater, using their bill, tail, and feet to change rapidly direction, pursuing fish or avoiding predators. During traveling long distances, penguin porpoise, leaping out of water, to reduce drag and conserve energy.

Most pygoscelid penguins are faithful to their mate and their nest site, returning to breed in the same place year after year. They join into small colonies, with a few breeding pairs, or millions of pairs. Males arrive first and build the nests with small rocks piled up in snow-free places. Females arrive after the males and locate their mate. Emperor and King Penguins carry eggs and young chicks on their feet which make it difficult for them to walk, to keep their eggs and young warm, and prevent them from freezing on cold Antarctic rocks. Emperor penguins breed in the Antarctic winter which allows the new chicks to be independent from their parents after 5 months, to adjust on their own during the mild Antarctic summer. Polar bears evolved white fur as it better conceals them in arctic. All other bear species are either black, or brown. The ancestors of today's polar bears, the white individuals probably had more hunting success .Perhaps their prey found it difficult to spot them against the snow and ice. Squirrels buried nuts during summer and fall for food through winter. Even the common dandelion has adapted to its environment by producing a characteristic, white fluff on its seeds to increase their spread, and thus their chances of survival in environment.

Most information isn't coded in the genes. It's coded in the way the genes interact with one another. Evolution allows drastic changes in a short period as exemplified in the evolution of the **FOXP2 gene**, which plays role in language. It is almost same in most animals. But the human version has changed dramatically in a brief period. Rather than slow, gradual change, species remain in a static state for long period. All the change occurs when DNA networks are reconnected and combined in new and interesting ways. A byeproduct of this is that every little evolutionary advantage no longer makes the great difference, reducing the emphasis on competition. DNA arranged in a vast intertwined network is influenced from the outside. Specialized proteins regulate access to the DNA. The environment plays a crucial part in whether or not a specific gene ever even does anything. Sometimes it. The cell needs the DNA to maintain itself. The DNA needs the cell in order to reproduce. The genes called **transposons** have the ability to leap from one sequence of DNA to other. Some sequences of DNA have evolved to create controlled mutations, where the DNA is shuffled around. This is true of genes in human brain cells. We literally have DNA designed to free ourselves from the constraints of DNA. No gene won't change the way it is expressed under an extreme environment. In the **polymorphic moths**, dark environment

favors the black variant over the white, which is easily discovered and consumed by predators. Both polygenic traits and polymorphism create phenotypic variations that act differently by selection.

## 4.11 Epilogue on Neo-Darwinian Theory

Darwin formulated his theory of natural selection before the birth of modern genetics. He assumed that acquired characteristic could be inherited. If we add this assumption to the operation of natural selection the picture changes. Now, every individual organism is the inactive participant in evolution. Any creative, or adaptive response of an organism can be passed onto subsequent generation. Since creativity tends to beget more creativity, one can see a steady upward movement of complexification. In such circumstances, an evolutionary force can originate as the total sum of individual adaptive responses. The advent of modern genetics refuted the Lamarckian theory of the inheritability of acquired characteristics, and dashed all hopes of explaining evolution by natural selection alone. This gave birth to the neo-Darwinian theory, which attributes the source of novel forms in evolution to mutation. Mutations are assumed to be totally random in the sense explained by Dicks. As per second law of thermodynamics, an infinitesimal number of mutations will be favourable, increase complexity. If typically neo-Darwinian materialistic reductionistic view of evolution is taken, even assumption of pure randomness is probably over-optimistic, as the known physical causes of mutations are events like incomplete chemical processes, or radiation which by their very nature tend to produce unfavorable mutations. Thus the distribution of mutations would, under such assumptions, most probably be skewed in direction of unfavourability.

Under the neo-Darwinian assumptions, mutations favourable to increased complexity would, at best, only be sporadic, i.e., insufficiently frequent to allow for any significant process of convergence towards greater complexity resulting from the operation of natural selection. Indeed, to achieve multigenerational convergence towards complexity, one needs much more than an occasional favourable mutation. One needs a certain minimal, tran generational rate of favourable mutations within same population for a considerable time. To have a complex process, a consistent string of favourable mutations within the same mutant **subpopulation** may be required. This requirement multiplies the probabilities for individual favourable mutation-events, thereby significantly reducing the probability that such process could occur spontaneously. Alternation between long periods of stasis and short periods of rapid change towards complexify as the fossil record show shortens the time gap during which successive complexification occurred. This decreases the probability in favour of spontaneous increase in complexity. Thus according to the neo-Darwinian theory, the only source of new forms in evolution is the mutation, which is assumed to be purely random. Despite the operation of natural selection which under certain circumstances positively selects new genotypes, the movement towards greater complexity in evolution is nevertheless confronted with the pure randomness of mutation. The neo-Darwinian theory does not really diminish the force of 'Abdul-Baha's argument.

Evolution towards complex form would have negative selective value during initial stages of process. A complex and flexible organ like the eye has a positive selective value only when it is about fully formed. Let us imagine the process of evolving an eye beginning with, say, a mutation-generated light-sensitive spot on the skin. Under most conceivable environmental circumstances, such a spot would increase the vulnerability of organism without conferring any immediate selective advantage. This would be the case for an unlimited number of generations, during which an incredible number of favourable mutations would have to occur among the already mutated population for any evolution towards higher complexity. This requirement increases the individual probabilities for mutation rendering such process even less likely. Recognizing the fundamental weakness in neo-Darwinism, some neo-Darwinian theorists have argued that a mutation generated change in physical characteristics must have been accompanied by a parallel mutation of the central propensity structure of organism's nervous system, thereby endowing the organism with capacity to use newly mutated characteristic in a positive way. Such assumptions do not buttress neo-Darwinian theory but rather are logically equivalent to postulating the existence of evolutionary force.

## 4.12 Modern Synthesis (MS)

The term, 'the modern synthesis' was coined by Huxley in 1942. At the start of the 20th century, the particulate views of inheritance (Mendelians) were incompatible with the models of gradual change. Mathematical model of the population dynamics of gene pools resolved it. Thus the first stage of the MS used the population genetics in a way to explain changes in **gene frequencies** within populations by the evolutionary forces. This led to the removal of the saltationism and orthogenesis Lamarckism, which are incompatible with gradual evolution. Drift could be seen as a significant force. The second stage of the synthesis dogmatically focuses on natural selection as the predominant evolutionary force (Gould, 2002). Gould cited Mayr's definition "… all evolution is due to the accumulation of small genetic changes … natural selection, and… transpecific evolution is nothing but an extrapolation and magnification of the events that occur within populations and species" (Mayr, 1963 ). Natural selection operating over genes within population is the predominant force of **micro-evolutionary change**. These micro-evolutionary changes can explain macro-evolutionary change. But after 23 years different view emerged that holds that populations contain genetic variation arises by random mutation and recombination. Populations evolve due to changes in gene frequency as the result of random genetic drift, gene flow, and natural selection. Most adaptive genetic variants have individually slight phenotypic effects so that phenotypic changes are gradual; diversification comes about by speciation, which entails the gradual evolution of reproductive isolation in populations. These processes continued for long to give rise to changes of great magnitude to warrant the designation of higher taxonomic levels (Futuyma, 1986). Natural selection is still predominant, but not the sole cause of evolution.

This form of analysis represented an extended synthesis as a multi-faceted research programme (Pigliucci and Müller, 2010) who presents interesting phenomena, with particular one area of focus, the epigenetic inheritance. Evolution of structural and functional complexity suggests that something may be increasing. Complexity increases when self-organizing systems organize themselves. **Genomic complexity** a consistent information-theoretic manner, which include intuitive notions of complexity used in the analysis of genomic structure and organization (Jablonka and Lamb, 2008) is found in functional complexity and vice versa. Evolution in an information-poor landscape leads to selection for replication only, and shrink genome size (Danchin et al, 2011). This allows us to observe the growth of physical complexity explicitly, and distinguish distinct evolutionary pressures on the genome and analyze them. DNA is the blueprint of an organism for building an organism that can best survive in its native environment, and passes on that information to its progeny which corresponds to Dawkin's view of selfish gene that use their environment, for their own replication (Williams, 1966). Those parts of the genome that corresponds to something, correspond in fact to the environment the genome lives in. This view referred to genes embody knowledge about their niches (Deutsch, 1997). An organism's DNA is not only a book about the organism, but is a book about its environment, including the coevolving species. Not all of the symbols in an organism's DNA correspond to something. These sometimes referred to as junk DNA, consisting of portions of the code that are unexpressed or untranslated. Unexpressed and untranslated regions in the genome can have a multitude of uses, such as satellite DNA near the centromere excised from *Tetrahymena* rRNA. *Mycoplasma mycoides* has a complexity of less than one million base pairs in our nasal passages, but close to zero complexity most (Adami et al, 2000) everywhere else, because it cannot survive (Adami et al, 2000) in other environment-meaning its genome does not correspond to anything there.

The genetic locus that codes for the essential information to an organism's survival will be fixed in an adapting population as all mutations of the locus result in the organism's inability to promulgate the tainted genome, whereas neutral sites is randomized by the constant mutational load. Examining an ensemble (Adami et al, 2000) of sequences large enough to obtain statistically significant substitution probabilities would thus be sufficient to separate information from entropy in genetic codes. The neutral sections (Adami et al, 2000) that contribute only to the entropy turn out to be exceedingly important for evolution to proceed (Kuhn, 1962). Sites are not independent. The probability to find a certain base at one position may be conditional on the probability to find another base at another position. Such correlations between the sites are called **epistatic**, which can render the entropy per molecule different from the sum of the per-site entropy (Pigliucci and Müller, 2010). This entropy per molecule takes into account all epistatic correlations between sites. **Epistatic effects** can be measured by estimating the entropy per molecule by creating mutational clones at several positions at the same time (Maher, 2008) to measure.

Weakest point in Darwin's theory of natural selection is that it lacks understanding about heredity. Darwin's theory has been enriched by population genetics and establishes the modern synthesis (Neo-Darwinism) which is combination of Darwinian selection and genetic theory. **Point mutation** and **genetic recombination** cause variation. Evolution which operates in small steps is resulted from natural selection that acts on genetic variation. Such processes explain the origin of higher taxa if they acted over long duration. This view is now called microevolution or **phylogenetic gradualism**. Modern understanding in genetics, population genetics, cytology and evolution formulated a coherent theory called modern synthesis around 1930's by Dobzhansky, Goldschimdt, Wright, Muller, Huxley, Fisher, Haldane, Mayr, and Stebbins. Synthesis theory explains transformation of a species by natural selection and splitting of species into reproductively isolated subgroups.

The latter process is the speciation. Modern synthetic theory recognizes five basic processes:

- **Gene mutation**
- **Alteration in chromosome structure and number**
- **Genetic recombination**
- **Natural selection**
- **Reproductive isolation (Figure 4.23)**

First 3 provides genetic variability for bring about changes. Natural selection and reproductive isolation guide population into adaptive channels. Mutation, genetic recombination, structural changes in chromosomes, and natural selection interact to produce a progressive change in population. Three accessory processes affect the working of these 5 basic processes. The genetic variability of a population increases through migration of members of one population to another, and hybridization between races. Effects of a chance, acting on small populations, may change the way in which natural selection acts. Mutation, genetic recombination, and natural selection are indispensable for evolution. Evolution of organisms can be compared to an automobile being driven along highway. Mutation can be compared to gasoline in tank. Mutation, the only possible source of new genetic variation is required for continued progress. But the source of motive power is genetic recombination as happens during sexual cycle. Since this process provides immediate source of variability on which selection exerts its primary action, it can be compared to engine of an automobile. Natural selection directing genetic variability for adaptation to environment, can be compared to driver of the vehicle. Structural changes in chromosome can have profound effects on interrelationships between genetic recombination and natural selection and can be compared to transmission

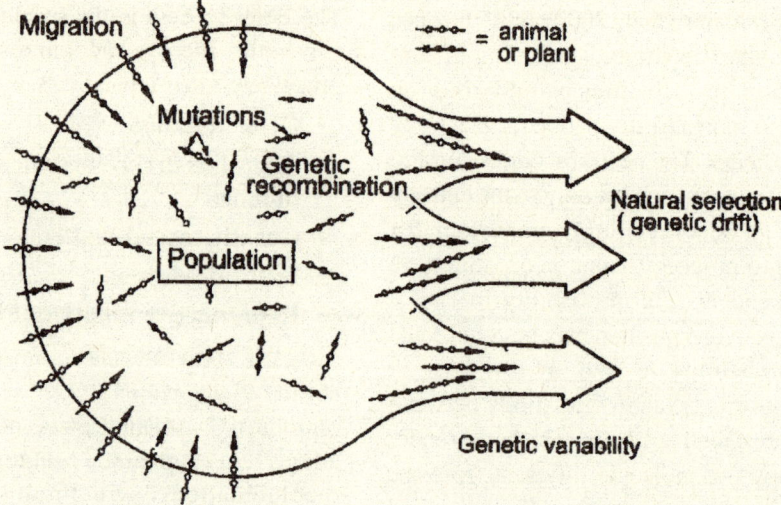

**Figure 4.23** Scheme to illustrate the interaction of the basic processes that bring about phenotypic evolution in a variable population of organisims (animals, or plants) (adapted from Stebbins, 1971).

and accelerator of automobile. Finally, reproductive isolation, which includes all the barriers to **gene exchange** between populations, has an effect like the highway, with its limits and directive signs, exerts on driver of automobile, thus allowing several vehicles to drive in same direction at same time.

Major Contributors of the modern synthesis were Ronald Fisher (1890-1962), J. B. S. Haldane (1892-1964), Sewell Wright (1889-1988), Theodosius Dobzhansky (1900-1975), Ernst Mayr (1904-2005) and Julian Huxley (1887-1975) .The Modern Synthesis could be summarized as follows :

1. The phenotype is a product of the interaction between genotype and environment.
2. Environment induced phenotypic differences are not passed to offspring, although the environment affect expression of genes.
3. Genes are particulate, but their effects are combined. They can have discrete effects on the phenotype. They can

produce continuous variation through interactions with other genes.
4. Genes form alternative alleles by mutations which have large, or small effects on the phenotype.
5. Environmental factors cause mutation. Mutations are not directed and not necessarily adaptive.
6. Evolutionary change is a population process.
7. The low rate of mutation does not cause significant evolutionary change. Genetic drift and natural selection change the proportions of alleles.
8. Slight phenotypic differences result in differences in fitness and rapid evolutionary change.
9. Selection produces phenotypic variation beyond the range of variation in the original population.
10. Natural populations are genetically variable.
11. Geographically discrete populations often have different allele frequencies.

12. Differences between populations usually have a genetic basis.

13. Natural selection can be observed in natural populations.

14. Differences among geographically discrete populations are often adaptive.

15. Species are reproductively isolated. Reproductive isolation in populations varies from slight to complete. Speciation usually occurs through geographic separation of populations.

18. Variation between species, genera, and families is graded.

19. Higher taxa appear to arise through successive accumulations of small differences.

20. The fossil record is consistent with the Darwinian model of evolution.

Let us summarize some assumptions of the late 20th century version of the Modern Synthesis:

1. Heredity through the transmission of germline genes are distinct units consisting of DNA on chromosomes.

2. Hereditary variation varies directly with change in DNA base sequence. Inheritance of acquired variations may be explained in terms of variations in DNA.

3. Hereditary variation is the result of the random combinations of pre-existing alleles that are generated by the sexual processes. New variations (mutations) are due to the accidental changes in DNA. The developmental history of the individual does not affect hereditary variation.

4. Selection occurs in individuals. Selection targets almost always the individual, rarely the community, which may co-evolve with its symbionts and parasites.

Some role for group selection is known with marginal significance in evolution.

5. Evolution is typically gradual. Heritable variations have small effects. Selection of individuals with phenotypes make them more adapted to the environment than other individuals due to abundance of some alleles in the population. **Mutation pressure** is not so important in evolution. With few exceptions, macroevolution continues with microevolution.

6. Evolution occurs through modifications from the common ancestor based on vertical descent. **Horizontal gene transfer** cannot alter the basic branching structure of phylogenetic divergence and has minor significance. The pattern of evolutionary divergence is tree-like.

This view is now being challenged. Biologists are now arguing that:

1. Heredity involves not only the DNA. The **non-DNA variations** can play role in evolutionary change and guide genetic evolution.

2. Soft inheritance including non-DNA variations and developmentally induced variations in DNA sequence is the important.

3. Many organisms contain symbionts and parasites that are transferred from one host generation to the next. Such communities could be the target of selection.

4. Saltational changes for evolution beyond the species level are common. Macroevolution may be the result of specific, stress-induced mechanisms leading to a re-patterning of the genome for the systemic mutations.

5. Pattern of divergence fails to explain all the sources of similarities and differences between the taxa. Sharing whole

genomes and partial exchange of genomes show web-like patterns of relations which are more evident in some taxa and in some periods of evolution. The soft epigenetic inheritance has implications.

Majority of geneticists thought the origin of species as gradual accumulation of small mutations. In 1940 this allowed for "The Synthetic Theory of Evolution" by Dobzhansky and others. This consensus or Synthesis was coined, forwarded and developed by Huxley, Mayr, Rensch, Simpson, and Stebbins. The word "gene" was coined in 1909 by Johanssen for heredity unit located on chromosome. Genes are the segments of DNA molecules, a self-replicating entity that along with others makes up the genome. Nathaniel C. Comfort pointed out, what DNA expresses is not a discrete chromosome part with fixed boundaries, but structural genes and regulator genes often act in clusters. All genes in an individual's genome are not expressed in phenomena but all, if the program remains successful, have a chance of being reproduced. Heredity traits are passed through generations by genes. DNA segments that constitute a single gene are sometimes far apart on chromosome. The same DNA segments can be combined in different ways to produce various expressions. Examples include overlapping genes, genes within genes, and genes that can be read forward or backwards. Coyne defines gene, as a piece of DNA that is translated into mRNA. Gene mutation produces new characteristics. J. M. Smith provided reductionist and holistic examples of use of genome information during development of individual. Natural nonrandom selection explains the incorporation of information in genome. It together with random genetic drift due to neutral mechanisms gives us neo-Darwinism.

Mendel in the 1860s discovered that parents can fail to always produce offspring with their own dominant features, one aspect of **sexual inheritance**. Mendel's yellow peas were hybrid forms with recessive green feature but able to reappear in later generations. Mendel's First Law of Inheritance tells that each organism inherits 2 copies of a trait, one from each parent. Each trait that each parent contributes is selected by chance from the 2-designated in his tabulations of dominant and recessive character pairs by capital- and lower-case letters, Aa, Bb. These traits are particulate. de Vries, Correns, von Tschermak and Bateson independently rediscovered Mendel's findings. Fisher, who developed probability theory, revealed that both partners in sexual union contribute to the offspring. This disproved the **spermatist dogma**, the notion that male contributes all. Mendel focused on formation of hybrids and did not know that character is expression of genes on chromosomes. Darwin claimed that individual is favored by selection owing to individual's overall quality. That is, natural selection acts on phenotype. Selection acts on particular alleles in relation to their overall contribution to all individuals that carry copies of them. G. C. Williams stated that evolution is not about survival of species but how natural selection operates on individuals and their genes? Neo-Darwinism makes explicit that as alleles cannot be individually exposed to selection, each rides on quality of whole genome which itself cannot ensure the quality of phenotype.

Within cells, proteins are found as enzymes, actin, hormones, antibodies, which are involved in transportation of other molecules. Any change in a gene called mutation occur by chance at any time due to non repair of lesions in DNA that arise from environmental agents like genotoxic chemicals, UV radiation, and ionizing radiation. The reactive oxygen species derived from oxidative respiration and products of lipid per oxidation products of normal cellular metabolism damage

DNA. The problem of **DNA damage** exists. DNA repair systems must have arisen early in evolution. At least 4 main, partly overlapping damage repair pathways operate in mammals are base-excision -, nucleotide excision repair, homologous recombination, and end joining. Enzyme repairs damaged RNA and proteins. Every day, in every cell of a human may occur up to 300,000 lesions that can stop the routine DNA-replication machinery. **SOS**, a type of unusual DNA polymerases, continues DNA replication which is efficient and imprecise, leaving mistakes to efficient error-correction enzymes as their substrates are specific mistakes made by other enzymes. Morgan proposed that embryonic growth results from genes switching on and off, in the right place, and at right time.

Proteins fold by formation of covalent bonds and by variety of non-covalent interactions among amino acids. Many proteins fold to become globular, some are fibrous; some have distinctive 3 dimensional shapes. The physiological environment works in protein folding. Some proteins fold to expose a hydrophobic region which helps them bind with membranes. Most proteins are modified after addition of sugars, phosphate, sulfate and other small molecules. Their function is modified and the same cause some to become active, or inactive. Certain proteins work alone and need no company. Others work when they are bound with other molecules of self, or other proteins or cellular components. Proteins form such complexes by binding along surface clefts created by folding in particular fashion, and by ionic and other non-covalent interactions. Less than 2% of our genome is genetic code for proteins that let us live. Much of that is repeated code of transposons, the jumping genes first described by McClintock that insert copies of themselves throughout the genome. Add regulatory sequences, and the total of

genes in chromosomes required to bring you into existence is 5% of genome. The conflicts between parental genes and childhood genes and between female genes and male genes are described. The rest, 95% of our genome, consists of nonfunctioning genes, HIV-like gene fossils that are mostly dormant, and huge stretches of no-function code rife with genetic parasites that, willy-nilly, will hitchhike into your children and their children's children. Cells with different functions produce particular proteins by turning on or turning off appropriate genes along DNA molecule. The complex hormonal, developmental and tissue-specific regulation influence gene expression. For prompting, the promoter, a gene is first transcribed into a tightly edited molecule of pure information called **mRNA**. Varieties of mRNA are different numbers and orderings of nucleotides U, A, G, C in a linear one chain. Each mRNA chain of nucleotides is the genetic code for a protein which is an ordered linear chain of amino acids. **Genetic code**, discovered by Nirenberg in 1961, is nucleotide triplets, called codons.

In RNA, U plays the role of T in DNA code. So mRNA is read out from DNA with uracil linking to adenine, and guanine links to cytosine. Proteins manufactured through code are assembled outside the nucleus. mRNA enters cytoplasm where it meets and bind with ribosome. Small variety of cytoplasmic RNA, tRNA, brings amino acids one at a time as called for to ribosome and links them into a strand. Ribosome which glides along mRNA is cued as to what calls to make by reading the mRNA's sequence of letter triplets. Nucleotide sequence present in DNA, via the mRNA intermediary, directs the synthesis of protein with a precise and predetermined sequence. As the words of the code along the mRNA chain are triplets of letters that follow each other end to end, the reading must be told where to start and stop. Initiation sequence includes the start word AUG. **Termination codes** are UAA, UAG, and UGA. In human cells, the genome size are 3,200,000,000 AT, GC, letter pairs and the chromosome number is 23 (Figure 4.24). The respective numbers for mouse *Mus musculus* are 3,454,200,000 and 20, for house fly : 900,000,000 and 5, and for amoeba *Amoeba dubia*: 670,000,000,000 and several hundreds for *Mycoplasma genitalium*: 580,000 and 1 (circular) has the smallest genome. We then differ from mice by 300 genes but number of human protein-coding genes has since been enhanced to 40,000 and in 2007 Whittled gave a firm count of 20,488. Venter estimates that of our genes about 16% play a role in our metabolism, 12 % are used for communication from cell to cell, 4% are devoted to help us. Our cells reproduce and of the remainder, most are active only in brain. He has found 3,000 of these already and planned to scoop the sequencing of entire human genome. Genome sequencing is fundamental to an understanding of how species cope. By comparing genomes, we reconstruct the evolution of one species from another and say what common ancestor species could have had. Such application of studying genes was proposed by Paul Linus. Genes controlling development is expected to have been conserved through evolution. Millions of zebra fish can be bred in a small space. The control genes of such fish are being

identified. Their genome has about 1.7 billion base-pairs and this has long regions of synteny-where order of genes is similar to that found in humans and likely plays same or similar functions in all other vertebrates (Figure 4.25).

A biopacemaker for circadian rhythms found in eukaryotic life as fungi, algae, higher plants, fruit flies, and mammals, has now been described. External environmental changes have only a secondary tuning input. Primary mechanism for each cellular beat of time begins in nucleus. There special, always "on" initiator genes make proteins that activate clock genes in another region of cell's DNA. mRNA output of these activates, outside the nucleus, is the construction of distinctive clock proteins that accumulate in cytoplasm. At a certain concentration the clock proteins bond in pairs and in this condition they gain entry into nucleus. There they block the operation of initiator genes. The **clock genes** close production. In time the blocking proteins are eliminated in nucleus. The cycle, which takes between 22and 26 hours, resumes. The pacemaker for circadian rhythms in mammals is the pinhead sized suprachiasmatic nucleus of about 20,000 neurons deep within brain. Light-detecting molecules that track the day and continually adjust the primary biological clock of animals and plants are proteins called **crypto chromes**. Our ungoverned wake-sleep cycle is 25 hours. Crypto chromes that detect

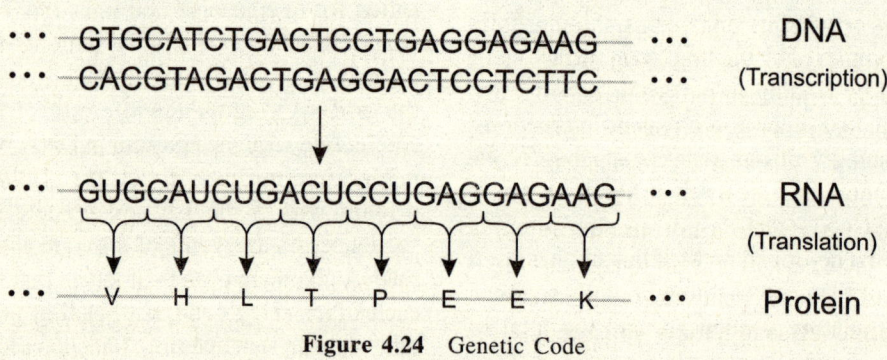

**Figure 4.24** Genetic Code

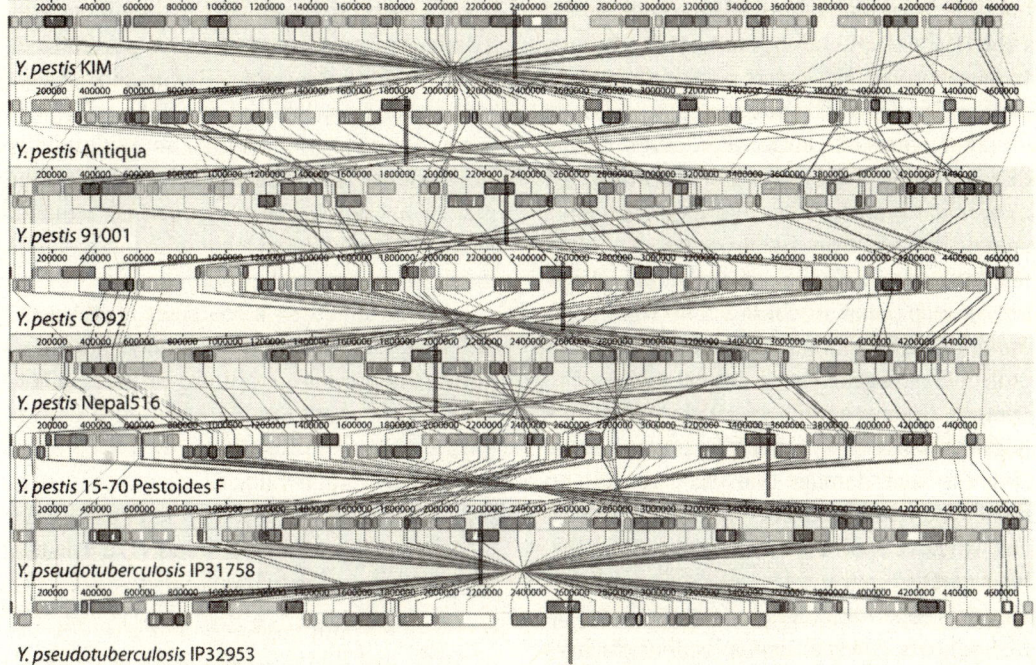

**Figure 4.25**  Genome alignment of eight *Yersinia* isolates

blue light have been found by Somers to govern the daily responses of weed *Arabidopsis thaliana*. Contrary to prior assumption that DNA contains but one copy of most genes, of the 25,500 genes of *A. thaliana*, two thirds are sequence-sortable, at one or more, into 11,600 families. Each *Arabidopsis* plant has 2 copies of each of it alleles. Hall and Rosbash have shown that fruit flies with mutations in a cryptochrome gene have altered circadian rhythms. Sancar has found that in mice lacking 1 of 2 mouse cryptochromes, their biological clocks run on a cycle 1 hour longer than those of normal mice. *A. thaliana* genome research is on patterning, the "dwarf" gene, hormone biosynthesis, ethylene mutations, and seed dispersal.

Normal strains of the nematode *C. elegans* pass through 4 larval stages to mature into fertile adults. Victor Ambros in 1993 came the surprise discovery that the gene that turns off other genes play a role in worm's development, encodes not a protein but a tiny molecule of RNA. In 1998, double-stranded RNA (dsRNA) was identified by Andrew Fire and Craig C. Mello as a key component of **gene-silencing**. Research has been rapid into understanding biological processes called RNA interference that involve either blocking the translation of specific messenger RNAs (post-transcriptional gene silencing; PTGS) or preventing transcription of specific regions of DNA into RNA (transcriptional gene silencing; TGS). One RNAi pathway control distinct aspects of development, tissue-specific differentiation, and maintenance of functions in both animals and plants. To explain how the number of protein-coding genes in an organism doesn't seem to reflect its complexity, Mattick with Gagen in 2001 proposed that small RNA genes may account for the diversity and complexity of eukaryotes.

Rapidly evolving non-coding regions could be what makes humans. Greatly different from our nearest primate relatives is the gene, called HAR1F, in a non-coding segment of genome. In the phenome, HAR1F is produced by brain cells called Cajal-Retzius cells and regulate the arrangement of 6 layers of the cortex during development. Free radicals are atoms or molecular units such as hydroxyl that have an unpaired electron. Electrons seek to travel in pairs. Free radicals are produced during aerobic respiration in cells. Each find numerous opportunities to cut an electron from another molecule. Creating a free radical of it and so damage chain reaction. In 1954, D. Harman began to consider that taken for granted aging is due to such damage that the body fails to repair. Evolutionary theory would predict that cell will have evolved to protect them. In 1967, biochemists discovered in cells the ubiquitous enzyme, super oxide dismutase that sops up free-radicals. Allergies and infectious diseases that, at their root, have vulnerability based in the proteins and RNA products of genes.

## 4.13   Evolutionary Rate

The common method to measure evolutionary changes in units is called Darwin. '**Darwin**' is a unit to measure evolutionary rates. One Darwin is a change by a factor of 2.718 (the base of the natural logarithm) in 1 million years. House sparrow introduced to North America about 100 years ago has spread across North America and diverged into local populations. Tibiotarsus bone has changed by about 5% in 100 years. The changes measured over short periods are greater than those in fossil record. In fossil record, rates are measured using pairs thought to be ancestor and descendant. If the change is extensive, one may not recognize the pair as ancestor and descendant. Thus the largest rates may not get measured. In selection studies, very small changes are ignored. The fossil record usually showed stasis for long periods, and

transitions between species are difficult to find. This is explained by gaps in fossil record which if filled, would show gradual evolution. Change occurs during speciation. Rate of evolution is a measurement of change in an evolutionary lineage over time. Method for measuring the rate of evolution can be gained by work of Mac Fadden on horse teeth which are classic materials in study of evolution.

Let assume that a character is measured at 2 times, $t_1$ and $t_2$. The $t_1$ and $t_2$ are times before the present in millions of years. Then, time interval between the 2 samples: Dt = $t_1$ – $t_2$, which is 1 million years, if $t_1$ = 15.2 and $t_2$ =14.2. Average value of character is $x_1$ in the earlier sample and $x_2$ in the later sample; then natural logarithms of $x_1$ and $x_2$ (the natural logarithm is the log to base e where e = 2.718).

Evolutionary rate then is $r$
$$r = (\log X_2 - \log X_1) / dt$$

Formula above for $r$ gives the rate in Darwin provided time interval in millions of years. An inverse relation between the rate of evolution and the time interval over which it was measured exist. The observed examples of rapid evolution tend to be shorter than the examples of slower evolution. Measurement of evolutionary rates in Darwin is appropriate for metrical changes, like a character evolving to be longer or shorter. For larger changes, like from leg to wing, this method is not useful. Evolutionary rates can be studied quantitatively in characters whose evolutionary changes are not simply metrical. Characters can be divided into discrete states; the states assigned arbitrary scores; and changes in those scores measured through time. An example is lungfish.

Westoll's classic study of lungfish that changed little from their fossil ancestors in distant past is mentionable. Westoll distinguished 21 characters of fossil Dipnoi. For each of the 21 characters, he identified several character states. The 21 characters were given scores between 0 and 4 with the highest score being the most primitive condition and the lowest the most derived and advanced state. Westoll made a total score, comprising of its total for all 21 characters for each fossil. The lungfish, with 21 characters would have a total score of zero.

**Carbon dating:** This method devised by W.F. Libby recognizes that living organism has $C_{14}$, the radioactive isotope of carbon in body, which disintegrates after their death into $C_{12}$ or $N_{14}$ by release of an electron, as it is no longer replenished after death of plant or animal. This radioactive carbon is absorbed by plants in the form of $CO_2$, and animals get it into their bodies from plants. As after death, animals and plants do not replenish $C_{14}$ through food, it only decays at constant rate. Rate of decay is calculated by counting the number of beta particles emitted by $C_{14}$ in unit time in Scintillation counter, or Geiger-Muller counter (that use $CO_2$ gas) and then incorporates the data into a computer.

**Fluorine dating:** When there is water seepage through bones, **fluorine** in water combines with calcium in bones to form fluorapatite, whose proportion in bones can be detected to find out the age of fossil.

**Nitrogen dating:** This method works on collagen degradation in bones and accumulation of other materials such as fluorine. How much nitrogen is retained in bone gives an idea of age of fossil.

**Uranium dating:** Bones buried in soil absorb **uranium** at a constant rate and longer they lay in those strata more uranium they will absorb. As uranium is radioactive, its presence in bones can be detected using scintillation counter.

**Fission-track dating:** Trace amount of Uranium238 impurities are found in nature and spontaneous fission of this radioactive element produces small but permanent damage

trails in insulating solids through long periods. Counting numbers of such trails gives the age of fossils.

**Potassium-Argon dating:** Potassium is abundant in nature and its isotope $K_{39}$ comprises over 93% of total, while $K_{40}$ is only 0.0118% and $K_{41}$ 6.8%. $K_{40}$ decays into its daughter isotopes, viz. $Ca_{40}$ by beta decay and $A_{40}$ by K capture, in a half life of 1.3 billion years. Ratio of daughter isotopes can be detected in a mass spectrophotometer. This method detect the age of old fossils but theoretically potassium rich samples of 5,000 year old.

**Amino acid racemization dating:** All animals have L-amino acids in their proteins which after death transform by racemization into D-amino acids, whose proportion increases with time. Ratio of 2 types of amino acids will give the age of skeletal material in fossil.

## 4.14 Shoot Vs. Ladder

Evolution is generally regarded as a progression from simple to complex, or ladder like or lineal and it would be possible to reconstruct a direct line of ancestors. The evolution of life is more similar to a branching tree. Each branch of a tree is a direct lineage. Multiple branches extending from a common point can joint by a set of characters present in the common ancestor. Lineages can acquire characters that are not shared in a common ancestor. Not all lineages persist in the future due to the extinction. Tree-thinking shift the focus to **synamorphies** that link collateral ancestors. The grandmother is a lineal ancestor while the great-aunt a collateral ancestor, but their lives and times were probably not that different (Padian and Angielczyk, 2007), which means that information about one provides information about the other(Mead, 2009)

## 4.15 Extinction

Extinction is the death, or disappearance of every member of a species. This is the eventual fate of all species. According to Raup, more than 99% of all species that have ever existed have gone extinct. Less than 1% of all species that have ever existed are still alive now. Many existing species are close to extinction .Thus the notion that evolution by natural selection ensures the survival of the species is wrong. Natural selection ultimately causes the extinction of about all species in the long run. An **"average" species** persists for 8 million years before becoming extinct. The elephant bird of Madagascar is an extinct species which was ten feet tall, weighed about 1,100 pounds, and laid 3-pound eggs. It survived for almost 60 million years before going extinct. Its extinction by human hunting is not unusual. Moa of New Zealand went extinct in 1835 which was 13 feet tall and flightless. It and about 15 other species of large flightless birds on islands of New Zealand were hunted to extinction .The common dodo; a flightless bird about the size of turkey of Mauritius went extinct in 1680. It did not taste very good and was hunted to extinction for sport. The passenger pigeon of North America was once the most abundant bird, numbering about a hundred billion went extinct in 1914 by hunters in less than 40 years. **Laughing owl** of New Zealand went extinct in 1914.Several species of macaws and parrots , large brightly colored birds of the tropics, went extinct during the 19th century, along with species of rails and the great auk , the largest flightless bird in northern hemisphere, which went extinct in 1844. Over 50 species of Hawaiian honey creepers, finches, flightless rails, and geese is now extinct. Ivory-billed woodpecker was last seen in 1951. The Round Island boa last observed in 1975 is now extinct. Many species of suckers were driven to extinction during 19th and 20th centuries. In Lake Victoria alone, over 200 species of **cichlid fish** became extinct since 1960, almost all as prey of an introduced predator, the Nile perch. Many orchid species have gone extinct, due to habitat destruction or from zealous collection by orchid fanciers.

## 4.16 Causes of Extinction

The fossil record shows many examples of extinction, but it rarely shows the cause of such extinction. An exception is the Cretaceous-Tertiary (K-T) mass extinction, which is linked with an asteroid collision, likely at the northern coast of Yucatan peninsula. The first evidence for the Cretaceous-Tertiary asteroid collision was an enrichment of iridium in rocks dating from **K-T boundary** which was puzzling to Walter Alvarez, who was studying iridium levels in rocks to track the rate of infalling space dust. Iridium is rare on earth's surface, but common in meteorites and asteroids. Louis Alvarez, Walter's father suggested that anomalous iridium enrichment in K-T boundary rocks as evidence of massive asteroid collision. The **K-T asteroid** collision is dated at 65 mya, and is linked to a drastic change in fossil record of dinosaurs. During Mesozoic, there was a steady extinction of dinosaurs, which began to accelerate during the Cretaceous. Dinosaur fossils completely disappear following K-T asteroid collision event. Thus the collision and its ecological after-effects might cause final mass extinction of dinosaurs. Fossil records shows background extinction, in which species disappear in a random fashion, one at a time. Species are the temporary, may last long time, or they may appear and disappear in short time, but finally all species go extinct. Some species last much longer, while others disappear quickly. Species with very small populations are easy to kill. Examples include whooping cranes, California condors, and ivory-billed woodpeckers. If a species is reduced to only a small number of individuals, even random chance events can drive it to extinction.

Widespread species are hard to kill but can be rapidly wiped out, if something prevents their reproduction. Passenger pigeon was reduced from several billion to zero in 40 years, due to market hunting of squabs. Stresses caused widespread species extinction not generally experienced by the species. Sometimes widespread success can be disastrous. When a species outgrows its environment, or so widespread that disease can easily spread from individual to individual causes it to extinct. American chestnut trees were once abundant and widespread in North America. During the early 20th century, they were faced to an introduced disease and are now at the verge of extinction throughout most of their former range. The same rapid change towards extinction has occurred in American elm trees, American beech trees, and other tree species. Simultaneous extinction of many species due to stresses cut across ecological lines.

## 4.17 Ordinary Extinction

Reasons for extinction are numerous. A species can be excluded by a closely related species by competition. In a habitat a species can disappear and/or the organisms that the species exploits could come up with an unbeatable defense. The majority believes that if the environment stays constant, a well adapted species could continue to survive indefinitely. **Mass extinctions** shape the overall pattern of macroevolution. History of life includes many episodes of mass extinction in which many groups of organisms were wiped off. Mass extinctions are followed by periods of radiation where new species evolve to fill the empty niches left behind. Surviving a mass extinction is mostly a function of luck.

Contingency plays a large role in macroevolution. The largest mass extinction happened at the end of Permian, about 250 mya. This coincides with formation of Pangaea II, when all the continents were brought together by plate tectonics. A worldwide drop in sea level occurred at this time. The well-known extinction at boundary between the K-T which is dated at around 65 mya eradicated the dinosaurs. The **K/T event** was perhaps occurred due to environmental hazards, due to impact of an asteroid on earth. After this extinction the mammals radiated and coexisted with the dinosaurs for long time but were confined mostly to nocturnal insectivore niches. With eradication of dinosaurs, mammals radiated to fill the vacant niches.

From time to time, our planet has experienced holocausts of gigantic magnitude which wiped out majority of life forms. The extinction events often engulfed the evolutionary lines defying the concept of natural selection and survival of the fittest. Study of 600 million years' history of animal reveals 5 major extinctions with periodicity of about 200 million years. Minor extinctions have occurred between showing periodicity of 50 million years. Marine animals exhibit periodicity of 26 million years. Extinction waves were so intense that out of 1800 families of marine animals that had evolved during 280 million years since the Permian, 970 have become extinct by now. Not much is known about the Cambrian extinction 600 million years ago but the Silurian-Devonian extinction (200 million years later) wiped out the abundant animals of that time-the Trilobites. Earth was quiet for the subsequent 150 million years when the Permian extinction wave removed 71 % of the tetrapod families that lived at that time. This followed Triassic extinction 50 million years later which killed 80% of tetropod families. End-Cretaceous extinction 70 mya is the best known of all which caused extinction of 61% of tetra pods and removed the dinosaurs. Another significant extinction occurred in the late Pleistocene when over 76% of large terrestrial animals vanished from, America, including horse. In Australia 86% of land animals became extinct. Tropical Africa suffered only 19% loss. The animal succumbed to this holocaust was woolly mammoth. The frozen corpses of these giants were discovered in Siberia. The animals were abundant before their sudden death and tusks of about 20,000 animals were taken out in 20 years. It is a strange coincidence that following every major extinction, there was sudden emergence of larger groups of organisms. After the Cretaceous extinction, suddenly the mammals, flowering plants, fruit trees and grasses evolved. Permian extinction led to the emergence of reptiles and dinosaurs, which ruled the earth for next 200 million years. **Triassic extinction** killed 80% of tetrapod families and caused the origin of birds (Figure 4.26, 4.27).

## 4.18 Mass Extinctions and Speciation

Massive/rapid environmental changes cause mass extinction. This wipes out many species, opens many **ecological niches**, which become occupied by new or existing species. Most mass extinction produces more species than they extinguish. Life on earth has witnessed mass extinctions since 542 mya. It sometimes accelerated the evolution of life. When dominance of particular ecological niches passes from one group of organisms to other, it is rarely due to the new dominant group's superiority to old and generally because extinction eliminates old dominant group and creates way for new one. Rate of extinction slowed down, with gaps between mass extinctions becoming longer and average and decreased background rates of extinction. It is not certain whether the actual rate of extinction has altered, since both observations

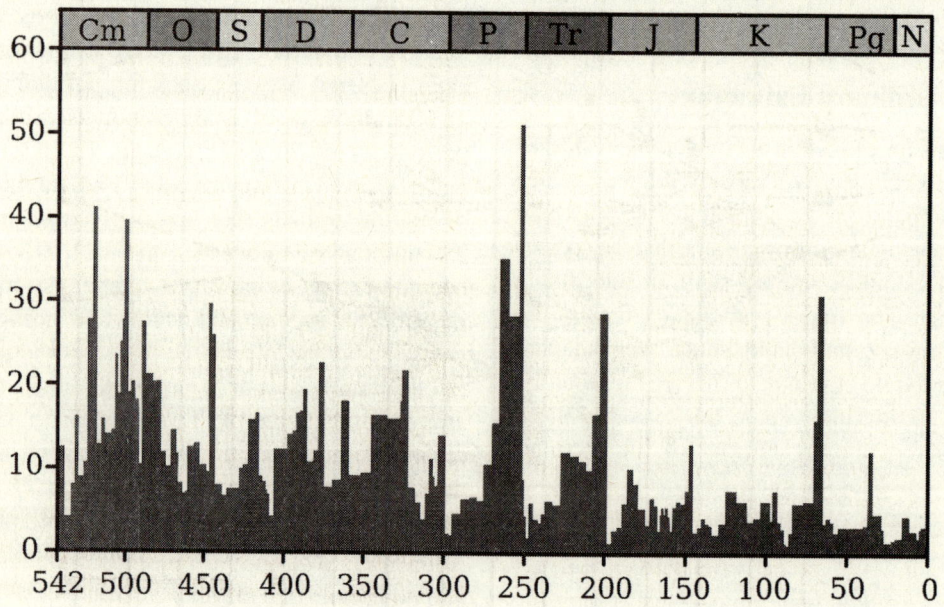

**Figure 4.26** Extinction intensity during various ages of earth

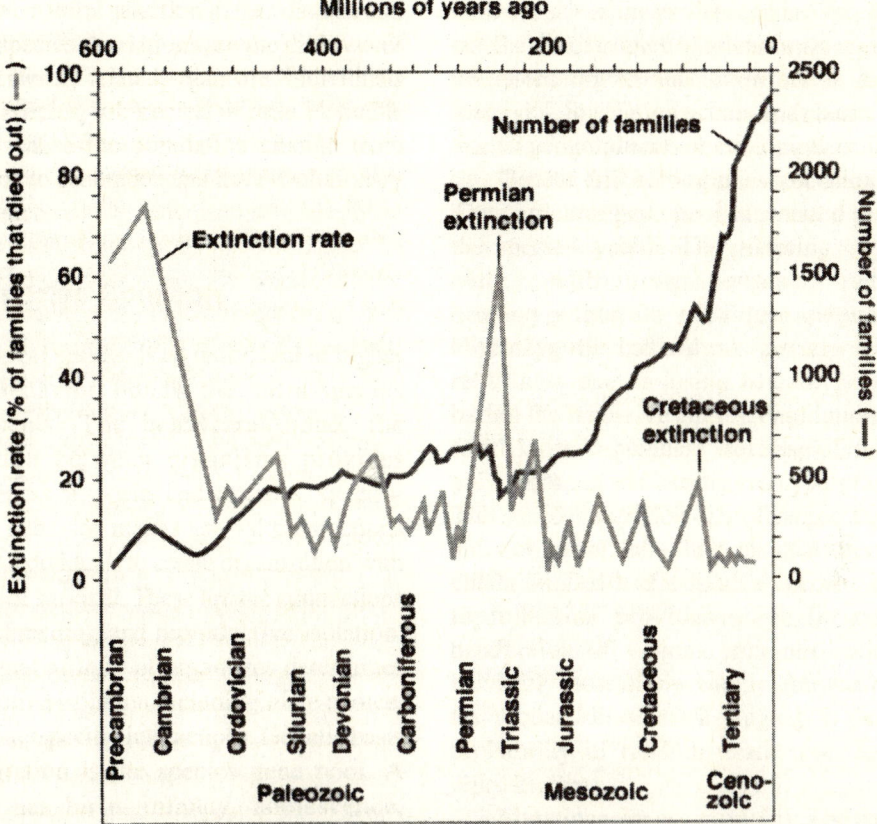

**Figure 4.27**  Understanding extinction

could be explained in several ways. Apparent extinction intensity, i.e. the rate of genera going extinct at any given time has been reconstructed from fossil record. The oceans may have become more hospitable to life over the last 500 million years and less vulnerable to mass extinctions. Dissolved oxygen became more widespread and penetrated to greater depths. The development of life on land reduced the run-off nutrients and risk of eutrophication and anoxic events. Marine ecosystems became more diversified so that food chains were less likely to be disrupted. **Complete fossils** are very rare. Most extinct organisms are found as partial fossils, and complete fossils are very rare in the oldest rocks. Many superfluous

genera are found as fragments that are not found again, and these superfluous genera become extinct quickly. Biodiversity in fossil record that is, those whose first occurrence predates and whose last occurrence postdates that time" shows a different trend: a swift rise from 542 to 400 mya, a slight decline from 400 to 200 mya, in which devastating Permian-Triassic extinction event was an important factor, and a swift rise from 200 mya to the present.

## 4.19 Big Five Mass Extinctions

In early modern Europe, systematic study of fossils emerged as integral part of changes in natural philosophy. At the end of 18th century Cuvier's work established comparative anatomy and that some fossil animals resembled no living ones, showed that animal could become extinct, leading to emergence of paleontology. Expanding knowledge of fossil record also played an increasing role in development of geology, particularly stratigraphy. The first half of 19th century saw geological and palaeontological activity which become increasingly well organized with growth of geologic societies and museums and an increasing number of professional geologists and fossil specialists. Interest increased was not purely scientific, as geology and palaeontology helped to find and exploit natural resources like coal. This contributed to rapid increase in knowledge about history of life on earth and to progress in definition of **geologic time scale**, largely based on fossil evidence. Paleontology refers to study of ancient living organisms through fossils which exhibit some kind of successive order to development of life. These encouraged early evolutionary theories on transmutation of species. After publication of *Origin of Species*, much focus of paleontology shifted to evolutionary paths, including human evolution. Last half of 19th century saw a tremendous expansion in palaeontological activity, especially in North America. Trend continued in 20th century with additional regions of the earth being opened to systematic fossil collection. Fossils found in China near end of 20th century are important as they have provided new information about earliest evolution of animals, early fish, dinosaurs and evolution of birds. Last few decades of 20th century saw a renewed interest in mass extinctions and their role in evolution of life on earth. There was also a renewed interest in the Cambrian explosion that apparently saw development of body plans of most animal phyla. Fossils of Ediacaran biota and

developments in paleobiology extended knowledge about the history of life back far before the Cambrian. Increasing awareness of Mendel's pioneering work in genetics led first to development of population genetics and then in mid-20th century to modern evolutionary synthesis, which explains evolution as outcome of mutations and horizontal gene transfer, which causes genetic variation. In the next few years, the role of DNA in inheritance was discovered, leading to Central Dogma of molecular biology. In 1960s investigation of evolutionary family trees by biochemical techniques, produced an impact, when it was proposed that human lineage had diverged from apes much recently than was generally thought at the time. Although this study compared proteins from apes and humans, molecular phylogenetics research is now based on comparisons of RNA and DNA.

Raup discusses both background and mass extinctions. The latter are the result of historical processes than former. **Background extinctions** are the result of random numerical fluctuations in population sizes that are limited in range and overall numbers. Such species are easy to kill. Mass extinctions are more indiscriminate. Fossil records show at least 5 major mass extinctions during past billion years, with many minor mass extinction. Mass extinctions perhaps set the stage for mass speciation. Permian/Triassic mass extinction killed off over 95% of all living species at that time. Rocks deposited just after the end of Permian are almost without fossils. About 15 million years later, new fossils begin to appear and proliferate. Cretaceous/Tertiary mass extinction caused due to an asteroid collision, wiped out all vertebrates larger than a turkey. This allowed mammals to proliferate into more species. All mass extinctions are followed by adaptive radiations of new types of organisms, so that species diversity of earth has increased. Such mass extinctions have wiped out almost all previously existing species. David Raup concludes that often, extinction is bad luck. The species become extinct as the environment changes faster than it can adapt to the change (Figures 4.28-4.39)

**The Sixth Extinction:** Bio-diversity is under threat. The earth has faced 5 major extinctions in the past. The first extinction occurred 450 million years ago, after the origin of the first land-based plants and 100 million years after the Cambrian Explosion of animal life. The 2nd extinction happened 350 million years ago (Doubleday, 1995). The earth faced 2 mass extinctions during the Triassic period. The 5th mass extinction occurred 65 million years ago, at the end of the Cretaceous period .The 6th extinction is going on due to our activities. Every year, up to 100,000 species become extinct. Fifty percent of the earth's species is expected to be lost in the next 100 years. We use about half the available energy and this figure will increase as our population will attain 10 billion in the next 50 years. Dramatic mass extinction threatens the complex fabric of life including our species. Arriving late on the evolutionary scene, our insights have not prevented us from exploiting earth's resources commercially.

We endanger the species in 3 principal ways; first through direct exploitation; second by introducing the alien species. In Africa's Lake Victoria, more than 200 fish species have lost within the past decade due to over fishing and pollution. The major culprit was the Nile perch, which was introduced to the lake. The 3rd is the destruction and fragmentation of habitat, especially the cutting of tropical rainforests. The forests cover 7% of the world's land surface and provide shelter to half of the world's species. The human populations encroaches wild habitats, through expanding agricultural land, or through building towns and cities. Habitat shrinkage limits capacity

to sustain biodiversity. The species extinction poses a serious threat to civilization. Human population explosion has accelerated the environmental change. **Current extinction** rates are equivalent to mass extinction. About 2% of standing forest being cut each year. About 80,000 square miles of forest lost each year. Tropical forests will be reduced to 10 % of their original cover after the turn of the

**Figure 4.28**   Edward's dodo

**Figure 4.29**   Armadillo-lives presently in Americas

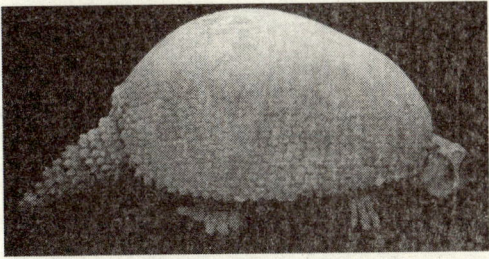

**Figure 4.30**   Glyptodon-lived 2,500,000 to 10,000 years ago

**Figure 4.31**   Elephant Bird

**Figure 4.33**   Extinct Passenger pigeon

**Figure 4.34**   Dodo

**Figure 4.32**   Mao

**Figure 4.35**   Extinet Golden Toad

**Figure 4.36**  Stegosaurus

**Figure 4.37**  Tyrannosaurus rea

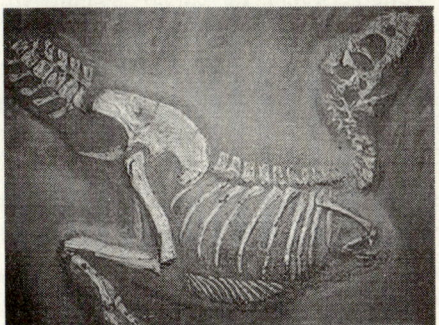

**Figure 4.38**  Tarbosaurus

**Figure 4.39**  Orange Bird

century at such rate of destruction. Forest is fragmented into small ecologically fragile islands varying in size from 2.5 acres to 25,000 acres. Species become extinct more rapidly in small patches than in larger ones. **Vulnerable species** require a large range. Extinction of them often causes other species to become extinct. Three species of frog vanished from one 250-acre plot, as the habitat was too small to support peccaries, whose wallowing in mud created ponds for (Doubleday, 1995) the frogs. Small populations can experience sudden bouts of disease or perturbations like storms. The large populations can tolerate such events. Even large forest patch is less sturdy than might be imagined for the so-called edge effect. Habitats deep in the forest are protected from external influences, but those at the boundary between forest and grassland are exposed to wind, varying microclimates over short distances, incursion by non-forest animals and human (Doubleday, 1995) hunters. Animals and plants are vulnerable to extinction for as much as a half a mile into the forest.

Edge effect is important even for large tracts of forest. Logging has been leaving a greater proportion of Amazonian tropical forests vulnerable to edge effects. Habitat loss is not confined to tropical forests. During this century, half the US natural ecosystems has been degraded. Entire communities are now on the brink of extinction. Human worldwide is encroaching the wild habitat, for constructing villages, towns, and cities, and the infrastructure, and for producing food, and livestock. To offer a standard of living for 10 billion people, the global economic activity will have to rise at least tenfold. Now we consume 40 % of **net primary productivity** (NPP) on land; the total energy trapped in photosynthesis worldwide, minus that required by the plants themselves (Doubleday, 1995) for their survival.

Faunas and floras of islands show a relation between the area of the islands and the number of species inhabiting them. The number of species about doubles with every 10 fold increase in area. The qualitative relationship between area (Doubleday, 1995) and number of species shows that the bigger the area supports more species. A thousand acres of flat terrain are likely to support fewer species than a thousand acres of varied topography as many **microhabitats** are present in the latter than the former. A thousand tropical acres will support more species than a similar area at high latitudes. If a significant proportion of species is restricted to small localities, then species loss will be higher than 50%. During periods of normal, or background, extinction, species loss occurs at an average of one every four years (http://www.mysterium.com/). Extinction at the rate of 30000 a year is easily comparable with the 5 biological crises of geological history, except that this is not being caused by global climate change, rise in sea level, or asteroid impact. It is caused by us. If background extinction rates applied to birds, a bird species could lost no more frequently than once every century. Stuart Pimm reports, in the Pacific alone, we see about one extinction per year." The Hawaiian Islands, a tropical paradise (Doubleday, 1995) to tourists, bear the scars of catastrophic extinctions. Since first human contact, as many as half the islands' bird species have gone extinct and the loss continues today. Of about 135 bird species there, only 11 thrive in numbers that ensure their survival well into the next century. Because of the rarity of a dozen there is little chance of saving them. A further dozen are classified as endangered. Almost a decade ago, 90 species of plants became extinct, when the forested ridge on which they grew was cleared for agricultural land. The ridge, in the western Andean foothills of Ecuador, (Doubleday, 1995) the Centinela has become associated with catastrophic **anthropogenic extinction**. Centinela an ecological island had developed a unique flora. Within 8 years the ridge had been transformed into farmland, and its endemic species became extinct. Each time an ecological island is cleared, species became extinct. Countless species become extinct before ecologists even know of their existence. Many species have very limited ranges, particularly in the tropics. Habitat destruction often results in the loss of species.

The list of such loss include half the freshwater fish of peninsular Malaysia, 10 bird species of Cebu in the Philippines, half the 41 tree snails in Oahu, 44 of the 68 shallow-water mussels of the Tennessee River shoals. Species introductions and physical habitat alteration are the serious factors for species loss. Current extinction is running at a rate some thousand to ten-thousand higher than background extinction. The bird's range covered much of the eastern seaboard of the United States. Hunting and habitat destruction (Doubleday, 1995) reduced the species' number to 50 individuals in 1908 and a reserve was established to save it from extinction. Over the next 2 decades the population's numbers began to rise, but finally the species became extinct, through calamities, including fire and pestilence. A small population is vulnerable to normal fluctuations due to disease and disasters. Becoming dominant as no other species in the earth, *Homo sapiens sapiens* is causing a major biological crisis.

## 4.20 Remarks

**Ice Ages:** General belief says that species become extinct as they are eliminated by natural selection, when environment changes and new ones emerge to take their place. These extinctions were apparently not the result of atmospheric changes or ice ages. Fossils point to sudden mass destructions. Woolly mammoth have died out by freezing in sudden cold wave. On islands, even small mammals became extinct. In oceans the destruction was of equal magnitude. Permian extinctions could be due to ice age but late Triassic and end-Cretaceous extinctions occurred without global refrigeration.

**Impact Hypothesis:** Alvarez in his Impact Hypothesis opined that iridium concentration and shock metamorphosed quartz on earth were created as impact of a giant asteroid or comet on the earth. In 1990, a crater 180 km in diameter was discovered near a town called Chicxulub in Yucatan peninsula in Mexico. It is believed to cause by the hit of a meteorite 10 km in diameter about 66 mya and caused extinction of dinosaurs. Our solar system, moving at a speed of 150 miles/second takes 240million years to complete one orbit around the nucleus of the Milky Way galaxy. It passes through spiral arms or density waves every 50 million years where interstellar planetismals are captured by the sun's gravity, causing cometary or asteroid bombardment on the earth. Sepkoski and Raup proposed that solar system also oscillates vertically every 67 million years, increasing the X-ray and UV radiation intensity on the earth, thereby causing mass extinctions and mega mutations. Through satellite images, a 480 km wide crater is discovered beneath the ice sheets in Antarctica, which is 250 million years old and perhaps was the cause of Permian mass extinction.

# SUMMARY

1. Darwinism mentions about survival of the fittest but failed to explain about arrival of the fittest. In fact character needs a long time to develop before it become useful or fittest. Bird's wings must have taken millions of years to develop and during this period of growth they were not useful organs and should have been eliminated by natural selection.

2. That all variations are heritable is not true. Only genetic variations are the heritable and not the somatic ones. Darwin tried to explain inheritance by theory of pangenesis.

3. Many useless and non-adaptive characters persist in many animals and are not eliminated. Small tails in giraffe or pig, ear muscles and appendix in man have no selective value.

4. Darwinism fails to explain over-specialization, which led to extinction of species. Dinosaurs became extinct due to overspecialization in body size. Saber tooth tiger (*Smilodon*) had oversized canines causing its extinction in Africa. Similarly Iris deer (*Megaloceros*) grew such huge antlers that they interfered in its movement through forest and ensured its extinction.

5. Struggle for existence is not physical one. Mostly it is passive. Many species of insects and other animals camouflaging against predators do not fight physically.

6. Most mutations that produce new characters are harmful, cause diseases and not useful in natural selection. Darwin explained sudden appearance of characters due to Spots or Sports.

7. Natural selection does not operate on one character.

8. Darwin believed in blending inheritance. Mendel's laws of heredity state that characters segregate in the 2nd generation.

9. No evidence contradicts the theory of evolution. Evolution is genetic change in population. Speed and direction of change varies in various species at different times. Continuous evolution through generations results in new varieties and species.

10. All life forms evolve from earlier species, and existing species continue to evolve today. Changing environment over a long period causes some organisms of a species to develop different forms and habits through accumulation of changes. Evolution is dominant mechanism behind the diversity of today's species. Large phenotypic changes due to genetic changes are the evolution.

11. Populations adapt to their surroundings by evolution .Any organism's success is related to behavior of its surrounding organisms. Each species modifies its own environment.

12. Comparative anatomy is one line of evidence which reveals that structure of organisms follows basic expected patterns if organisms are related to each other. The overall anatomical structure of whale fins is similar to structure of other mammal forelimbs than to structure of penguin wings or fish fins. Structure of penguin wings is similar to structure of bird wings than to structure of fish fins or whale fins.

13. Gene structure and gene function show relationships in ways similar to anatomical structure and function. Biogenetic law states that organism developed through their ancestral forms.

If human were descended from fish, and then reptiles, and primates, than human embryo would go through fish stage, reptile stage, and finally its human form.

14. Vestigial organs are less developed with reduced utility compared to equivalent organs in related organisms. All life is related, and organisms should only be present where their ancestors were present. Fossil record confirms that types of organisms are geographically constrained and related to locations of their ancestors. Any given life forms may survive and thrive in a variety of places, but life forms present where they are related to other life forms. Cuvier the 'Father of paleontology' studied fossils to develop phylogenies. Remains of animals and plants found in sedimentary rock give the idea of past changes through vast periods.

15. Some extinct species were transitional between major groups of organisms. Many transitional forms show development of amphibians from fish, of birds from dinosaurs, of mammals from early tetrapods, of humans from a common ape ancestor. Majority of life forms in fossil record are probably not ancestors of modern life forms. They are organisms in lineages that went extinct, but they are nevertheless members of populations and can tell us about types and combinations of features that existed and were likely to exist among ancestors of modern organisms. Fossil record is sparse by nature as an extremely small number of things that ever lived became fossilized. So we do not have a complete fossil record. Discoveries have shed light on evolution of land dwelling organisms from fish.

16. Index fossils in undisturbed sedimentary rocks and in short geological time period lie in recognizable strata of older rocks below and formed layers above. Based on presence of such fossils in rocks, age of other fossils in same rock can be determined without dating, because such fossils are an index to particular geological period. Ammonites are good index fossils as different species represent specific geological periods in rocks.

17. Evolution is the continual, random, change in the genetic code. Evolution creates different life forms naturally. Organisms are well-suited to their environment as they go through natural selection.

18. Weakest point in Darwin's theory of natural selection is that it lacks understanding about heredity. Darwin's theory has been enriched by population genetics and establishes the modern synthesis (Neo-Darwinism) which is combination of Darwinian selection and genetic theory.

19. Darwin suggested that modern species had descended from other similar species which are now extinct. When the environment changes, new traits get the chance of being selected. Vast changes in population results from the cumulative natural selection. Natural selection lacks any foresight. It allows organisms to adapt to the existing environment. Natural selection favoured dark moths, which camouflage on bare tree trunks. The moth has only one generation per year, and in less than 50 generations, natural selection replaced gray population with black population.

20. Natural selection forces the species to keep improving generation after generation so that they remain in the fittest state to survive in a given environment. Random genetic changes provide raw material for variations and gives natural selection a chance to operate. Natural selection produces systematic heritable changes in population from generation to generation

and continues evolution. Thus it is directional phenomenon, producing changes in definite direction, producing new species. Such selection is associated with change in environment, due to which average character becomes non-adaptive. It favours individuals with non-average or extreme characters, which may be useful in changed environment. This selection is directional and progressive, which produces new types from original population with ability to survive the change in environment.

21. By keeping the gene from disappearing in heterozygous state, natural selection provides the species with reserves on which it may draw when condition changes. Species with lots of heterozygote and plenty of variations are the fittest ones in changing environments. The "survival of the fittest" is used synonymously with natural selection.

22. Survival is one component of selection and perhaps the less important ones in many populations. In polygynous species, several males survive to reproductive age, but only few ever mate. Natural selection is the engine of organic change, driving different variants of the same species to diverge until they become new species. Variations are differences among closely related individuals.

23. Heritable variations are caused by aneuploidy, polyploidy, mutations, chromosomal aberrations, and hybridization. Somatic variations are short-lived without influence on natural selection. No 2 individuals are alike within a species, race or cohort. Difference in reproductive capability is natural selection, the only mechanism of adaptive evolution. It is differential reproductive success of pre- existing classes of genetic variants in gene pool.

24. Natural selection removes unfit variants as occurs via mutation. Selection removes deleterious alleles and depletes genetic variation. When heterozygotes are fit than homozygote, selection maintains genetic variation called balancing selection as found in sickle-cell alleles in malaria patient. Natural selection occurs as a result of interactions of organisms with both biotic and abiotic factors of environments.

25. Major agent of natural selection in biotic environment is competition with other members of same species. Sexual selection can be separated from natural selection, as the 2 processes have separate effects. In many bird species, the grotesquely exaggerated character of males lower the male's chances of survival. If any species evolved a trait exclusively for benefit of another species, it could not have done so by natural selection.

26. Adaptation is based on the concept that population change over time due to natural selection. Increased survivorship and/or increased reproductive success cause adaptive evolution when group of individuals in population gain advantage because of sharing the special traits

27. Nearly-neutral theory of molecular evolution states that many neutral mutations will have small effect. The rest of mutations are deleterious, with only small mutation being beneficial.

28. Point mutation and genetic recombination cause variation. Evolution which operates in small steps is resulted from natural selection that acts on genetic variation. Natural selection and reproductive isolation guide population into adaptive channels. Mutation, genetic

recombination, structural changes in chromosomes and natural selection interact to produce a progressive change in population.

29. G. C. Williams stated that evolution is not about survival of species, but how natural selection operates on individuals and their genes.

30. Cells with different functions produce particular proteins by turning on, or turning off appropriate genes along DNA molecule. The complex hormonal, developmental and tissue-specific regulation influence gene expression.

31. Genome sequencing is fundamental to an understanding of how species cope. By comparing genomes, we reconstruct the evolution of one species from another and say what common ancestor species could have had. Such application of studying genes was proposed by Paul Linus. Genes controlling development is expected to have been conserved through evolution. The number of protein-coding genes in an organism doesn't seem to reflect its complexity. Mattick with Gagen in 2001 proposed that small RNA genes may account for the diversity and complexity of eukaryotes.

32. Many existing species are close to extinction. Thus the notion that evolution by natural selection ensures the survival of the species is wrong. Natural selection ultimately causes the extinction of about all species in the long run. An "average" species persists for 8 million years before becoming extinct.

33. Evolutionary rates can be studied quantitatively in characters whose evolutionary changes are not simply metrical.

34. Dinosaur fossils completely disappear following the Cretaceous-Tertiary asteroid collision event. Thus the collision and its ecological after-effects might cause final mass extinction of dinosaurs.

35. Fossil records shows background extinction, in which species disappear in a random fashion, one at a time. Species are the temporary, may last long time, or they may appear and disappear in relatively short time, but finally all species go extinct. Most mass extinction produces more species than they extinguish.

36. Life on earth has witnessed mass extinctions at least 5 times since 542 mya. It sometimes accelerated the evolution of life.

## REVIEW QUESTIONS

### Short Answered Questions

1. Define homologous structure.
2. Define analogous structure.
3. Define Hackle's theory of ontogeny.
4. What do you understand by vestigial organ?

5. Define continental drift.

6. Define biogeography.

7. Define Lamarckism.

8. Define Modern Synthesis.

9. Define extinction.

10. What do you understand by mass extinction?

## Long answered questions

1. Describe the modern synthetic theory of evolution.

2. Describe various causes of extinctions.

3. Write a note on the importance of vestigial organs in evolution.

4. Discuss the continental drift in relation to evolution.

5. Describe the importance of fossils in evolution.

# REFERENCES

Danchin, E., Charmantier, A., Champagne, F. A., Mesoudi, A., Pujol, B., Blanchet, S. 2011 Beyond DNA: integrating

Dennett, D. C. 1995. Darwin's dangerous idea. London, UK: Allen Lane.

Deutsch, 1997. The fabric of reality ( Penguin, New York) p.179

Futuyma, D. J. 2005. Evolution. Sinauer and Associates,.

Godfrey-Smith, Peter 2009. Darwinian Populations and Natural Selection. New York: Oxford University Press.

Gould, S. J.2002. The structure of evolutionary theory. Cambridge, MA: Belknap Press of Harvard University Press.

Hone D.W.E., Benton, M.J. 2005 The evolution of large size: how does Cope'srule work? Trends Ecol Evol;20:4-6.

Hull, David L., Rodney E. Langman, and Sigrid S. Glenn. 2001. A General Account of Selection: Biology, immunology, and behavior." Behavioral and Brain Sciences 24: 511-528.

Huxley, J.1942. Evolution: the modern synthesis. London, UK: Allen and Unwin.

Jablonka, E., Lamb, M. J.2008. Soft inheritance: challenging the modern synthesis. Genet. Mol. Biol. 31: 389-395.

Jablonka, E., Lamb, M. J.2010. Transgenerational epigenetic inheritance. In Evolution: the extended synthesis (eds Pigliucci M., Müller G. B.), pp. 137-174. Cambridge, MA: MIT Press.

Kerr, Benjamin, and Peter Godfrey-Smith.2002 "Individualist and Multi-level Perspectives on Selection in Structured Populations." Biology and Philosophy 17 : 477-517.

Kuhn, T. S.1962 The structure of scientific revolutions. Chicago, IL: University of Chicago Press.

Lewens, T. 2010 "The Natures of Selection." British Journal for the Philosophy of science 61: 313-333.

Lewontin, R. C.1970. "The Units of Selection." Annual Review of Ecology and Systematics 1 : 1-18.

Lieberman, B.S. 2010. Paleobiogeoghraphy: using fossils to study globalchange, plate tectonics, and evolution. New York: Kluwer Academic Press.

Lieberman, B.S. 2012. The geography of evolution and the evolution of geography. Evo. Edu. Outreach,5: 521-525

Maher, B.2008. Personal genomes: the case of missing heritability. Nature ,456, 18-21.

McShea, D. W.1994. Mechanisms of large-scale evolutionary trends. Evolution;48:1747-63.

McShea, D.W.2005 The evolution of complexity without natural selection, a possible large-scale trend of the fourth kind. Paleobiology 2005; 31 (Suppl.) : 146-56.

Mead, L.S. 2009. Transforming our thinking about transitional forms. Evo. Edu. Outreach, 2: 310-314

Morrone, J.J. 2008. Evolutionary biogeography: an integrative approach with case studies. New York: Columbia University Press.

Myers C., Lieberman, B.S.2011. Sharks that pass in the night: using GIS to investigate competitionin the Cretaceous Western Interior Seaway. Proc. R. Soc. Lond B Biol. Sci., 278(1706): 681-689

Padian, K. , Angielczyk, K. D. 2007. "Transitional "forms versus transitional features. In: Petto, A. J. , Godfrey, L. R. , editors. Scientists confront intelligent design and creationism. New York: Nortonp. 197-230.

Pigliucci, M., Müller, G. B.2010. Evolution: the extended synthesis. Cambridge, MA: MIT Press.

Ross, M.I. , Scotese, C. R. 2000. PaleGIS/Areview 3.5, PALEOMAP project. Arlington: University of Texas.

Sober, E. 1984. The Nature of Selection. Cambridge, MA: The MIT Press

Sober, E.2006 "Models of Cultural Evolution." In Conceptual Issues in Evolutionary Biology, Third Edition, edited by E. Sober, 535-551. Cambridge MA: The MIT Press.

Stigall, A. L. 2010. Speciation decline during the Late Devonian biodiversity crisis related to species invasion. PLoS One, 5(12):e15584

Valentine, J.W., Collins, A.G., Meyer, C.P.1994 Morphological complexity increase in metazoans. Paleobiology 1994;20:131-42.

Walsh, D. M.2004 "Bookkeeping or metaphysics? The units of selection debate." Synthese 138 : 337-361.

Wiley, E.O., Lieberman, B.S. 2011.Phylogenetics: theory and practice of phylogenetic systematics. 2nd ed.New York. : Wiley

Williams, G. C.1966 Adaptation and natural selection: a critique of some current evolutionary thought. Princeton, NJ: Princeton University Press.

# CHAPTER 5

# Species and Speciation

A Phylum is the highest level in Linnaean system for classification. Phyla are group of animals based on general body plan. Despite varied external structure of organisms, they are grouped into phyla based on internal and developmental organizations. Despite their differences, spiders and barnacles belong to Phylum Arthropoda. The earthworms and tapeworms although similar in shape belong to different phyla. The Phylum is not a fundamental division of nature. It is a high-level grouping in classification to describe all living organisms. The Phylum accommodates extinct organisms poorly; if at al .The concept of **stem groups** cover evolutionary aunts and cousins of living groups. The crown group is a group of closely related living animals plus their last common ancestor plus all its descendants. A stem group is a set of shoots from the lineage at a point earlier than last common ancestor of crown group. Tardigrades form a crown group in their own right, but Budd regarded them as a stem group relative to arthropods. **Triploblastic animals** consist of 3 layers, which are formed in embryo early in development from a single-celled egg to larva. The innermost layer produces digestive tract; the outermost forms skin; and middle one forms muscles and all internal organs except digestive system. Most living animal are triploblastic except Porifera and Cnidaria. The **Bilaterians** have right and left sides at some point in their life history. They have top and bottom surfaces and distinct front and back ends. Al known bilaterians are triploblastic and all known triploblastic animals is bilaterian. Living echinoderms are radially symmetrical, but their larvae show bilateral symmetry. Some earliest echinoderms might have been bilateral symmetry. Porifera and Cnidaria are radially symmetrical, non-bilaterian, and non-triploblastic. A coelomate possesses a gap between gut and outer tubes. Body cavity of coelomate contains internal organs. Most phyla featured in debate about the Cambrian explosion are coelomates. The priapulids are non-coelomate. All known coelomate are triploblastic bilaterians, some without coelom. In flatworms, organs are surrounded by unspecialized tissues.

Individual organism can be recognized. Populations, subspecies, or species are not easily identifiable. Species are grouped into kingdom, order, family, genera, or into structure like ecosystem and community (Figure 5.1). **Species** are similar in kind to grouping at lower and higher taxonomic levels (Darwin, 1859). Species are more objectively identifiable than population, or genus. Now ecology and biodiversity consider species as fundamental taxon. The International Commission of Zoological Nomenclature considers species as basic rank of classification. In **taxonomy**, the species is used to collect and document information on individual specimen. Speciation produces qualitatively different from in population. Species is countable both regionally and globally and can compete. Apparently distinct species in different areas often intergraded where they overlap. Such replacement species are combined as subspecies within polytypic species, the idea given for **geographical variety** (Wallace, 1865). The rank below the species level like subspecies was about complete by 1920s and 1930s. Other infraspecific taxa like local varieties, or forms were not named in the Linnaean taxonomy. Such changes are now incorporated in International Code of Zoological Nomenclature.

Speciation is the origin of new species from existing one through reproductive isolation between 2 populations. Speciation can be classified into 2 categories. In **convergent speciation**, 2 species have not developed complete reproductive isolation, but maintains their separate existence due to geographical barriers. Such species can produce new species by hybridization, if the geographical barrier between them is removed. Sometimes migrations merge 2 species into one species. In **divergent speciation** (Figure 5.2), 2 or more species are produced by splitting of single species, due to migration or adaptive radiation. In phyletic speciation (sequential evolution), the species change over long period to become different from the ancestor. Evolution of horse from *Hyracotherium* and elephant from *Moeritherium* are the phyletic speciation. Darwin's theory is based on such idea of gradualism. In quantum speciation, sudden formation of new species occurs by mega mutation and disruptive selection, hybridization, or polyploidy. In small, scattered populations or populations that migrate to new area, genetic drift, rather than natural selection plays role to change the species quickly.

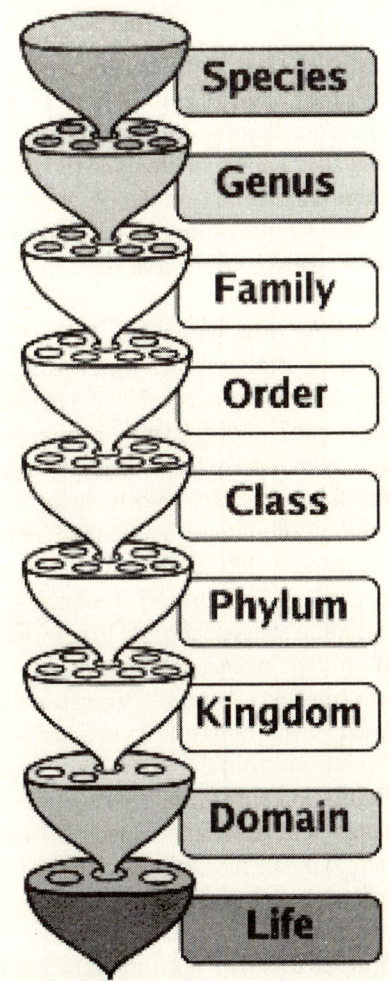

**Figure 5.1** Biological classification

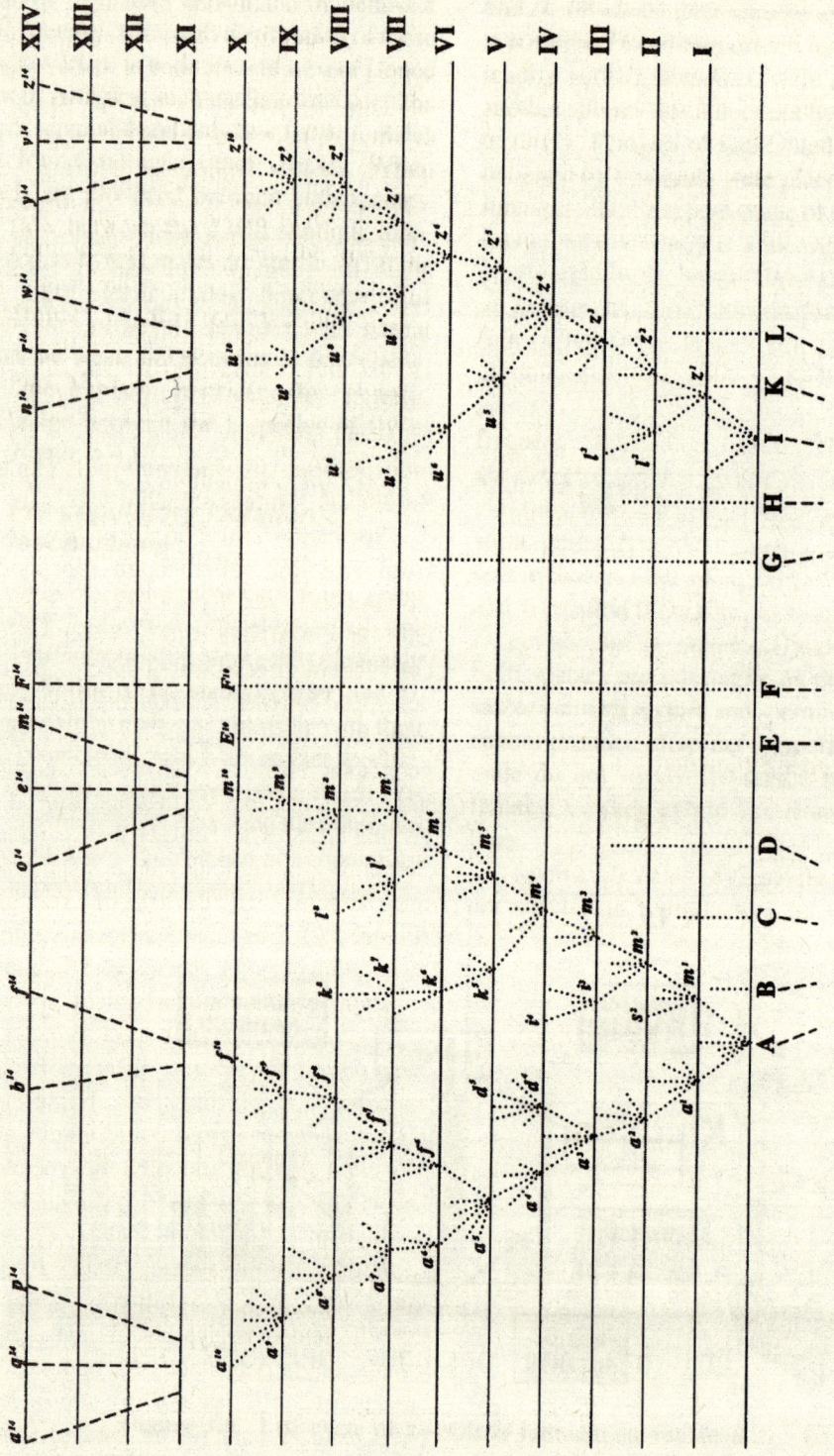

**Figure 5.2** Appearance of Divergence according to Darwin

Disruptive natural selection makes distinct and diverse species. Variations are the differences between closely related organisms, individuals of same species, subspecies, or race. Heritable changes cause the population change from generation to generation and have evolutionary significance.

## 5.1 Species Concepts

A **biological species** is an individual of higher rank. Behavior bonds the intra-species components. The behavioral bond, the interaction between organisms provides cohesiveness of species and societies. Species-specific behavior makes special connections between individuals to create organization with function of survival. These are the connections for interbreeding and reproductive isolation. The mutual affinity of organisms determines reproductive isolation including mate choice, and all interspecific interactions. Genetic basis of integration is the species gene pool. A species has birth, infancy, adolescence, maturity, aging, and death. The essence of reproduction is not just increasing organism's number. It is the renewal, replacement of the old bodies by new ones with restoration of live systems by genetic information (Shcherbakov, 2010). Multiplication causing biological expansion is largely a matter of contingency. **Reproduction** is a creative force of ontogenetic programs.

Speciation differs from reproduction in one important aspect. Asexual organisms reproduce their exact copies. With the very high fidelity of DNA replication in simple organisms, they attain eternity. Sexual organisms fail to reproduce their copies; exist as constituents of the species body. During reproduction, gene combinations are tested and selected for and put in species gene pool. In biological species, speciation is not a way of its own salvation as descendant species are not the exact copies of ancestor. A species is interested not in speciation but in its own longevity. Speciation occurs as a chance event due to geographical, or ecological isolation.

Genetic drift accompanies speciation event. The particular gene pool of isolated group is the species wealth. The offspring of group mate with one another causing homozygotization of future population. **Homozygotic individuals** possess reduced robustness due to losing heterozygosis and baring the recessive alleles. Populations they form loose robustness because of missing polymorphism and loosing genotypic plasticity. This, besides a still low size of incipient species and new niche, lefts the population in critical condition. Death of individuals occurs. Sexual reproduction provides scope for random production of various genomes including inevitably those of low vitality. Another reason for production of extra progeny is the high probability of random death (r-strategy of reproduction).

Mutations are accumulated. The isolated population become about different from species to which isolated group had belonged. During initial period of rapid evolution, accumulation of new beneficial alleles and those compensating the harmful mutations enrich gene pool for creating new species robustness. This is a complex phenomenon, including capability of individual organisms to perform development in background of mutational and environmental perturbations and genotypic plasticity of populations (Shcherbakov, 2010). During this short period new form is created with special behavior. The crucial moment for emerging species occurs, when a contact of isolated population with ancestor species is restored. If a reproductive barrier against the ancestor species was put up during isolation, the new species will be fixed and creation of unique co adapted gene pool (Mayr, 1963) becomes possible. The new species will not come into life if reproductive barrier was not put up, being dissolved in maternal species (Williamson, 1981).

**Reproductive isolation** is an important homeostatic mechanism. Sexual reproduction is necessary as a higher organism is unable to cope with mutation flood by acting alone. Promiscuous crosses are lethal for species because of its dissipation. Isolating mechanisms are ethological, anatomical obstacles for mating, sterility, or low vitality of hybrids. The important is the ethological mechanisms of isolation. All animals have complex systems for identification of individuals of same species. Such psychological incompatibility is found in wolf and dog. Physiologically they are completely compatible, hybrids are fully vital and fertile, but they hate one another. Such hatred maintains themselves as separate species. Reproductive isolation is necessary step in creating an individual of a higher rank - the biological species. The well-known **ethological process** of sexual and filial imprinting (Irwin and Price, 1999) is actually the teleonomic mechanisms for maintaining reproductive isolation and behavioral integration of species.

As the size of population increases and its gene pool gets better co adapted, the probability of successful fixation of novelties becomes lesser, and evolution rate becomes zero. The species reaches stasis which does not imply complete cessation of genetic alterations. The species struggle against mutational and environmental perturbations. A species may acquire adequate adaptation to varying environment by creating intra-species **genotypic diversity**, enabling it to support suitable phenotypic configurations without committing irreversible evolutionary steps, but it cannot tolerate mutation flow once and for all. **Purifying selection** works permanently to the same goal, removing organisms and populations of low vitality.

There is no selection other than selection for survival. It leads automatically to robustness of individual development and genotypic plasticity of populations.

The prokaryotes are probably potentially immortal. They are capable of reproducing with high fidelity, causing about one change in DNA sequence per 300 progeny cells. For higher organisms, the answer is not so evident. Apparent **evolutionary stasis** of higher species is multifaceted. They may remain phenotypically unchanged during millions of years, but genotypically they continuously change, and mutation flow cannot be stopped, or slowed down. New mutations enter the gene pool, and these are neutralized by genetic changes. This financial pyramid of evolution does not seem to work endlessly. It may possibly crash down when species capacity to resist entropy pressure becomes inadequate. New small group may undertake an attempt to create another variant of species, which may happen to be more perfect than its predecessor and live longer. G. G. Simpson (1944) construed direct interspecies competition as rarely the cause of extinction of species.

Rules of classifying a group of individuals as species are called species concept (Ereshefsky, Marc, Species, The Stanford Encyclopedia of Philosophy, Spring 2010 Edition). Reproductive barriers are semi permeable to gene flow. Species can differentiate instead of continuing interbreeding. **Parallel evolution** produces single species polyphyletically. Uniparental organisms are organized in units that resemble biparental organisms. **Genic concept** defines species as reciprocally characterized groups of individuals with probable negative fitness effects in other groups that cannot be exchanged between groups on contact. This differential fitness species concept classifies groups that keep differentiated and keep on differentiating despite interbreeding. It is not limited to specific mutations causing speciation, and is not applicable to uni- and bi-parental organisms. Though at least 25% of plant species and 10% of animal species hybridize with other species, most species concept does not consider gene exchange between the species.

Persistence of differentiated species and ongoing differentiation despite gene exchange challenge such concepts. In **parallel speciation**, similar traits confer reproductive isolation originate in separate closely related lineages and results in polyphyletic entity that evolves independently of ancestral species (Nosil et al., 2002). Traits determining reproductive isolation may evolve as by-product of adaptation to different environments. They may originate by process like independent polyploidization. Most polyploid entities originate by recurrent polyploidization, often in different ancestral populations (Soltis and Soltis, 2000). Polyploid individuals may interbreed with each other and form coherent polyphyletic entities as species. **Uniparental organisms** are organized in units resembling the species of biparental organisms (Cohan and Koeppel, 2008). Different species concepts often refer to different stages during speciation (de Queiroz, 1998). Almost every species concept is usable. Group of polyploid individuals should not be considered as species which is distinct from otherwise similar diploid individuals.

Darwin (1859) argued against using hybrid sterility and zygote inviability as a cut-and-dried characters of species. He did not mention pre-mating isolation. Darwin argued that mate choice, like hybrid sterility and inviability, is found within and between species. He treated species purely typologically as characterized by degree of difference and had strong, perhaps unconscious motivation to show that species lack constancy and distinction, (Mayr 1942). Mayr's proposition that interbreeding is the true reality of species immediately opens debate.

## 5.2   Morphological Species Concept

Before Darwin, each species was believed to have an Aristotelian form or essence. Each species was defined by its static essence, which was inherently different from essences of other species. Such thought viewed that the God created each form separately and precluded change of one species into another. Species were little more than varieties with great reality only when intermediates died out leaving the morphological gap (Darwin, 1859). Early evolutionists accepted this gap (e.g., Robson, 1928). Darwin used the gaps in morphology to define species which is the morphological species concept. His species criterion is easily extendable to ecology, behavior, or genetics.

## 5.3   Biological Species Concept

This concept defines species as "*groups of actually or potentially interbreeding populations, which are reproductively isolated from other such groups*" (Mayr, 1963). The concept has originally been formulated exclusively for biparental organisms, not for uniparental organisms like **parthenogenetic species**, although they are organized in units that resemble species of biparental organisms (Cohan and Koeppel, 2008). Modification of concept can be applied to uniparental prokaryotes (Dykhuizen and Green, 1991) as prokaryotes are not simple clonal organisms, but frequent gene exchange occurs in some groups. Gene exchange between prokaryotes is not limited to closely related species (Ochman et al., 2005) and modification of biological species concept requires lumping many well-differentiated and generally accepted species.

## 5.4   The Interbreeding Concept

Poulton (1904) proposed **syngamy** (interbreeding) as the true meaning of species. Swallowtail butterflies (Papilionidae) exhibit strong sexual dimorphism. Females are often polymorphic and their color pattern often mimics unrelated and unpalatable butterflies. Male is nonmimetic. Based on morphology, each form should be treated as different species, but mating pattern showed that forms were part of same interbreeding group. J. P. Lotsy termed the interbreeding species a **syngameon**. In morphologically indistinguishable sibling species of *Drosophila*, a species rarely interbreed with its sibling; rather each chooses mates from within its own species. Dobzhansky proposed interbreeding species concept. Mayr later popularized it as the biological species concept. Interbreeding within species, coupled with reproductive isolation between species, is the single reality for species (reviewed by Mayr, 1982).

Instead of species being defined simply, using man-made criteria like morphology, species defined by characteristics that is by means of their biological function became important (Mayr, 1982). The species defined by biological concept are protected from gene flow by isolating mechanisms, including pre-zygotic - and post-zygotic factors. Later on Dobzhansky simply concluded that asexual forms could not have species. Hybridization intermediates, or inapplicability to many plants and asexual forms caused taxonomic problems, but did not disprove the underlying truth of concept. Mayr claimed that biological concept in many ways can be seen as a reversal to new kind of essentialism, where evolutionary maintenance through interbreeding is the cause of reality of species.

## 5.5   Ecological Species Concept

Asexual forms like the bdelloid rotifers can be clustered into species, as competition made intermediates extinct (Hutchinson, 1968). Distinct forms like oaks (*Quercus*) with high rates of hybridization, remain distinct even where they co occur. van Valen and others (1976) suggested that true meaning of species is the occupancy of an ecological niche , called ecological species concept. Gene flow could not unite each population in a polytypic, biological species range. Stabilization of phenotype might be effected by ecologically mediated stabilizing selection rather than solely because of **gene flow** (see Mayr, 1982).

## 5.6   Recognition Concept of Species

Recognition concept differs from biological species concept by focusing on subset of isolating mechanisms occurring before fertilization. **Interbreeding concept** always stressed common gene pool, compatibility within species and isolation between species. Isolating mechanisms implied that reproductive isolation was adaptive, which Paterson felt as unlikely. True reality behind designating a species was the pre-zygotic compatibility. Compatibility is strongly conserved by stabilizing selection. Isolating mechanisms like hybrid sterility, or inviability are non-adaptive, can be a result rather than cause of species separateness (Paterson, 1985). True reality of species must be adaptive. Paterson proposed specific **mate recognition systems** (SMRSs). Species are the most inclusive population of individual biparental organisms sharing common fertilization system (Paterson, 1985).It is pointed out that SMRSs are about the inverse of prezygotic isolating mechanisms.

## 5.7  Monophyly

Cladistic methods states that all classification should be based on monophyly. This system states paraphyletic and polyphyletic taxa were unnatural groupings and should not be used in taxonomy. **Monophyly** criterion is a type of phylogenetic species concept (Hennig, 1966). Species forms when a single interbreeding population split into 2 branches, or lineages that did not exchange genetic material. In cladistic concept, species are considered as branch segments in phylogeny, with every branching event resulting to a new pair of species (Ridley, 1996). If one of the 2 branches were considered as new, other branch would be paraphyletic. This idea causes nomenclatural instability. Monophyly shows fractal self-similarity and exist at very high, or low levels of the phylogeny. Thus, exact level at which species taxa exist becomes unclear.

**Genealogy:**  A single, true phylogeny of taxa rarely exists. Hybridization and occasional horizontal gene transfer selectively flow genetic material between unrelated forms. Multiple gene lineages exist in population. If such a population split into 2 distinct forms, some time would be required before fixing of each branch for various, reciprocally monophyletic gene lineages at single gene, especially near species boundary. In **genealogical concept**, species should be defined when a consensus between multiple gene genealogies shows reciprocal monophyly (Baum and Shaw, 1995). Geographic forms that have become isolated in small populations rapidly become fixed for gene lineages, and viewed as separate species without any important evolution. Distinct sister taxa like human and chimpanzee share gene genealogy polymorphisms at genes like the human leukocyte antigen (HLA) complex and may be classified as same species based on genealogical consideration.

## 5.8  Diagnostic or Phylogenetic Species Concept

An alternative to biological species concept applicable to uni-parental organisms is the phylogenetic species concept. Here a species is a diagnosable cluster of individuals within which there is a parental pattern of ancestry and descent, which shows phylogenetic ancestry and descendent among units of like kind (Eldredge and Cracraft, 1980). This concept is recommended for prokaryotes (Staley, 2006) and is referred in several DNA barcoding /DNA taxonomy studies (Monaghan et al., 2009). In uniparentals in which there is no or little gene exchange, each clone with a mutation would be classified as separate species (Coyne and Orr, 2004). In uniparentals with higher levels of gene exchange and in biparentals, each substitution will have its own particular distribution and little or no concordance might exist among the sets of individuals with independently derived mutations except those bounded by barriers to gene flow (Avise and Ball, 1990).

Eldredge and Cracraft (1980) restricted the clusters of individuals that are ranked as species by including condition within which there is a parental pattern of ancestry and descent, beyond which there is not (Davis, 1997). This condition restricts the cluster of individuals that are ranked as species to those that are bounded by barriers to gene flow. **Hybridization** between closely related species is frequent. Species can differentiate despite continuing interbreeding means. Many recognized species do not show separate "parental patterns of ancestry and descent," now, but that some descendants belong to one species may have ancestors belonging to other co existing species and vice versa (Rajagopalan and Fujimura, 2012). These insights have similar consequences for application of phylogenetic species concept as for biological species concept. **Polyphyletic species** originating by parallel speciation do not show parental patterns of ancestry, descent separate from that of ancestral species and may be clumped under phylogenetic species concept.

Diagnostic concept called phylogenetic species concept attempts to incorporate phylogenetic thinking (Hennig, 1966) into species-level taxonomy. Cases of hybridization between taxa on various branch of species-level phylogeny suggest that interbreeding and phylogenetic reality conflict. Many bird taxa, believed as **subspecies**, are recognizable and stable nomenclaturally than polytypic species to which they supposedly belong (Cracraft, 1989). Cracraft argued for using a diagnostic criterion in form of fixed difference at one, or more inherited characters. A phylogenetic species is an irreducible (basal) cluster of organisms, distinct from other such clusters, and within which parental pattern of ancestry and descent (Cracraft, 1989) exists. If diagnostic criteria are strictly applied, small group of individuals, or single specimen, might be defined as separate species, leading to taxonomic inflation. Cracraft argued that such groups have no parental pattern of ancestry and descent.

Characters used to identify phylogenetic species may be primitive characters, or they would have evolved several times. Phylogenetic species thus need not be monophyletic and could probably be paraphyletic and polyphyletic. Phylogenetic species will never be non monophyletic except through error (Cracraft, 1989), although "their historical status may be unresolved because relative to their sister species they are primitive in all respects. Whether they [are] truly paraphyletic is probably unresolved (Cracraft, 1989). Some genuine problems noted with polytypic application of interbreeding concept and as a result phylogenetic species concept became influential. Davis(1997), following Cracraft's

suggestion and using diagnostic differences between **geographic populations** showed, in some cases at single DNA base pairs, as evidence that 2 forms are separate species even if they intergrade freely at boundaries of their distribution. Ornithologists and primatologists have used diagnostic characters to reassign many taxa long thought of as subspecies to level of full species, resulting in taxonomic inflation (Isaac et al., 2004).

## 5.9   Phenetic Species Concept

This concept is based on presence, or absence of shared derived characteristics and is highly dependent on information about characteristics of organisms. If 2 organisms share the same set of derived characteristics, they belong to same species. Characteristics are quantified and then results are used to construct a **cladogram**, the fit of which is either accepted, or rejected using statistical analysis. If there is insufficient data to perform such an analysis, an organism simply can't be classified. Species are reproductively isolated populations.

Numerical methods in taxonomy now referred as phenetics states that taxonomy and systematics should based on multivariate statistical analysis of characters rather than on underlying evolutionary process information. Species, like other taxa, would be defined in **numerical taxonomy** on the basis of multivariate statistics, as clusters in phenotypic space (Sokal and Crovello, 1970). **Phenetics** is closely similar to intuitive methods adopted by most taxonomists, who use multiple (morphological or genetic) characters to sort individual specimens into discrete groups between which there are few intermediates. Some large areas of practical taxonomy are based on phenetic approach. Bacterial systematists use multiple biochemical tests to assign microbes to species taxa. Phenetic classifications based on morphology possess the risk that, if convergent characters are used as data, one can classify unrelated forms into polyphyletic or paraphyletic taxa. Single gene polymorphisms and sexual dimorphism can affect multiple external characters leading to designation of multiple species within polymorphic populations. Sibling species could be lumped into same species using aphenetic approach, unless a set of highly diagnostic characters could be found. These problems are due to the lack of characters found in morphological datasets. Phenetics is more successful in distinguishing unrelated forms, although cryptic taxa from polymorphic form in combination with molecular genetics techniques (Avise, 1994).

## 5.10 Genic Species Concept (Differential Fitness Species Concept)

Wu (2001) developed a model of speciation which states that speciation depends primarily on genes responsible for differential adaptation to different natural or sexual environments, the speciation genes. Differential adaptation is a form of divergence in which alternative alleles of a gene have opposite fitness effects in 2 groups of individuals. Speciation genes account for a small fraction of genome. Gene exchange is restricted at these loci, whereas gene exchange at other loci, the marker loci, could persist after speciation. In speciation, the groups of individuals do not lose their divergence on contact and continue to diverge. Species are differentially adapted groups which on contact are unable to exchange genes by controlling adaptive characters, by direct exchange or through hybrid population (Wu, 2001). It was originally created for **biparental species**, but in principle it is applicable to uniparentals. *Wolbachia* infection may be about neutral in host species, but have a negative fitness effect in noninfected species. Groups

of individuals become evolutionary independent only if gene flow is restricted in both directions. Each group must have features that have negative fitness effects on other species. The concept is criticized as it only focuses on differential adaptation caused by mutations in genes (Noor, 2002). Genetic factors that may result in reproductive isolation like chromosomal changes are designated as special cases (Wu, 2001).

This concept considers mutations in genes, and any differences including chromosomal changes (Rieseberg, 2001), *Wolbachia* infections (Werren et al., 2008), other selfish genetic elements like transposable elements (Hurst and Werren, 2001), or niche-specifying genes are acquired by horizontal transfer in prokaryotes (Cohan and Koeppel, 2008). Differences may result from differential adaptation due to natural or sexual selection, but may be the result of genetic drift, or polyploidization, or infections by symbionts. Features may be adaptive for one group's niche, but maladaptive for another's, or may be incompatible in the context of genetic background of another group. Such incompatibilities could result in lowered fertility, or in unbalanced physiological function. Differential fitness species concept differs from biological species concept in that the exchange of species-specific features may not only be restricted by reproductive isolation, but by divergent selection. Differential fitness species concept is thus close to Darwin's (1859) understanding of species than to biological species concept.

Darwin (1859) argued that difference between species and well-marked variety is that the latter are linked at the present day by intermediate gradations, whereas species were previously connected. Lack of intermediate gradations in first difference characterizes nascent species that results from inability to exchange features between groups and is not connected with reproductive isolation. One shortcoming of a purely phenotypic assessment of ability to exchange potentially species-specific features is that groups of individuals characterized by different discrete polymorphisms may be taken for separate species, because of lack of intermediate gradations. This does not affect differential fitness species concept that does not focus on lack of intermediate gradations as such, but is based on inability to exchange species-specific features between groups. As long as groups of individuals characterized by different discrete polymorphisms are able to exchange these polymorphisms, that is the underlying genes; these groups are not considered species under differential fitness species concept.

## 5.11   Lineage Species Concept

Modern species concepts explicitly, or implicitly equate species with separately evolving (segments of) metapopulation lineages (de Queiroz, 2005). This concept, the re-formulation of evolutionary species concept (Wiley, 1978), became popular. Lineage is defined as an ancestral-descendant sequence of populations (Simpson, 1961). Population is a group of con-specific organisms that occupy well-defined geographical region and exhibit reproductive continuity from generation to generation (Futuyma, 1998). In some definitions of population, the con-specific organisms or individuals of same species are replaced by interbreeding individuals (de Queiroz and Donoghue, 1988). Such definition of population introduces biological species concept by including it into definition of population.

de Queiroz (1998) used population as an organizational level above that of organism rather than specific sense of community of sexual organisms. He did not specify property of this organizational level above the organism,

and exact definition of population, lineage and species remain unclear. **Metapopulation** is an inclusive population made up of a set of connected subpopulations (de Queiroz, 2005a, b). It is circular to define a species as a sequence of (meta-) populations. The term **population** must not be used in defining species unless it is defined without referring to con-specific organisms. Notion of separately evolving units is the core of all modern species concepts. Classification of populations connected by limited gene flow is not clear. Biparental taxa may not evolve separately at some loci, but do at most others (Mallet, 2005). Coherent polyphyletic entities that originate by parallel speciation and evolve separately of ancestral species cannot be classified as separate species by lineage-based species concepts, which shift the problem of defining the term species to problem of defining the term population and result in an intricate re-formulization of biological species concept . Thus, lineage-based species concepts present only limited progress toward a generally applicable species concept. Simpson's (1951) evolutionary species concept defines species as a lineage evolving separately from others and with its unitary evolutionary role and tendencies. This concept combines idea that species are historical lineages with concept of their evolutionary and ecological role. de Queiroz (1998) proposes the general lineage concept which states that species are segments of population-level lineages. But all the species concept acknowledge implicitly, or explicitly that evolutionary separateness of lineage is the primary concept.

## 5.12   Cohesion Species Concept

Evolutionary processes other than reproductive isolation contribute to formation and maintenance of coherent entities, especially in uni-parentals (Templeton, 1989). Templeton compiled these processes and classified these cohesion mechanisms. He defined the limits of spread of new genetic variants through gene flow and mechanisms affecting demographic exchangeability by defining fundamental niche and limits of spread of new genetic variants through genetic drift and natural selection. The term "cohesion mechanism" may be misleading, as most biological properties that confer cohesion probably did not arise for that purpose (Harrison, 1998). Thus, **cohesion** is an effect, not a mechanism.

**Cohesion species concept** states that species is the most inclusive group of organisms having potential for genetic and/or demographic exchangeability. This concept fails to provide decision about species status of 2 groups when criteria of genetic and demographic exchangeability conflict (Coyne and Orr, 2004). If 2 groups of biparental individuals are genetically non-exchangeable but demographically exchangeable, all individuals linked by any cohesion are classified as one species in cohesion concept. If there are groups of biparental individuals that are genetically **non-exchangeable, but demographically exchangeable, the most inclusive group includes all individuals. Thus, groups of** individuals may be con-specific under cohesion concept, if they occupy the same fundamental niche, even if they are reproductively isolated. Cohesion concept does not specify the kind of gene flow that affects genetic exchangeability and result in lumping species that differentiate despite ongoing gene exchange.

Cohesion concept combines interbreeding species concept with ideas from ecological-, recognition-, and genealogical concepts. Combination of ecological and reproductive cohesion is important for maintaining a species' evolutionary unity and integrity. This idea is applied to species like oak that often undergo hybridization and gene flow. Species evolve and are maintained as cohesive whole by all such multifarious processes. Species can be

seen as separate from its history of origin and from current reasons for its integrity. When groups with different and conflicting biological and evolutionary characters are considered species, there should simple criteria for uniting them.

## 5.13 Genotypic Cluster Definition

Mallet (1995) formulated the **genotypic cluster definition** which states that a species is a genotypic cluster that can overlap without fusing with its sibling. This concept is applicable for uni-parental organisms. In this definition, each genetically different clone will be identified as a separate species (Coyne and Orr, 2004). The prediction that species sooner, or later form genotypic and phenotypic clusters can be derived from most species concepts. Thus, this is useful criterion for delimiting provisional species. Incipient species might not yet be recognized as distinct cluster based on random sample of genetic markers. In peripatric speciation, the peripheral species will initially often form a cluster with neighboring populations of the widespread species. The more widespread species does not form a genotypic cluster distinct from peripheral species. At least in their initial stages, coherent entities originating by parallel speciation from different populations of an ancestral species cannot be recognized as separate genotypic clusters and would have to be lumped with ancestral species under genotypic cluster definition.

DNA has a digital, rather than analog code. There are genetic gaps between virtually any pair of individuals. Any discreteness at gene level can not be used to define species. Separate sexes and polymorphic female forms of mimetic *Papilio* (butterfly) have gaps between them. A genetic element, which may be a sex chromosome, a single base pair, the entire mitochondrial genome, a rearrangement of chromosome may determine genetic or morphological differences between such polymorphic forms.

Polymorphic genetic elements like **mimicry genes** and sex chromosomes would randomly combine with polymorphisms at genetic elements found on other chromosomes or extra chromosomal DNA. Each individual may be a distinct multilocus-genotype, but single grouping of genotypes as polymorphism at one genetic element is independent of polymorphism at others. Some proposed a genotypic cluster criterion for species (Mallet, 1995). Species are recognized by morphological and genetic gaps between populations in a local area, rather than by phylogeny, cohesion, or reproductive isolation that create gaps (Mallet, 1995). In a local area, separate species are recognized if there are several clusters separated by multilocus phenotypic or genotypic gaps. A single species is recognized if there is a single cluster in frequency distribution of multilocus phenotypes and genotypes. **Genotypic gaps** may be entirely vacant, or may contain low frequencies of intermediate genotypes. Genotypic clusters are neither profound nor original. Multi-locus genotypic clusters are almost universally considered as criterion of speciation in theoretical models of sympatric speciation (Kondrashov and Kondrashov, 1999). Most genotypic cluster species can be recognized morphologically. Minor pattern elements in *Papilio* can unite the various polymorphic forms.

Asexual forms can be clustered and classified as genotypic clusters in exactly the same way as sexual species. Exact level of species clustering for asexual is arbitrary. Asexual forms like bdelloid rotifers have distinguishable species taxa (Hutchinson, 1968), probably due to ecological selection for distinct characters. Bacteria are largely asexual where populations are not recognizable in genetic clusters (Cohan, 1994). Thus, reproductive isolation is not required for genotypic clustering. Genotypic cluster criterion in sexual species is nothing other than a gene flow concept of species. This holds true for specialized interpretation of gene flow in sexual population. Random associations of genes within genotypes in genotypic cluster approach will be methodologically the same as genotypic analysis to determine whether a population is interbreeding, but the latter process needs extra assumptions. In sexual species, genotypic cluster criterion could be viewed as a practical application of the biological species concept. Genotypic cluster criterion is character-based, instead based on interbreeding, and is thus applicable to a sexuals and sexual species. If a single geographic race, which previously intergraded at all its boundaries with other geographic races, splitted into 2 forms that coexist as separate genotypic clusters, the original polytypic species became paraphyletic in changed situation. The new species is derived from one component subspecies. Thus, paraphyly of species should be considered as a possibility under this definition, as in interbreeding and diagnostic concepts.

## 5.14  Comparison of Species Concepts

Biological species concept differs from other concept of species. It is formulated only for biparentals and requires **intrinsic reproductive isolation**. Phylogenetic species concept differs from other concepts and classifies groups that are only extrinsically isolated as species. This results in considerable taxonomic inflation (Agapow et al., 2004). It classifies diagnosable groups of uni-parentals that are not characterized by features and would have negative fitness effects in other groups as

**Table 5.1** Alternative contemporary species concepts (i.e., major classes of contemporary species definitions) and the properties on which they are based (modified from de Queiroz, 2005). Properties (or the converses of properties) that represent thresholds crossed by diverging lineages and that are commonly viewed as necessary properties of species are marked with an asterisk (*). Under the proposal for unification the various ideas summarized in this table would no longer be considered distinct species concepts (see de Queiroz, 1998, for an alternative terminology). All of these ideas conform to a single general concept under which species are equated with separately evolving metapopulation lineages, and many of the properties (*) are more appropriately interpreted as operational criteria (lines of evidence) relevant to assessing lineage separation-

| Species concept | Property(ies) | References |
| --- | --- | --- |
| Biological | Interbreeding (natural reproduction resulting in viable and fertile offspring) | Wright, 1940; Mayr, 1942; Dobzhansky, 1950 |
| Isolation | *Intrinsic reproductive isolation (absence of interbreeding between heterospecific organisms based on intrinsic properties, as opposed to extrinsic [geographic] barriers) | Mayr, 1942; Dobzhansky, 1970 |
| Recognition | *Shared specific mate recognition or fertilization system (mechanisms by which conspecific organisms, or their gametes, recognize one another for mating and fertilization) | Paterson, 1985; Masters et al., 1987; Lambert and Spencer, 1995 |
| Ecological | *Same niche or adaptive zone (all components of the environment with which conspecific organisms interact) | Van Valen, 1976; Andersson, 1990 |
| Evolutionary (some interpretations) | Unique evolutionary role, tendencies, and historical fate | Simpson, 1951; Wiley, 1978; Mayden, 1997; Grismer, 1999, 2001 |
| | *Diagnosability (qualitative, fixed difference) | |
| Cohesion | Phenotypic cohesion (genetic or demographic exchangeability) | Templeton, 1989, 1998a |
| Phylogenetic | Heterogeneous (see next four entries) | (see next four entries) |
| Hennigian | Ancestor becomes extinct when lineage splits | Hennig, 1966; Ridley, 1989; Meier and Willmann, 2000 |
| Monophyletic | *Monophyly (consisting of an ancestor and all of its descendants; commonly inferred from possession of shared derived character states) | Rosen, 1979; Donoghue, 1985; Mishler, 1985 |
| Genealogical | *Exclusive coalescence of alleles (all alleles of a given gene are descended from a common ancestral allele not shared with those of other species) | Baum and Shaw, 1995; see also Avise and Ball, 1990 |
| Diagnosable | *Diagnosability (qualitative, fixed difference) | Nelson and Platnick, 1981; Cracraft, 1983; Nixon and Wheeler, 1990 |
| Phenetic | *Form a phenetic cluster (quantitative difference) | Michener, 1970; Sokal and Crovello, 1970; Sneath and Sokal, 1973 |
| Genotypic cluster | *Form a genotypic cluster (deficits of genetic intermediates; e.g., heterozygotes) | Mallet, 1995 |

Source : Syst. Biol. 56(6) : 879-886, 2n7

species which would result in larger inflation in prokaryote taxonomy. Such groups might be classified as species using genotypic cluster definition which is most similar to differential fitness species concept. Its outcome cannot be predicted unequivocally because genotypic clusters are expected to emerge under all specified conditions, but may come about only in later stages of differentiation. Genotypic clusters arising in only extrinsically isolated groups do not qualify as species under genotypic cluster definition, as they will fuse on contact. In cohesion concept, the groups of bi-parental individuals are considered conspecific if they occupy the same fundamental niche, even if they are reproductively isolated. Genic and differential fitness species concepts do not demand complete reproductive isolation, but tolerate the exchange of features except those are causal for differentiation. Groups are classified as species under genic species concept under general differential fitness species concept. The same is true for groups classified as species under **biological species concept**, because features that cause reproductive isolation in bi-parentals have negative fitness effects in other groups and cannot be regularly exchanged. (Table 3.1)

## 5.15　Species Delimitation

Formulation of species concept and development of approaches for delimiting species are 2 tasks (de Queiroz, 2007). A species concept needs to be formulated in a way that cover the earliest stages after speciation, which may differ only in few properties. As divergence differentiation of resulting species continues, it becomes easier to distinguish later stages. Approaches for delimiting species should be based on specified conditions and correspond exactly to species concept. Often properties acquired later during

differentiation are observed easily than few properties acquired during speciation. **Delimitation approaches** are often based on such properties. If approach for delimiting species is properties of later stages, the approach may result in false negatives. Species in early stages of speciation are not recognized. If the properties acquired prior to speciation, false positives may result. Methods not directly based on specified conditions in species concept sometimes fail to delimit species boundaries properly, an eclectic approach to delimiting species is necessary (Sites and Marshall, 2004).

If species delimitation is based on specified conditions of differential fitness concept, groups of individuals should be reciprocal by features that have negative fitness effects on other groups and that cannot be regularly exchanged between groups on contact. In many cases, features that have **negative fitness effects** on other groups, like different beak sizes, or forms in birds that allow individuals belonging to different species to use different food, or different ploidy levels are known and are used for species delimitation. Discontinuous variation of such features cannot be regularly exchanged between groups. Such groups are considered species under the differential species concept. In principle, fitness effects of genes of other species be tested by experimental transformation in uni-parentals. If nothing is known about the fitness effects of features and their exchangeability, a standard approach for delimiting species is to determine phenotypic, or genotypic clusters (Hausdorf and Hennig, 2010). This approach is justified at least in biparental species by differentiation processes. After speciation , more alleles become restricted to one of the descendant species, and the individuals belonging to one species form genotypic and phenotypic clusters soon, or late even if they were initially paraphyletic, or polyphyletic.

## 5.16 Taxonomic Practice

Geographic variants blended together at their boundaries are united in a single, polytypic species, unless differences (morphological, or genetic) cause reorganization of 2 species. Whenever 2 divergent forms differ at several unrelated traits, overlapped spatially, they are recognized as separate species, even if few intermediates suggest some hybridization or gene flow. Subspecies as artificial taxa may be avoided and may ignored geographic variation of polytypic species to the status of species. Diagnostic version of phylogenetic species concept is making in roads into **zoological nomenclature**, resulting in increase of species number as former subspecies are designated as species, in spite of inter-gradation at their boundaries (Isaac et al., 2004). Many *Heliconius* butterflies have over 30 geographic diagnosable subspecies per species. Numbers of bird and butterfly species could easily increase 2-10 times in some groups if the diagnostic criteria were adopted. In well-known groups like primates, a doubling of species number is noted. Most increase has resulted from reclassification of known subspecies, or populations rather than from the discovery. The species multiplication is termed **taxonomic inflation** (Isaac et al., 2004). Thus, the species is not real enough to remain at the same taxonomic rank while fashions in species concept change.

## 5.17 Populations: The Evolutionary Units

Gene flow, even in mobile animals like birds may not unite local populations into common gene pool. If local populations rarely exchange genes, gene flow through range of a continental species does not explain species integrity, as it would be outweighed by weak local patterns of adaptation, or genetic drift (Endler, 1977). This input led to proposal that species are not real biological units. Local populations are the only real groupings united by gene flow within common gene pool, which adapt to local conditions and compete. **Homogeneity** of ecological niche, or genetics over the range of species might be due to simple evolutionary inertia, or to similar stabilizing selection everywhere. Species exist and are real in local communities.

## 5.18 Speciation

Two individuals who do not interbred are the members of different species. To designate a species criteria like reproductive incompatibility and reproductive isolation are most important.

## 5.19 Reproductive Isolating Mechanisms

Reproductive isolation maintains the identity of a species. Mechanisms preventing gene exchange between populations called isolating mechanisms which are maintained by pre- and post- mating isolating mechanisms. Incompatibilities preventing mating between species called pre-mating isolating mechanisms include geographical-, ecological-, temporal- and behavioral isolation and mechanical incompatibilities. When pre-mating isolation fails, or has not evolved members of different species mate, and then post mating isolating mechanisms operates. If resulting hybrid offspring die during development, than 2 species are reproductively isolated from one another.

When viable hybrid offspring are produced, they are less fit than their parents or are themselves infertile, and then 2 species, may remain separate, with little, or no gene flow between them. Incompatibilities preventing formation of fertile hybrids between species called post mating isolating mechanisms include hybrid inviability, gametic incompatibility, and hybrid infertility. Reproductive isolation, or hybridization barriers is a collection of mechanisms, behaviors and physiological processes that prevent the members of 2 species to cross or mate , or to ensure that produced offspring are fertile. Such barriers keep the integrity of species over time, reduce, or impede gene flow between individuals of various species, allowing retention of each species' characters. Two populations become sufficiently distinct when interbreeding is difficult, or impossible between them. However, then there must be relatively little gene flow (migration) between them. Genetic changes in one population will soon become widespread in other, when there is a great deal of gene flow. Mechanisms of reproductive isolation are genetically controlled and evolve in species with overlapping geographic distribution, or due to adaptive divergence that accompanies allopatric speciation

**Selection for reproductive isolation:** Selection can increase reproductive isolation between 2 *Drosophila* species (Koopman, 1950) as found in *D. pseudoobscura* and *D. persimilis*. Generation percentage of hybrids persimilis was calculated. When these fly species were kept at16°C about a third of mating are interspecific. In experiment equal number of males and females of both species were placed in containers suitable for their survival and reproduction. Progeny of each generation were examined to determine the interspecific hybrids which were then removed. Equal number of males and females of resulting progeny were selected as progenitors of next generation. As hybrids were destroyed in every generation flies that solely mated with members of their own species produced more surviving descendants than flies that mated solely with

individuals of other species. For each generation, number of hybrids continuously decreased up to 10th generation when hardly any interspecific hybrids were produced.

Selection against the hybrids was very effective in increasing reproductive isolation between these species. From the 3rd generation, proportions of hybrids were less than 5% confirming that selection acts to reinforce reproductive isolation of 2 genetically divergent populations if hybrids formed by these species are less well adapted than their parents. From such discovery, assumptions were made on origin of reproductive isolation mechanisms in nature. If selection reinforces the degree of reproductive isolation between 2 species due to poor adaptive value of hybrids, the populations of 2 species located in same area should show a greater reproductive isolation than populations that are separated geographically.

Reproductive isolation mechanism can be a consequence of allopatric speciation. A population of *Drosophila* was divided into sub pupulations that were selected to adapt to different food types. After a number of generations the two sub populations were mixed again. It was observed that the subsequent matings occured between individuals belonging to the same adapted group.

Reproductive isolation mechanisms can be the result of allopatric speciation. A *Drosophila* population was divided into sub populations and was selected to adapt to various food (Figure 5.3). After several generations, the 2 sub populations were mixed again. Subsequent mating between individuals belonging to same adapted group reinforce hybridization barriers in sympatric populations called **Wallace Effect**, as it was proposed by Wallace for the first time. Sexual isolation between *D. miranda* and *D. pseudoobscura*,

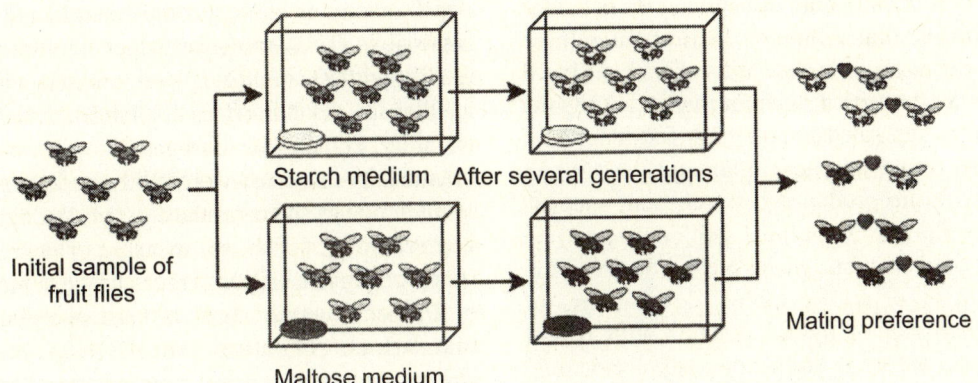

Initial sample of
fruit flies

Starch medium   After several generations

Maltose medium

Mating preference

**Figure 5.3**  Speciation in *Drosophila*

is pronounced according to the geographic origin of flies. Flies from regions where distribution of species is superimposed show a greater sexual isolation than exists between populations originating in distant regions.

Interspecific hybridization barriers can arise due to adaptive divergence that accompanies allopatric speciation which is proved on **D. pseudoobscura**. A single fly population was divided into 2, with one population fed with starch-based food and other with maltose- based food. Thus each sub population was adapted to each food type over several generations. After populations had diverged over many generations, groups were again mixed. Flies mated only with others from their adapted population showing that mechanisms of reproductive isolation can arise even though the interspecific hybrids are not selected against. Issue of speciation is very contentious. Any factors that prevent potentially fertile individuals from meeting will reproductively isolate the members of distinct species. Types of barriers that cause this isolation include different habitats, physical barriers, and a difference in time of sexual maturity, or flowering. When factors change, species may branch off. An example of ecological or habitat differences that impede the meeting of potential pairs occurs in 2

sticklebacks. One species is found all year round in fresh water, mainly in small streams. Other species lives in the sea in winter, but in spring and summer they come to river estuaries to reproduce. Members of 2 populations are reproductively isolated due to adaptations to distinct salt concentrations. **Reproductive isolation** due to differences in mating season is found in *Bufo americanus* and *B. fowleri*. Members of these species can be crossed in laboratory producing healthy, fertile hybrids. Mating does not occur in the wild, even though geographical distribution of 2 species overlaps. Reason for lack of inter-species mating is that *B. americanus* mates in early summer while *B. fowleri* in late summer.

Plants like *Tradescantia canaliculata* and *T. subaspera* are sympatric, yet they are reproductively isolated, as they flower at different times of year. One species grows in sunny areas and other in deeply shaded areas. Potential mates do not come in contact with one another because of differences in breeding seasons of 2 species, due to different flowering seasons in plants. *Bufo americanus* breeds in early monsoon (May), while *Bufo fowleri* breeds in late rainy season. Barriers that separate species do not consist of one mechanism. *Drosophila pseuoobscura* and *D. persimilis* is isolated from each other by habitat, by timing of mating season and by behavior during mating. Although distribution of these species overlaps in wide areas of west of United States of America, such mechanisms keep the species separated and only few fertile females were found amongst other species. When hybrids are produced between both species, gene flow between the 2 will continue to be impeded as hybrid males are sterile. With the great vigor sterile males, descendants of backcrosses of hybrid females with parent species are weak and notoriously non-viable. This mechanism restricts the genetic interchange between the 2 species of fly in wild (Figure 5.4.)

## A. Pre-copulatory isolation mechanisms

Pre-copulatory isolation occurs when genes necessary for sexual reproduction of one species differ from equivalent genes of another species. Males of **D. melanogaster** and *D. simulans* exhibit elaborate courtship with their female counterparts, which are species specific. *D. simulans* male hybridize with *D. melanogaster* female. Assessing the minimum number of genes involved in pre-copulatory isolation between *D. melanogaster* (Figure 5.5)

and *D. simulans*, their chromosomal location is possible. *D. melanogaster* which hybridizes readily with *D. simulans*, were crossed with another species that it does not hybridize with, or rarely. Females of segregated populations obtained by this cross were placed next to *D. simulans* males and percentage of hybridization was recorded, which is a measure of degree of reproductive isolation. Three chromosomes of *D. melanogaster* carry at least one gene that affects isolation, substituting one chromosome from a line of low isolation with another of high isolation reduces hybridization frequency. Interactions between chromosomes are detected, so that certain combinations of chromosomes have a multiplying effect. In some cases, reproductive isolation between species occurs after a long time of fertilization and formation of zygote, as in twin species, *D. pavani* and *D. gaucha*. Hybrids between both species are not sterile as they produce viable gametes, ovules and spermatozoa. They cannot produce offspring as sperm of hybrid male do not survive in semen receptors of females, be they hybrids, or from the parent lines.

Sperm of males of 2 parent species do not survive in reproductive tract of hybrid

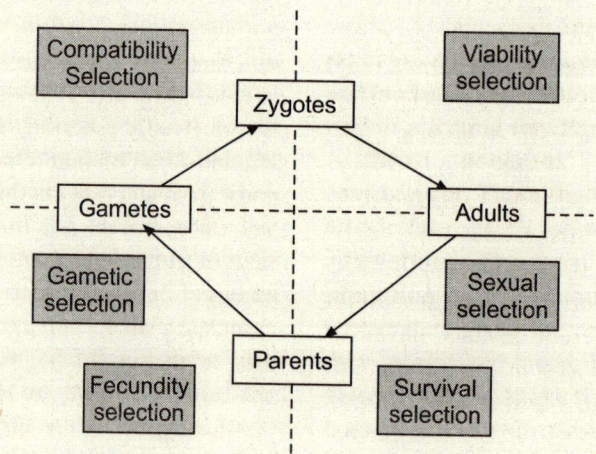

**Figure 5.4**  Life-cycle of a sexually reproducing organism

**Figure 5.5** *Drosophila melanogaster*

female. This post copulatory isolation is the efficient system to maintain reproductive isolation in many species. Development of zygote into adult is complex involving interactions between genes and environment. Any alteration in usual process due to absence of necessary gene, or presence of different one, arrest normal development causing non-viability of hybrid, or its sterility. Half of chromosomes and genes of a hybrid come from one species and other half come from other. If 2 species are genetically different, there is less chance that gene from both acts harmoniously in hybrid. Only few genes would be required for post copulatory isolation, as opposed to situation in **pre-copulatory isolation**.

In species without pre-copulatory reproductive isolation, hybrids of only one sex are produced as in hybridization between females of *D. simulans* and *D. melanogaster* males. Hybridized females die in their early development and only offspring males are seen. *D. simulans* has genes for development of adult hybrid females, that is, viability of females is rescued. These speciation genes inhibit expression of genes that allow growth of hybrid. There are regulator genes. Several genes are found in melanogaster species group. The first to be discovered was *Lhr* (Lethal hybrid

rescue) located in Chromosome II of *D. simulans*. This dominant allele allows development of hybrid females from cross between simulans females and melanogaster males. A gene located on Chromosome II of *D. simulans* is Shfr that allows development of female hybrids and its activity is dependent on temperature during development. Other similar genes are located in distinct populations of species of this group. Few genes are needed for an effective post copulatory isolation barrier mediated through the non-viability of the hybrids.

*Hmr* gene, linked to the X chromosome is related with viability of male hybrids between *D. simulans* and *D. melanogaster* originates from proto-oncogene family *myb*, which codes for a transcriptional regulator. Two variants of this gene work perfectly in separate species, but in hybrid they do not function correctly, due to different genetic background of each species. Examination of allele sequence of 2 species shows that change of direction substitutions are abundant than synonymous substitutions, indicating the exoposure of this gene to intense natural selection. Dobzhansky-Muller model proposes that reproductive incompatibilities between species are caused by interaction of genes of respective species.

*Lhr* has functionally diverged in *D. simulans* and interact with *Hmr* which has functionally diverged in *D. melanogaster* causing lethality of male hybrids. *Lhr* is found in a heterochromatic region of genome. Its sequence has diverged between these 2 species in consistent with mechanisms of positive selection. *OdsH* (abbreviation of Odysseus) gene causes partial sterility in hybrid between *D. simulans* and related species, *D. mauritiana* and is of recent origin. This gene is monophyletic in both species and subjects to natural selection, intervenes in initial stages of speciation, while other genes differentiating 2 species are polyphyletic. *Odsh* originated by duplication in genome of *Drosophila* has evolved at higher rate in *D. mauritania*, but its paralogue, unc-4, is about identical in species of group melanogaster. All such cases show the way of speciation mechanisms in nature. They are called **speciation genes**, gene sequences with a normal function within populations that diverge rapidly in response to positive selection forming reproductive isolation barriers with other species. All these genes have functions in transcriptional regulation of other genes.

*Nup96* gene is example of evolution of genes implicated in post-copulatory isolation which regulates the production of one of about 30 proteins required to form a **nuclear pore**. In each of simulans groups of *Drosophila*, the protein from this gene interacts with protein from another, as yet undiscovered, gene on X chromosome to form a functioning pore. In a hybrid the pore that is formed is defective and causes sterility. Differences in sequences of *Nup96* have been subject to adaptive selection, similar to other examples of speciation genes described above. **Pre-zygotic isolation** mechanisms are the most economic in terms of biological efficiency of population, as resources are not wasted for producing weak, non-viable, or sterile descendent. Songs of insects, birds, and many other animals are part of a ritual to attract potential partners of their own species. The song presents specific patterns for members of the same species, and show a mechanism of reproductive isolation.

**Geographical isolation:** When the populations are separated by **geographical barrier** like the sea, mountain, Deserts, and river, and for aquatic animals land physically prevented them from interbreeding, such populations are termed allopatric, which are forced to evolve freely and accumulate genetic differences. **Geographical isolation** may be different for various species. A small stream

may be an effective barrier for land insects and small mammals, while for birds even mountain and oceans may not be barriers. Members of different species cannot mate if they never get near one another by geographical barrier, for example intervening river. Such isolation typically provides condition for speciation in first place and prevents interbreeding through development of different courtship rituals. This isolation is considered to be a mechanism that allows new species to from mechanism that maintains reproductive isolation between species.

**Temporal or habitat isolation:** Here potential mates do not meet one another due to their differences in habitats, space requirements, climate, and food. This is called **ecological isolation**. Potential mates inhabit different areas but do not come in contact with each other. Spawning grounds of riverine fishes in different tributaries prevent interbreeding. When 2 species occupy similar habitats, but cannot mate due to different breeding seasons, then it is called temporal isolation. Hawthorn-liking and apple-liking fruit files are partially isolated from one another because they emerge from their host fruits and breed at slightly different time of year.

**Behavioural isolation:** Different signals and behaviors create behavioral isolation. Striking colors and calls of male songbirds may attract females of their own species, while females of other species treat them with utmost indifference. Among frogs, males impressively indiscriminate, jumping in every female in sight, regardless of species, when spirit moves them. Females approach only male frogs that croak the Ribbet appropriate to their species. If they find themselves in an unwelcome embrace, they emit release call to cause the male to let go which resulted in few hybrids. It is also called **behavioral isolation**. Here the potential mates meet but cannot mate, due to differences in their courtship displays, or specific signals required for rituals before mating. Signals may be of following 3 types, which stimulate opposite sex for mating:

(a) **Visual stimuli:** Feather displays and dancing in male birds attract the female, e.g. pheasants, birds of paradise and peacocks. Colour, shape of feathers and display pattern is so unique for each species that mating between 2 different species is not possible. Collection of nest material and construction of the nest by weaverbird male is very specific that cannot be imitated by other species.

(b) **Auditory stimuli:** Songbirds like cuckoos, mynas, nightingales, parakeets use auditory signals to attract the sex partner. Sometimes the singing continues for several days before the pair can come together for mating. *Auditory communication* is used by toads, cicadas, frogs, gibbons, monkeys, and jackals.

(c) **Chemical stimuli:** This includes odors of animals that attract opposite partner for mating. Scent of musk deer and musthing in elephants help to attract the females. In insects, particularly in Lepidoptera, females produce highly specific pheromones that are detected by the specialized antennae of males from a distance of about 2 kilometers. Different mating rituals create powerful reproductive barriers called sexual, or behavior isolation that isolates similar species in most animal groups. In dioecious species, males and females search for partner for complex mating rituals and finally copulate, or release their gametes to breed.

Mating dances, songs of males to attract females or mutual grooming, are the examples of courtship behavior that allows both recognition and reproductive isolation. Each stage of courtship depends on behavior of partner. Male only move onto second stage of exhibition if female shows certain responses in her behavior, onto third stage when she displays a second key behavior. Behaviors of both interlink are synchronized in time and lead to copulation, or liberation of gametes. No animal that is not physiologically suitable for fertilization can complete this demanding chain of behavior. The smallest difference in courting patterns of 2 species is enough to prevent mating. A specific song pattern acts as isolation mechanism in distinct species of *Chorthippus*. Even where there are minimal morphological differences between species, behavioral differences can be enough to prevent mating.

*D. melanogaster* and *D. simulans* as twin species do not mate when they are kept together in laboratory. *D. ananassae* and *D. pallidosa* are twin species from *melanesia*. In the wild, they seldom produce hybrids, although in laboratory it is possible to produce fertile offspring. Male's court the females of both species . Females show marked preference for mating with males of their species. A different regulator region found on chromosome II of both species affects selection behavior of females. **Pheromones** play an important role in sexual isolation of insects. These compounds identify individuals of same species and of same or different sex. Evaporated molecules in volatile pheromones serve as wide-reaching chemical signal. Pheromones may be detected only at a short distance or by contact. In *D. melanogaster*, pheromones of females are mixtures of different compounds. There is clear dimorphism in type and/or quantity of compounds present for each and differences in quantity and quality of constituent compounds between related species. It is assumed that pheromones serve to distinguish between individuals of each species.

An example of role of pheromones in sexual isolation is found in corn borers (*Ostrinia*). The 2 twin species in Europe occasionally cross. Females of both species produce pheromones that contain a volatile compound which has 2 isomers, E and Z; 99% of compound produced by females of one species is in E isomer form, while females of other produce 99% isomer Z. Production is controlled by one locus and interspecific hybrid produces an equal mix of 2 isomers. Males, for their part, almost detect the isomer emitted by females of their species, such that hybridization although possible is scarce. Perception of males is controlled by one gene, distinct from one for production of isomers. Heterozygous males show a moderate response to odour of either type. Here 2 loci produce the effect of ethological isolation between species that are genetically very similar. Sexual isolation between 2 species can be asymmetrical when mating that produces descendants only allows one of the 2 species to function as female progenitor and other as male, while reciprocal cross does not occur. Half of wolves tested in the Great Lakes area of America show mitochondrial DNA sequences of coyotes. **Mitochondrial DNA** from wolves is not recovered in coyote populations reflecting an asymmetry in inter-species mating, due to difference in size of 2 species as male wolves take advantage of their greater size to mate with female coyotes, while male coyotes and female wolves do not mate.

**Mechanical isolation:** Here mating is attempted but is not successful, due to mechanical problems like differences in structure of genitalia. Dufour described Lock and key mechanism in genitalia of insects. In *Drosophila* spp., genitalia are so different that copulation is mechanically not possible. Mating pairs may not be able to couple successfully if their genitals are not compatible. Relationship between reproductive isolation of species and form of their genital organs was signaled for the first time in 1844 by Léon Dufour. Insects' rigid carapace sact in a manner analogous to a lock and key, as they allow mating between individuals with complementary structures, that is, males and females of same species. Evolution has led to development of genital organs with increasingly complex and divergent characteristics, which cause mechanical isolation between species. Certain characteristics of genital organs often have converted them into mechanisms of isolation. Organs that are anatomically different may be functionally compatible, showing that other factors influence these complicated structures. **Mechanical isolation** occurs in plants which is related to adaptation and co evolution of each species in attracting certain type of pollinator through a collection of morphophysiological characters of the flowers, happens in such a way that transport of pollen to other species does not occur.

**Gametic Isolation:** Synchronous spawning of many species of coral in marine reefs means that inter-species hybridization can take place as gametes of hundreds of individuals of tens of species are liberated into same water at the same time. About a third of the possible crosses between species are compatible as gamete's fusion leads to individual hybrids. Such hybridization apparently plays a role in evolution of coral. The other two-thirds of possible crosses are icompatible. In sea urchins, *Strongylocentrotus* concentration of spermatocytes that allow 100%fertilization of ovules of same species is able to fertilize 1.5% of ovules of different species.

Inability to produce hybrid offspring is due to the fact that gametes are found at the same time and in same place. Angiosperm has evolved due to gamete incompatibility, which often attract and reward. In plants, the pollen grains of a species can germinate in stigma and grow in co evolution with pollinator species. Sometimes pollinator detained at some point between stigma and ovules, in a way that species fertilization does not take place. Such reproductive isolation is common barrier in the Angiosperms and is called **cross-incompatibility**, or incongruence. A relationship exists between self-incompatibility and phenomenon of cross-incompatibility. In crosses between individuals of a self-compatible species (SC) with individuals of a self-incompatible (SI) species give hybrid offspring. A reciprocal cross (SI x SC) do not produce offspring, because pollen tubes do not reach ovules. This is known as **unilateral incompatibility**, which occurs if 2 SC or 2 SI species are crossed. They prevent site-specific crosses in sympatric populations.

**Ecological isolation:** Lack of interbreeding between populations that occupy distinct habitats within same general area is called ecological isolation. White throated sparrow frequents dense thickets, whereas white crowned sparrow inhabits fields, meadows, and seldom penetrates far into dense growth. Two species may coexist within a few hundred yards of one another yet seldom meet during breeding season. In case of 750 species of fig wasp, each species of fig wasp breeds in fruits of a particular species of fig and each fig species hosts only one species of pollinating wasp. Ecological isolation may slow down interbreeding, but may not prevent gene flow entirely. Other mechanism also contributes to **inter-specific isolation**.

## B. Post-copulatory isolating mechanisms

This can arise between chromosomally differentiated populations due to chromosomal translocations and inversions. If a reciprocal translocation is fixed in a population, hybrid produced between this population and one that does not carry translocation will not be able to complete meiosis. This results in the production of gametes containing unequal numbers of chromosomes with reduced fertility. In some cases, **complete translocations** involving more than 2 chromosomes exist, so that meiosis of hybrids is irregular and their fertility is zero or nearly zero. Inversions also give rise to abnormal gametes in heterozygous individuals but this effect has little importance compared to translocations.

An example of chromosomal changes causing sterility in hybrids comes from the study of *D. nasuta* and *D. albomicans*, which are twin species from Indo-Pacific region. Sexual isolation between them is absent and the F1 hybrid is fertile. F2 hybrids are relatively infertile and leave few descendants which have a skewed ratio of sexes. X chromosome of albomicans is translocated and linked to an auto some which causes abnormal meiosis in hybrids. **Robertsonian translocations** (Figure 5.6) are the variations in the chromosome numbers that arise from either the fusion of 2 acrocentric chromosomes into a single chromosome with 2 arms, causing a reduction in haploid number, or fission of one chromosome into 2 acrocentric chromosomes, in this case increasing haploid number. Hybrids of 2 populations with differing numbers of chromosomes experience a certain loss of fertility, and poor adaptation, because of irregular meiosis which reduce the success of crosses between species. If premating mechanisms fail to prevent mating, then several post mating mechanisms prevent the success of mating and hybridization. Four such mechanisms are given below:

**Gamete mortality/incompatibility:** Mating and sperm transfer takes place but egg is not fertilized. In inter-specific crosses in

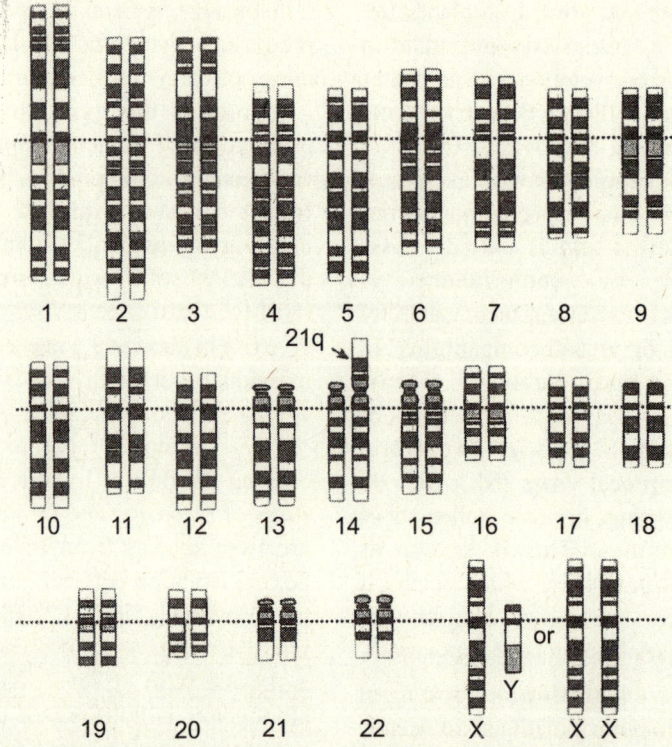

**Figure 5.6**   Translocation karyotype in Down syndrome

*Drosophila* vaginal wall swells killing spermatozoa. When mating occurs between *Bufo fowleri* and *B. valliceps*, sperms fail to penetrate the egg membrane of each other, leading to of gamete mortality. Even if a male inseminates a female, sperm may not fertilize eggs, a situation called **gametic incompatibility**. For example, fluids of female reproductive tract may weaken, or kill sperm of other species.

**Zygote mortality and non-viability of hybrids:**   A type of incompatibility that is found often in plants when the ovule is fertilized but zygote fails to develop, or it develops with reduced viability in resulting individual. This is the case for crosses between species of frog, where differing results are found based on species involved. In some crosses, segmentation of zygote does not occur. In others, normal segmentation occurs in blastula, but gastrulation does not start. In some crosses, initial stages are normal but errors occur in last phases of embryo development showing variation in differentiation of embryo development genes in these species. Such differences determine the non-viability of hybrids. In coral reefs, gamete incompatibility prevents the formation of numerous inter-species hybrids. Similar results are observed in *Culex*, but differences are found in reciprocal crosses. The same effect occurs in interaction between genes of cell nucleus as occurs in genes of cytoplasmic organelles, which are inherited solely from female progenitor through cytoplasm of ovule.

In angiosperms, the successful development of embryo depends on normal functioning of its endosperm. Failure of endosperm development and its subsequent abortion is found in many interploidal crosses, and in certain crosses in species with same level of ploidy. Collapse of endosperm and abortion of hybrid embryo is the most common post-fertilization reproductive isolation mechanism found in angiosperms. Egg is fertilized but zygote dies.

Eggs of many species of fishes may be present in spawning grounds and some may be fertilized by sperms of different species forming zygote, but such zygotes fail to develop due to differences in chromosomes. Zygote develops and hybrid is produced, but is physically weak and inviable due to physiological disturbances in body. It fails to survive for long and dies untimely as found in different duck species. If fertilization occurs, resulting hybrid may be weak, or even unable to survive, a situation called hybrid inviability. Gametic programs directing development of 2 species may be so different that hybrids abort early in the development. Hybrids, if survive as found in certain love birds have great difficulty in learning to carry nest materials during flight and probably could not reproduce in wild.

**Hybrid sterility:** A hybrid has normal viability but is deficient for reproduction, or sterile as found in mule and in many other hybrids. **Sterility** is due to interaction between genes of 2 species involved; or chromosomal imbalances due to the different number of chromosomes in the parent species or nucleus-cytoplasmic interactions as in *Culex*.

Both horses and donkeys belong to (Genus) *Equus*. *Equus caballus* has 64 chromosomes, while *Equus* has 62. A cross between them produce offspring (mule or hinny) with 63 chromosomes that not form pairs as their germ cells do not divide at meiosis. They can cross with each other but

mule or hinny are created by human. Sterility of many inter-specific hybrids among angiosperms is widely recognized phenomenon. A variety of causes can determine the inter-specific sterility of hybrids in plants. These may be genetic, related to genomes, or interaction between nuclear and cytoplasmic factors. Hybridization in plants creates new species. Although hybrid may be sterile, it can continue to multiply in wild through asexual reproduction, be the vegetative propagation or apomixes, or production of seeds. Inter-specific hybridization can be associated with polyploidia and origin of new species are called allopolyploids. *Rosa canina* is the result of multiple hybridizations. A type of wheat is an allohexaploid that contains genomes of 3 different species.

**Hybrid** is viable, physically strong and physiologically sound, but is sterile due to differences in chromosomes and different gene arrangements. Mule is a product of cross between male donkey and female horse and Hinny between female donkey and male horse and both are sterile, albeit physically strong. Sometimes all isolating mechanisms break causing fertile hybrids, which are not reproductively isolated from parents and produce fertile offsprings by introgression. This is instant speciation. Most animal hybrids, like mule or liger (a zoo based cross between a male lion and a female tiger) are sterile. **Hybrid infertility** prevents hybrids from passing on their genetic material to offspring. A common reason is failure of chromosomes to pair properly at meiosis, and the eggs and sperm never develop (Figure 5.7, 5.8).

**Significance of isolating mechanism:** If isolating mechanisms are distinct and specific, only individuals of same species indulge in courtship. Isolating mechanism protects gene pool of a species and prevents hybridization. It prevents wastage of gametes and energy.

**Figure 5.7**    Zonkey: A hybrid

**Figure 5.8**   Liger : A hybrid animal

A weak isolating mechanism leads to production of new species through hybridization. Absence of isolating mechanism leads to production of new species by instant speciation. Geographical isolation followed by reproductive isolation leads to production of new species. Isolating mechanisms protect the identity of a species.

## 5.20   Origin of Species

Speciation is the process of a single species becoming 2 or more species. Some think that certain evolutionary phenomena apply only at speciation and macro evolutionary change cannot occur without formation of new species.

Other believes that major evolutionary change occur without speciation. Changes between lineages are the extension of changes within each lineage. Paleontologists belong to former group and geneticists in the latter. Two types of speciation are **allopatric** and **sympatric** speciation which differs in geographical distribution of populations. Allopatric speciation (Figure 5.9) is the most common which occurs due to splitting of a population into 2 geographically isolated subdivisions that organisms cannot bridge. The 2 populations' gene pools change until they could not interbreed, even if they were brought back together. **Sympatric speciation** occurs when 2 subpopulations become reproductively isolated without first becoming geographically isolated. Insects found in a single host plant provide a model of sympatric speciation. When a group of insects switched host plants, they would not breed with other members of their species still living on their former host plant. The 2 subpopulations could diverge and speciate. *Rhagolettis pomenella* started to infest apples in 1860's. Prior to that it only infested hawthorn fruit.

Feder and others found that 2 races of *R. pomenella* have become behaviorally isolated. Allele frequencies at 6 loci are diverging. Significant linkage disequilibrium is found at such loci, showing that they are hitchhiking on some allele under selection. Sympatric speciation is called microallopatric speciation to emphasize that subpopulations are physically separate at an ecological level. Some think a series of small changes in each subdivision gradually lead to speciation. Founder effect prepares the stage for rapid speciation. Few key genes could change and confer reproductive isolation called genetic transilience. Margulis thinks most speciation is caused by changes in internal symbionts. In the plant *Tragopogon*, *T. mirus* and *T. miscellus* have originated in past 60 years as

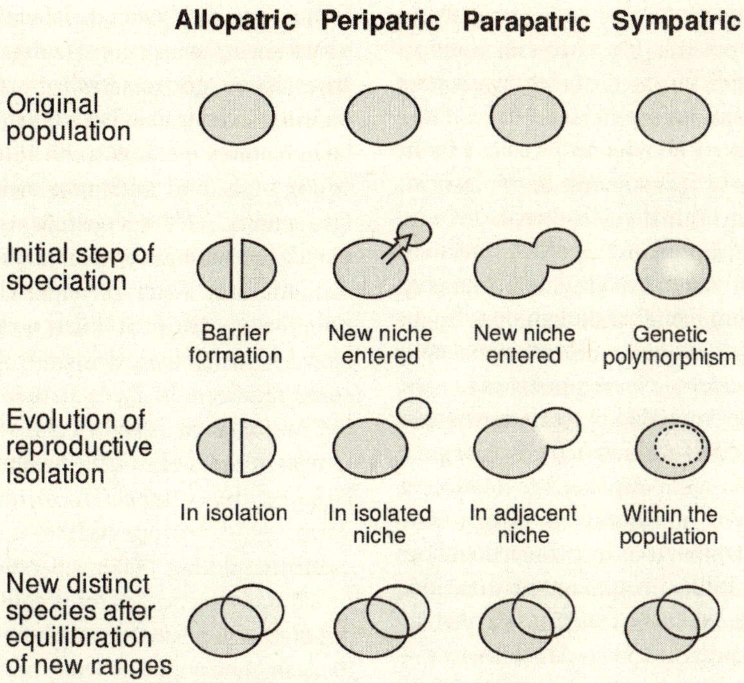

## Allopatric   Peripatric   Parapatric   Sympatric

Original
population

Initial step of
speciation

| Barrier formation | New niche entered | New niche entered | Genetic polymorphism |

Evolution of
reproductive
isolation

| In isolation | In isolated niche | In adjacent niche | Within the population |

New distinct
species after
equilibration
of new ranges

**Figure 5.9**   Various modes of speciation

2 new species. New species were formed when one diploid species fertilized a different diploid species and produced a tetraploid offspring which could not fertilize or be fertilized by either of its 2 parent species. This is reproductively isolated.

## 5.21   Basic Modes of Speciation

Evolutionary process leading to production of new species involves some reproductive isolation between populations of single species until they become unable to interbreed under natural conditions and produce viable or reproductively capable offspring. Evolution of life on earth from simple matter is accepted. Natural selection supports the belief that all life has emerged from a common ancestor. Natural selection acts to decrease, or conserve the amount of genetic information in a population. For transforming an amoeba into

an ape, an increase in genetic information is required. Biologists insist on increasing complexity of life over time. The change evidenced in fossil record is the result of change by natural selection within the created kinds. Dogs exhibit different types, but never produce anything other than dogs. The process leading to formation of new species is the speciation. Four basic modes of speciation are allopatric, parapatric, peripatric, and sympatric.

In adaptive radiation one species gives rise to many. Derived species do not interbreed, or interbreed with limited degree of genetic exchange. Divergence of signaling and response system involved in mate choice occurs as a correlated effect of adaptive divergence. A sympatric phase of the speciation process then follows the allopatric phase, when a derived species disperses into the environment of another. Coexistence in the same habitat relies on ecological and reproductive differences arisen largely, or completely in allopatry. If this was the universal route to species multiplication, those environments without physical barriers should have fewer species. Relatively homogeneous lakes exhibit high diversity of fish species. In the African Great Lake, cichlids diversity traces back to one, or a few colonizing species, and they must have evolved in broad scale geographic sympatry within the lake, though possibly in local parapatry, or allopatry. How a population splits into 2 under such circumstances is not clear. How does reproductive isolation evolve through disruptive selection in the face of gene flow counteracting divergence? A solution in plants, but rare in animals is polyploidy: **autopolyploid** through chromosomal doubling, or allopolyploid through hybridization. Adaptive radiation implies speciation, whereas speciation implies neither adaptation nor a radiation.

**A. Allopatric speciation:** In Greek, allopatric means with a different fatherland. This common speciation involves physical separation of a species into 2 groups due to environmental changes, **tectonic plate** movement, or eruption of land mass, creation of waterways, or due to mountain range. Due to such a barrier, the 2 separated groups fail to interact with each other. Each group accumulates changes to adapt to the environmental changes and the 2 groups become distinct from each other and cannot breed with each other. Formation of Isthmus of Panama about 3.5 million years ago separates the Pacific Ocean and the Caribbean Sea causing formation of some new aquatic species. Formation of this land mass blocked the gene flow between populations of marine animals of same species. Two pork fish species *Anisotremus virginicus* found in the Pacific Ocean, and *A. taeniatus* found in the Caribbean Sea are the results of the **allopatric speciation**.

Formation of Hawaiian Islands resulted in evolution of Canada geese (*Branta canadensis*) from Nene goose (*Branta sandvicensis*), endemic to Hawaiian Islands. The Nene geese have features like less webbed feet, long and strong nails, and thick foot pads to walk on lava plains. This speciation occurs when 2 populations are geographically isolated from one another. After physical separation of population, little or no migration occurs between them. Physical separation occurs when some members of a population drift, swarm or flow to remote island, or part of a population remain in patch of suitable habitat that became isolated by change in climate, or by geographical process like volcanism, or **continental drift**. After separation, pressures of natural selection differ in 2 locations, or if populations are small enough for genetic drift to occur, then population may accumulate large genetic differences to become separate species. Allopatric speciation also called geographical speciation is a slow process.

Mexican platy fish, *Xiphophorus conchianus* inhabits Northern most part is restricted to a single river system and is sterile in crosses with other species. *X. variatus* with 3 distinct populations are found in 3 rivers, but can produce fertile hybrids with one another. *X. maculatus* is found in many streams and hybridize with those from other populations. In the past, perhaps an ancestral platy fish colonized all these streams. Subpopulations, isolated in different streams have been diverging since then. Some have attained full reproductive isolation, others have not. Large areas are required for geographical speciation to separate different population from one another for a time which is sufficient to evolve reproductive isolation. Smallest islands from where a single species have given rise to 2 species are known for several organisms. Birds and mammals require larger areas than fish, amphibians, and reptiles. Isolation of a subpopulation from remainder of ancestral species is the first prerequisite for allopatric speciation. Secondly, divergent selection pressure establishes new barriers of reproductive isolation based on number of genetic differences. Physical barrier divides the population into small units. Separated populations on islands evolve differently by constant genetic changes, translocations, and inversions. They change into races, subspecies and if time are long, into species. Sometime the population separates, when they are carried by chance into new area, or some individuals cross over barriers and did not return back. Fourteen **Darwin finches** (Figure 5.10) species found in various islands of Galapagos (Figure 5.11) archipelago originated from single population blown with storm from Ecuador in South America by chance.

**Simple model of allopatric speciation:** Speciation occur when a small subset of a large, interbreeding population becomes reproductively isolated from large population. In a large population its members freely interbreed. Any new allele that appears within this population can spread through population. When the new allele is deleterious mutation it is removed from population rapidly and completely if mutation is dominant, or slowly if mutation is recessive .If new allele is neutral or beneficial, then it spread throughout the population, so long as all individuals are freely interbreeding. If a small sub-population of this original population is reproductively isolated from it, then any new allele that occurs in either population do not spread to other. So the longer these 2 sub-populations are kept reproductively isolated from each other, the more different alleles accumulate in each. Accumulation of such non-identical genetic elements result in partial, or complete genetic incompatibility between 2 sub-populations, and

they then qualify as 2 species. Mayr's definition of a species advocates that key to speciation is anything that cause reproductive isolation. For example in **geographic isolation** organisms are so far apart that probability of their mating with each other is zero. This is called allopatry and speciation that results from allopatry is allopatric speciation in which reproductive isolating mechanisms are incidental, but not the result of natural selection.

**B.    Sympatric    speciation:** This speciation occurs due to genetic divergence among few members inhabiting the same geographic area, not due to increase in geographic distance. While sharing a common range, certain members adapt to particular aspect of range leading to development of new behavioral, or genetic traits. Such adaptations include specific host preference, shelter, food or diurnal to nocturnal habits, and mutations. Over a period of time, local selective pressure results in huge difference to an extent that they cannot interbreed and form new species. Cichlids inhabiting Lake Victoria shows a diverse range of species that have formed through such speciation. Here, the clarity of water and color of ambient light varies with varying depths of lake. In shallow water, the ambient light is blue, whereas in deep water red light dominates. Genetic variation increases sensitivity to red and blue light enabling some fish to see clearly in red light, and some to see clearly in blue light. Fish with high sensitivity to blue light remains in shallow waters, and those with high sensitivity to red light restrict themselves to deep regions. **Cichlids** exhibit coloured patterns. Female cichlids prefer bright males of their own species. Female cichlids inhabiting the shallow region, being more sensitive to blue light prefer blue- coloured male. Female cichlids in deep region perceive red coloured males over blue-coloured cichlids. Thus, reproductive isolation

1. *Geospiza magnirostris.*
3. *Geospiza parvula.*

2. *Geospiza fortis.*
4. *Certhidea olivasea.*

**Figure 5.10**   Finches from Galapagos Archipelago

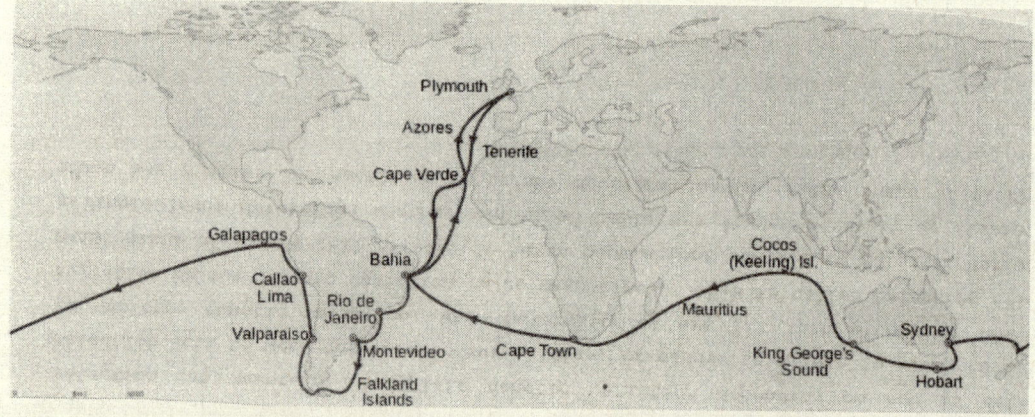

**Figure 5.11**   Map showing the route of travel of Darwin

happens between the 2 populations giving rise to 2 different species. The North American apple maggot fly (*Rhagoletis pomonella*) (Figure 5.12) perhaps diverged from its hawthorn-feeding co-members through sympatric speciation. These flies exhibit host fidelity and specifically mate on or near the fruit of host plant. Ancestor of these flies fed on hawthorns. Some members feed on apples, when apples were available in North America. These individuals prefer to mate with other individual that fed on apples and did not prefer the counterparts that fed on hawthorns. The 2 groups of flies thus developed adaptations specific for feeding on specific fruit, and failed to feed on both. This led to formation of new population that is reproductively isolated from hawthorn-feeding maggot. **Sympatric** means with the same fatherland which occurs within the single geographical area, as in Hawaiian land snails. This requires limited gene flow. Two mechanisms that can reduce gene flow between members of single population in a given area are ecological isolation and chromosomal aberrations. In chromosomal aberration, gametes and somatic cells containing multiplies of haploid number of chromosomes are the **polyploids**. Polyploidy is common in plants than in animals. About 47% of all angiosperms are polyploids. Most

**Figure 5.12** *Rhagoletis palmonella*

animals cannot self fertilize, or reproduce asexually. If an animal produces a tetraploid offspring, this would have to mate with a member of diploid parental species and would produce all triploid offsprings. Which are almost sterile. Polyploidy is known from asexually reproducing groups like isopods, bagworm moth, weevils, and flies. This may be a result of duplication of chromosome of a single species (autopolyploidy), or by combination of chromosomes from 2 species (**allopolyploidy**). This speciation occurs in freely interbreeding populations without geographical isolation, but sometimes host preferences may create pockets within same area. Reproductive isolation is produced by polyploidy or hybridization. Hybridization normally produces sterile offspring, but sometimes fertility is possible by introgressive hybridization. Also if hybrids become polyploids, they possess full chromosome complements and will be fertile and can create a new species instantly.

**C. Parapatric speciation:** This speciation occurs due to partial spatial isolation of populations combined with small overlap in their ranges and significant gene flow among the populations. The gene flow reduces due to changes in the local conditions and the reproductive isolation between 2 populations. Such process occurs when populations belonging to same species inhabit adjacent areas experience changes in climatic, or geographic conditions and results in varying selection pressures in 2 regions. One or both populations may diverge from one another for this event. Members of a population prefer to mate with members of the same region, though mating with members of adjacent region is possible and divergence occurs to extent of reproductive isolation. A population of Buffalo grass (*Anthoxanthum odoratum*) developed tolerance to heavy metal and became divergent from adjacent population. A group of such plants near mines and area of mining activity cause contamination of soil with heavy metals like zinc and lead, which occurred in soil within certain distance. Plants present within this distance developed tolerance to heavy metals. Neighboring plants not subjected to such selection pressure are intolerant, although 2 populations are continuous and gene flow is possible among them. The populations evolved different flowering times, limiting the gene flow. **Ring species** of *Ensatina* salamander (*Ensatina eschscholtzii*) found in the Pacific coast has evolved from common ancestor due to varied selection pressure through adjacent region.

**D. Stasipatric speciation:** This chromosomal model of speciation proposed by White and others suggests chromosomal rearrangements and polyploidy as a driving force for speciation. This may be a separate mode of speciation, or a special case of sympatric speciation. Here, members of a population in particular range face uniform selection pressure, undergo spontaneous chromosomal changes like polyploidy. Polyploidy is common in some plants, in some races of grasshopper and stick insect. Here new species arises sympatrically, first by rearrangement of chromosome within geographical range of parent species. New population spreads within range of parent species, by parapatric distribution. Seven species of Australian grasshopper having parapatric distribution, with zones of hybridization and perhaps all species are chromosomally distinguished and that their differences arose sympatrically. (Figure 5.13-5.16)

**E. Peripatric speciation:** Proposed by Ernst Mayr, in such speciation, few members found in periphery of range undergo reproductive isolation to form new species. It is sometimes considered as a variation of **allopatric speciation**. Mayr observed that populations inhabiting the peripheral regions of a range often showed certain variance with respect to other members of its species and spread in range. This divergence further increased if they enter the niche of parental population. Members in periphery remain as distinct and reproductively isolated. Peripatric speciation occurs when a small group colonizes new habitat, or if periphery of range fragments off to become isolated. Both these events cause development of new characters due to unique selection pressure of that region, forming new species. Speciation because of fragmentation of peripheral part is similar to allopatric speciation, but isolated population is small in comparison to ancestral one. Evolution of polar bears (*Ursus maritimus*) from brown bears (*Ursus arctos*) is the best examples. During the **Pleistocene**, glaciation caused the isolation of small population of brown bears. The individuals of this small population acquired physical and physiological characters like white fur for camouflaging, swimming ability in extreme cold water and over long distances, and tolerance to extreme cold. **Polar bear** is not reproductively isolated from brown bear and 2 species breed to produce hybrid bear called grizzly bear. *Culex pipiens f. molestus* presents a case of peripatric speciation. This mosquito has evolved from *C. pipiens*. Some *C. pipiens* members migrated into London and adapted to subterranean environment. They acquired characters like adaptation to warm conditions, loss of cold tolerance and hibernation, mating all year round and unlike their ancestral species. *C. pipiens f. molestus* developed the ability to bite rats, mice, and humans, the ancestral species bite only birds.

**Directional Speciation:** Directional change within populations can result if offspring tend to differ from their parents

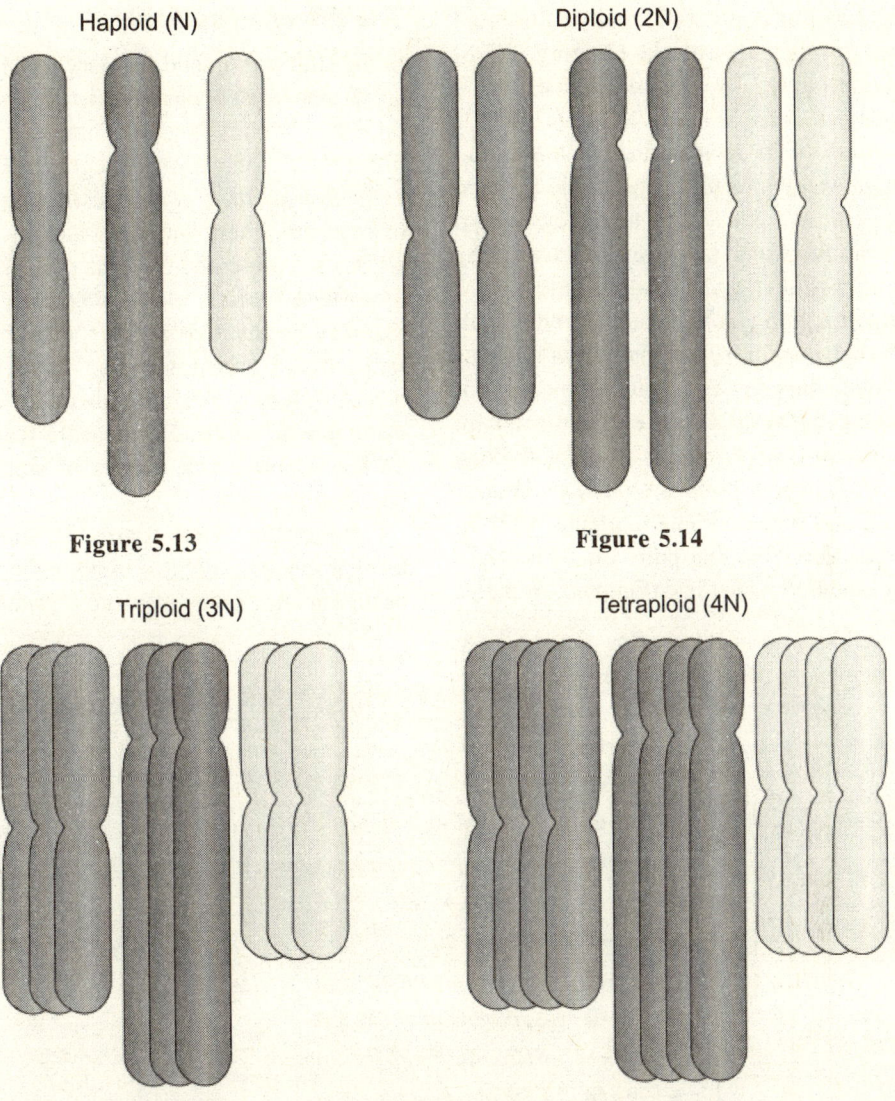

Haploid (N)

Figure 5.13

Diploid (2N)

Figure 5.14

Triploid (3N)

Figure 5.15

Tetraploid (4N)

Figure 5.16

Figure showing N, 2N, 3N, 4N number of chromosomes

frequently in one direction than another and species-level trends can derive from a tendency for new species to differ from their ancestors in a biased manner (Finarelli, 2007). Certain types of changes are inherently more likely than others to occur when new species arise, as some mutations are more likely than others to appear in offspring. In this sense, Gould (2002) envisions the analog of "**mutation pressure**" at species level as a cause of trends. Similar pattern may result from external factors, if the niches available for new species to occupy consistently differ in same way relative to niches occupied by existing species (Grant, 1989). Once a change occurs in a new species, it cannot be undone. According to "Dollo's

Law,"(Lonning e al., 2007) many substantial changes during the course( Gregory, 2008) of evolution are irreversible. Once lost, a complex feature cannot be regained as probability of its reemergence is too slight. Possible exceptions to this principle is noted (Domes et al., 2007), but tendency toward irreversibility would produce a trend resulting from *directional speciation*. Changes in a certain direction would create a moving wall under such condition, making further changes possible in only one direction. Wagner (1996) proposed that reversals of evolution remain possible until a certain threshold has been crossed, at which point the *lineage* remains trapped and may continue to change only in one direction from that point on. This could result in a large-scale trend (Figure 5.17, 5.18).

## The Effect Hypothesis

Some authors contend that large-scale trends are the end result of directional natural selection operating within species. Others argue that differential speciation (Gregory, 2008), or extinction perhaps even constituting a form of more important species selection. A third alternative presented by Vrba (1983), "**effect hypothesis**" which view, anagenetic change that may be adaptive within species can have incidental consequences for species diversification, or extinction, thereby generating cladogenetic trends. **Large-scale trends** can be non-adaptive side effects of small-scale, adaptive processes.

**Remarks:** Both constraints and higher-level processes may cause trends either passive or driven and maybe influenced by a number

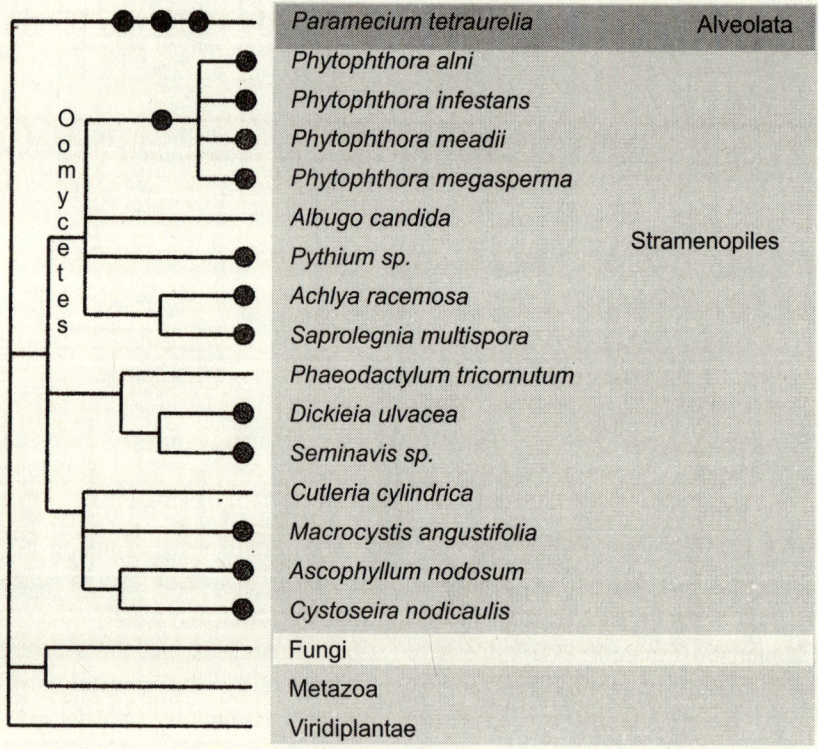

**Figure 5.17**  Polyploidy in fungi

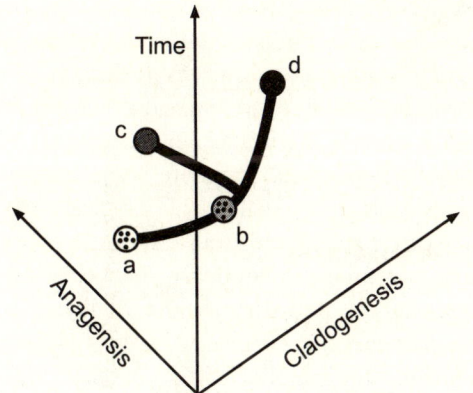

**Figure 5.18** Anagenesis and cladogenesis species a anagenetically changes to b specis, and b species cladogenetically splits into species c and d

of factors. Many trends are localized either taxonomically or in time. There is no evidence to support popular conceptions of evolution as an inexorable march in any direction, be it toward larger size, greater complexity, heightened intelligence, or any other trait. Rather, processes and patterns of evolution are, like its products, intriguingly diverse.

**Genetic divergence:** Two populations become separate species if during the period of isolation, they evolve large genetic differences, so they can no longer interbreed, or produce vigorous fertile offspring if they are reunited. In both small and large populations, different environmental pressures in separate environment may favour evolution of large **genetic differences**. Origin of new species falls into following categories:

**Operation of reproductive isolation:** Genetic divergence during period of isolation is a necessary condition for origin of new species, but it is not sufficient unless part of that genetic divergence happens to cause the development of something that ensures reproductive isolation. Structural and/ or behavioral modifications that prevent

interbreeding are called isolating mechanisms. Once such mechanisms have arisen, they have a clear **adaptive value**. Separate species tend to be genetically distinct in ways that adapt them to different environments. Any individual that mate with a member of another species probably produces unfit or sterile offspring, thereby, wasting its genes and contributing nothing to future generations. Thus, there is strong environmental pressure to avoid mating between species.

**Ecological Opportunity:** Adaptive differentiation depends on the availability of ecological resources and properties of the organisms that facilitate their evolution. **Adaptive landscapes** provide visualizing opportunities for diversification in the environment. Adaptive landscape first developed by Sewall Wright in 1932 explores possibilities and limits to change of genotype frequencies in a population. An adaptive landscape represents variation in fitness in relation to combinations of traits, or environmental conditions. There are hills and valleys in landscape. **Fitness** is the maximum when a population occupies the top of a hill. Each hill can be equated with ecological opportunity. Spatial distribution of hills reflects another property of environments. Some are closer and more within reach of evolving organisms than others. The proximity governs the sequence in which hills can be climbed through natural selection. Neither the environment nor the landscape it represents is static. Flowering plants from diverse families have invaded freshwater from the land at least 50 times, with repeated evolution of floating, bladder like structures in the leaves. Dioecy, the presence of 2 separate sexes in a population, has evolved repeatedly in various lineages of plants colonizing the Hawaiian archipelago. Algal-scraping, mollusc crushing, fish-scale-scraping cichlid fish and other ecomorphs seen in Lake Malawi are seen in Lakes Victoria and Tanganyika, though their origins differ and their evolution is independent.

Organisms diversify in response to ecological opportunities. Coping with ultraviolet radiation and predation from fish and invertebrates is a challenge to *Daphnia* in upper waters of lakes. Repeated response of different *Daphnia* lineages has been the evolution of melanism in the face of the first challenge, and head shields and spines when confronted by the second. Evolution in similar environments is predictable to a degree, because ecological opportunities are similar. History of the Caribbean *Anolis* lizards has been reconstructed from a molecular phylogeny. It shows that the 4 largest islands of the Greater Antilles were colonized by different species, which then underwent parallel **evolutionary diversification** into same **ecotypes** occupying same spatial niches. Subsequent evolutionary diversification on a given island took the form of variation on these 4 themes. Evolutionary history repeats itself at the higher taxonomic levels. As pointed out by Osborn, marsupial radiation led to convergence in many phenotypic and ecological traits with eutherian mammals: browsing, burrowing, and gliding habits and associated morphological adaptations evolved independently in 2 groups, on separate continents. **Parallel radiation** within 2 continents produced convergence between them. This is strong indirect evidence for driving force of environmental factors in adaptive radiations.

Ecological opportunity is one side of a coin; organism responsiveness or intrinsic evolvability is the other. Importance of intrinsic factors is hinted at by differences in taxonomic groups in their radiation. Speciation in lacustrine fish appears to be faster than speciation in terrestrial birds and arthropods. Time required for speciation is estimated to be 15,000 to 300,000 years for fish, and half a million to1 million years for birds and arthropods on islands. Diversification of cichlid fish has been rampant, but most other fish taxa in similar circumstances in African lakes have failed to diversify beyond a few species. Since various environmental factors are not clearly implicated, there are some intrinsic genetic, developmental, or physiological factors that differ among the groups. Intrinsic factors may potentiate evolutionary change, or may constrain it. Radiating groups of these fish cluster together in the grand African phylogeny, implying a common inheritance of one, or more predisposing factors for radiating. These factors are poorly known. Evidence of intrinsic potential is correlative and indirect. It is the association between evolution of a novel trait and several related species that share the trait. Prominent novel traits are the pharyngeal jaws of cichlid fish which are plate's with tooth like projections in roof of mouth that enable them to split the functions of procuring and processing food into oral and pharyngeal regions of mouth. Fish with pharyngeal jaws have diversified greatly in oral jaws and associated diets, as different as grazing algae and catching other fish. Pharyngeal jaws are not sufficient to explain cichlid radiation, as groups that have not radiated also possess them.

Evolution of antifreeze glycoproteins by fish in the suborder Notothenioidei is another example of a key innovation that facilitated invasion of Antarctic waters and diversification in a relatively empty environment. Detailed comparative work is needed to determine whether a key innovation resulted in an enhanced diversification. Nectar spurs in columbines (*Aquilegia*) vary in shape and color, which affect reproductive isolation by attracting different pollinators, and facilitated speciation, as there are more species in clades with than without spurs. The evolutionary invention of resin canals by some plants, constituting a defense against chewing insects, is another example.

**Species Interactions:** An important factor in rapid diversification in a new environment is the absence of competitors- the related species. During development of concept of **adaptive radiation**, absence of competitors was emphasized as a key facilitating factor. Islands were viewed as empty environments when the first colonists arrived, and diversification proceeded until all ecological niches were filled. The **Gaussian principle** that no 2 species can occupy the same ecological niche is, in modern language, no 2 species can occupy the same adaptive peak. If all peaks in a landscape are occupied, there is no ecological opportunity for a radiation to occur, or there is no opportunity for its continuation. Objective reality of adaptive peaks can be shown only with environmental data. For Darwin's finches in the Galapagos, the peaks have been quantified on several islands by measuring food supply. The fitness has been estimated for each island with a given seed size profile by measuring population sizes in relation to beak sizes. The adaptive landscape for the finches is rugged. Only one species is ever associated with one peak. Different species are sometimes associated with the same peak on different islands. They are interchangeable, although ecologically incompatible. Competitive interactions between species lead to exclusion of one species by another, or to evolutionary adjustments. Competitive displacement in food-related body size and shape among sticklebacks has been shown in controlled conditions in ponds.

Limnetic species are small and slender than benthic species. A solitary species intermediate in size and shape between them suffered slower growth in the presence of a limnetic species. As a result, their average body sizes diverged. Under natural conditions character displacement has been observed in Darwin's finches. During drought large members of the medium ground finch population died at a high rate. They were out competed by a larger efficient species, the large ground finch, when feeding on a reduced supply of large and hard Prey size. Mean fitness for the phenotype in a given combination of body size and prey size is indicated by height of surface.

**Hybridization:** One possible potentiating factor is introgressive hybridization. An exchange of genes can result in enhanced genetic variation under the right ecological condition and lead to formation of a new species. Study of Loren Rieseberg on sunflowers (*Helianthus*) and Tom Whitham on poplars (*Populus*) in western North America has shown that introgression of certain genes is selective according to nature of environment. Populations of sunflower hybrids in peripheral and ecologically extreme environments undergo large-scale genome reorganization, leading to reproductive isolation from parental populations. These findings help explains early stages of major radiations. Introgressive hybridization is widespread in many young radiations, including Darwin's finches of the Galápagos, African cichlids, and *Heliconius* butterflies, but is almost absent in older radiations such as Hawaiian honeycreeper finches and Caribbean *Anolis* lizards. Seehausen suggested that gene mixing through hybridization does more than accompany a radiation. It creates a hybrid swarm and facilitates a radiation. Cichlids of Lake Victoria, with a history of 15,000 years, are far younger than fish in Lake Malawi and have less mitochondrial gene diversity, yet they have same level of nuclear diversity as their Lake Malawi counterparts. Similar evidence of incongruence between nuclear and chloroplast genes is found in one lineage of Hawaiian silvers words. Pattern is even repeated in human history.

**Timing of Speciation and Diversification:** Adaptive radiation starts with an increase in number of species, or higher taxa, at a diminishing rate with an increase in **phenotypic disparity**, or morphological differences among the species. Fossil record shows high extinction rates at times of geophysical perturbation, followed by speciation that is initially rapid and than slower. Fossils are missing for most living taxa. Molecular phylogenies have been used to see whether the early burst phenomenon with subsequent decline is consistent in adaptive radiations. A phylogeny describes the increase in number of species at any one time from the starting point. The same pattern species diversity results from a declining speciation rate, as expected, or an increasing extinction rate. Daniel Rabosky and Richard Glor constructed 12 ways in which species multiplication might be expected to occur as continuous processes of birth and death. Using molecular phylogeny of Caribbean *Anolis* lizards to reconstruct the pattern of diversification, they found that one model out performed others in fitting the data. In this model, speciation rates declined toward a presumed equilibrium on 3 out of 4 islands. Agamid lizards in Australia, warblers and Lampropeltine snakes in North America, and pythons in the Australo-Papuan region all declined in rate of diversification. Jonathan Losos and colleagues found that most of the groups they studied did not fit the early burst pattern. Darwin's finch species increased in parallel with an increase in the number of islands of the Galapagos archipelago. In this case, diversification was facilitated, if not actually driven, by expanding spatial and temporal ecological opportunities. Seehausen found a different pattern of diversification in a changing environment.

Species diversity of the rapidly evolving cichlid fish in Africa increased in pulses interspersed with periods of relative stasis. Thus speciation and extinction may not vary in coordinated manner through time. With **ecological opportunity** being maximal at beginning of a radiation, initially high rates of phenotypic diversification is expected, which then decrease as the environment becomes progressively filled with a large number of ecologically diverse exploiters. This expectation has been tested with large number of species that make up the Caribbean *Anolis* faunas. Expectation was met with rates of diversification of 2 ecological traits, body size and limb length, decreased with increasing radiation, and decreased as number of inferred potential competitors increased. These results matched **paleontological evidence** over larger time spans and taxonomic categories. Morphological disparity increased among hard-bodied invertebrates in so-called Cambrian explosion and in certain groups of organisms following the end-Permian and end-Cretaceous mass extinctions. Branching points in a phylogeny can be estimated, but morphological change independent of speciation can be dated only with fossils, and then with difficulty. Morphological change is likely to accompany speciation and be a vital part of it, but change may continue without further speciation. Anagenesis might contribute to diversification of Lampropeltine snakes in North America as morphological disparity increased prior to the Pliocene, about the time when rates of speciation decreased. Selective extinction through competitive interactions is another process that leads to the same increase in morphological disparity.

## 5.22  Ring Species

When the species are present in circular arrangement in a central region they are called ring species. The original *Ensatina* spreads southward, from the Oregon and Washington, along 2 continuous regions of the Central valley, the coastal arm and inland arm, and evolved based on selection pressures. About 20 *Ensatina* species are known, and interbreeding among them occurs in about 13 zones. The 2 arms of region meet in Southern California, with diverged *Ensatina* that interbreeding is not possible. This is a variation of sympatric speciation and does not require geographic isolation. When structural rearrangements of chromosomes, or other genetic changes occur in small number of individuals, reproductive isolation may develop. Individuals carrying new rearrangements become isolated from other members of population and become new species as reported in *Spalax ehrenbergi*. Its 2 northern subspecies have 52 and 54 chromosomes, central Israeli subspecies has 58 and Southern subspecies has 60 chromosomes. Hybrids of these subspecies rarely occur and when occurs, fertility is reduced. **Fossil evidences** show that all subspecies are descended from an ancestral species *S. mimtus* lived in said area about 500000 years ago. This speciation occurs in widely distributed species with continuous distribution without geographical barrier separating populations. Large populations are broken into clines, without effective geographical barrier between them, but distance and habitat differences themselves may form barriers. Such populations with overlapping boundary serve as zones for hybridization. Eleven Partula species on the island of Moorea have originated by parapatric speciation.

**Ring species** (Figure 5.19) provide a unique glimpse of speciation. At least 23 cases of ring species have been proposed. Most of them are not so clear as in salamanders and warblers. Most cases have major gaps in distribution in chain of populations connecting the terminal forms. Some cases have more than one species boundary in the ring of populations. Most cases have one thing in common. In one place; there are 2 species, and in another area the boundary between species is difficult to detect. Ring species are rare for several reasons (Wake, 2001) as follows:

1. Their formation needs unusual geographic conditions, in which a species could expand around a geographic barrier through a continuous ring of suitable habitat. Range expansion must occur slowly enough that the 2 expanding fronts have time to diverge before they meet on other side of barrier and size of barrier must be large compared to distance that individuals disperse.

2. Taxonomic rules that are used by biologists to classify organisms create a bias against recognizing ring species. Under these rules, a ring species must be classified either as a single species, or as 2 species. Both schemes of classification conceal the fact that there is gradual variation between reproductively isolated forms.

3. Ring species might be rare because many of them were destroyed before they could be discovered. Greenish warblers and *Ensatina* salamanders show three fundamental ways that ring species tell us about evolution.

   (a) Ring species provide strong evidence for evolution causing appearance of new species, demonstrating that much small change can eventually accumulate into large differences between distinct species. Some think that evolution can only cause limited change within a species and cannot lead to formation of new species. Ring species shows that variation between species is qualitatively similar though different in degree within the species.

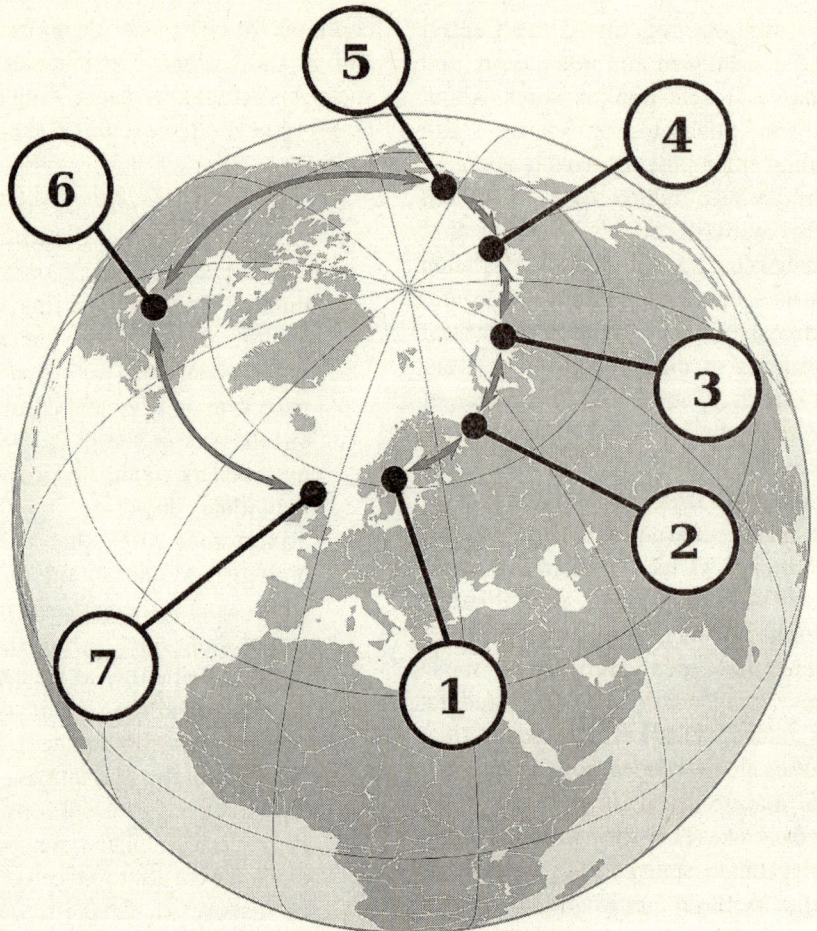

**Figure 5.19**   The Larus gulls inbreed in a ring
1. *L. Fuscus*, 2. S. Siberian population of *L. fuscus*, 3. *L. leuglini*, 4. *L. vegal*,
5. *L. smithsanianus*, 6. *L. argentatus*

(b) Ring species allow a reconstruction of history and causes of divergence during speciation, as spatial variation may show change through time .Without the rings of populations connecting the terminal forms; we would have little understanding of history of divergence of greenish warbler songs or *Ensatina* color patterns.

(c) Ring species provide evidence that how speciation could result from geographic isolations (Figure 5.20-5.22).

## 5.23   Hybrid Gender: Haldane's Rule

**Haldane's Rule** states that when one of 2 sexes is absent in interspecific hybrids between 2 species, then gender is not produced, or is

**Figure 5.20** Greenish warbler ring

**Figure 5.21** Herring gull and Lesser backed gull with clear differences

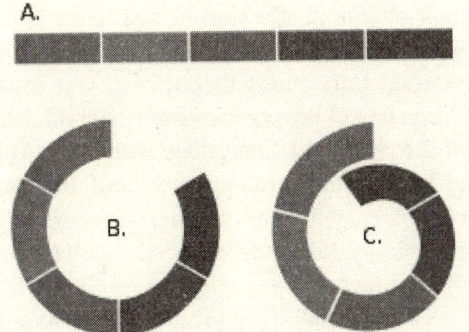

**Figure 5.22** The interbreeding populations in this circular breeding sroup are referred to as ring specis

rare or is sterile or is the heterozygous (or heterogametic) sex. In mammals, there is evidence to suggest that this is due to high rates of mutation of genes determining masculinity in the **Y chromosome**. Haldane's rule simply reflects that male gender is more sensitive than female when sex-determining genes are included in a hybrid genome. But there are also organisms in which heterozygous sex is female. Birds and butterflies follow this law. It is not a problem related to sexual development, nor with the sex chromosomes. Stability of hybrid individual development requires the full gene complement of each parent species, and hybrid of the heterozygous sex is unbalanced. Hybrid non-viable male lacks the X chromosome of *D. simulans* which was found by mating *D. melanogaster* females with *D. simulans* males.

## 5.24 Incompatibility Due to Microorganisms

Presence of microorganisms in cytoplasm of certain species can cause post zygotic isolation. Presence or absence of these organisms causes non-viability of corresponding hybrids. For example, in *D. paulistorum* the hybrid female is fertile, but males are sterile due to presence of mycoplasma in cytoplasm, which alters spermatogenesis leading to sterility. Incompatibility or isolation can arise at an intra-specific level. Factor determining sterility is found to be present or absent of a *Wolbachia* and the population's tolerance, or susceptibility to these organisms. Such inter population incompatibility may be eliminated in laboratory through administration of a specific antibiotic to kill **microorganism**. Similar situations are known in several insects, as around 15% of species show infections caused by this symbiont. In some cases, speciation process has taken place because of incompatibility caused by these bacteria. Two wasp *Nasonia* *giraulti* and *N. longicornis* carry 2 strains of *Wolbachia*. Crosses between infected population and noninfected population produce a nearly total reproductive isolation between semi-species. If both species are free from bacteria, or both are treated with antibiotics there is no **reproductive barrier**. *Wolbachia* induces incompatibility due to weakness of hybrids in populations of spider mites (*Tetranychus urticae*), between *Drosophila recens* and *D. subquinaria* and between species of *Diabrotica* (beetle) and *Gryllus* (cricket).

## 5.25  Speciation by Polyploidy and/or Hybridization

Sexually reproducing organisms generally have a set of 2 for every chromosome; one came from mother and another from father. This makes most sexually reproduced organisms diploid, meaning that they have 2 copies of each chromosome. Organisms can have more than 2 copies of a chromosome which are called polyploids, meaning that organism have many copies of each chromosome. Polyploid condition can arise instantly during formation of gametes and through variety of other ways, including hybridization .When a ployploid organism is produced, it is unable to reproduce with other organisms of its kind unless they also have the same number of chromosomes. Polyploids with an odd number of chromosomes, such as 3 or 5, are not able to reproduce under any circumstance.

The condition of polyploidy arises commonly in plants for a variety of reasons, including how plants make gametes, and volume of gametes that plants produce, and frequency of hybridization. Between 20% and 30% of all plant species are polyploids. Polyploids also exist among animals, but they are rarer. **Polyploidy** is significant in plants as many plants can self-fertilize, so even when a spontaneous occurrence like this happens

and there are no other polyploids around, the plant can at least mate with itself. High occurrence of polyploidy in plants causes its multiple independent occurrences, so that the plant can find a mate. Trees that lives for hundreds or thousands of years, if a polyploid is created and no other polyploidy mates for the tree exists immediately, since another polyploid is crated in the region from the same original species sometime over the next hundreds or thousands of years the 2 can potentially mate. This makes the chances for polyploids finding mates pretty high. As polyploids only breed with plants that have same chromosome count, each time a polyploid plant is created with a new chromosome count that did not previously exist among its parent species and a new species is formed, because new plant can no longer mate with any of other plants of its kind, aside from other polyploids with same chromosome count. This new polyploid now goes down its own road of evolution. Genetic alterations in new polyploid lineage accumulate distinctly from genetic alterations in parent lineage, leading to further divergence over time. Polyploidy occur among plant hybrids, both in wild and under human control. Here, 2 different species of plants are crossed with one another and none of diploid offspring are able to reproduce, but some of the polyploids are capable of reproduction among themselves. This results in creation of new species from a hybrid. None of the polyploids reproduce with plants from either of ancestral species, but they can reproduce with one another. Speciation by polyploidy has been recoded among plants many times.

Karpechenko produced a new species by crossing a cabbage with a radish. Although belonging to different genera (*Brassica* and *Raphanus*), both parents possess a diploid number of 18. Mostly infertile hybrids are produced due to fusion of their respective gametes (n=9). A few fertile plants are probably formed, through the spontaneous doubling of the chromosome number in somatic cells that form gametes (by meiosis). Thus these contained 18 chromosomes - a complete set of both cabbage (n=9) and radish (n=9) chromosomes. Fusion of such gametes produced vigorous, fertile, polyploids with 36 chromosomes. These plants could breed with each other, but not with cabbage or radish ancestor, Thus Karpechenko produced a new species.

*Tragopogon mirus* provides an example of a reproductive hybrid species that has recently developed in wild. Polyploidy is rare among animals. Several species of polyploid insects, amphibians, reptiles, fish, and birds are observed, and recently a polyploid species of rat being the first known example of a polyploid mammal is discovered. Every case of polyploidy presumably results in instant speciation by the fact that polyploid organisms are not capable of mating with their parent species. Recently a diploid hybrid cross between 2 species of fly was believed to have developed. *Rhagoletis lonicera* was discovered to be a new hybrid species in 2004. Genetic tests show that lonicera fly is a hybrid cross of *R. zephyria* and *R. mendax*. *Lonicera* fly is believed to be a recently developed hybrid species in part because its host plant is a recently introduced invasive species from Asia. Study on Lonicera fly was first published in the journal *Nature* in 2005.

## 5.26 Controlled Speciation without Polyploidy (Standard forms of Sexual Speciation):

Erneso Paterniani conducted a 5 year speciation experiment using maize. Paterniani reduced the intercrossing of 2 varieties of maize from 46.7% to 3.4% by planting 2 different varieties of corn side by side in a field and then replanting only the non-crossed seeds for each generation. He never planted the seeds that were the product of crosses between the 2 different varieties. In only 5 years he almost completely eliminated interbreeding even though the 2 different types of corn were planted right next to each other. The experiment was not continued past the 5th year, but it clearly shows the ability for barriers to breeding to develop among populations.

*Drosophila* is commonly used for genetic study. The closest known relative to *D. melanogaster* is *D. simulans* which looks very similar and is believed to have diverged from *D. melanogaster* about 5 million years ago. These 2 species produce sterile hybrids in the laboratory showing the difficulty of production of distinct species over short period. If 2 species of *Drosophila* look similar and can produce sterile offspring, and these are believed to be divergent by 2+ million years, the challenge of producing 2 distinct species from one population in a lab concerning tens of years is obvious. Indeed *D. melanogaster* is capable of producing hybrids with species that it is believed to be separated from by as much as 20 million years. While *D. melanogaster* is a great specimen for genetic studies, it is a poor specimen for speciation research, because the qualities that make *D. melanogaster* good for genetic study are the same qualities that reduce the likelihood of speciation events within the species. The fly only has 4 pairs of chromosomes, compared to 24 in humans, and it has a very compact genome without much junk DNA.

Organisms with very short life spans high reproduction rates and intense reproductive competition tend to have very efficient reproductive systems. Evolutionary theory explains this by stating that the size of the genome as a trait is selected for by natural selection in these cases, causing such organisms to have genomes without much excess DNA. This quality makes *D. melanogaster* a good subject for genetic study, but reduces the amount of genetic material that is likely to undergo successful mutations, because there is not much excess DNA in *D. melanogaster* and DNA is very finely tuned. Small number of chromosomes reduces the likelihood of the development of chromosomal breeding barriers. So, fruit fly populations are not very susceptible to genetic change. Organisms with large genomes are better suited to speciation experiments than *D. melanogaster*.

## 5.27  Wild Speciation without Polyploidy (Standard forms of Sexual Speciation)

It is difficult to observe speciation events in laboratory and in wild. Conditions for **speciation** are more prevalent in wild, where there are large diverse populations. In wild, the populations are much larger and there is much more variability, both genetically and often environmentally as well. Several different speciation studies on wild populations have been published. *Anolis* lizards are one subject of major speciation studies both for field experiments and genetic and morphological studies. Of about 138 recognized species of *Anolis* lizards among the islands of the Caribbean, phylogenic studies show that these 138 species have developed relatively recently from as few as 2 colonizing species from mainland that initially colonized the islands less than 20 million years ago. Diversity of *Anolis* in the Caribbean is the result of recent speciation from an initially small variety of species; 2-4 species have become 138 species.

### *Anolis sagrei*

Comparative studies of these lizard populations began reporting significant morphological, behavioral, and genetic divergence among the populations and from ancestral population. Differences were adaptive. They followed patterns of adaptive differentiation, though to a lesser degree, found among other species of *Anolis* in Caribbean. Studies of these populations include research to make associations between the morphological changes and genetic changes, and speciation studies. Major examples of observed speciation in action in wild are examples of ring species. One ring species is (salamander) *Ensatina eschscholtzii* from the California, other is (song bird) the *Phylloscopus trochiloides* from central Asia.

Many cats can interbreed. A dozen different combinations of different cat species can breed and produce offspring. Several crosses produce fertile offspring. Female Ligers are usually fertile, male Ligers are not. Many canines can interbreed. Wolves, Coyotes, and Dogs can interbreed, producing fertile offspring. Zebras and Horses can interbreed, and Bison, Yaks, and domestic Cows can interbreed. Bison and Cows produce fertile offspring when they interbreed and widely farmed breed are called Beefalo. One most interesting crosses is the recently produced cross of a Camel and a Llama, called a Cama. The Llama from South America weighs an average of about 165 pounds, whereas Camel from Middle East, weighs an average of almost 1,000 pounds. Evolutionary model, states that all organisms are related. Evolutionary biologists determined through study of fossil evidence, comparative morphology, and genetics that Llamas and Camels are closely related. Genetic studies also show that both Camels and Llamas have 74 chromosomes.

Based on knowledge that these animals have a recent common ancestor, scientists believed it would be possible to cross breed them, (Figure 5.23).

**Figure 5.23** *Anolis sageri*

### *Real Animal Hybrids*

The proper way to view organisms is not as distinct species, but as genetic islands. Islands and continents can be viewed as completely separate, isolated bodies of land, unconnected to one another, but we can trace foundations of land under the water to prove all land is connected. The same is the case with life. Time and death creates the isolation of species. We use the term species to discuss genetically isolated populations. An allele is a variant of a gene that controls eye color. Different alleles of this gene cause eyes to be blue, brown, and green, grey. Race is a collection of alleles.

## 5.28  Kinds of Variations

1. **Group variations:**  When population of a species differs from other. African human population is black while European population is white in colour. Insect populations exhibit group variations in small range of distribution.

2. **Individual variations:**  Differences among individuals of same population called individual variation is important in taxonomy, when variation within a deme is considered for comparing and assigning the population to a taxon. Variations can be classified based on type of character considered:

3. **Meristic variations:**  Man has 12 pairs of ribs but if some individual has 13 pairs of ribs, it is a meristic variation. Some people possess 6 fingers instead of 5 due to trisomy. A starfish may have 6 arms instead of 5.

4. **Quantitative variations:**  Variations that are measured in size or weight, like tall versus dwarf, light versus heavy body, long versus short tail.

5. **Qualitative variations:**  Characters which depict identification quality of an individual like presence or absence of spots, colour, stripes and hairs. Based on continuity of character, variations can be classified as follows:

(a) **Continuous variations:** Variations when fluctuate above or below the average, with intermediate stages. In a population, some individuals are large, some small and some intermediate. Some people have light skin, some have dark and intermediate shades are also found.

(b) **Discontinuous variations:**  Such variations deviate mostly from the average individuals. Major mutation and disruptive selection produce distinct individuals as found in white tiger; hornless calf and albino peacock. Based on inheritance, variations are classified as follows:

**(a) Somatic variation:** This variation produced due to environmental effect is not heritable, for example, muscles of a wrestler.

**(b) Genetic variation:** This variation occurs in genes, heritable, blastogenic, and cause phenotypic change in populations. If somatic variations exist for a long, they become genetically important. Variations not classified in above are as follows:

**(a) Age variations:** When young ones differ from adults, as the difference between caterpillars and adult butterfly. Such variations are important from the view of adaptation value, but not evolutionary significance, although they aid in identification.

**(b) Seasonal variations:** When animals change their appearances in different seasons. Winter fur in snowshoe hare becomes white in winter and brown in summer. Such variations are seen in grasshoppers, butterflies, and bugs.

**(c) Habitat variations:** In sedentary animals like sponges variations are due to local influences of habitat. Mobile grasshoppers, locusts and plant bugs show habitat variations. Chameleon changes its colour according to habitat.

**(d) Castes in social insects:** Division of labour in social insects like termites, honey bees creates castes with specialized organs to carry out particular job.

**(e) Polymorphism:** When different individuals occur in single interbreeding population of a non-social animal as in butterflies and beetles, which exhibit dry and wet season forms.

## 5.29    Sources of Variations

Somatic and genetic variations are produced in following ways:

**Environment:** **Environmental factors** like heat, cold, rain, drought, food influence populations and cause somatic variations in short period, which are not heritable and disappear if environmental conditions revert back. When tadpoles are fed on thyroid extract, they metamorphose rapidly.

**Endocrine glands:** Optimum activity and balance in functioning of **endocrine glands** is important for normal growth. Over or under functioning of pituitary, thyroid, adrenal produce variations in body, which are somatic and not heritable. Such variations are due to environmental effects.

**Blending inheritance:** Darwin emphasized blending inheritance as major source of variation in animals. Characters of both parents merge in first generation offspring, due to crossing over. But characters segregate in subsequent generations according to Mendel's laws, but still produce variations.

**Mutation:** Mutation is the change in sequence of nitrogenous base pairs in DNA. If AGC is a codon on DNA, its complementary codon on m-RNA is UCG, which code for serine. If a mutation replaces guanine with adenine in codon of DNA, complementary codon in m-RNA changes to UUG, which synthesize leucine. This point mutation alters relevant character, produces sickle-cell anemia in man, when CTT codes for glutamic acid is changed to CAT, which codes for valine. Mutations occur by environmental conditions or during duplication of DNA, when new bases are produced. As such mutations are recessive. They do not express for a long but play role in evolution. Mutations may form multiple alleles. New alleles change gene frequency in population that upsets genetic equilibrium, causing microevolution. In diploid organisms, mutations are recessive, but their spread and accumulation in population cause hybrid vigor (**Heterosis**) and variations. In haploid organisms, they affect the character immediately and expose it to natural selection. Recombination of mutant genes occurs in crossing over, which causes the population heterozygous and much fit when exposed to natural selection. Due to genetic drift, advantageous mutation may be quickly fixed helping in natural selection or sometimes harmful mutations may be fixed, causing extinction of population.

**Chromosomal aberrations:** This is the breaking and rejoining of chromosomal segment during prophase of meiosis, by deletion, duplication, inversion, or translocation.

1. **Deletion:** When a small part of chromosome breaks apart at the time of crossing over by enzyme endonuclease and lost. In man, **terminal deletion** of chromosome 21 causes granulocytic leukemia. In *Drosophila*, deletion in X chromosome causes notched wings.

2. **Duplication:** When one gene is present more than once, and if the deleted portion of one chromosome gets attached with other chromosome, the zygote contains 3 doses of genes and crossing over is unequal. Bar-eye in *Drosophila* arises due to duplication of a small section of X chromosome.

3. **Inversion:** When a chromosome breaks and joins at the same place after rotation, chromosomes form a loop and the sequence of genes is altered in meiosis. The breakage of chromosome at center and reattach after reversing cause abnormal synopsis and crossing over. There is no loss, or gain of genes but resultant effect is due to new position of genes.

4. **Translocation:** This is transfer of one gene block from one linkage group to another in non-homologous chromosomes, which do not produce abnormality in carriers as deletion in one is balanced by other. Fifty percent of offspring of carriers (heterozygote) are grossly abnormal. **Translocations** do not produce immediate effects, but accumulate over long periods to introduce reproductive isolation in the allopatric populations. Six species of *Drosophila* have been produced by translocation from the ancestral *D. virilis* with 6 pairs of chromosomes.

5. **Aneuploidy:** In aneuploidy, the entire chromosome is either lost or duplicated. In *Drosophila*, the gene for white eye is X chromosome. When 2 X chromosomes do not separate during oogenesis, female offspring with 3X chromosomes produce super females with low viability and sterility, and some eggs do not possess a X chromosome. OY individuals do not survive because of upset in gene balance. In man, non-disjunction in sex chromosomes influences secondary sexual characters. **Trisomic** XXX produces super females, while XXY causes Klinefelter's syndrome. **Monosomic** (2n-1) XO causes Turner's syndrome and the YO is non-viable. Down's syndrome due to

duplication of chromosome No. 21 causes mental retardation, malformed ears, susceptible lungs, and sexual immaturity. Aneuploidy produces abnormality in organisms which are sometimes useful in different environmental conditions ( Figure 5.24-5.26).

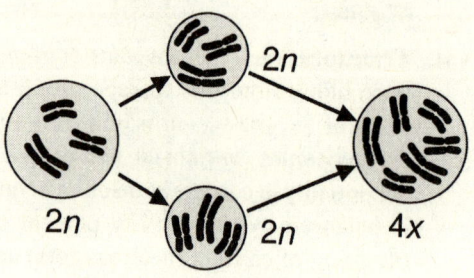

**Figure 5.24**   Polyploidization

6. **Polyploidy:** Duplication of entire haploid set of chromosomes due to abnormal mitosis or meiosis, produce triploid or tetraploid organisms. Such offspring are normally sterile. Polyploidy is not successful in animals, except in some parthenogenetic shrimps, isopods, moths, beetles, and flies.

7. **Hybridization:** When isolating mechanisms between 2 species break down, they produce normally sterile offspring but they may **back cross** with parents to produce second generation individuals (introgressive hybridization). *Raphanobrassica* is produced by crossing *Raphanus* (Radish) and *Brassica* (Cabbage). Both have 18 chromosomes. Hybridization is the mechanism for producing variation in non-chordates, particularly the insects.

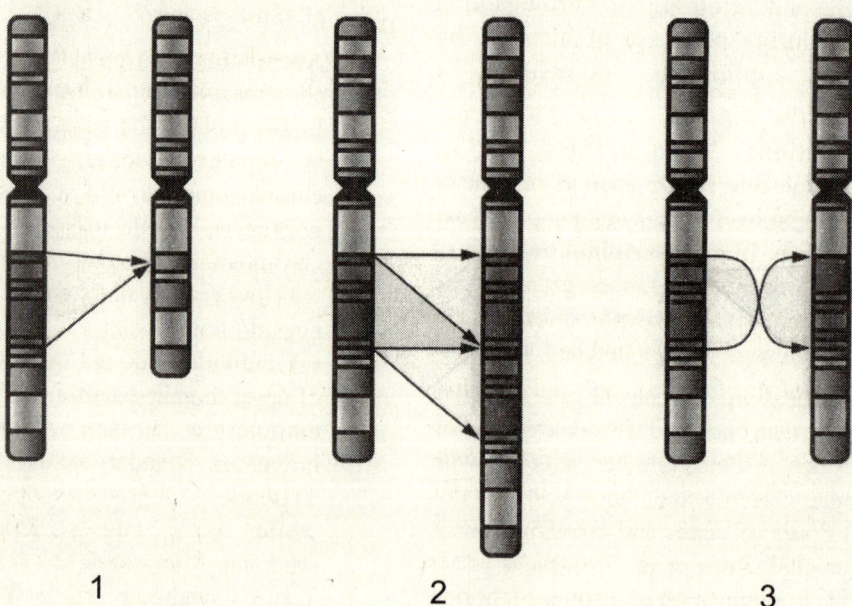

**Figure 5.25**   Three mutations involving single chromosome 1. Deletion, 2. Duplication, 3. Inversion

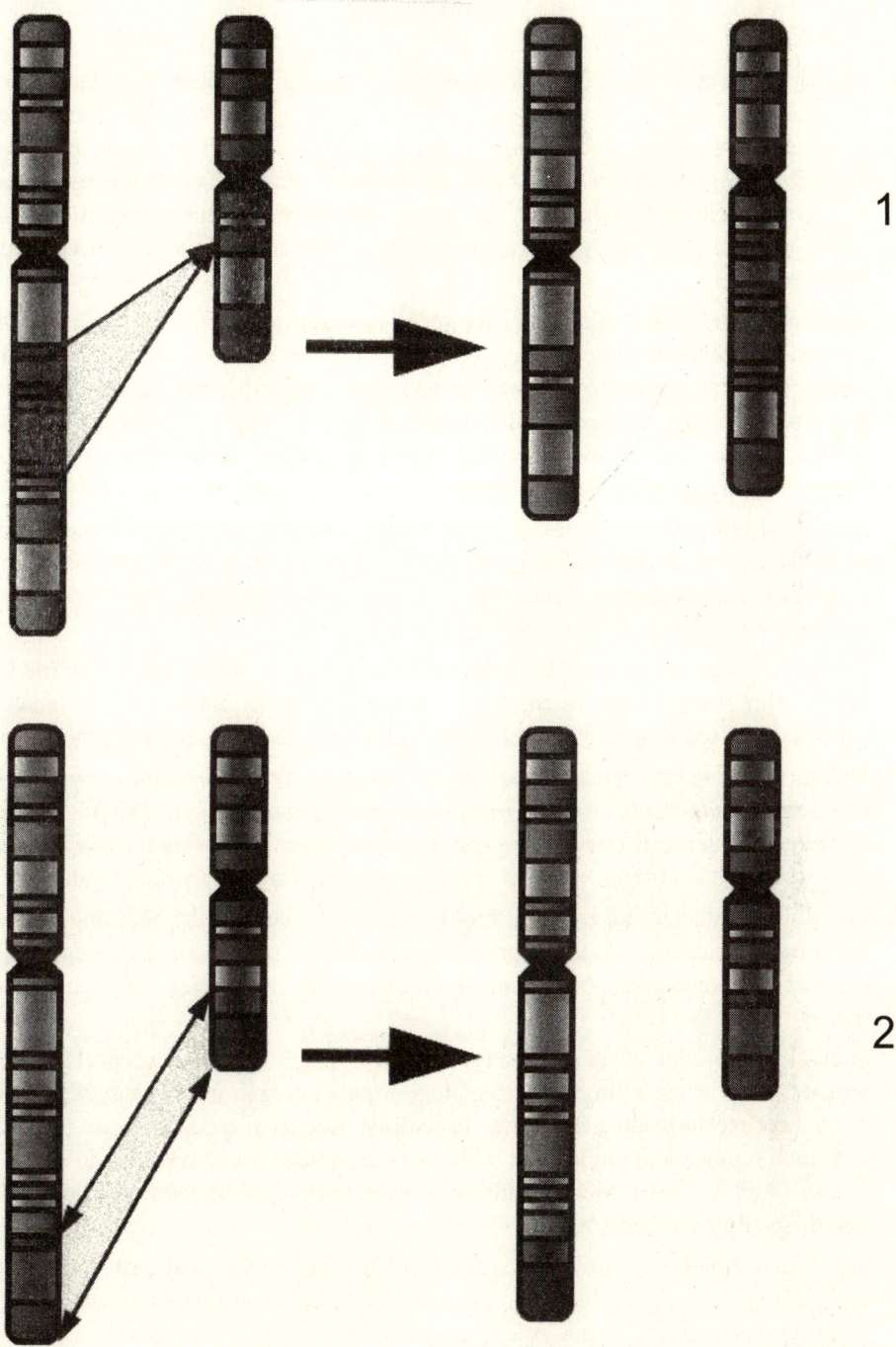

**Figure 5.26** 1. Insertion, 2. Translocation

# SUMMARY

1. In convergent speciation, 2 species have not developed complete reproductive isolation, but maintains their seperate existence due to geographical barriers. Such species can produce new species by hybridization if the geographical barrier between them is removed. Sometimes migrations can make 2 species merge into one species. In divergent speciation, 2 or more species are produced by splitting of single species, due to migration or adaptive radiation. In phyletic speciation, a species changes over long period to become different from ancestor.

2. In quantum speciation, sudden formation of new species occurs by mega mutation and disruptive selection, hybridization or polyploidy. In small, scattered populations, or populations that migrate to new area, genetic drift, rather than natural selection plays role in quickly changing the species. Disruptive natural selection makes species distinct and diverse. Differences between closely related organisms, individuals of same species, subspecies or race are called variations.

3. Genic concept defines species as reciprocally characterized groups of individuals with probable negative fitness effects in other groups that cannot be exchanged between groups on contact. This differential fitness species concept classifies groups that keep differentiated and keep on differentiating despite interbreeding.

4. Different species concepts often refer to different stages during speciation (de Queiroz, 1998). Almost every species concept is usable. Groups of polyploid individuals should not be considered as species, which is distinct from otherwise similar diploid individuals.

5. Biological species concept states that "*species are groups of interbreeding natural populations that are reproductively isolated from other such groups* (Mayr, 1970)." This concept answered problems of Darwin's morphological approach by using a naive interpretation of morphological criteria, mutants and polymorphic variants within populations.

6. Recognition concept differs from biological species concept by focusing on subset of isolating mechanisms occurring prior to fertilization. Interbreeding concept had always stressed a common gene pool and compatibility within the species, and isolation between species

7. Monophyly criterion of species is a type of phylogenetic species concept (Hennig, 1966). Species forms when a single interbreeding population split into 2 branches, or lineages that did not exchange genetic material. In cladistic concept of species, species are considered as branch segments in phylogeny, with every branching event resulting to a new pair of species (Ridley, 1996). Mechanisms preventing gene exchange between populations are called isolating mechanisms.

8. Reproductive isolation between species is maintained by pre- and post- mating isolating mechanisms. Incompatibilities that prevent mating between species are called premating isolating mechanisms, which include geographical-, ecological-, temporal- and behavioral isolation and mechanical incompatibilities. When premating isolation fails or has not evolved members of different species mate.

9. Polyphyletic species originating by parallel speciation will not show parental patterns of ancestry and descent separate from that of ancestral species and would have to be clumped under phylogenetic species concept.Single gene polymorphisms and sexual dimorphism can affect multiple external characters leading to designation of multiple species within polymorphic populations. Sibling species could be lumped into same species using aphenetic approach, unless a set of highly diagnostic characters could be found.

10. Opposite fitness effects of alternative alleles causing speciation do not necessarily have alternative alleles and their fitness effects do not necessarily have to be opposite in different species, that is positive in the species that they characterize and negative in other species.

11. Speciation is the process of a single species becoming 2 or more species. Some think certain evolutionary phenomena apply only at speciation and macro evolutionary change cannot occur without formation of new species. Two populations become separate species if, during period of isolation, they evolve large genetic differences so they can no longer interbreed or produce vigorous fertile offspring if they are reunited. Ring species provide a unique glimpse of speciation. At least 23 cases of ring species have been proposed. The diversity of *Anolis* in Caribbean is the result of relatively recent speciation from an initially small variety of species; 2-4 species have become 138 species.

# REVIEW QUESTIONS

## Short Answer Questions

1. Define species.
2. Comment on "biological species concept".
3. What are the criteria of "ecological species concept"?
4. Define taxonomy.
5. Define ring species.
6. What do you understand by the speciation?
7. Define polyploidy.
8. Define population.
9. What do you understand by the reproductive isolation?
10. Define hybrid.

## Long Answer Questions

1. Describe the biological species concept with its merits and demerits.
2. Compare various concepts of species in brief.
3. Describe pre-mating isolating mechanisms.
4. Describe post-mating isolating mechanisms.
5. Write a note on allopatric speciation.

6. Write a note on sympatric speciation.

7. Examine the importance of ring species in evolution.

# REFERENCES

Agapow, P. M. et al. 2004. The impact of species concept on biodiversity studies. Q.Rev. Biol., 79: 161-179

Avise, J. 1994. Molecular Markers, Natural History and Evolution. Chapman & Hall, New York.

Avise, J.C. and Ball, R.M.1990. Principles of genealogical concordance in species concepts and biological taxonomy. Oxford surv.Evol. Biol., 7:45-67.

Barraclough, T.G. et.al, 2009. Accelerated species inventory on Madagascar using coalescent-based models of species delineation. Syst.Biol, 58:298-311.

Baum, D. And Shaw, K.L. 1995. Genealogical perspectives on the species problem. In: Koch, P.H. and Stephenson, G.D. eds. Experimental and Molecular Approaches to plant Biosystematics. Missouri Botanical Garden, St Louis, pp.289-303

Cohan,F. M. 1994. Genetic exchange and evolutionary divergence in prokaryotes. Trends ecol. Evol., 9: 175-180.

Cohan, F.M., and Koeppel, A.F.2008. The origins of ecological diversity in prokaryotes, Curr. Biol., 18:R1024-R1034.

Coyne, J.A., and Orr, H.A. 2004. Speciation, Sinauer, Sunderland, MA.

Cracraft, J. 1989. Speciation and its ontology: The empirical consequences of alternative species concepts for understanding patterns and processes of differentiation. In: speciation and its Consequences( D. Otte and J. A. Endler, Eds) , pp 28-29. Sinauer Asociates, Sunderland, M.A.

Darwin, C. 1859. On the origin of species by means of natural selection, or the preservation of favoured races in the struggle for life. John Murray. London.

Davis, J. I. 1997. Evolution, evidence, and the role of species concepts in phylogenetics. Syst. Bot 22: 373-403.

deQueiroz, K. 1998. The general lineage concept of species, species criteria, and the process of speciation : a conceptual unification and terminological recommendations, Pp 57-75 in D.J. Howard and S.H. Berlocher, eds. Endless forms : Species and speciation. Oxford Univ. Press, New York, Oxford.

deQueiroz, K. 2005. Ernst Mayr and the modern concept of species. Proc.Natl.Acad.Sci. USA, 102:6600-6607.

deQueiroz,K. 2005b. Different species problems and their resolution. BioEssays 27:1263-1269.

deQueiroz, K. 2007. Species concepts and species delimitation, Syst. Biol. 56:879-886.

deQueiroz,K.,and.Donoghue, M.J .1988. Psylogenetic systematics and the species problem Cladistics, 4:317-338.

Domes K, Norton, R.A, Maraun, M, Scheu, S. 2007.Reevolution of sexualitybreaks Dollo's law. Proc Natl Acad Sci U S A ,104:7139-44.

Dykhuizen,D.E.and Green. L.1991. Recombination in *Escherichia coli* and the definition of biological species.J.Bacteriol.,173:7257-7268.

Eldredge, N. and Cracrft, J. 1980. Phylogenetic patterns and the evolutionary process. Columbia University Press: New york.

Endler, J. A. 1977. Geographic variation, Speciation , and Clines. Princeton University Press, Princeton, NJ.

Ereshefsky, M. 1992. Eliminative pluralism.Phil.sci., 59, 671-690

Ereshefsky, M., E D. 1992. The Units of Evolution. Essays on the Nature of Species. MIT Press, Cambridge,

Finarelli, J.A. 2007.Mechanisms behind active trends in body size evolution of the Canidae (Carnivora: Mammalia). Am Nat ,170:876-85.

Futuyma, D.J. 1998. Evolutionary biology. Third edition. Sinauer, Sunderland, MA.

Gould, S.J. 2002.The structure of evolutionary theory. Cambridge, MA:Harvard University Press.

Grant ,V. 1989.The theory of speciational trends. Am Nat ,133:604-12.

Gregory, T.R. 2007. Evolutionary Trends. Evo.Edu. Outreach,1: 259-273

Harrison, R.G. 1998. Linking evolutionary pattern and process. The relevance of species concept for the study of speciation. Pp-19-31 in D.J. Howard and S.H. Berlocher, eds, Endless forms: species and speciation . Oxford Univ.Press, New York. Oxford.

Hausdorf,B., and Hennig. C. 2010. Species delimitation using dominant and codominant multilocus markers. Syst.Biol., 59:491-503

Hennia, W. 1966. Phylogenetic systematia (trans.D.D. Davis and R. Zangerl) Urbana : University of Illinois press.

Hurst, G.D.D and Werren, J.H. 2001. The role of selfish genetic elements in eukaryotic evolution. Nat. Rev.Genet., 2:597-606.

Hutchinson, G.F. 1968. When are species necessary? In: Population Biology and Evolution( R.C. Lewontin, Ed.) pp. 177-186. Syractional Symposium , sponsored by Syracuse University and the New york State Science and Technology Foundation. June7-9, 1967, Syracuse, New York.

Irwin, D.. Price, E .1999 Sexual imprinting, learning and speciation. Heredity, 82:347-354.

Isaac, N.J.B., Mallet, J and Mace. G.M. 2004. Taxonomic inflation : its influence on macroecology and conservation. Trends Ecol.Evol., 19:464-469.

Kondrashov, A.S. and Kondrashov, F. A. 1999. Interactions among quantitative traits in the course of sympatric speciation. Nature, 400: 351-354

Lönnig ,Wolf-Ekkehard , Stüber, K, S, H., Kim, J.H. 2007 Biodiversity and Dollo's Law: To What Extent can the Phenotypic Differences between *Misopates orontium* and *Antirrhinum majus* be Bridged by Mutagenesis? Bioremediation, Biodiversity and Bioavailability, 1(1): 1-30

Mallet, J.1995. A species definition for the Modern Synthesis. Trends Ecol. Evol., 10:294-299.

Mallet, J., 2005. Hybridization and an invasion of the genome. Trernds Ecol.Evol.20:229-237.

Mayr, E. 1942. Systematics and the Origin of Species. Columbia University Press: New York.

Mayr, E. 1963. Animal species and evolution. Cambridge, Massachusetts: Harvard University Press.

Mayr. E. 1982. The growth of biological thought. Combridege, Harvard University press.

Monaghan, M.T.R., Wild, M., Elliot, T., Fujisawa, M., Balke, D.J.G., Inward, D.C., Lees, R. Ranaivosolo, P.Eggleton, Noor, M.A.F. 2002. Is the biological species concept showing its age ? Trends Ecol.Evol., 17:153-154.

Nosil, P.B.J. Crespi. and Sandoval, C.P. 2002. Host-Plant adaptation drives the parallel evolution of reproductive isolation. Nature, 417:440-443.

Ochman, H. , Lerat, E. and Daubin, V. 2005. Examining bacterial species under the specter of gene transfer and exchange, Proc. Natl. Acad.Sci. USA, 102:6595-6599.

Paterson, H.E.H. 1985. The recocgnition concept of species. In: Vrba, E.S. ed. Species and speciation( Transvaal Museum Monoigraph 4). Transvaal Museum , Pretoria, pp.21-29.

Poulton, E. B. 1904. What is a species. Proc. Entomol. Soc. Lond. 1903, lxxxvii-cxvi

Rajagopalan, R. and J.H. Fujimura .2012.Will Personalized Medicine Challenge or Reify Categories of Race and Ethnicity?" Virtual Mentor. American Medical AssociationJournal of Ethics , 14(8): 657-663.

Ridley, M. 1966. Evolution, Blackwell Science, Oxford.

Rieseberg, L.H.2001. Chromosomal rearrangements and speciation. Trends Ecol. Evol., 16:351-358.

Robson, G. C. 1928. The Species Problem. An Introduction to the Study of Evolutionary divergence in Natural Populations. Oliver and Boyd, Edinburgh.

Shcherbakov, V.P. 2010. Stasis is inevitable consequence of every successful evolution. Biol Philos( in press)

Simpson, G.G. 1944. Tempo and Mode in Evolution. New York: Columbia University Press.

Simpson, G.G. 1951. The species concept. Evolution, 5:285-298.

Simpson, G.G. 1961. Principles of animal taxonomy, Columbia university press. New york.

Sites, J.W. Jr. and Marshall, J.C. 2004. Operational criteria for delimiting species. Annu.Rev. Ecol. Evol. Syst., 35: 199-227

Sokal, .R. and Crovello, T.J. 1970. The biological species concept: A critical evaluation. Am. Nat. 104: 127-153

Soltis, D.E., and Soltis, P.S, 1999. Polyploidy : recurrent formation and genome evolution. Trends Ecol. Evol., 14:348-352.

Soltis, P.S. and Soltis, D.E. 2000. The role of genetic and genomic attributes in the success of polyploids. Proc. Natl. Acad. Sci. USA, 97:7051-7057.

Staley, J. T.2006. The bacterial species dilemma and the genomic-phylogenetic species concept. Phil.Trans. R. Soc. Lond.B., 361: 1899-2009.

Templeton, A.R.1989. The meaning of species and speciation,Pp3-27 in D.Otte and J.A. Endler, eds, speciation and its Consequences. Sinauer, Sunderland, MA.

Van valen, L. 1976. Ecological species, multispecies and oaks. Taxon, 25; 233-239

Vrba, E.S. 1883. Macro evolutionary trends: new perspectives on the roles of adaptation and incidental effect. Science,221:387-9.

Wagner, P.J. 1996. Contrasting the underlying patterns of active trends in morphologic evolution. Evolution, 50:990-1007.

Wake, D. B. 2001. "Speciation in the round." Nature, 409: 299-300.

Werren.J.H., Baldo, L. and Cleark, M.E. 2008. Wolbachia : Master manipulators of invertebrate biology. Nat.Rev. Microbiol., 6:741-751.

Wiley, E.O. 1978. The evolutionary species concept reconsidered.Syst. Zool., 27:17-26

Williamson, P.G. 1981. Paleontological documentation of speciation in Cenozoic molluscs from Turkana Basin. Nature., 293:437-443.

Wu,C-I, 2001. The genic view of the process of speciation. J.Evol.Biol., 14:851-865.

# CHAPTER 6

# Soft Inheritance

The term soft inheritance was coined by Ernst Mayr (Figure 6.1) to include ideas of **Lamark** and Geoffroy which contrasts the modern ideas of the hard inheritance. Modern genetics view the hereditary material as impervious to environment. The soft inheritance advocates that the genetic basis of characters could be modified directly by environment induced factors, by use and disuse, or by an intrinsic failure of constancy. Such modified genotype is then transmitted to the next generation, which is a type of soft inheritance. Genetic and environmental variation was traditionally thought to direct phenotypic variation. Now the line between these 2 components is blurred by inherited epigenetic variation. **Epigenetic** variants and inheritance provide the missing puzzle for the complex phenotypes. Every gene affects several different processes and work as gene networks. During development, the networks perhaps changes in competence. **Waddington theory** tells that a collection of genes play the basic role. Bacterial research shows that genetic processes are not singular. At least 7 genes control the synthesis of argenine in *Neurospora*. More than 40 genes are involved in wing development of *Drosophila melanogaster*. There is no one-to-one relation between a gene and a phenotypic character.

A new extended evolutionary synthesis (**EES**) has been the necessity (Pigliucci and Müller, 2010) to explain specific phenomena. Many of them can be grouped as soft inheritance, which refers inheritance of the

**Figure 6.1** Ernst Mayr

resulting variations of non-genetic effects. Soft inheritance combines the evolutionary developmental biology, epigenetics, niche construction, and cultural transmission. Each of such fields deals with the non-genetic processes that introduce heritable phenotypic variation. There is heritable variation that is developmentally induced and not caused by the underlying DNA sequence as in the licking and grooming studies in rats (Jablonka & Lamb, 2008). The level of licking and grooming experienced by a pup becomes an epigenetically inherited part of maternal behavioural phenotype and is passed on several generations (Youngson and Whitelaw, 2008) associated with other phenotypic changes. Here, no new mutation is required to lead to significant inherited change. The MS could not account for such phenomenon satisfactorily. Soft inheritance, saltational changes due to systemic mutation, genetic exchange and cooperation are now the topics of evolutionary biology. We now need new evolutionary theory incorporating Darwinian, Lamarckian, and saltational processes (Jablonka and Lamb, 2008). The **soft inheritance** refers to the proximate process, with the function to calibrate behaviour in the uncertain environment.

## 6.1  Non-genetic Inheritance

Various non-genetic mechanisms of inheritance in parallel with Mendelian-genetic inheritance are known. **Non-genetic inheritance** comprises all vertical mechanisms of inheritance, including transgenerational epigenetic inheritance, somatic inheritance, environmental inheritance, and behavioral or cultural inheritance (Jablonka and Lamb, 2005). The non-genetic mechanisms are mediated by the transmission of elements of a parent's extended phenotype to offspring like components of parent's body, behavior or ambient environment. Non-genetic inheritance is soft as some traits acquired by parents during their lifetime can be transmitted to offspring via such mechanisms which function via factors transmitted in the gametes at conception. The other mechanisms operate via post conception interactions. The role of soft inheritance as a general factor in evolution was primarily inspired by Jablonka and Lamb (2005). Non-genetic inheritance differs from the hypothetical mechanism of genetic encoding. The genetic encoding involves the transmission of DNA sequence (allelic) variation, and can be viewed as biased mutation (Steele, 1979). Non-genetic inheritance involves the transmission of other factors alongside alleles. The extension of the Mendelian model of heredity to encompass non-genetic inheritance represents a synthesis of the 19th-century idea that parents manufacture their offspring with the 20th-century belief that offspring develop autonomously following genetic blueprint. The offspring genome guides development, although many facets of this process are influenced by the developmental environment and resources, which are determined by parental phenotypes. Any form of inheritance, either hard or soft, must occur via the transmission of DNA. Mendelian inheritance was the sole mechanism of heredity. The multiple mechanisms of **heredity** could operate in parallel as a heterodox position within mainstream genetics (Sapp, 1987). Acquired somatic changes amplified by somatic selection can be encoded in the germ-line DNA through **reverse transcriptase**, which catalyzes the synthesis of DNA from an RNA template.

## 6.2  Diversity of Inheritance Mechanisms

Non-genetic inheritance comprises inheritance that operates in parallel with genetic inheritance. Human cultural and linguistic variation transmits from parents to offspring via non-Mendelian mechanisms. Beginning in the 1970s, these ideas developed into cultural evolution and ultimately gene-culture co evolution. **Cultural inheritance** occurs in other animals (Danchin et al., 2004). Niche construction theory proposed that any species can undergo analogous processes, where the activities of organisms lead to modification of their environment affecting selection and thereby influences evolution (Odling-Smee et al., 2003). The parental environment, genotype or phenotype affects offspring phenotype, the parental effects or indirect genetic effects. Transgenerational epigenetic inheritance (Youngson and Whitelaw, 2008) accounts for a variety of parental effects. The parental effects mediated by substances in the gametes, parental glandular and somatic donations. Behavior and environment affect a range of offspring traits (Bonduriansky and Day, 2009). Although **maternal effects** were thought as far more common than paternal effects on offspring development, this view is now challenged by the increasing number of examples of **paternal effects** (Bonduriansky and Head, 2007). Non-genetic inheritance is attracting increasing attention in medicine, ecology and evolutionary biology.

## 6.3 The New Debate

Efforts to extend the Mendelian model of heredity to include non-genetic inheritance are important change in the nature of the inheritance debate. Throughout the past century, the debate was centered on the existence of soft inheritance. The existence of non-genetic inheritance is established by diverse empirical research, although some putative mechanisms of non-genetic inheritance remain poorly understood. Debate continues on the importance of non-genetic inheritance for phenotypic variation and evolution (Sterelny, 2007). The Mendelian-genetic model

is adequate in most contexts and need to be replaced by a complex pluralistic model. Proponents of this view argue that outside of some specialized domains like human culture, the importance of non-genetic inheritance has not been shown (Various, 1998). This position is reinforced by the practical difficulties involved in extending population and quantitative genetic approaches to encompass the diverse non-genetic mechanisms of inheritance (Tal et al., 2010). However, non-genetic inheritance should not be ignored as studies have provided many large effects of non-genetic inheritance on offspring phenotype (Jablonka and Lamb, 2005) and such effects can influence the dynamics and course of evolution (Pal and Miklos, 1999) .Debate on the importance of non-genetic inheritance will ultimately be resolved empirically.

## 6.4 Pluralistic Model of Heredity

The model of heredity now emerging is pluralistic or inclusive or extended combining the genetic and non-genetic mechanisms of inheritance (Jablonka and Lamb, 2005). The pluralistic model recognizes the reality of both hard and soft inheritance and the potential for a range of intermediate phenomena. A corollary of the **pluralistic model** is variation in the nature of inheritance among different traits and taxa, spanning a continuum from purely genetic to purely non-genetic. Variation, in the relative importance of genetic and various non-genetic mechanisms of inheritance, has important implications for evolution. Two human traits - eye color and language illustrate opposite ends of the continuum. Eye color may be determined largely by allelic variation (Sturm and Larsson, 2009), whereas mother tongue is determined by non-genetic (cultural) factors. Both traits transmit from parents to offspring and respond to natural selection, although the dynamics and endpoints of change in mean phenotype over generations may be different (Richerson and Boyd, 2005) .Many traits may fall in between these extremes. Both genetic and non-genetic inheritance influences the life history, physiology and morphology. The relative importance of non-genetic inheritance may be greater in some taxa than in others. **Transgenerational epigenetic** inheritance plays important role in plants and unicellular organisms (Jablonka and Lamb, 1995). Most animals are highly plastic in behavior and in some species variation in behavioral traits may be shaped considerably by parental behavior (Avital and Jablonka, 2000). Variation among traits and taxa in the nature of inheritance presents an important novel research problem arising from the pluralistic model of heredity.

## 6.5    Soft Inheritance as Genetic Encoding

Prior to the 20th century, heredity concept advocated the transmission of parental features to offspring. During the 1st decades of the 20th century, proponents of hard heredity redefined heredity narrowly as the transmission of genes or the presence of identical genes in ancestors and descendants (Johannsen, 1911). The gene eventually acquired a material basis in the DNA, and inheritance was interpreted as the transmission of germ-line DNA sequences (Bowler, 1989).Acceptance of this new definition required a change to conceptualize the soft inheritance. If heredity is mediated by the transmission of **DNA sequences**, then soft inheritance occurs via this mechanism of transmission. It must be possible for the environment or soma to modify the germ line DNA sequence to operate the inheritance of acquired traits.

## 6.6    Soft Inheritance in Animal Social Evolution

Many traditions in animals are associated with observational learning. Several ways exist in which **learning** through social interactions lead to similarity between generations and to animal traditions. Learning through vocal imitation is common in song birds and marine mammals. Few cases of motor imitation are reported. The imitative learning of songs in communities of birds leads to local dialects. Group-specific dialects are found in whales. The socially learnt and transmitted habits affect many aspects of an animal's life like habitat choice, food preferences and handling, predation and defense, mating, parenting and social interactions (Figure 6.2) Social learning, especially early learning, can have strong long-term effects. Some traditions are very stable and can evolve through cumulative additions and modification, but one behavior serve as the foundation on which another is built.

**Figure 6.2**   Social learning in orangutan

## 6.7 Soft Inheritance and Human Cultural Evolution

Humans transmit information through the symbolic system, which is a rule-bound system in which signs refer other symbols within the same system, objects, processes and relations (Jablonka and Lamb, 2005). The symbol-based system is found only in human, although precursors are found in other mammals. **Human language** is the clearest example of symbolic communication. The process of cultural evolution based on the transmission of symbols is accepted. The complexity of such type of evolution is enormous. Clearly animal and human traditions based on learning cannot be explained by the gene-based Modern Synthesis, as learning requires developmentally-mediated acquisition and transmission of variations.

## 6.8 Soft Inheritance Involving DNA

In long-term evolution, it is difficult to isolate one particular type of heritable variation from the others, as the different types of variations and processes underlying their origin and transmission interact. The interactions between epigenetic inheritance and genetic system based on DNA replication is good example of this and explains the nature of genetic mutation (Figure 6.3, 6.4). The origin of many genetic variations, especially in conditions of stress was believed as not random, often predictable and could result in saltational changes. New genetic variation has different nature from that of **modern synthesis**. The rates of mutation, recombination and transposition are lower when chromatin is condensed (Jablonka and Lamb, 1995). The movement of transposable elements as major cause of genomic change (Kidwell and Lisch, 1997), is influenced by

various stress that affect DNA methylation and other components of chromatin.

**Figure 6.3**   Mutation causing the garden moss rose to produce flowers of different clours

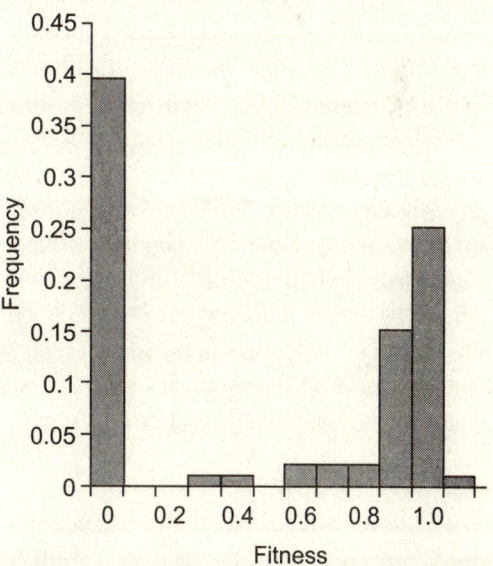

**Figure 6.4**   The distribution of fitness effects of mutations in vesicular stomatitis virus

Clearly epigenetic variations can bias when genetic changes occur. The effect of epigenetic control can go beyond the localized mutational changes induced by local chromatin variations. Developmentally regulated genome rearrangements under epigenetic control are an ancient feature of eukaryotes (Zufall et al., 2005). Global epigenomic macro variations occurring during stress are inherited between generations and lead to macro evolutionary changes. The epigenetic control mechanisms bring out genome wide changes that under normal environmental conditions act in limited and specific manner. During animal and plant phylogeny, developmental genes are duplicated and re-used (Gu et al., 2004). Epigenetic silencing may play a role in this regard (Rodin et al., 2005). Epigenetic control plays a significant role in speciation through polyploidization and hybridization (Rapp and Wendel, 2005).Auto- and allo-polyploidization dramatically alters DNA methylation. The genes in some duplicated chromosomes are heritably silenced. The burst of new variation becomes the raw material for natural selection (Fontdevila, 2005).

In ciliates, where epigenetic control systems cause deletions and amplifications of genes in developing macronucleus, operate in other organisms under conditions of genomic and ecological stress. Mechanisms based on DNA-RNA, DNA-DNA, and RNA-RNA pairing involves enzymatic alterations of chromatin and result in changes in DNA which may underlie the systemic changes seen in stress conditions. Such mechanisms are responsive to external and internal conditions. The genetic variations go through the natural selection. They are not the random mutations that are central to the synthesis. Goldschmidt (1940) promoted the idea that systemic changes in the genome underlie macroevolution. But his views were supposed to be wrong. Recent data from biological fronts is changing this attitude. It seems that **systemic epigenetic** and genetic heritable variations occur and lead to macro evolution (Figures 6.5, 6.6).

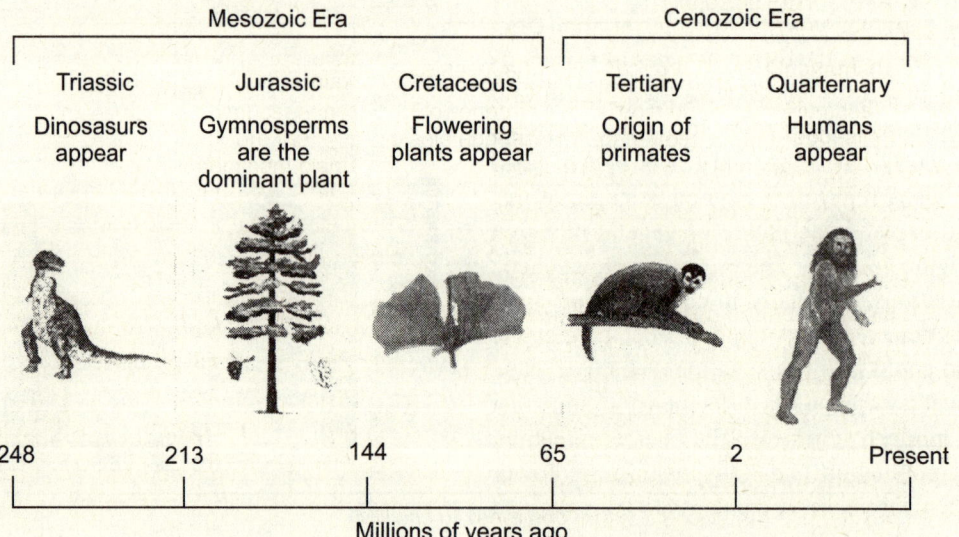

**Source:** https://wikispacer.psk.edu/display/110master/machanism+of+macroerolution

**Figure 6.5**  Mechanism of Macroevolution

**Figure 6.6** Homeosh A mutation in the bithorax locus causes four winged fly

## 6.9 Epigenetics

Epigenesis was championed by the developmental biologist C.H. Waddington (Figure 6.7) throughout the last century (Waddington, 1957). The epigenetic rule is a gene-directed process in the presence of environmental events, for the normal development of the brain and the mind. Three broad classes of **epigenetic rules** are the primary rules, secondary rules, and the reification rules. The primary rules are active in the development of brain systems for events ranging from sensory reception to the early stages of perception. The secondary rules are active in the life history of mental processes ranging from the later stages of perception through conscious thought and experience. The reification rules, which act in beliefs about the reality of what we imagine or conceive by the system of beliefs and imaginative identifications we attribute reality to the minds of others based on our experiences of what they say and do (Seager, 1999). Another is moral realism, the characteristically human tendency to believe that rules and laws distinguishing right from wrong, good from evil, exist outside us in the Universe, quite apart from ourselves and our beliefs about them (Greene, 2003). Up to about the year 1980, a second classification of the epigenetic rules conjectured that an evolved strategy of gene action in epigenesis could take one of the following 3 forms:

- **Pure genetic transmission:** The outcome of brain and mind development is tightly specified by gene activity. Culture and social environment either have no effect, or always lead to the same outcome across the population. **Pure genetic transmission** with negligible impact of the social environment is the closest to the innately hard-wired "genetic robot". Thus, for pure genetic transmission, one would predict zero correlation between changes in the sociocultural environment and the directions taken by mental development.

- **Pure cultural transmission:** Gene activity creates nerve circuits in the brain that force the organism to acquire general capacity for learning. This activity forces a range of possible outcomes in temperament, personality, emotional valence, knowledge, and morality. Events in the culture and social environment set the course taken by such capacities.

**Figure 6.7** C.H. Waddington

Individuals like this are the blank slates or tabula rasa minds, which are molded by the forest of symbols into which they are born. Thus, in pure cultural transmission the properties of individual minds, together with the differences among individual minds, arise solely from differences in their sociocultural environment.

- **Gene-culture transmission:** A category stretching from the pure genetic transmission to **pure cultural tranmission**. Gene action plays a subtle role, does not commit the mind and brain to one innately programmed result and does not force a flat or unbiased general learning potential. The potential field is bent into a landscape of hills and valleys. The hills are linked with outcomes that are partially blocked, or less likely to occur. The valleys mark outcomes to which the mind's development is more likely to flow. **Gene-culture transmission** can hug the extremes defined by pure genetic transmission and pure cultural transmission, or take the form of complex transitional patterns, in which some outcomes for the brain and mind are more likely than others. The intermediate mode of gene-culture transmission predicts that the mind, in effect, shapes itself. Genes act not to set the final behavior, or intervening mental action, but to endow selectivity toward the learning environment. In pure genetic transmission, changes in mind and brain are uncorrelated with changes in the social and culture environment. If the necessities of mental survival are supplied, a singular outcome is certain. In **pure cultural transmission**, differences among individual minds and within each mind's life history are correlated with matching differences in

social and cultural upbringing. In gene-culture transmission, some inputs have higher salience and receive more attention from the developing mind. While certain primary epigenetic rules act via pure genetic transmission, the majority of epigenetic rules from language acquisition to schemas for mating seem to act via gene-culture transmission, an evolutionary game of mind against culture in which genes set odds that some paths of learning are better than others.

For epigenetic switches, turning the single gene on or off can produce a major change in the nerve cell function. In mice and rats, the gene **Dasml** has been shown to modulate the dendrite arbor to similar ends. Blocking production of the cell membrane protein coded for by *Dasml*, or manipulating the gene so that an altered protein is made by the cell, blocks the outgrowth of the all-crucial dendrite tree (Shi et al, 2004). In the brain of *Apis mellifera*, increases in the activity of the gene for foraging occur at the time, a bee switches from maintenance duties of the hive to food gathering roles (Ben-Shahar et al., 2002). Honeybee behavior is molded by an epigenetic rule that ratchets for activity up at a point partway through a bee's adult life. Neutralization of the *Drosophila* gene period, for example, yields flies in which the male's normal courtship song is replaced by a less effective roughened sound pattern (Greenspan, 1995). In the land snail *Euhadra*, a single genetic change switches the spiral twist of the animal's shell and body layout from clockwise to anti-clockwise, and disrupts the geometry and positioning needed in normal mating behavior. Clockwise and anti-clockwise snails belong literally to different *Euhadra* species

C. H. Waddington in 1942 introduced the term epigenetics for the interactions of genes and environment that bring the phenotype.

Epigenetic means the phenotypic changes that are inheritable without **DNA mutation**. The chromatin structure plays role in the regulation of gene expression. The tail modifications in histones are important in chromatin structure. Epigenetic inheritance is the transmission of information without the information being encoded in nucleotide sequence as found in the development of multicellular organisms. That epigenetics applied only to developmental processes was undoubted because all cells of multicellular eukaryote are genetically identical. They start as fertilized unicellular zygotes, which divide through mitosis to produce all cells, tissues, and organs. Mitosis produces genetically identical daughter cells. The cells of a multicellular eukaryote contain the same genetic information. Cells have different structures and functions as genetic material is expressed variously in different cells.

Development in multicellular eukaryote proceeds by successive up- or down-regulation of the genes in a given cells, producing different structures and functions in an organism without changing its genetic material. New study suggests that **epigenetic effects** are not restricted to intra-individual phenotypic changes at the individual cell level. During morphogenesis, totipotent stem cell changes into neurons, epithelium muscle cell etc. by activating some genes, while inhibiting others through epigenesis. The dividing fibroblasts create new fibroblasts, though their genome is identical to other cells. **Epigenetic transmission** of traits first found in maize is comparatively rare. Epigenetic effects occur by several self-reinforcing and inter-related covalent modifications of DNA and/or chromosomal proteins like **DNA methylation**, histone modifications, and by chromatin remodeling like repositioning of nucleosomes. Such heritable modifications are all together called the epigenetic codes.

Epigenesis believes that the fertilized egg contains few elements which react together during development to produce several adult features that were not present before. A fertilized egg produces something new and contains some preformed elements, the genes and different regions of cytoplasm. During development, these interact to produce adult features. Thus preformation and epigenesis are involved in **embryonic development**. The differences between fully differentiated cells relate not to single substance, but to many. The differences arise progressively. The tissue of an early neural plate is distinct from that of the contemporary epidermis, but both undergo change as the former develops into fully functional nervous tissue and the latter into the adult skin. Such changes are not the uniform continuation of the first stages of differentiation. Thus, of the early neural plate, some become eye-cup and some do not. The whole process is not the simple unfolding of a single trend.

As such epigenetic processes have a 2nd order effect on DNA expression. The inclusive inheritance model views phenotypic variation as explicandum and inclusive inheritance as explicans. Epigenetic mechanisms address the missing heritability problem in genome-wide association studies (**GWAS**), showing most traits with high heritability can not be explained by association with common genetic variants (Maher, 2008). This is viewed as evidence of inclusive inheritance. Several possible explanations for the poor link between DNA sequence variants and heritability include imprecise heritability estimates, and ambiguity in modern behavioural genetics about the sources of non-genetic influence (Turkheimer and Waldon, 2000) exists. Epigenetic processes allow non-genetic information to be inherited across generations. Multiple epigenetic mechanisms are proposed to affect development and disease. The erasure of epigenetic marks at the germ line and **embryogenesis** in vertebrates is extensive and makes it a difficult topic (MacKay et.al.2007). The potential for epigenetic trans-generational inheritance appears limited.

Quantitative genetic evidence shows that epigenetic variation is influenced by genetic variation as shown from twin studies which offer a way of estimating heritability as a proportion of phenotypic variation. They compare similarity between monozygotic (Dickins and Rahman, 2012) and dizygotic twins and are unbiased by age effects. The twin approach with a narrow definition of heritability includes anything that does not involve DNA sequence variation in its broad definition of environment. The most heritable epigenetic sites are correlated with functional genome regions, showing that function-specific **epigenetic indicators** work under the strongest genetic control. This evidence for genetic control of epigenetic signals is consistent with several approaches, including population-level findings from quantitative trait loci studies (Gibbs et al., 2010) and familial clustering of methylation profiles (Johannes et al., 2009). Heritable effects are known for other epigenetic processes like individual- and allele-specific chromatin signatures (McDaniell et al., 2010). The epigenetic processes as a challenge to the MS suggests that epigenetic systems are themselves phenotypes and their genes are subject to natural selection and drift allele. Twin and family studies do not claim that all epigenetic variation can be accounted for by genetic variation. Animal studies also show that epigenetic variation can be separated from genetic variation (Sollars et al., 2003) (Figures 6.8, 6.9).

By the 1940s, combination of the Darwinian view of evolutionary change and the Mendelian genetics produced the modern synthesis (MS) (Ruse, 2003). Evolution describes a kind of change caused by natural selection, drift etc (Webb, 2011). The general theory of evolution including the basic Darwinian dynamics of variation, inheritance, competition and selection are applicable across several classes of systems. The MS is a mature theory with clear cut criteria of operation. As per the inclusive inheritance, evolution acts on phenotypes, which can be inherited by genetic and non-genetic means and the phenotypic change occurs due to epigenetic processes. The **EES** does not deny hard inheritance. However, hard and soft contributions affect the evolutionary dynamic (Jablonka and Lamb, 2008) differently. Some equate phenotypic variation as a result of variation of DNA sequence and DNA expression (Danchin et al., 2011).

## 6.10  Origin of Epigenetics

Epigenetics or heritable changes in gene expression patterns occur by mechanisms not concerning the general nucleotide sequence (Mehta et al., 2006). The heritable changes occur due to mechanisms other than changes in the general DNA sequence. The environment influences development and evolution. Chromatin, the unit formed by DNA, histones and associated proteins undergo remodeling leading to facilitation or hindrance of transcription. **Chromatin** is not so loosely bound that transcription can easily occur. DNA must be uncoiled from the histones for the transcription factors to bind to their matching sequences (Rocha, 2007). Tighter winding of promoters or enhancers can hide them from transcription factors or polymerases. DNA methylation, **DNA demethylation**, and histone acetylation are the methods of chromatin remodeling (Wilson et al., 2009).

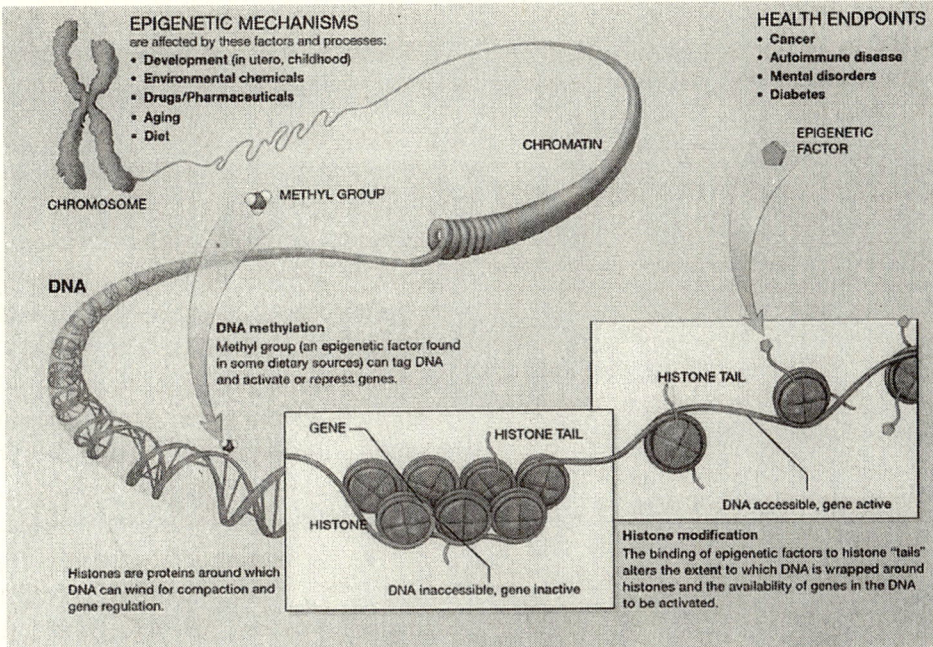

**EPIGENETIC MECHANISMS**
are affected by these factors and processes:
- **Development** (in utero, childhood)
- **Environmental chemicals**
- **Drugs/Pharmaceuticals**
- **Aging**
- **Diet**

CHROMATIN

CHROMOSOME

METHYL GROUP

DNA

**HEALTH ENDPOINTS**
- Cancer
- Autoimmune disease
- Mental disorders
- Diabetes

EPIGENETIC FACTOR

**DNA methylation**
Methyl group (an epigenetic factor found in some dietary sources) can tag DNA and activate or repress genes.

HISTONE TAIL

GENE

HISTONE TAIL

HISTONE

DNA accessible, gene active

DNA inaccessible, gene inactive

Histones are proteins around which DNA can wind for compaction and gene regulation.

**Histone modification**
The binding of epigenetic factors to histone "tails" alters the extent to which DNA is wrapped around histones and the availability of genes in the DNA to be activated.

**Figure 6.8** Epigenetic mechanism

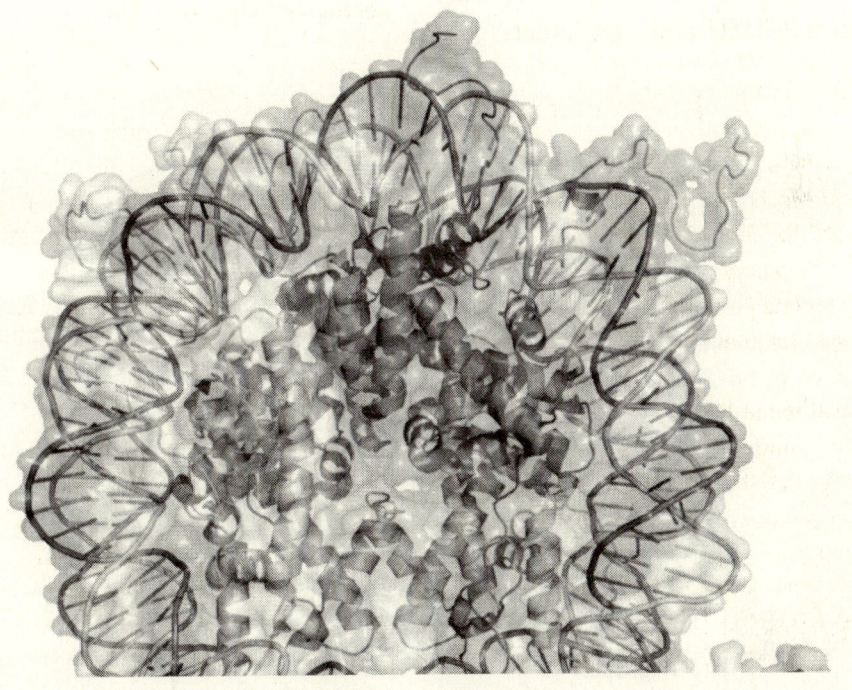

**Figure 6.9** Nucleosome

**Methylation:** Alteration of the DNA methylation pattern is a major epigenetic change. DNA methylation pattern sequences are a part of the epigenetic code of a cell. DNA methyl transferases (DNMT) are vital for survival in mammals (Okano et al., 1999). Methyl groups are added post synthetically to the 5-position on cytosines in methylation resulting in modified major groove on the DNA double strand, making it harder for DNA binding proteins to dock. This phenomenon can be copied after DNA synthesis, making it hereditary (Singal and Ginder, 1999). But the frequent occurrence is the non-hereditary gene silencing. Most methylations occur at CpG dinucleotides, a rare dinucleotide in mammals (Lander et al., 2001). As CpG dinucleotides are often present in genes (Lander et al., 2001), methylation of these dinucleotides modify gene expression (Weber et al., 2005). This modification by methylation of CpG dinucleotide-containing promoters has

importance as synthetic DNMTs or DNMT inhibitors can aid in therapeutic applications where gene expression is aberrant.

**Acetylation:** DNA acetylation of histone-proteins making DNA more available to transcription factors. Histone acetyl-transferases mask the positive charge of the histone lysine tails, loosening the strength of the histone-DNA interaction, enabling docking of transcription factors. The process for obtaining tightening of (Hagmann, 2010) histone-DNA binding is done by proteins called **histone deacetylases** (HDAC). This modulation of chromatin structure by compression has set up HDAC as welcome targets for pharmacological intervention (Kelly et al., 2005). These modifications, albeit a completely natural and widely occurring process of cellular development and evolution, are subject sometimes to pathological aberrancies (Figure 6.10).

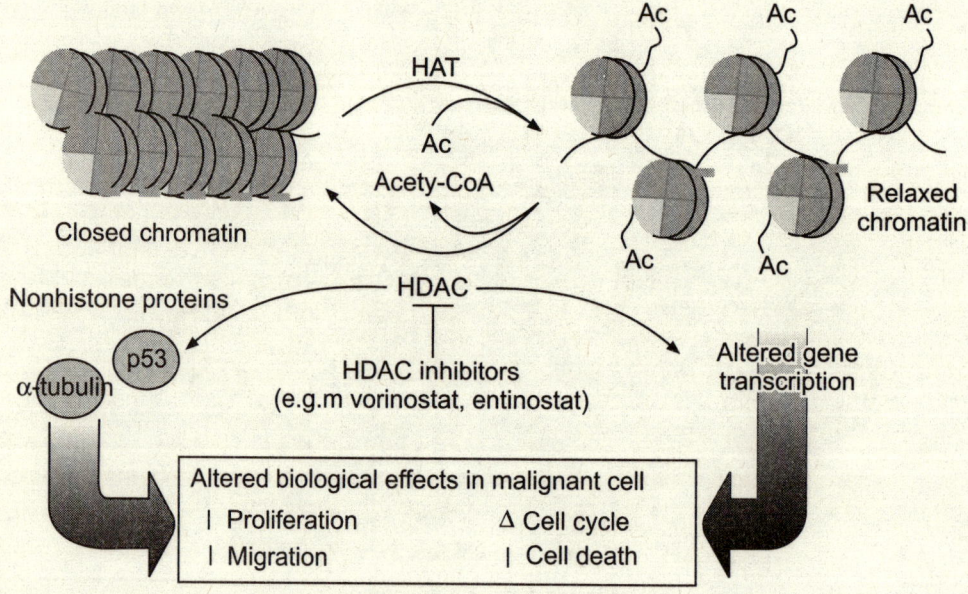

**Figure 6.10** Histone acetylation and deacytylation

## 6.11 Mechanism for Producing Phenotypic Variation

Waddington recognized the importance of the Mendelian laws and marked phenotype and genotype as characters of an adult individual. The characters present in the germ-cells pass on into the next generation (Waddington, 1956 (1939) .This distinction was made after the development of basic theory of genetics. Waddington considers the genotype as the whole genetic system of the zygote as a set of potentialities for developmental reactions and set of hereditable units. It includes the sum of the genes, their arrangement, as expressed in position effects, translocations, and inversions (Waddington, 1956 (1939). The phenotype includes developmental processes. The whole set of characters of an organism is viewed as a developing entity. Waddington argued that when brought together, such terms can be used for the differences of development in a single organism. **Crossing over**, gene-linkage, translocation, duplication are based on the study of chromosomes, or the study of phenotypic effects after crossing. The phenotype-genotype distinction is linked to the theory of the germ track, which states the continuous succession of germ line cells through the generations. This idea is echoed in the selfish gene (Dawkins, 1989).

As Per Waddington's germ-soma distinction, the temporary body and the immortal germ cells are not always distinct. **Germ cells** often belong to the body, as they develop in the same way as other differentiated cells in response to some organizer. Regarding immortality of the cell, Waddington told that all young embryonic cells are potentially immortal. He considered development as an epigenetic process. The phenotype is the result of the interrelations among genetic processes and environment. Waddington expands the classical genetic vision on the phenotype by including the epigenotype, the genetic and non-genetic developmental interactions that organize the organic substances and tissues. The genetics only deals with the behavior of genes during inheritance and needs to be supplemented by an **epigenetic theory**, which deals with their behavior in the developmental processes by which the fertilized egg becomes the adult. Epigenetics thus becomes the turning point between heredity, evolution, and development.

Some phenotypic changes in adults are not related with underlying genetic changes, and are due to changes in expression of particular genes. Some purely phenotypic changes are heritable from parents to offspring. An example is the agouti gene/phenotype in mice. Pregnant female mice that are fed a diet rich in proteins that contain methyl groups produce offspring that express the agouti phenotype. They have yellowish fur, tend to be obese, and have a tendency to develop cancer. These agouti offspring are genetically identical to non-agouti mice. Their phenotype is the result of change in expression of agouti gene, a heritable change from mothers to their offspring for many generations.

Paternal grandsons of Swedish boys who were exposed during preadolescence to famine in 19th century were less likely to die due to cardiovascular disease. When food was sufficient then diabetes mortality in grandchildren increased, suggesting a trans-generational epigenetic inheritance, although the opposite effect was found in females. The paternal granddaughters of women who experienced famine while in womb lived shorter lives on average showing **Lamarckian inheritance**, and suggest that something like Lamarkian evolution is possible, at least for those phenotypic changes that are produced by epigenetic mechanisms. Epigenetic changes are completely outside the scope of modern evolutionary synthesis, and have generated controversy. Epigenetic inheritance affects many phenotypic traits of living organisms and the post-modern evolving synthesis has departed significantly from the modern synthesis to incorporate such findings.

**Developmental explanations:** Several theories are based on concept that minor modifications in animals' development as they grow from embryo to adult may cause large changes in the adult. **Hox genes** control the development of the specific part. Certain Hox gene expression results in development of limb. The expression of different Hox gene in that region could develop into an eye. Such system allows large range of disparity to appear from limited set of genes and raise the question regarding the cause of origin of increased diversity or disparity. Support for the existence of the Precambrian metazoans combines with molecular data show that much of genetic architecture could have played role in explosion already well established by the Cambrian. This apparent paradox is addressed focusing on physics of development. Simple multicellular forms provided a changed context and spatial scale in which novel physical processes and effects were mobilized by products of genes that had previously evolved to serve unicellular functions. Morphological complexity arose, in this view, by self- organization.

**Ecological explanations:** These focus on interactions between different types of organism. Some hypotheses deal with changes in food chain; some suggest arms races between predators and prey, and others focus on more general mechanisms of co evolution. Such theories are well suited to explain a rapid increase in both disparity and diversity.

## 6.12  Organization, Contexts, and Holism

The expansion of the **genotype-phenotype distinction** shows somewhat holistic nature

of Waddington's program on embryology. The single harmonious living organism under the action of causes brings about the unifying arrangement of separate processes. The organizational principles are still studied via reductionistic means. As organization is a rather abstract concept, guidance lies in its exploratory epistemic value than in its ontological or direct experimental value (Waddington, 1947 (1940). The real value of organization comes with quantifiability: that is, the degree of organization depends on the (Speybroeck, 2002) degree of specific part. Organization needs to be defined in relation to its context. Through interactions of elements, new relevant contexts or a new level of organization can occur, excluding a total reduction of the hierarchy of organizational levels, because new level of organization cannot be accounted for the properties of its elementary units as they behave in isolation. Waddington's goals were 2-sided as he wanted not merely to explain the complex by the simple, but to discover about the simple by studying the complex. He held it possible to study the problem of causal embryology experimentally, while remaining holistic in interpreting the results.

## 6.13  Cellular Epigenetic Action System

Waddington studied induction in animal embryology, the development of embryonic cells triggered by and conforming to other elements to gain insight into the organizational principles of development that lead to flexibility and stability. The work of Spemann on the presence of internal organizing determinants in embryonic organo- and histogenesis, especially the experience that early embryonic development could recover after cutting away certain regions of cells, stimulated Waddington's views on the existence of harmonious equipotential, individuation fields, developmental fates, evocation and self-differentiation. The idea behind these terms is that a given stimulus can induce a change which only occurs as the reacting tissues or cells have the potentiality for the change. The focus was given previously often on the stimulus, but Waddington highlighted potency and competence. This idea is fundamental to embryonic cell differentiation and development. Cells react to chemical and organic stimuli as allowed by their state at that time. Each new reaction differentiates the cell further and brings it under more constraints, but opens new possibilities or potentialities. Waddington did not agree with Driesch (Speybroeck, 2002) that these cellular potentialities should be called a developmental soul or non-material entelechiem Potency or the ability of future. With induction as part of a developing epigenetic system, Waddington brought principles to account for the role of genes in it. The development of an organ takes place in a series of steps following epigenetic path which is governed by instructions in the genotype. Genetic action forms a major influence, making development the result of all gene-influenced tendencies. The diverse paths in organismic development are protected or canalized by threshold reactions and by the interrelations between elements and processes of the living system. This canalization is based on the genetic capacity to buffer developmental pathways against mutational or environmental perturbations.

The system is to some extent self-righting. The stability results from canalization arise due to natural selection and inevitable consequence of a system in which many genes interact with one another. Epigenetic crises or instability can arise due to small change in the normal conditions during development with major impact on stadia. Waddington introduced a gene-protein system, the biochemical processes leading from a gene to its protein

level and a gene action system, the biochemical processes leading from a gene to a phenotypic character. The gene action system makes up the epigenetic action system of the cell, in which many feedback loops should be stressed. These feedback relations admit a flexible view on gene action. Genes have the stability to be transmitted through generations. The cell as a double-cyclic system allows genes to make gene products, which elaborate the final cytoplasmatic elements. The cytoplasmatic elements feed back to the genetic level to control gene-activity. Genes regulate and are regulated by non-genetic factors. A gene itself promoted linear reaction pathways. Waddington included branching pathways, analogous to embryological bifurcation pathways in which an external factor push a competent cell line into new track. Such modified genes can have secondary effects. Genes could not be seen as a unit of developmental activity as it is subjected to many influences. **Gene-activity** is not attributable to circumscribed particles .The basic elements are short stretches of chromosome, which are not sharply bounded off against each other, but shade into or overlap one another. A change in the chromosome thread in that case alters the character of fundamental reactions carried out by it (Waddington, 1956).

## 6.14 Embryonic Cell Differentiation

A true epigenetic theory gives a clue about cell differentiation. As per the mutational theory during cell differentiation only those genes responsible for the differentiation are activated. Studies of coelenterates show that after cell division, new differentiation patterns could appear despite a similar genetic makeup. The idea of **plasma genes** with long-lasting genetic continuity of character seemed inconsistent with the facts. Waddington speculated in the direction of cytoplasmic non-homogenity, the location of the nucleus in a specific cytoplasmic region, differential protein synthesis due to small changes in chemical gradients or to genetic mutations, and the existence of autocatalysis. He thought for expanding his theory to an intercellular, organismic, and environmental level. But his approach could not continue. It looked as if cell-cell interactions are not abundant, as muscle cells and nerve cells lie next to each other without mutual interference with protein synthesis. The main accent of Waddington's epigenetics lies on the interconnectivity and continuous feedback relations within a cell and its direct progeny. Waddington was inspired by Volterra's mathematical approach on population dynamics to develop an epigenetic thermodynamics as statistical tool for open biological systems and to provide mathematical foundation for embryology.

Plasmagenes, also called cytogenes, blastogenes, and **proviruses**, were thought to be cytoplasmic elements with a stable nature similar to genes in the nucleus. They were viewed as determinants of cytoplasmic inheritance. The proof for the existence of plasma genes relied on multiple backcrosses of the F1 generation with the parental male. If the presence of (Speybroeck, 2002) parental female chromosomes slowly diminishes in each generation, and if a character of the parental female persists through the generations, it could only be inherited via the cytoplasm of the fertilized egg (Waddington, 1962). Waddington leaned toward the molecular biology to account for non-genetic components that may be effective at intermediate levels of complexity. A classical genetic approach would miss out on these components. If in a system involved in pattern formation, an important part is played by factors like the permeability or tension of a cell surface, or resistance to bending of an epithelium (Speybroeck, 2002). We should not be brought to realize this by genetic analysis. Many genetic factors were active which are yet to be discovered (Speybroeck, 2002). Many of them are operative because they affect cell membranes. The genetic analysis of pattern reveals about their complexity in terms of ultimate units, and about the degree of integration of ultimate units into stabilized creodes, but it is less informative about the intermediate steps between the genes and final patterns. A certain gene substitution in *Drosophila* causes the appearance of a 4-segmented leg in place of a 5-segmented one should be regarded as a problem. The intertanglement of genetics and development became clear with the emergence of **bacterial genetics** in which the genotype-phenotype distinction is less evident and cytoplasmic inheritance is more obvious. The expansion to eukaryotic and multicellular genetics made the study of developmental cellular changes during embryogenesis. The cells of higher organisms exhibit phenomena which may be essential for the whole process of genetic determination. One such phenomenon is the almost universal occurrence of feedback relations between cytoplasm and genes in higher organisms, such that the nature of the cytoplasm determines the intensity of the syntheses, controlled by the various genes in nucleus (Waddington, 1962).Control of gene expression was particularly interesting to the new causal **embryology**.

## 6.15  Waddington's Evo-Devo Program

Waddington incorporated evolutionary aspects into epigenetics and believed that in gene-cytoplasm regulatory interactions, a key can be found to explain limited number of clear developmental pathways in a species. He posstulated the idea that natural selection does not work directly on gene pools, nor on adult organisms, but on the characters of organisms during their lifetime development. The natural selection of chance mutations could not account for many natural adaptations and that evolutionary theory would miss out many biological aspects without developmental theory in which the genotype of evolving organisms respond to the environment in a co-ordinated way. This led to the speculative model on genetic assimilation, exemplified in the case of the callosities of the ostriches and supported by an experiment on the size of the anal papillae of *Drosophila*. When environmental conditions are normal, the fruit fly's papillae do not deviate in size. Once the conditions change to abnormal on an adult, ostrich embryos develop callosities on the exact spot where they are needed in adult life in absence of a stimulus. Genetic assimilation show such a trait which only has an adaptive function after birth can be selected for. Although a physiological or developmental adaptation to an environmental stimulus is not heritable, the capacity to do so is, as it is under genetic control. As natural selection works indirectly on any genetic control, it offers benefits to those organisms that show the capacity to develop a trait sooner than others. As long as the pathway that develops the trait before its adaptive functional time does no harm to the organism, selection is in no way work an early development of the trait away. The development of a trait becomes canalized, once it comes under the control of mutant genetic variation.

Where skin cells have the competence to develop callosities after induction by friction, they can be induced by a new factor, like (Speybroeck, 2002) an embryonic gene product. Being controlled by a heritable factor, the induction becomes fixated and shows itself apart from the **environmental stimulus** without loosing much of the normal effect that the original stimulus triggered. By performing constant positive selection per new generation for those flies, Waddington succeeded in creating a *Drosophila* stock that showed the larger papillae under normal conditions. The new stock was better adapted to high salt concentrations in food than the original stock, leading to the conclusion that selection ameliorate the rate of adaptability and genetically assimilate an environmental reaction by selecting the appropriate genetic lines.

Natural selection works on complex, developing phenotypes and on epigenetic processes that build these phenotypes in interaction with the environment. Stressing the developing phenotype in evolutionary theory expands the classical focus on the transmission of genetic information with a second focus on gene regulation on how to use the genetic information. To bring evolution and development to full synthesis, a developmental theory is needed. Waddington's epigenetics situates itself on the developmental plane as a model to link the genotype and the phenotype during development in specific environment. Waddington's epigenetic originated in an embryological era when the gene concept had no strict definition. As unit of heredity, it could mean anything that fell under the denominator of being heritable.

The **Baldwin effect** was coined at the end of the 19th century. The Baldwin effect means that under stress certain habits are developed by an active organism. Such habits spread initially through imitation. But a genetic mutation can appear that makes the genetic heritability of the habit possible. Function comes first, structure later, an idea basic to Lamarckism. Under the environmental stress, organisms can survive long enough to passively await the appearance of a random genetic mutation that provides the **phenotypic modification** to resist this stress. Waddington claims that the basic idea of genetic assimilation is forgotten, physiological adaptations are under genetic control and a switch is necessary to make this control heritable. One does not have to await a totally new (Speybroeck, 2002), random genetic mutation causing a specific context. Waddington defined a **gene** as a DNA sequence, making epigenetics necessary, as it denoted interactions of genes with their environment that bring the phenotype into being.

Waddington approached the problem within experimental embryology to analyze the developmental processes of fertilized egg. The genomic contexts he studied restricted themselves to intracellular components and networks. Today, in the molecular age, not only the gene concept, but the entire phenotype is often reduced to the DNA. This extreme gene centrism is refuted by the complex molecular networks.

Currently epigenetics is defined as the study of mitotically and/or meiotically heritable changes in gene function that cannot be explained by changes in DNA sequence, but are important for understanding the developmental processes and phenotypic traits of the organism (Henikoff and Matzke, 1997). However, it does reveal Waddington's legacy in several links (Jablonka and Lamb, 1995). First, there is the renewed interest in the existence and generation of non-random and even **non-DNA variation** in biology. Although Waddington focused more on stability and canalization as constraining the possible variation caused by the genome and the environment, he oriented his epigenetic theory on themes like genetic assimilation, which is not about the preservation of developmental changes but about their origin. He was interested in the canalizing constraints exhibiting their evolutionary role in guiding the direction of possible changes in evolution. Second, it shows that not all information

leading to the phenotype is inscribed in the DNA sequences, but that regulation and the developmental context of these sequences should be taken into account. Third, a strong link between genetics and molecular biology was viewed by Waddington as most promising for studying epigenetics.

Evolutionary developmental biology (**evo-devo**) deals with bases of developmental changes in body plan evolution of complex organisms on premise that evolution cannot be fully understood without knowing the evolution of developmental programmes( Raff ,2000). Several novel conceptual frameworks supplement traditional evolutionary biology, like developmental reprogramming (Wilkins, 2002) which describes the acts between mutation and selection on level of organism leading from an altered gene product to new ontogeny and phenotype. Reprogramming is proposed to constitute an extra evolutionary mechanism as some ontogenetic changes may be promoted by existing developmental mechanisms while other alterations are prevented (Arthur, 2000). Evolution could be biased by development, and this would have profound impact on direction of evolution (Arthur, 2001).

Most animals share specific gene families like homeobox-containing genes that regulate major aspects of body pattern (Scott and Weiner, 1984), which are present in Cnidaria (Finnerty and Martindale, 1999). Morphologically simple organisms often possess genes like members of pax gene family that are homologous and show high level of sequence similarity to higher vertebrates (Raible and Arendt, 2004). Despite such evolutionary conservation in developmental **regulatory genes** across major taxonomic groups, cases exist where gene expression patterns differ markedly among closely related taxa, as found in molecular mechanisms determining spatial axes of tetrapod limb

(Christen and Slack, 1998).Evo-devo search for putative Phylum-specific genes producing Phylum-specific evolutionary novelties. That new phyla arose in concert with advent of novel genes has been challenged (Levine and Tjian, 2003). Evolution of lineage-specific body plans does not primarily depend on invention of new genes but on deployment of new gene regulatory circuitries. Changes in **transcriptional regulation** of genes may be significant than changes in gene number or protein function (Caroll, 2000). The use of old genes for novel structures is shown in several instances (Wang et al., 2004). Controversies exist that whether or not the utilization of conserved molecular components in developmental programmes across animal phyla can be taken as evidence for shared developmental function in their recent common ancestor. The alternative co-option of conserved genes along with gene pathways for new functions would likely work in non-homologous structures. It is thus fundamental to employ phylogenetic methods and homology criteria to resolve issues, including idea of retention of genetic programmes or re-awakening (Abouheif , 1997).

Common molecular aspects of segment ontogenesis between insects and annelids (Prud'homme et al, 2003), and between arachnids and chordates (Stollewerk et al., 2003), are used to reinforce the hypothesis that last common ancestor of all bilaterian animals was segmented ( Seaver, 2003). One must be critical about such deep homologies since probability of gene recruitment to non-homologous roles grows with phylogenetic distance (Holland, 2004). Use of gene expression patterns to establish homologies between morphologically similar features among distantly related organisms is another matter of ongoing debate (Locascio et al., 2002).

**Morphological versus Genetic Complexity:** Little correlation is found so far between genetic complexity and degree of morphological organization. Large similarities in genetic make-up between morphologically complex organisms and relatively simple forms (corals) are found. The most examples are the cnidarians, which exhibit genes like wnt, pax, and genes organizing bilaterian head ( Kusserow et al., 2005). About 12 % genes found in EST library of **Acropora** were shared with vertebrates without match with *Caenorhabditis elegans* or *Drosophila melanogaster* (Kortschak et al, 2003). Until this finding, many vertebrate genes were presumed to be lineage-specific, e.g., Churchill and Tumorhead(Kortschak et al., 2003), which are functionally associated with highly differentiated nervous system. In cases where particular gene sequence was present in all 3 animals, the coral sequence matched the human counterpart strongly than any corresponding *Caenorhabditis* or *Drosophila* sequences (Kortschak et al., 2003). This suggests higher rates of divergence in these invertebrate model systems. Analysis of **Puffer fish** genome likewise shows that many fish proteins have diverged faster than their mammalian homologues (Málaga-Trillo and Meyer, 2003). The sequence was most similar to that of human estrogen receptor. Evolution of genes for major receptor like thyroid hormone and steroid receptors occurred very early in metazoan history. Nuclear receptor genes would be far older than thought, with an estimated origin for protostome-deuterostome split of about 960 mya (Hedges et al., 2004). These genes were probably lost independently in several bilaterian lineages. The extensive **gene losses** of ancestral gene families are reported in *D. melanogaster* and *C. elegans* with 25.8% and 31.0% respectively (Hughes and Friedmann, 2004). The sparse taxon sampling now hampers a thorough analysis and informative interpretation of changes in character like some critical nodes in metazoan phylogeny which are still unresolved (Philippe et al., 2005). Similarity in genomic repertoire between some invertebrate and human sequences can be explained by atypical condition of commonly used model organisms (Raible and Arendt, 2004). In corals, many genes in simple organisms are probably confined to a single ancestral role, during acquisition of extra functions later in evolution, which allowed for more complexity in organism. Many nuclear receptor genes did probably not yet have their present function in organisms in which they first occurred (Bertrand et al., 2004).

**Evolutionary Genetics of Morphological Novelties:** Morphological novelty at the level of different animal phyla is not always reflected in similar changes of gene composition and sequence divergence of genes controlling development. **Evolutionary novelties** appear to be based on co-option of already existing genes for new functions. One molecular mechanism acting at gene level is the acquisition of new regulatory sequences that leads to novel patterns of transcriptional activation (True and Carroll, 2002). As a result new genes can be recruited into existing regulatory gene networks and result in functional changes to network. Genes may gain novel expression domains by chance mutations, or recombination events in their cis-regulatory elements. Changes in expression of upstream transcription factors themselves can initiate activation of target genes in new domains. When combined with an increase in gene family size through duplications, the acquisition of novel functions in duplicates by co-option (Lynch and Force, 2000) is a powerful mechanism to increase the pool of developmental building blocks.

Mechanism of co-option is facilitated by modular character of gene interactions. Modularity, one of central paradigms of molecular evolutionary biology (Duboule and Wilkins, 1998) proposes the use of pre-existing building blocks in novel ways, rather than origin of new elements (Bonner, 1998). In a gene-based, developmental context individual genes together perform a given network function (Gilbert et al., 1996). Such modules can be visualized as being composed of a set of interacting genes that can associate in novel ways with other modules, producing networks of higher organization. This gene-set called Genetic Toolkit, determines the overall body plan and number, identity and pattern of body parts (Carroll et al., 2001). Evolution of metazoan development and body plans is based on increase in complexity of control circuitry regulating an ancestral toolkit of genes, rather than on invention of novel developmental genes (Peterson et al., 2000).

Major transitions in evolution are often accompanied by phylum-specific innovations like occurrence of mineralized tissue in vertebrates (Shimeld and Holland, 2000), which helped to radiate the modern vertebrates. Its development changed the body armour, teeth and endoskeleton, providing adaptive phenotypes for developed locomotion, predation and protection. Members of secretory calcium binding phosphoprotein (**SCPP**) gene family at the heart of such tissue produce the specific ionic conditions in extra cellular matrix required for skeletal mineralization. The SCPP gene family has been traced back to an ancestral gene, **SPARC** (encoding a non-collagenous bone matrix protein), present in protostomes and deuterostomes (Kawasaki et al., 2004). Tetrapod SCPPs arose from SPARC by several gene duplications followed by co-option for new functions, and include genes for dentin/ bone, enamel, milk caseins and salivary proteins in mammals (Kawasaki and Weiss, 2003). Birds have an eggshell SCPP but lack genes for enamel. Milk and salivary-associated proteins, probably reflects the loss of teeth in birds, and evolution of mammary glands in mammals (International Chicken Genome Sequencing Consortium, 2004). SPARC homologues are identified in a variety of invertebrates, where they influence cell behaviour and interactions with extra cellular matrix, rather than involving in production of mineralized tissues. An evolutionary modification of these proteins seems likely to have occurred through a functional shift from facilitating stabilization of structural proteins towards enabling more diverse interactions between cells and proteins of extracellular matrix in course of vertebrate evolution.

**Evolution of Regulatory DNA in Invertebrates:** Significant morphological transformations in body plan of invertebrates are found to correlate with developmental changes of Hox gene expression patterns (Lee et al., 2003). As 2 examples in molluscs demonstrate, some changes are not related to characteristic Hox function of establishing pattern along anteroposterior axis. In *Haliotis asinina*, 2 Hox genes (Has-Hox1 and Has-Hox4) are expressed in mantle margin, where they have been co-opted into new developmental role in shell formation (Hinman et al., 2003). Patterns of Hox expression in squid *Euprymna scolopes* strengthen the argument that co-option events that are often associated with origin of new morphological structures (Lee et al., 2003). Acquisition of 3 innovations derived from ancestral molluscan foot, namely brachial crown, funnels tube, and stellate ganglia, could be ascribed to Hox gene recruitment during **cephalopod evolution**. The large diversity of molluscan body plans suggests that morphological flexibility may result from relaxation of regulatory constraints on recruitment of morphological patterning genes (Lee et al., 2003). Variation at the species level is often based on changes to gene regulation. Expression of yellow gene at the wing tips of *Drosophila biarmipes*, a closely related species to *D. melanogaster*, resulting black spots. The gene's regulatory sequences had gained additional binding sites for highly conserved transcription factors (Gompel et al., 2005). An appealing concept is that combinatorial nature of transcriptional regulation creates a large reservoir for diversity in morphology, and provides variation for natural selection than changes in gene product alone. Changes in cis-regulatory systems of genes may be more significant than changes in gene number or protein function (Caroll, 2000) (Figure 6.11).

**Developmental Innovations in Protein Sequence:** Examples in which evolutionary changes in gene regulation lead to morphological changes are numerous (Carroll, 2005) in which changes in protein sequence have been linked to new adaptations. Many genes have **pleiotropic functions** and changes to their protein sequence are deleterious. Most cases involve gene duplications as in SPARC gene family, or splicing, causing one copy to retain its function, while the other acquires a new function. In the latter case, mutations resulting from detrimental mechanisms, like frame shift mutations, retain for up to hundreds of millions of years and have evolved new protein functions (Raes and Van de Peer, 2005). Changes to **protein-protein interactions** can lead to alterations in developmental mechanisms, by integrating novel regulators into existing pathways, or by eliminating old ones. **Transcription factor** brachyury is expressed in circumference of blastopore of most animal phyla and its orthologues from most Bilateria are capable of inducing

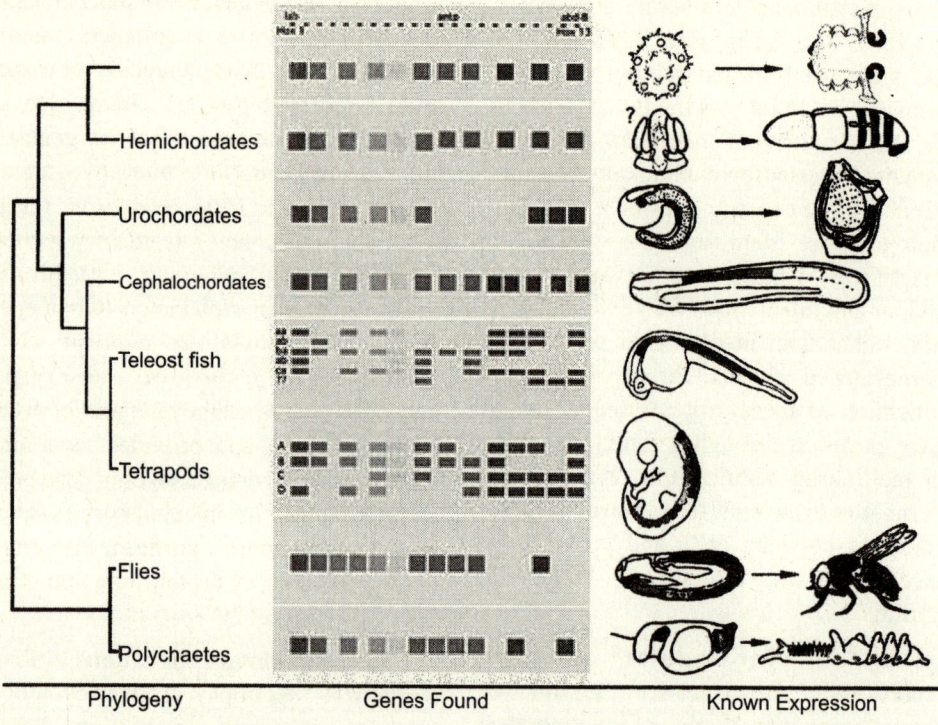

Phylogeny                    Genes Found                    Known Expression

**Figure 6.11**   Hox gene clustes in several phyla.

mesoderm when assayed in *Xenopus* animal caps. Brachyury orthologues from tunicates and *Drosophila* induce formation of mesoderm and endoderm, and this is correlated with loss of a short protein-protein interaction motif, N-terminal of the DNA-binding domain (Marcellini et al., 2003). Insects and tunicates have lost independently circumferential blastopore brachyury expression, but have derived mode of gastrulation which is mostly independent of brachyury. Messenger et al. (2005) have identified. The cofactor that binds to conserved brachyury N-terminal peptide and inhibits endoderm induction the possibility exists that this repression module, which is absent in *Hydra*, evolved in bilaterians to separate mesoderm .In Tunicates and *Drosophila*, these tissue types are derived from topologically separate regions, and derived

mode of development may have relaxed the selective pressure required to maintain this motif.

**The Basic Concepts:**   Homology as a historical concept is defined as shared inheritance of a trait from a most recent common ancestor, though plain similarity has inadequately been used in several instances. **Historical homology** can be applied to different levels including morphological structures, developmental processes, and genes ( Abouheif et al., 1997). Abouheif (1999) suggested a **hierarchical analysis** of homology relationships to reveal alternative evolutionary hypotheses, like the recruitment of homologous genes and networks to work in convergent embryonic and morphological structures. Gene family phylogenies can be used to show if genes are orthologous or paralogous. The concept of generative homology, or syngeny, uses shared developmental pathways to imply that a given character is generated by same genetic machinery inherited from a common ancestor (van Valen , 1982). Wilkinson (2002) has expanded this idea to use a combination of shared **key genes** plus a shared biological or developmental function for which those genes are crucial as indicator of homology. Both these concepts view that continuity of inheritance of potential to make a particular trait rather than continuity of its appearance matters most in homologous relationships (Tautz, 1998). Phylogenetic re-appearance of characters is often observed in well-established phylogenies, and genetic potential to create those characters is probably retained, a phenomenon referred to as **latent homology** (Minelli, 1998) or reawakening (Meyer, 1999). The concept of homology is undoubtedly challenged when it comes to tracking down the evolutionary origins of developmental programmes (Nielsen and Martinez, 2003). Use of gene expression patterns to infer homology is a matter of intense debate (Haszprunar, 1992). Comparisons of gene expression patterns to assess homology hypotheses should be restricted to orthologous gene copies, since new expression domains evolving among paralogues are likely to be convergent. These gene relationships may be complex and need to be tested based on a phylogenetic tree including whole gene family by identification of timing of speciation events relative to gene duplications (Abouheif et al., 1997).

An analysis of snail/slug gene family revealed functional modifications after original copy had become duplicated during vertebrate evolution. The study demonstrated unexpected evolutionary changes in snail/slug expression domains, which imply a higher plasticity and highlight the risk of using expression or function as homology indicators when studying the evolution of gene families (Locascio et al., 2002). Patterns of **gene expression** are complicated to interpret in terms of homology as they interconnect different levels of biological organization, here morphological structures and genes (Abouheif et al, 1997). Conserved expression patterns do not show conservation of gene function. Now it is known that homologous genes can be expressed in structures with different evolutionary origin, such as distal-less in animal appendages (Haszprunar, 1992). Similar developmental roles may be the result of convergent evolution (Mindell and Meyer, 2001).

**Gene Network Evolution:** Behind the scenes are complex interacting gene networks that form the genetic machinery required for origin and functioning of morphology. Such networks may be considered as distinct level of biological organization, homology relationships of which differ from other such levels (Butler and Saidel, 2000). Some have dismissed strictly hierarchical view of biology calling for a combinatorial approach to homology (Hillis, 1994), because interactive combinatorial processes, like co-option and modularity bears a role. Homology assessments cannot be reduced to a yes-or-no question even at the molecular level (Hinman et al., 2003). The obvious problem is how much of a given entity, a gene network, must be continuous between lineages. Two networks are homologous if all genes and their interactions are derived from an identical network in recent common ancestor. Often networks share some elements but differ in others, and they are not fully homologous, or independently evolved. They are partially homologous. Recruitment of **novel genes** into existing regulatory gene network would lead to partial homology. As a result, gene expression patterns need evaluation at the level of gene networks to know their true evolutionary relationships (Lowe et. al, 2003). Studying closely related

species in a phylogenetic context is helpful
in this case, because smaller changes are
expected. This knowledge can then be used
to connect more ancient characters.

Wing patterning network in *Drosophila*
is well studied. when expression patterns of
its constituent genes in wingless castes of ants
are compared, the results give insights into
evolution of wing polyphenism. Several closely
related species do not share common
mechanism to interrupt wing development in
wingless castes, which is unexpected given
that a common origin of all wingless castes
can be assumed (Abouheif and Wray, 2002)
(Figure 6.12).

**Similar Expression Patterns do not
Necessarily Indicate Homology:** Gene
expression patterns should not be used to infer
morphological homology of structures without
employing phylogenetic criteria to test
hypotheses about orthology of genes and partial
homology, or convergence of gene networks.
Genes expressed during basic cellular
processes, like cell proliferation or epithelial-
mesenchymal transitions are likely to be
utilized in **non-homologous organs**. With high
developmental and evolutionary constraint,
they do not support homology of structures.
If homology of structures is to be based on
expression of genes, members of genetic toolkit
require to be consulted, as they control general
and diverse patterning processes. Lowe et al.
(2003) proposed that a comparable expression
pattern in nervous system of *Drosophila*,
hemichordates, and chordates indicates
homology, suggesting that ancestor of
deuterostomes had a diffuse nervous system
that was centralized independently in
arthropods and chordates. This idea disagrees
with prevailing theory of a single origin of
central nervous system with dorsoventral axis
inversion, which is supported by inversion of
TGF-signalling in chordates and *Drosophila*
(De Robertis and Sasai, 1996). The example

(a) *Junonia almana,* wet season form

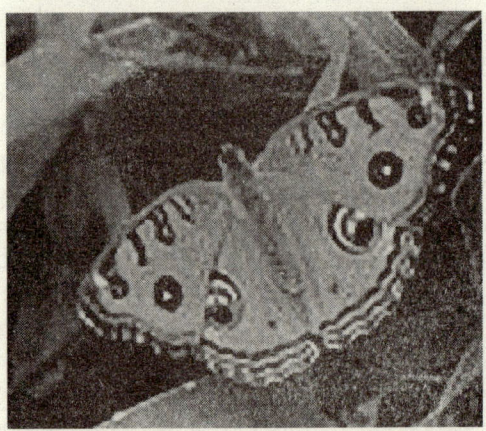

(b) *Junonia almana*, dry season form

(c) *Junonia almana*, wet season form

**Figure 6.12**   Wing Polyphenism

illustrates that there is no obvious boundary as to number of genes need to be co-expressed to proclaim a structure homologous, i.e., would 10 of 22 genes justify homology of such nervous systems. Like morphological analyses, similarity cannot be used as proof to show common descent. Such approaches are clearly phenetic and by no means phylogenetic.

The use of homologous genes to achieve bilateral symmetry in larvae of sea anemones was used to infer that bilaterality originated before the split of Cnidaria and Bilateria (Finnerty et al., 2004). But is bilateral symmetry of **sea anemones** really homologous to that observed in more derived animals or did it arise by convergent evolution (Martindale et al., 2003)? If **orthologous genes** are used to control the development of similar structure in 2 otherwise morphologically different animals, this lends some support to hypothesis of homology, but it is not sufficient and requires rigorous tests. Early metazoan interrelationships have remained difficult to understand, once more stressing the importance of phylogenetic certainty in relation to co-option or homology question. One adequate method of testing morphological homology hypotheses by using gene expression data has recently been proposed (Svensson, 2004). This approach involves investigation of changes in developmental systems in a parsimony analyses by mapping gene expression data onto a phylogenetic tree along with other characters.

Different homology hypotheses can be evaluated in terms of number of evolutionary steps required for each hypothesis to be valid, and the most parsimonious solution can be identified. An example can be seen in depicting a single co-option event leading to shared gene expression in organ 2 in frog and mouse which supports homology of organ 2 in these 2 species. But if instead we were to assume organ 2 in frog as being homologous to organ 3 in mouse, then 2 independent changes in

expression of gene would be required , which would be less parsimonious and thus would not support this homology relationship. Several homology hypotheses can be compared in combined analysis including many genes and other characters (Svensson, 2004). Whenever a co-option event can be identified as a **synapomorphy** for a set of taxa, it could support the homology of a structure. Gene expression data could distinguish between convergence and re-awakening, as in previous case distinguishing features to reflect different origins of 2 independently evolved structures is expected while in latter we do not (Figure 6.13).

**Conserved Developmental Programs or Repeated Evolution?** Despite growing evidence for widespread conservation of genetic toolkit that is used to produce complex body plan of bilaterian animals, there is reason to believe that last common ancestor (Urbilateria) did not necessarily employ same developmental programmes. That function of many homologous developmental genes in last common ancestor was of same kind, but in different developmental context. **Conserved genes**, or entire networks, might have been co-opted repeatedly into new regulatory regions or morphological structures. If available genetic toolkit is of limited size, then possibility of co-option for similar functions, i.e., repeated evolution of **analogous** developmental processes does not appear unlikely (Nielsen and Martinez, 2003). The **homeobox gene** distal-less (dll) and its vertebrate homologue dlx are expressed across extant animals in various types of appendages that are clearly non-homologous (Panganiban et al., 1997). Shared functions of dll genes among animal phyla are few and consensus function is reduced to a role in regulating cell proliferation (Nielsen and Martinez, 2003). The bilaterian ancestor probably had no legs but perhaps some inconspicuous body wall outgrowths triggered

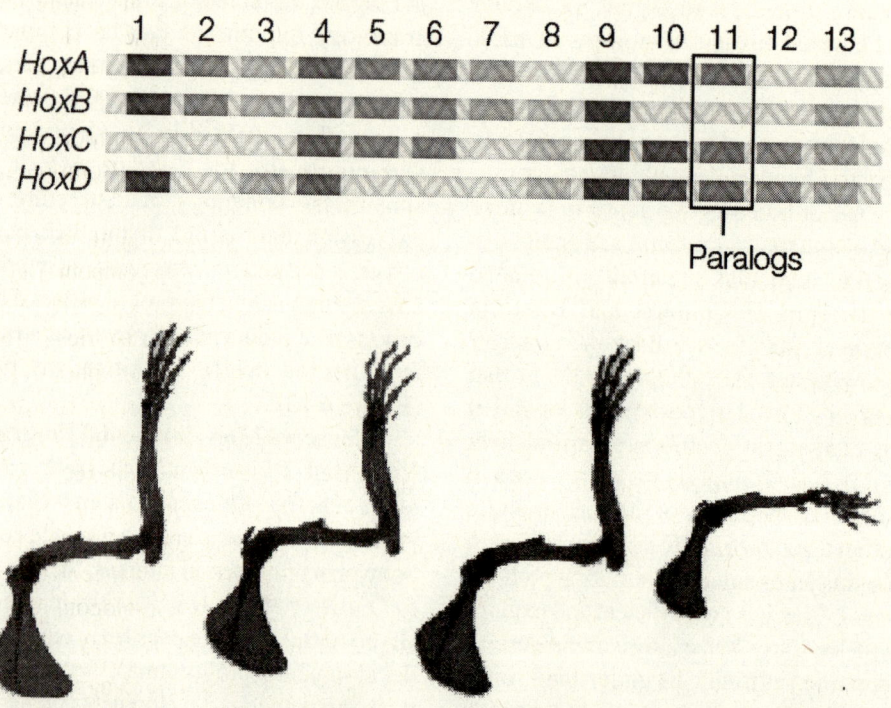

In vertebrates the entire *Hox* cluster has been duplicated multiplen times. Mice and other mammals have 4 *Hox* clusters. All 4 are similar, but each is different. Similar genes in different clusters are called paralogs.

The figures on the left, show mouse forelegs. Inactivating one paralog or the other has subtle effects (middle two images). But inactivating both makes a dramatically different limb (right). This experiment and others have shown that Hox genes in mice work much the same way as they do in fruit flies.

**Figure 6.13**   Vertebrate *Hox* genes

by dll expression. Co-option of this pre-existing process into more specific building blocks could have subsequently occurred during evolution of different animal phyla. It is debated whether segmentation in different animal phyla has had a common origin or not (Minelli et al., 2003). Whether last common protostome-deuterostome ancestor was already segmented or whether segmentation arose on 3 separate occasions in arthropods, annelids and vertebrates? Current 'single origin of segmentation' hypothesis finds that notch and delta genes participate in segmentation of both spiders and vertebrates (Prud'homme et al.,

2003). Segmentation in *Drosophila* is notch and delta independent. The basal phylogenetic position in spiders in relation to *Drosophila* suggests a derived mode of **segmentation** in latter (Telford and Budd, 2003) and thus allows evaluation of more distant ancestors within arthropod clade. Arthropod-like expression pattern of engrailed and wingless genes in segment formation of the annelid *Platynereis* points to a segmented ancestor of all protostomes (Stollewerk et al., 2003).

There are objections to 'single origin of segmentation' hypothesis. Part of the problem might reside in definition of segmentation, since some relax the definition to include repeated structures, like paired coeloms in echinoderms, which render the absence of segmentation, an uncommon trait throughout animal kingdom (Telford and Budd, 2003). Segmentation is a mesodermal process in annelids and vertebrates, whereas in arthropods it is primarily ectodermal (Minelli, 2003). It is conceivable that the same deeply conserved modules have been co-opted for similar functions many times, producing not only to morphologically different types of animal segmentation, but to segmented tissues of different embryonic origin found in metazoans. Segmentation may be a case of repeated evolution not implying existence of a segmented Urbilateria. The way in which *Pax6* and its associated genes are involved in eye development across Metazoa suggests shared genetic potential for occurrence of eyes (Fernald, 2000). Yet the phylogenetic pattern of distribution of eye structures is polyphyletic, which shows multiple independent origin from forms lacking eye development (Salvini-Plawen and Mayr , 1995). The current situation most likely reflects successive losses and gains of use of *Pax6* network during evolution of metazoan animals (Wilkins, 2002).

Some have suggested that Pax6 has become integrated into several independently evolved genetic programmes to regulate particular aspects of eye development (Nielsen and Martinez, 2003), rather than being a master regulator of eye development (Haider et al. ,1995). Repeated utilization of similar genetic pathways involving pre-existing building blocks may emerge as common theme in animal evolutionary history. Completely new morphological structures can evolve by integration of independent anatomical entities. Vertebrate eye is a compound structure comprising 2 types of light-sensitive cells with independent evolutionary histories (Arendt et al., 2004). Presence of ciliary photoreceptors containing an opsin similar to those of vertebrates in brain of ragworm *Platynereis* (Annelida) suggests that both receptor types were present in Urbilateria. It is straightforward to propose that vertebrate and invertebrate eyes are partially homologous, since they contain homologous and non-homologous cell types derived from different germ layers.

**Retention of Genetic Programs:** Certain structures may be lost and later re-appear yet genetic potential to produce those structures may be retained, even though structures are not continuously present in all ancestors. This observation is known as latent homology or re-awakening (Tautz, 1998). Re-awakening may be a valid hypothesis where phylogenetic character distributions suggest reappearance of character that would be considered homologous on morphological grounds. Instances of reversals that appear in robust phylogenies might be hidden cases of latent homology (Wake, 2003). If dormant genes can be re-activated after further speciation, we expect high similarities in developmental systems between lost and regained structure. The convincing way of testing for re-awakening would be to show functionality of genetic programme by experimental induction of re-evolved trait in species that do not exhibit the trait. Several species of *Xiphophorus* display a so-called sword, an extension of ventral tail fin. The sword was lost once and later re-evolved at least twice in different branches, suggesting re-awakening of '**sword-developing**' programme (Meyer, 1999). Swords could be induced in some sword-less species through testosterone treatment, making a strong support for this hypothesis. Swords could be induced by artificial selection in distantly related species.

There is molecular phylogenetic evidence for independent origin of compound eyes in arthropods. Myocopids have compound eyes which are nested phylogenetically within several groups that lack this kind of eye (Oakley and Cunningham, 2002). Ancestral-state reconstruction was highly significant in supporting independent origin of compound eyes from eyeless ancestors. The ommatidia of arthropods have an arrangement of photoreceptive cells different from regained eyes in ostracods, which cast doubt on their homology from a morphological point of view. Whether the same genetic pathways are used to produce these slightly different types of compound eyes is interesting. A recent example from stick insects suggests that wings have re-evolved as many as 4 times during radiation of this group (Whiting et al., 2003). An alternative interpretation would involve 13 independent occasions of wing loss, which is less parsimonious solution. But accepting the alternative hypothesis would not appear implausible if different method of character optimization were used, i.e., one that assumes loss to be more likely than re-appearance (Telford and Budd, 2003) while establishing the basic requirement for re-awakening, which is a robust phylogenetic hypothesis of character evolution. Phylogenies based on morphological characters will be affected in negative way by re-awakening as reappearing character

causes homoplasy and will be incongruent with other characters. Although re-appearance of character using same genetic machinery is evolutionarily truly parsimonious, conflicting hypotheses may arise when applying parsimony principle for phylogenetic tree construction in such cases.

## 6.16   Epigenetic Links to Disease

Epigenetic mechanisms like methylation are linked to various phenotypes in humans and animals. *Daphnia atkinsoni* develops defensive spines when live in predator-rich habitats. This phenotype for stress-intense situations (Hagmann, 2010) is then pass to offspring even when the ring is no longer reared in the same environment (Laforsch et al., 2009). Trans-generational epigenetic changes are visible in mammals. Pregnant agouti mice were fed with variable (Hagman, 2010) amounts of folic acid, choline, which are important in the production of S-adenosyl methionine, a player in methyl group transfers. Agouti mice give birth to sickly and overweight agouti offspring. The treated mice then bore pups of fur colors directly tied to the amount of supplements the mothers received; mice given supplements produced healthy wild-colored pups, and mice without supplements produced agouti pups. The agouti fur color and its associated symptoms are linked to alterations in DNA methylation. The offspring of mothers fed with the above-mentioned supplements did not exhibit obesity in adulthood, indicating possible solutions to developed-world obesity epidemics (Dolinoy et al., 2007).

In humans, the Angelman syndrome is the result from aberrant methylation patterns of the maternal allele on chromosome 15 (Weeber et al., 2002). The Prader-Willi syndrome, a sister syndrome to the Angelman syndrome occurs when the paternal allele on chromosome 15 is silenced (Sahoo et al., 2008,

Hagmann, 2010). Non-functioning methylation results in grave consequences, even if only one allele is afflicted. **Environmental toxins**, endocrine disruptors showed the effect on epigenetic transgenerational phenotypes.

Male infertility in rats resulting from the anti-androgenic fungicide vinclozolin was transferred to the F1 generation. The offspring of rats exposed to vinclozolin had higher rates of immune abnormalities, kidney disease, prostate disease, and breast cancer (Skinner et al., 2010, Hagmann, 2010). Yehuda et al. (2009) found that low basal cortisol levels were indicator of vulnerability to post traumatic stress disorder. The children of sufferers with post traumatic stress disorder had low basal cortisol levels. Several abnormal regulations and diseases are based on epigenetic changes. To study the molecular mechanisms of epigenetically caused diseases and (Hagmann, 2010) to figure out which ailments can be traced back to these alterations is interesting.

## 6.17 Epigenetic Change and Transplantation

Epigenetic changes during transplantation may contribute to post-transplant gene dysregulation, influencing transplantation outcome (Schildberg et al., 2010). These findings could possibly help to improve the acceptance of drafts and widen the donor pool. After optimizing **immunosuppressive therapy** postoperative care, and surgical techniques (Hagman, 2010), the present major problem in organ transplantation is the shortage of donor organs. Although organ transplantation is a life-saving operation, few patients can be treated, as the organ demand is higher than the supply. Different strategies are now in use to improve the quality of donor organs after cold storage during transplantation. Such innovations are new formulations of organ

preservation solutions, the stimulation of the cAMP second-messenger (Hagmann, 2010) signal pathway (Akbar and Minor, 2001), or venous systemic oxygen persufflation (Hagmann, 2010) during cold preservation (Treckmann et al., 2008), for restoring the viability of is chemically pre-damaged organs. Such processes improve the quality of existing organs (Wei et al., 2007) by reducing the incidence of **primary graft dysfunction**. They enlarge the donor pool by facilitating the use of marginal donors like non-heart-beating donors (Reddy et al., 2004b).

## 6.18 Transplantation: Self versus Non-self

The immunity is a defender of the infection. van Swieten used the term immunity in 1775 to describe the effects induced by an early attempt at vaccination. Transplantation protocols and immunosuppressive care are now highly developed. But transplantation of tissues is still the challenging and complex area. The act of defense requires distinction between the defending and attacking fractions: self and non-self. An organism must distinguish self from non-self tolerating the former and engaging the latter (Medzhitov and Janeway, 2002). Higher organisms need an innate immune system, represented by germ line-encoded receptors, recognizing staple pathogen-associated molecular patterns like lipopolysaccharide and flagellin, and an adaptive immune system. The WBC of the adaptive immune system possesses the unique receptor with a single specificity. To prevent recognition of self antigens by auto reactive lymphocytes and attack on the body's own cells, WBC must go through the clonal deletion process (Hagmann, 2010). Passing this selection, lymphocytes circulate the body up to the moment when their receptor binds to specific **foreign antigens**, which are presented by highly polymorphic proteins of MHC (Williams et al., 2002). This antigen presentation leads to T cell activation, proliferation, and differentiation (Marrack and Kappler, 1997). Many other triggers, so-called danger signals namely **heat-shock proteins**, nucleotides, interferon, and reactive oxygen species, start immune reactions without previous activation of antigen-presenting cells (Barchet et al., 2008).All these mechanisms allow the body to differentiate between self and non-self to combat foreign antigens with a potent immune response.

**Rejection Mechanisms:**    The rejection of tissues is a response of the adaptive immune system to donor antigens based on antibodies, which are produced by B cells and cell-mediated immunity, mainly dependent on T cells. Immune cells recognize **alloantigens**, like foreign major histocompatibility complex (Hagmann, 2010) and minor histocompatibility antigens from the donor organ to become activated. Host T cells then attack the antigen-bearing( Hagmann, 2010) cells. B cells produce specific antibodies to allow specific killing via macrophages or complement lyses, with resulting in graft rejection (Arakelov and Lakkis, 2000).Since host T cells need to recognize the specific antigen in professional antigen-presenting cells, there are 2 main mechanisms for allorecognition. If graft antigen-presenting cells, like (Hagmann, 2010) dendritic cells and macrophages, migrate into recipient lymphatic tissues and activate recipient T cells. This is called direct allorecognition. With the help of T cell receptor, alloreactive T cells recognize processed (Hagmann, 2010) peptides from graft antigens on the MHC molecules of donor cells (Warrens et al., 1994). In the 2nd mechanism, the indirect allorecognition, donor derived antigens are presented by (Hagmann, 2010) professional

antigen-presenting cells from the recipient (Caballero et al., 2006). Thus **allorecognition** and graft rejection are not restricted to MHC-mismatched grafts, but are also an issue in transplants with (Hagmann, 2010) differences in minor histocompatibility antigens. It is possible to prevent grafts from acute rejections using immunosuppressive drugs like cyclosporine A, tacrolimus, azathioprine, mycophenolate-mofetile,( Hagman, 2010) sirolimus, and antibodies like anti-CD25. These drugs have side effects like hypertension, dyslipidemia, hyperglycemia, and peptic ulcers (Ferrara et al., 2009, Hagmann, 2010)).As most of them act non-specifically, under their influence the immune system( Hagmann, 2010) as a whole is less resistant to pathogens, and the risk of post transplant cancer( Hagmann, 2010) is increased.

**Problems in Transplantation:**    The increased sensitivity of immunosuppressed patients is one problem in transplantation. Major challenges are balancing of immunosuppressive drug side effects and rejection. To prevent rejection, high doses of these drugs are used with critical and long lasting side effects. Krischnan et al. (2008) reported a case of a kidney transplantation between monozygotic twins, which allowed a discontinuation of immunosuppressive (Hagmann, 2010) treatment due to the optimal genetic background. This change allowed the usage of a better-fitting therapy for focal segmental glomerulosclerosis, a disease already existing in recipient before transplantation. This balancing act between low- and high-dose immunosuppressive medications is the same in cell transplantations like autologous hematopoietic stem cell transplantations. Such therapy is established to treat hematopoietic and oncological diseases, like multiple myeloma after high-dose chemotherapy (Harousseau, 2009).

Leukemias are curable using chemo- and radiotherapy, followed by allogeneic hematopoietic stem cell transplantations (Litzow et al., 2009). Immunosuppressive drugs are remedies after autologous cell transplantations. In some cases, based on the genetic background of donor and recipient and T cell amount in the graft, donor cells can attack the host cells, called graft-versus-host disease (**GVHD**). T cells mediate beneficial graft-versus-tumor effect, so that T cell-depleted or pharmacologically treated transplants lead to reduced risk of GVHD but increase the risk of cancer relapse (Apperley et al., 1986). Since it is difficult to calculate a risk balance between immunosuppressive drug dosages and their side effects, it is important to recognize early immunological responses imperiling transplant integrity. The clear and meaningful biomarkers, or diagnostic tools to predict potential rejections or manifest immune responses are required.

## 6.19 Epigenetic Changes as Biomarkers

It would be helpful if tailored epigenetic changes could be used as diagnostic tools to non-invasively determine transplantation outcome at increasingly early time. **Biomarkers** for transplantation outcome should be able to predict severity and onset of acute and chronic graft rejection, and exactly foretell the response to immunosuppressive medication. The fact that epigenetic changes are easily (Hagmann, 2010) categorizable makes them favorable candidates for a type marker. The global histone modification levels identify adverse prognoses in prostate cancer (Ellinger et al., 2010). Mehta el al. (2006) used quantitative detection of **CALCA** promoter hypermethylation as an indicator of post transplant kidney injury. Investigation into therapeutic uses for epigenetic changes, or

markers for diagnostic tools concerning graft and transplantation-related complications extends into the domain of graft arteriosclerosis. Arteriosclerosis in transplanted hearts, as with atherosclerosis of native coronary arteries, is accompanied by activation of the innate and adaptive immune responses. Use of pharmacological agents like **HDAC** (Hagmann, 2010) inhibitors for restriction of proinflammatory cytokine production such as IFN- would, aid in diminishing the inflammatory milieu of arteriosclerosis (Ahmad et al., 2010).

## 6.20 Epigenetic Inheritance

Trans-generational epigenetic effects are those where the DNA sequence remains unchanged but higher-level systems are altered, leading to differences in (Dickins and Rahman, 2012) phenotypic expression in the next generation. These systems include DNA methylation and histone modification. The inheritance of the epigenetic modification of gene expression through processes like methylation has been interesting for advocates of the EES, because they are regarded as the challenge to the hard inheritance (Pigliucci, 2007).

Several hypotheses about experience-dependent epigenetic inheritance are as follows:

### Licking and grooming: Adapted Parental Effect

Within rodents there are stable and heritable individual differences in the grooming offered by mothers. This inheritance is not genetic but it is modified by behaviour. Several integrated **proximate mechanisms** are involved (Champagne, 2008). Low levels of oxytocin receptor binding in the medial preoptic area of hypothalamus are associated with low levels of grooming (Dickins and Rahman, 2012). Oxytocin is implicated in facilitating bonding (Campbell, 2007). In high-grooming females, dopamine levels in **nucleus acumbens** increase before grooming. When grooming stops the dopamine level returns to baseline. This effect is not found in low-grooming mothers. The hypothalamic oxytocin neurons mediate this response, which leads to behavioural differences between mothers. Pups of mothers who are low in grooming have a prolonged passive stress response after the removal of the stressor. This response has behavioural effects, such as reduction in exploratory behaviour and increase in inhibition (Dickins and Rahman, 2012), sometimes classified as behavioural depression. The pups of low-grooming mothers have low oestrogen sensitivity. **Oestrogen** possesses an epigenetic effect as it increases the expression of the oxytocin receptor gene. Birth is associated with high oestrogen exposure for pups, which lead to increased **oxytocin** receptor binding in the hypothalamus. Low oestrogen sensitivity (Dickins and Rahman, 2012) is equivalent to that of mice lacking a functioning copy of the oestrogen receptor alpha, which affects gene expression, thus reduces the ability of

oestrogen to affect expression (Dickins and Rahman, 2012). In the offspring of low-grooming mothers, the expression of oestrogen receptor alpha is reduced relative to that of high-grooming mothers, which potentially affect the oxytocin activity. This alteration to gene expression is affected by postnatal maternal care-low-grooming mothers cause reduction in the level of oestrogen receptor in the medial preoptic area of the hypothalamus. This is linked to high levels of methylation at several sites in the promoter region of oestrogen receptor alpha. This presents a highly integrated set of proximate mechanisms. Low levels of grooming set in motion a particular stress response, and inhibition of bonding in turn leads to low grooming investment when those pups become mothers. Stressing high-grooming females lead to reduced grooming, and this is then epigenetically inherited (Figure 6.14).

**Learning Biases:** Epigenetic modification of genes that are expressed in the brain affect learning and memory (Wells, 2003) suggesting that specific learning biases should emerge under specific developmental conditions. Not

**Figure 6.14**    Genetically identical, but epigenitically different agouti mouse (modified after Duhl et al, 1994)

all organisms calibrate epigenetic mechanisms. Birds deposit hormones in egg yolk affecting chick development and adult reproductive success (Zhang and Meaney, 2010). The existence of calibrating mechanisms proves that organism lives in a stochastic ecology, moves between varieties of environments. If they move from environment to environment, then they are generalist. Selection shapes the calibration. Generalist strategies present a great deal of moment-to-moment and day-to-day change that would be served by learning mechanisms. The extended synthesis has packaged learning as soft inheritance at the cultural level (Dickins and Rahman, 2012), but paid little attention to individual learning. Even cultural learning (Dickins and Rahman, 2012) processes are found within individuals. There is variation in learning and teaching.

**The Hard/Soft Dichotomy:** How traits like personality, facial features and certain diseases run in families, can be traced back to the 18th century (Lo´pez-Beltra, 2007). Two views of heredity - hard versus soft (Mayr, 1998) were crystallized by the late 19th century, into a dichotomy. Proponents of **hard heredity** believed that parents transmit the developmental blueprint to offspring at the conception. Proponents of soft heredity believed that parents transmit their features to their offspring, including features acquired during their lifetime, and transmission occurs via subsequent interactions between parent and offspring (Churchill, 1987).

## 6.21  Epigenetic Variations and Soft Inheritance

Epigenetic inheritance used in 2 overlapping ways means transmission of environmentally-induced and environmentally-regulated variations to subsequent generations of cells

or organisms (Jablonka and Lamb, 2007a). The first is the inheritance of any developmental variations that do not stem from differences in DNA sequence including cellular inheritance through the germ line, and soma to soma transfer of information that bypasses the germ line. **Soma-to-soma transmission** through developmental interactions between mother and embryo exists (Clark et al., 1993). The 2nd, the transmission from mother cell to daughter cell of variations that are not the result of DNA differences is cellular epigenetic inheritance. It occurs during cell division in prokaryotes and mitotic cell division in eukaryotes, or during the meiotic divisions in the germ line, so that offspring may inherit epigenetic variations through the germ line. The mechanisms causing cellular epigenetic inheritance can lead to cell memory. The early maternal behavior in rats has long-term effects that are associated with chromatin marks in the key gene in their offspring's brain cells (Weaver et al., 2004). The changes in marks are associated with fear conditioning in rats (Miller and Sweatt, 2007). The mechanisms that underlie cellular epigenetic inheritance is called epigenetic inheritance systems (**EISs**), and 4 distinguished types of EISs are :

(i) Self-sustaining feedback loops: When gene products act as regulators to control the genes own transcriptional activity, the transmission of such products during cell division results in the same gene activity being reconstructed in daughter cells. Such positive feedback lead to alternative and heritable cell phenotypes is found in fungi (Malagnac and Silar, 2003), bacteria and other microorganisms (Smits et al., 2006). It plays an important role in the development of multicellular organisms (Ferrell, 2002).

(ii) Structural inheritance: Existing cellular structures act as templates for the production of similar structures, which become components of daughter cells. Such templating involves prion-based inheritance in fungi, inheritance of cortical structure in ciliate, and the reconstruction of genetic membrane.

(iii) Chromatin marking: Chromatin marks are proteins and small chemical groups attached to DNA those control gene activity. Relicts of such marks segregate in the DNA strands after replication, and reconstruct similar marks in daughter cells (Henikoff and Smith, 2007). Chromatin marks are modifiable histone and non-histone proteins and methyl groups.

(iv) RNA-mediated inheritance: The interactions between transmissible, RNA molecules and the mRNAs or the DNA chromatin (Jablonka and Lamb, 2008) regions with which they pair (Matzke and Birchler, 2005) control the transcriptional states. Such interactions are transmitted between cell and organism generations through an RNA-replication system, or via the interaction of the small RNAs with chromatin leading to heritable modifications of chromatin marks. RNA-DNA and RNA-RNA pairing interactions lead to silencing, target gene deletion and gene amplification (Mochizuki and Gorovsky, 2004). The cellular epigenetic inheritance is ubiquitous. Over a hundred cases of inherited epigenetic variations in bacteria, protists, fungi, plants and animals are reported.

Over 250 years ago, Carl Linnaeus described epigenetic inheritance in a variant of *Linaria vulgaris* with a different floral structure that of the normal toadflax and named the variant 'Peloria', the Greek word for 'monster'. Coen and others studied the molecular basis of Peloria by looking at *Lcyc*, the homologue of a gene that control dorso-ventral asymmetry in other plants and when mutated, results in similar morphological phenotype (Cubas et al., 1999). The DNA sequences of the normal and peloric forms of *Linaria* were the same, but the pattern of DNA methylation differed (Jablonka and Lamb, 2008). In the Peloria variant, the result of epimutation, the gene was highly methylated and transcriptionally silent. Peloria strains are not completely stable, and rarely branches with wild-type flowers develop on peloria plant. The epigenetic marks transmit to progeny for at least 2 generations. Many epimutations found in plants appear in conditions of genomic or chemical stress. Examples of epigenetic inheritance are also found in animals. Vastenhouw and others found that feeding *Caenorhabditis elegans* with bacteria expressing double-stranded RNA that targets specific nematode genes causing morphological and physiological variations that transmits for at least 10 generations (Vastenhouw et al., 2006).

Epimutations are studied in *Drosophila*. The geldanamycin, a drug that inhibits the activity of the heat shock protein on the phenotype of an isogenic *D. melanogaster* strain that carried a mutant allele of the Krüppel gene affecting eye morphology (Sollars et al., 2003). Adding geldanamycin to the flies' food increased the development of the abnormal eye phenotype. Addition of the drug to the food for one generation, followed by 6 generations of selective breeding, increased the proportion of flies with anomaly from over 1% to above 60%. As the strains used were isogenic, the selectable variation probably resulted from new heritable epiallelic differences. Epigenetic inheritance in mammals comes from studies of mice and rats. The epigenetic inheritance is found in every taxon which affects every type of locus in the genome. The conditions inducing cellular epigenetic variations and the stability of their inheritance depend on the type of **EIS** and organism. Often, the generation and transmission of epigenetic variations are responsive to external conditions and are developmentally controlled. In other words, it is soft inheritance.

The soma-to-soma transmission is included under the epigenetic inheritance. Soma-to-soma routes of transmission include those involved in transmitting or acquiring symbionts and parasites; transmitting products of development and soma-dependent deposition of specific chemicals in the eggs of oviparous animals and plants. Morphological affordances or constraints (Jablonka and Lamb, 2008) could lead to heritable developmental effects. Transmission of ecological legacies through ecological niche construction is often part of soma-to-soma transmission (Turner, 2000). The parent's behavior is often instrumental in the developmental reconstruction of similar phenotype in offspring in animals. But the distinction between behaviorally and non-behaviorally transmitted information is often not clear. The soma-to-soma transmission through maternal behavior is studied in rats. The **behavior** and stress resistance in rats is influenced by the mother's care for their young during the early period. Less maternal licking and grooming reduce stress resistance in the offspring. Increased licking and grooming makes the offspring more resistant to stress. The behavioral phenotype is incorporated in the lineage. When the daughters become mothers, they exhibit the maternal-care style (Weaver et al., 2004).

## 6.22 An Extended View of Heredity

This view of heredity that incorporates non-DNA variations and the epigenetic mechanisms of their production have many implications for evolutionary biology. Some are as follows:

1. Adaptation can occur through the selection of **heritable epialleles**, without any genetic change. This is important when populations are small and lack variability, as in cases of intense inbreeding following isolation, or changes in reproductive strategies. The discovery of much epigenetic variation in natural populations strengthens the view that they may play important role in evolution (Bossdorf et al., 2008).

2. Heritable non-DNA variations enhance the effectiveness of accommodation processes, genetic assimilation and enhance adaptive evolution.

3. Heritable epigenetic variations may start the process of reproductive isolation. Differences between chromatin structures inherited from father and mother may result in hybrid offspring which fails to develop normally or being sterile, because the 2 sets of parental chromosomes carry incompatible chromatin marks. **Behavioral differentiation** between populations cause differences in courtship behavior, like the preferred time or place at which it occurs, or the song dialect used, may start tradition-based pre-zygotic isolation.

4. Epigenetic inheritance and epigenetic control mechanisms probably played role in the major transitions in evolution particularly during the **symbiogenesis**, evolution of eukaryotic chromosome, and evolution of language through culturally guided genetic assimilation (Jablonka and Lamb, 2006).

5. Epigenetic inheritance has constrained the evolution of **ontogeny**. Several developmental phenomena, like the difficulty of reversal determined and differentiated cell states found in animal

groups may be the result of selection against transmission to offspring the **epigenetic memories** associated with the parents' development and chance epimutations. All of which would prevent zygote to start its development from a totipotent epigenetic state. Selection may favour the enhancement of germ line transmitted epigenetic memory (Jablonka and Lamb, 1995) in the evolution of genomic imprinting.

6. The origin of mechanisms of epigenetic and behavioral transmission, and evolution of different strategies of transmitting information are neglected areas of evolutionary biology.

The view of heredity now emerging is challenging. The tree metaphor based on the assumption that the pattern of evolution is the branching one, with each branch start from the single common ancestor. If hybridization and infection are common in evolution, the tree metaphor is not suitable. There are strong arguments in support of the conjecture that in early evolution **horizontal gene transfer** (Figure 6.15) may have been the rule than the exception, and that it may still be of major importance( Jablonka and Lamb, 2008) today in evolution of microorganisms and plants (Goldenfeld and Woese, 2007). The target of selection (may be the community rather than individual) may alter our understanding of evolutionary dynamics (Rosenberg et al., 2007). Gene expression in the eyespot model is mentionable in this regard (Figure 6.16).

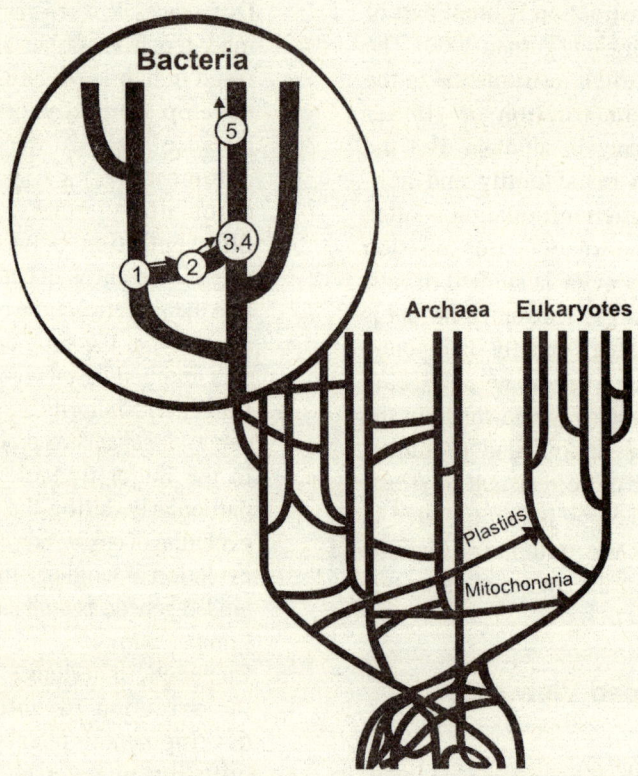

**Figure 6.15**   Horizontal gene transfer

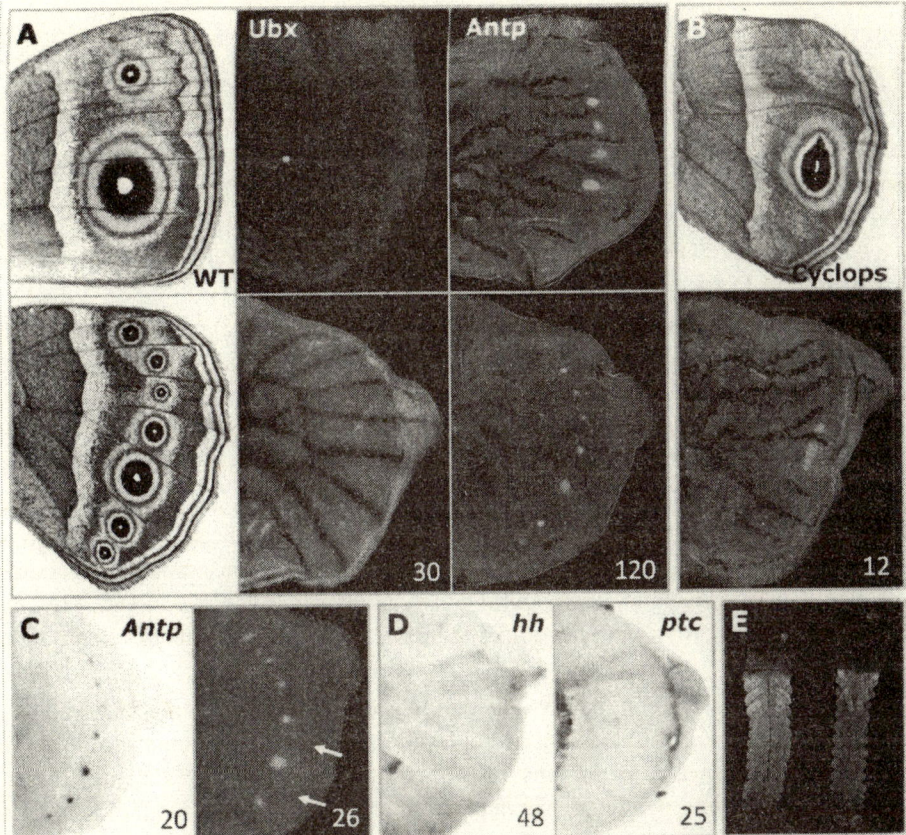

**Figure 6.16   Gene expression in the eyespot model *Bicyclus anynana* (Nymphalidae, Satyrinae))**

(A) Fore-(top) and hindwing (bottom) of 'wild-type' adult and larval wing discs visualized for Ubx and Antp. Ubx is detected throughout the hindwing, but not in the forewing (as is characteristic of insects), and is not associated with any colour pattern element (the bright spot visible in the forewing is an artefact). Antp is detected in the presumptive eyespot organizers in both fore- and hindwing. (B) Adult (top) and larwal (bottom) hindwing of *Cyclops* venation mutant with altered eyespot number and shape: *Antp* is upregulated in a single elongated organizer, matching the morphology of the adult eyespot centre. (C) Larval hindwings stained for *Antp* mRNA (left) and protein (right). Antp is detected in eyespot centres shortly after the last larval molt, prior to the extension of trachea into the vein lacunae (arrows) and before the upregulation of otheir organizer proteins. (D) Larval hindwings stained for hh and ptc mRNA. Absence of both transcripts in *B. anynana* eyespot fields reveals genetic divergence in organizer determination. (E) Immunostainings in embryos of *B. anynanm* (left) and *J. coenia* (right) at 30 to 40% development show the typical pattern of Antp in thorax and abdomen. Proteins are shown in colours, mRNA in gray; numbers indicate individuals examined.

Saenko *et al. EvoDevo* 2011 **2**:9 doi:10.1186/2041-9139-2-9

# SUMMARY

1. The term "soft inheritance" coined by Mayr, refers to the inheritance of variations that are the result of non-genetic effects. Transgenerational epigenetic effects are those where the DNA sequence remains unchanged but higher-level systems are altered, leading to differences in phenotypic expression in the next generation. These systems can include DNA methylation and histone modification. Several researchers equate phenotypic variation as a consequence of DNA sequence variation, with phenotypic variation as a consequence of variation in DNA expression

2. The epigenetic processes have a second-order effect on DNA expression. The inclusive inheritance model sees phenotypic variation as explicandum and inclusive inheritance as explicans. Epigenetic processes are a potential answer to this problem because they allow non-genetic information to be inherited across generations. But there is a great deal of variation in epigenetic mechanisms. They are present in many taxa, and multiple epigenetic mechanisms are proposed to affect development and disease.

3. Jablonka & Lamb in their comprehensive case for soft inheritance, argue that there are 4 dimensions or sources of evolutionary change-genetic inheritance, and then 3 sources of soft inheritance: epigenetic, behavioural and symbolic inheritance systems. They regard soft inheritance as providing phenotypic tailoring during the lifetime development of an organism.

4. Epigenetic changes are completely outside the scope of modern evolutionary synthesis, and have therefore generated considerable controversy within scientific community. Epigenetic inheritance affects many phenotypic traits of living organisms, and so the post-modern evolving synthesis has departed significantly from "modern synthesis" to incorporate such findings.

5. C. H. Waddington in 1942 introduced the term epigenetics for the interactions of genes and environment that bring the phenotype into being. Epigenetic means the phenotypic changes that are inheritable without DNA mutation. The chromatin structure is important in regulation of gene expression. Tail modifications in histones are important in chromatin structure. Epigenetic inheritance is the transmission of information from a cell or multicellular organism to its offspring without the information being encoded in nucleotide sequence. Such inheritance is found in the development of multicellular organisms.

6. Waddington considers development to be an epigenetic process. Phenotype is the result of interrelations among genetic processes, their potentialities and constraints, cytoplasmic differentiations, and environment. Waddington further expands the classical genetic vision on phenotype by including epigenotype, genetic and non-genetic developmental interactions that organize organic substances and tissues. Genes have sufficient stability to be transmitted through generations. The cell has been described as a double-cyclic system in which genes make gene products, which elaborate the final cytoplasmatic elements, while the cytoplasmatic elements feed back to genetic level to control gene-activity.

7. Genes not only regulate, but are regulated by non-genetic factors-relative concentrations of substances. A gene itself regarding genetic influences, promotes linear reaction pathways.

8. Natural selection work on complex, developing phenotypes and on epigenetic processes that build these phenotypes in interaction with the environment. Stressing the developing phenotype in evolutionary theory expands the classical focus on transmission of genetic information with a second focus on gene regulation or instructions on how to use the genetic information. To bring evolution and development to full synthesis, a developmental theory is needed.

9. Various non-genetic mechanisms of inheritance that operate in parallel with Mendelian-genetic inheritance have explored. Non-genetic inheritance comprises all vertical mechanisms of inheritance, including transgenerational epigenetic inheritance, somatic inheritance, environmental inheritance, and behavioral or cultural inheritance. So epigenetic, behavioural and symbolic systems are viewed as proximate mechanisms that provide a phenotypic tailoring service across the life span, and this action introduces new variation into the phenotype, changing frequencies at the population level, transmitted through generations and seems like evolutionary change.

# REVIEW QUESTIONS

## Short Answer Questions

1. What do you understand by the soft inheritance?
2. What is the full form of EES?
3. What do you understand by non-genetic inheritance?
4. What is the "pluralistic model of heredity"?
5. Define epigenetics.
6. What do you understand by "pure genetic transmission"?
7. What do you understand by "pure cultural transmission" ?
8. Define phenotypic variation.

## Long Answer Questions

1. Discuss the mechanism for production of phenotypic variation.
2. Discuss the epigenetic link to diseases.
3. Discuss in brief the core ideas of evo-devo biology.

# REFERENCES

Abouheif, E. 1997. Developmental genetics and homology: a hierarchical approach.Trends Ecol Evol, 12:405-408.

Abouheif, E. 1999. Establishing homology criteria for regulatory gene networks: prospects and challenges. In : Homology. Symposium on Homology held at the Novartis Foundation (Symposium 222); London. Edited by Brock G R, Cardew G. Wiley: Chichester, UK; pp. 207-225.

Abouheif, E., Akam, M., Dickinson, W.J., Holland, P.W.H., Meyer, A., Patel, N.H., Raff, R.A., Roth, V.L., Wray, G.A.1997. Homology and developmental genes.Trends Genet , 13:432-433.

Abouheif, E., Wray, G.A.2002. Evolution of the gene network underlying wing polyphenism in ants. Science, 297:249-252.

Ahmad, U., Ali, R., Lebastchi, A.H., Qin, L., Lo, S.F., Yakimov, A.O., Khan, S.F., Choy, J.C., Geirsson, A., Pober, J.S., Tellides, G. 2010. IFN-gamma primes intact human coronary arteries and cultured coronary smooth muscle cells to double-stranded RNA- and self-RNA-induced inflammatory responses by upregulating TLR3 and melanoma differentiation-associated gene 5. J Immunol, 185(2):1283-1294

Akbar, S., Minor, T.2001. Significance and molecular targets of protein kinase A during cAMP-mediated protection of cold stored liver grafts. Cell Mol Life Sci ,58(11):1708-1714

Apperley, J.F., Jones, L., Hale, G., Waldmann, H., Hows, J., Rombos, Y., Tsatalas, C., Marcus, R.E., Goolden, A.W., Gordon-Smith, E.C., et al. 1986. Bone marrow transplantation for patients with chronic myeloid leukaemia: T-cell depletion with Campath-1 reduces the incidence of graft-versus-host disease but may increase the risk of leukaemic relapse. Bone Marrow Transplant, 1(1):53-66

Arakelov, A., Lakkis, F.G. 2000. The alloimmune response and effector mechanisms of allograft rejection. Semin Nephrol ,20(2):95-102

Arendt, D., Tessmar-Raible, K., Snyman, H., Dorresteijn, A.W., Wittbrodt, J.2004,: Ciliary photoreceptors with a vertebrate-type opsin in an invertebrate brain. Science, 306:869-871.

Ariew, A.2003. Ernst Mayr's 'ultimate/proximate' distinction reconsidered and reconstructed. Biol. Philos. 18: 553-565.

Arthur, W.2000. The concept of developmental reprogramming and the quest for an inclusive theory of evolutionary mechanisms.Evol Dev , 2:49-57.

Arthur, W. 2001. Developmental drive: an important determinant of the direction of phenotypic evolution.Evol Dev , 3:271-278.

Avital, E. and Jablonka, E. 2000. Animal Traditions: Behavioural Inheritance in Evolution, Cambridge University Press

Barchet, W., Wimmenauer, V., Schlee, M., Hartmann, G. 2008. Accessing the therapeutic potential of immunostimulatory nucleic acids. Curr Opin Immunol ,20(4):389-395

Beckman, M. 2004. Crime, culpability and the adolescent brain. Science, 305: 596-599.

Ben-Shahar, Y, A., Robichon, Sokolowski, M. B., and Robinson, G. E. 2002. Influence of gene action across different time scales of behavior. Science, 296: 741-744.

Bertrand, S., Brunet, G.F., Escriva, H., Parmentier, G., Laudet, V., Robinson-Rechavi, M. 2004. Evolutionary genomics of nuclear receptors: from twenty-five ancestral genes to derived endocrine systems.Mol Biol Evol, 21:1923-1937.

Bonduriansky, R. and Day, T. 2009. Nongenetic inheritance and itsevolutionary implications. Annu. Rev. Ecol. Evol. Syst., 40: 103-125

Bonduriansky, R. and Head, M. 2007. Maternal and paternalcondition effects on offspring phenotype in *Telostylinus angusticollis* (Diptera: Neriidae). J . Evol. Biol. ,20: 2379-2388.

Bonner, J.T. 1998. The Evolution of Complexity. Princeton: Princeton University Press.

Bowler, P.J. 1989. The Mendelian Revolution: the Emergence of Hereditarian Concepts in Modern Science and Society, The Athlone Press

Boyd, R. & Richerson, P. J. 1983.Why is culture adaptive? Q. Rev. Biol., 58: 209-214 .

Boyd, R., Richerson, P. J.1985. Culture and the evolutionary process. Chicago, IL: University of Chicago Press.

Butler, A.B., Saidel ,W.M.2000. Defining sameness: historical, biological, and generative homology.

Caballero, A., Fernandez, N., Lavado, R., Bravo, M.J., Miranda, J.M., Alonso, A. 2006.Tolerogenic response: allorecognition pathways. Transpl Immunol, 17(1):3-6.

Campbell A. 2007. Attachment, aggression and affiliation: the role of oxytocin in female social behavior. Biol. Psychol., 77: 1-10.

Caroll, S. 2000. Endless forms: the evolution of gene regulation and morphological diversity.Cell , 101:577-580.

Carroll ,S.B. 2005. Evolution at two levels: on genes and form. PLOS Biology, 3:e245.

Carroll, S.B., Grenier, J.K., Weatherbee, S.D.2001. From DNA to Diversity: Molecular Genetics and the Evolution of Animal Design. Malden, MA: Blackwell Science.

Champagne F. A.2008. Epigenetic mechanisms and the transgenerational effects of maternal care. Front. Neuroendocrinol., 29: 386-397.

Christen, B, Slack, J. 1998. All limbs are not the same.Nature, 395:230-231.

Churchill, F.B.1987. From heredity theory to vererbung: thetransmission problem, 1850-1915. ISIS 337-364

Cubas, P., Vincent, C. and Coen, E.1999. An epigenetic mutation responsible for natural variation in floral symmetry. Nature, 401: 157-161.

Danchin, E ´. et al. (2004) Public information: from nosy neighbors tocultural evolution. Science, 305: 487-491

Danchin, E., Charmantier, A., Champagne, F. A., Mesoudi, A., Pujol, B., Blanchet, S.2011. Beyond DNA: integrating inclusive inheritance into an extended theory of evolution. Nat. Genet., 12: 475-486.

Dawkins, R. 1989. The Selfish Gene: New Edition (Oxford: Oxford University Press).

De Robertis, E.M., Sasai ,Y.1996. A common plan for dorso- and the origin of chordates ventral patterning in bilateria. Nature, 380:37-40.

Dennett, D. C.1995. Darwin's dangerous idea. London, UK: Allen Lane.

Dickins, T. E., Dickins, B. J. A.2007 Designed calibration: naturally selected flexibility, not non-genetic evolution. Behav. Brain Sci. 30, 368-369.

Dickins, T. E., Dickins, B. J. A.2008 Mother nature's tolerant ways: why non-genetic inheritance has nothing to do with evolution. New Ideas Psychol, 26: 41-54.

Dickins, T.E. and Rahman, Q. 2012. The extended evolutionary synthesis and the role of soft inheritance in evolution. Proc.R.Soc. B: Biological Sciences , rspb20120273

Duboule, D., Wilkins ,A.S. 1998. The evolution of 'bricolage'.Trends Genet , 14:54-59.

Ellinger, J., Kahl, P., von der, Gathen J., Rogenhofer, S., Heukamp, L.,C., Gutgemann, I., Walter, B., Hofstadter, F., Buttner, R., Muller, S,C., Bastian, P. J., von Ruecker, A.2010 Global levels of histone modifications predict prostate cancer recurrence. Prostate, 70(1):61-69.

Fernald, R.D. 2000. Evolution of eyes. Curr Opin Neurobiol , 10:444-450.

Ferrara, J,L.., Levine, J.E., Reddy ,P., Holler, E. 2009.Graft-versus-host disease. Lancet, 373(9674):1550-1561.

Finnerty, J.R., Martindale, M.Q. 1999. Ancient origins of axial patterning genes: Hox genes and para Hox genes in the Cnidaria.Evol Dev, 1:16-23.

Finnerty, J.R., Pang, K., Burton, P., Paulsen, B., Martindale, M.Q.2004 Origins of bilateral symmetry: Hox and Dpp expression in a sea anemone. Science, 304:1335-1337.

Fontdevila, A. 2005. Hybrid genome evolution by transposition. Cytogenet genome Res, 110: 49-55.

Fraga, M. F., Fraga, M. F.2005 .Epigenetic differences arise during the lifetime of monozygotic twins. Proc. Natl Acad. Sci. USA, 102: 10604-10609.

Futuyma, D. J.1986. Evolutionary biology. Sunderland, MA: Sinauer.

Gibbs, J. R., et al.2010 Abundant quantitative trait loci exist for DNA methylation and gene expression in human brain. PLoS Genet. 6: e1000952.

Gilbert, S.F., Opitz, J., Raff, R.A. 1996. Resynthesizing evolutionary and developmental biology.Dev Biol 1996, 173:357-372.

Goldschmidt, R. B. 1940. The material Basis of evolution. Yale University Press, New Haven, 438pp.

Gompel, N., Prud'homme, B., Wittkopp, P.J., Kassner, V.A., Carroll, S.B. 2005. Chance caught on the wing: cis-regulatory evolution and the origin of pigment patterns in Drosophila. Nature , 433:481-487.

Gould, S. J.2002. The structure of evolutionary theory. Cambridge, MA: Belknap Press of Harvard University Press.

Greene, J. D. 2003. From neural 'is' to moral 'ought': what are the scientific implications of neuroscientific moral psychology? Nature Reviews Neuroscience, 4: 847-850.

Greene, J. D., and 4 others. 2001. An fMRI investigation of emotional engagement in moral judgment. Science, 293: 2105-2108.

Greenspan, R. 1995. Understanding the genetic construction of behavior. Scientific American, 272(April): 72-78.

Gu,Z. Rifkin, S.A. , White, K.P. and LiW-H. 2004. Duplicate genes increase gene expression diversity within and between species. Nat Genet, 36: 577-579.

Hagmann, C. A. and Schildberg, F.A. Epigentics and Transplantation: Clinical Application of Chromatin Regulation. Discovery Medicine, 10(55): 511-520

Haider, G., Callaerts, P., Gehring, W.J. 1995. Induction of ectopic eyes by targeted expression of the eyeless gene in Drosophila. Science , 267:1788-1792.

Haig, D. 2007. Weismann rules! OK? Epigenetics and the Lamarckian temptation. Biol. Phil., 22: 415-428.

Harousseau, J.L. 2009. Hematopoietic stem cell transplantation in multiple myeloma. J Natl Compr Canc Netw ,7(9):961-970.

Haszprunar, G.1992. The types of homology and their significance for evolutionary biology and phylogenetics. J Evol Biol , 5:13-25.

Hedges, S.B., Blair, J.E., Venturi, M.L., Shoe, J.L. 2004. A molecular timescale of eukaryote evolution and the rise of complex multicellular life.BMC Evol Biol , 4:2.doi: 10.1186/1471-2148-4-2

Henikoff, S. and Smith, M. 2007. Histone variants and epigenetics. In: Allias, D.C. , Jenuwein, T, Reinberg, D. and Caparros, M-L( Eds)Epigenetics. Cold Spring Harbor Laboratory Press. Cold Spring Harbor, pp. 249-264.

Hillis, D.M. 1994. Homology in molecular biology. In The Hierarchical Basis of Comparative Biology. Edited by Hall BK. San Diego, New York, Boston, London, Sydney, Tokyo, Toronto: Academic Press:339-366.

Hinman, V.F., Nguyen, A.T., Cameron, R.A., Davidson, E.H.2003. Developmental gene regulatory network architecture across 500 million years of echinoderm evolution. Proc Natl Acad Sci USA , 100:13356-13361.

Hinman, V.F., O'Brien, E.K., Richards, G.S., Degnan, B.M.2003. Expression of anterior Hox genes during larval development of the gastropod Haliotis asinina. Evol Dev , 5:508-521.

Holland ,P.2004. The ups and downs of a sea anemone.Science , 304:1255-1256.

Hughes, A.L., Friedmann ,R. 2004. Shedding genomic ballast; extensive parallel loss of ancestral gene families in animals.J Mol Evol , 59:827-833.

Huxley, J.1942. Evolution: the modern synthesis. London, UK: Allen and Unwin.

International Chicken Genome Sequencing Consortium: Sequence and comparative analysis of the chicken genome provide unique perspectives on vertebrate evolution. 2004, Nature 432:695-716.

Jablonka, E. and Lamb, M.J. 1995. Epigenetic Inheritance and Evolution, Oxford University Press

Jablonka, E., Lamb, M. J.2005 .Evolution in four dimensions: genetic, epigenetic, behavioral, and symbolic variation in the history of life. Cambridge, MA: MIT Press.

Jablonka, E., Lamb, M. J.2008. Soft inheritance: challenging the modern synthesis. Genet. Mol. Biol. 31, 389-395.

Jablonka, E., Lamb, M. J.2010 Transgenerational epigenetic inheritance. In Evolution: the extended synthesis (eds Pigliucci, M., Müller, G. B., pp. 137-174. Cambridge, MA: MIT Press.

Johannes F., et al.2009 .Assessing the impact of transgenerational epigenetic variation on complex traits. PLoS Genet. 5: e1000530.

Johannsen, W. 1911 .The genotype conception of heredity. Am. Nat.45: 129-159

Kawasaki, K., Weiss, K.M. 2003. Mineralized tissue and vertebrate evolution: The secretory calcium-binding phosphoprotein gene cluster.Proc Natl Acad Sci USA , 100:4060-4065.

Kelly, W.K., O'Connor, O.A., Krug, L.M., Chiao, J.H., Heaney, M., Curley, T., Macgregore-Cortelli, B., Tong, W., Secrist, J.P., Schwartz, L., Richardson, S., Chu, E., Olgac, S., Marks, P.A., Scher, H., Richon, V.M. 2005.Phase I study of an oral histone deacetylase inhibitor, suberoylanilide hydroxamic acid, in patients with advanced cancer. J Clin Oncol, 23(17):3923-3931

Kidwell, M.G. and Lisch, D. 1997. Transposable elements as source of variation in animals and plants. Proc. Natl. Acad. Sci.USA, 94: 7704-7711

Kortschak, R.D., Samuel, G., Saint, R., Miller, D.J.2003. EST analysis of the cnidarian, *Acropora millepora*, reveals extensive gene loss and rapid sequence divergence in the model invertebrates.Curr Biol , 13:2190-2195.

Krishnan, N., Buchanan, P.M., Dzebisashvili, N., Xiao, H., Schnitzler, M.A., Brennan, D,C. 2008.Monozygotic transplantation: concerns and opportunities. Am J Transplant, 8(11):2343-2351

Kuhn, T. S.1962 The structure of scientific revolutions. Chicago, IL: University of Chicago Press.

Kusserow, A., Pang, K., Sturm, C., Hrouda, M., Lentfer, J., Schmidt, H.A., Technau, U., von Haeseler, A., Hobmeyer, B., Maretindale, M.Q., Holstein, T.W. 2005. Unexpected complexity of the Wnt gene family in a sea anemone.Nature, 433:156-160.

Lander, E.S., Linton, L.M., Birren, B., Nusbaum, C., Zody, M.C., Baldwin, J., Devon, K., Dewar, K., Doyle, M., Fitzhugh, W., Funke ,R., Gage, D., Harris, K., Heaford, A., Howland, J., Kann, L.,

Lehoczky, J., Levine, R., McEwan, P., McKernan, K., et al.2001. Initial sequencing and analysis of the human genome. Nature, 409(6822):860-921.

Lee, P.N., Callaerts, P., de Couet, H.G., Martindale, M.Q. 2003. Cephalopod Hox genes and the origin of morphological novelties. Nature , 424:1061-1065

Levine, M., Tjian, R.2003. Transcription regulation and animal diversity.Nature , 424

Litzow, M,R., Tarima, S., Perez, W.S., Bolwell, B.J., Cairo, M.S., Camitta, B.M., Cutler, C.S., Lima, M.D., Dipersio, J.F., Gale ,R.P., Keating, A., Lazarus, H.M., Luger, S., Marks, D.I., Maziarz, R,T, McCarthy, P.L., Pasquini ,M.C., Phillips, G.L., Rizzo, J.D., Sierra, J., et al. , 2009.Allogeneic transplantation for therapy-related myelodysplastic syndrome and acute myeloid leukemia. Blood, 115(9):1850-1857

Lo´pez-Beltra,´ N, C. 2007. The medical origins of heredity. In Heredity Produced: at the Crossroads of Biology, Politics, and Culture, 1500-1870 (Mu¨ller-Wille, S. and Rheinberger, H-J., eds), pp. 105-132, The MIT Press

Locascio, A., Manzanares, M., Blanco, M.J., Nieto, A. 2002. Modularity and reshuffling of Snail and Slug expression during vertebrate evolution.Proc Natl Acad Sci USA , 99:16841-16846.

Lowe, C.J., Wu, M., Salic, A., Evans, L., Lander, E., Stange-Thomann, M., Gruber, C.E., Gerhart, J., Kirschner, M. 200 3.Anteroposterior patterning in hemichordates and the origins of the chordate nervous system. Cell , 113:853-865.

Lynch, M., Force, A. 2000. The probability of duplicate gene preservation by subfunctionalization.Genetics , 154:459-473.

MacKay, A. B., Danzmann, R., MacKay, A. B., Mhanni, A. A., McGowan, R. A., Krone, P. H.2007. Immunological detection of changes in genomic DNA methylation during early zebrafish development. Genome, 50: 778-785.

Maher, B.2008 Personal genomes: the case of missing heritability. Nature, 456: 18-21.

Málaga-Trillo, E., Meyer, A. 2003: Genome duplications and accelerated evolution of Hox genes and cluster architecture in teleost fishes.Am Zool ,41:676-686.

Malgnac, F. and Silar, P. 2003. Non-Mendelian determinants of morphology of fungi.Curr. Opin. Microbiol, 6: 641-645

Mameli, M. 2005. The inheritance of features. Biol. Philos. 20: 365-399

Marcellini, S., Technau, U., Smith, J.C., Lemaire, P. 2003. Evolution of Brachyury proteins: identification of a novel regulatory domain conserved within Bilateria.Dev Biol , 260:352-361.

Marrack, P, Kappler, J. 1997. Positive selection of thymocytes bearing alpha beta T cell receptors. Curr Opin Immunol 9(2):250-255.

Martindale, M.Q., Finnerty, J.R., Henry ,J.Q. 2002 The Radiata and the evolutionary origins of the bilaterian body plan. Mol Phyl Evol , 24:358-65.

Mayr, E.1961. Cause and effect in biology. Science, 134: 1501-1506.

Mayr, E.1963. Animal species and evolution. Cambridge, MA: Harvard University Press.

Mayr, E. 1982. The Growth of Biological Thought: Diversity, Evolution, and Inheritance, Belknap, (Harvard)

Mayr, E. 1998. Prologue: some thoughts on the history of theevolutionary synthesis. In The Evolutionary Synthesis: Perspectiveson the Unification of biology (Mayr, E. and Provine, W.B., eds), pp. 1-48, Harvard University Press

McDaniell ,R., et al.2010 Heritable and individual-specific and allele-specific chromatin signatures in humans. Science, 328: 235-239.

Mehta, T.K., Hoque, M.O., Ugarte, R., Rahman, M.H., Kraus, E., Montgomery, R., Melancon, K., Sidransky, D., Rabb, H. 2006.Quantitative detection of promoter hypermethylation as a biomarker of acute kidney injury during transplantation. Transplant Proc, 38(10):3420-3426.

Messenger, N.J., Kabitschke, C., Andrews, R., Grimmer, D., Nunez Miguel, R., Blundell, T.L., Smith, J.C., Wardle, F.C.2005. Functional specificity of the *Xenopus* T-domain protein Brachyury is conferred by its ability to interact with Smadl.Dev Cell , 8:599-610.

Meyer, A. 1999. Homology and homoplasy: the retention of genetic programs. In Homology. Symposium on Homology held at the Novartis Foundation (Symposium 222); London. Edited by Brock GR, Cardew G. Wiley: Chichester, UK :141-157.

Miller, C. A. and Sweat, J.D. 2007. Covalent modification of DNA regulates memory formation. Neuro, 53: 857-869.

Mindell, D.P., Meyer, A. 2001. Homology evolving.Trends Ecol Evol , 16:434-440.

Minelli, A. 1998. Molecules, developmental modules, and phenotypes: a combinatorial approach to homology. Mol Phyl Evol , 9:340-347

Mochizuki, K. and Gorovsky, M.A. 2004. Small RNAs in genome rearrangement in *Tetrahymena*. Curr. Opin. Genet. Dev, 14: 181-187.

Oakley, T.H., Cunningham, C.W. 2002. Molecular phylogenetic evidence for the independent evolutionary origin of an arthropod compound eye. Proc Natl Acad Sci USA , 99:1426-1430.

Odling-Smee, F. J., Laland, K. N., Feldman, M. W.2003 Niche construction: the neglected process in evolution. Princeton, NJ: Princeton University Press.

Okano, M., Bell, D.W., Haber, D.A., Li, E. 1999.DNA methyltransferases Dnmt3a and DNMT3b are essential for de novo methylation and mammalian development. Cell 99(3):247-257.

Pal, C and Miklos, I.1999. Epigenetic Inheritance, Genetic Assimilation and Speciation J. theor. Biol. 200: 19-37

Panganiban, G., Irvine, S.M., Lowe, C., Roehl, H., Corley, L.S., Sherbon, B., Grenier, J.K., Fallon, J.F., Kimble, J., Walker M, Wray, G., Swalla, B.J., Martindale, M.Q., Carroll, S.B. 1997. The origin and evolution of animal appendages. Proc Natl Acad Sci USA , 94:5162-5166.

Peterson ,K.J., Davidson, E.H. 2000. Regulatory evolution and the origin of the bilaterians.Proc Natl Acad Sci USA, 97:4430-4433

Philippe, H., Lartillot, N., Brinkmann, H. 2005. Multigene analysis of bilaterian animals corroborate the monophyly of Ecdysozoa, Lophotrochozoa, and Protostomia.Mol Biol Evol .

Pigliucci, M.2007 Do we need an extended evolutionary synthesis? Evolution, 61: 2743-2749.

Pigliucci, M., Müller, G. B.2010 Evolution: the extended synthesis. Cambridge, MA: MIT Press.

Pigliucci, M., Murren C. J., Schlichting C. D.2006. Phenotypic plasticity and evolution by genetic assimilation. J. Exp. Biol., 209, 2362-2367.

Prud'homme, B., de Rosa, R., Arendt, D., Julien, J.F., Pajaziti, R., Dorresteijn, A.W.C., Adoutte, A., Wittbrodt, J., Balavoine, G. 2003. Arthropod-like expression patterns of engrailed and wingless in the annelid Platynereis dumerilii suggest a role in segment formation.Curr Biol, 13:1876-1881.

Raes, J., Van de Peer, Y. Functional divergence of proteins through frameshift mutations.

Raff, R.A. 2000. Evo-devo: the evolution of a new discipline.Nat Rev Genet 2000, 1:74-79.

Raible F, Arendt D: Metazoan evolution: some animals are more equal than others.Curr Biol 2004, 14:R106-108.

Rapp, R.A. and Wendel, J.F. 2005. Epigenetics and plant evolution. New Phytol., 168, 81-91

Richerdson, P.J. and Boyd, R. 2005. Not by Genes Alone: How Culture Transforms Human Evolution. The University of Chicago Press

Rocha, S.2007. Gene regulation under low oxygen: holding your breath for transcription. Trends Biochem Sci, 32(8):389-397.

Rodin, S. N. , Parkhomchuk, D.V. and Riggs, A.D. 2005. Epigenetic changes and repositioning determine the evolutionary fate of duplicated genes. Biochemistry, 70: 559-567.

Ruse M.2003. Darwin and design: does evolution have a purpose? Cambridge, MA: Harvard University Press.

Sapp, J. 1987.Beyond the Gene: Cytoplasmic Inheritance and the Struggle for Authority in Genetics, Oxford University Press

Scott, M.P., Weiner, A.J. 1984. Structural relationships among genes that control development: sequence homology between the Antennipedia, Ultrabithorax and fushu tarazu loci of *Drosophila*. Proc Natl Acad Sci USA, 81:4115-4119.

Scott-Phillips, T. C., Dickins, T. E., West, S. A.2011. Evolutionary theory and the ultimate-proximate distinction in the human behavioral science. Perspect. Psychol. Sci. 6: 38-47.

Seager, W. 1999. Theories of consciousness: an introduction and assessment. Routledge, New York, x + 306 pp.

Seaver, E.C. 2003. Segmentation: mono- or polyphyletic?Int J Dev Biol, 47:583-595.

Shi, S.-H., and 4 others. 2004. Control of dendrite arborization by an Ig family member,dendrite arborization and synapse maturation 1 (Dasml). Proceedings of the NationalAcademy of Sciences of the United States of America, 101: 13341-13345.

Shimeld, S.M., Holland ,P.W.H. 2000. Vertebrate innovations.Proc Natl Acad Sci USA , 97:4449-4452.

Sollars, V., Sollars, V.,Lu, X., Xiao, L.,Wang X., Garfinkel, M. D., Ruden, D. M.2003. Evidence for an epigenetic mechanism by which Hsp90 acts as a capacitor for morphological evolution. Nat. Genet. 33, 70-74.

Speybroeck, L.V. 2002. Erom epigenesist to epigenetics: The case of C.H. Waddington. Ann.N.Y.Acad. sci.981: 61-81

Steele, E.J. 1979. Somatic Selection and Adaptive Evolution: on the Inheritance of Acquired Characters, Williams and Wallace

Stollewerk ,A., Schoppmeier, M., Damen, W.G.M. 2003. Involvement of Notch and Delta genes in spider segmentation.Nature , 423:863-865.

Sturm, R.A.and Larsson, M. 2008. Genetics of human iris colourand patterns. Pigment cell Melanoma Res.22: 544-562

Svensson, M.E. 2004. Homology and homocracy revisited: gene expression patterns and hypotheses of homology. Dev Genes Evol , 214:418-421.

Tal, O., Kisdi, E. & Jablonka, E. 2010.Epigenetic contribution to covariance between relatives. Genetics 184: 1037-1050

Tautz D. 1998. Debatable homologies. Nature , 395:17-19

Telford, M.J., Budd, G.E. 2003. The place of phylogeny and cladistics in Evo-Devo research.Int J Dev Biol , 47:479-490.

True, J.R., Carroll, S.B. 2002. Gene co-option in physiological and morphological evolution.Annu Rev Cell Dev Biol , 18:53-80.

Turkheimer, E., Waldon, M.2000 Nonshared environment: a theoretical, methodological, and quantitative review. Psychol. Bull. 126: 78-108.

Ueshima, R., and Asami, T.2003. Single-gene speciation by left-right reversal. Nature,425: 679.

Van Valen ,L.M. 1982. Homology and causes. J Morphol 1982, 173:305-312.

Vastenhouw, N.L., Brunschwig, K., okihara, K. L. , Muller, F., Tijsterman, M. and Plasterk, R.H.A. 2006. Long-term gene silencing by RNAi. Nature, 442: 882

Waddington, C. H. 1942. Canalization of developmentand the inheritance of acquired characters. Nature 251,

Waddington, C.H. 1947 (1940). Organizers and Genes. (Cambridge: Cambridge University Press), 143.

Waddington, C. H.1953. Genetic assimilation of an acquired character. Evolution 7, 118-126.

Waddington, C. H. 1957. The strategy of the genes: a discussion of aspects of theoretical biology. George Allen & Unwin, London, x + 262 pp.

Waddington, C.H. 1956 (1939). An Introduction to Modern Genetics (London: George Allen & Unwin Ltd.),

Waddington, C.H. 1962. New Patterns in Genetics and Development (New York and London: Columbia University Press), 84.

Wake, D.B. 2003. Homology and homoplasy. In Keywords and Concepts in Evolutionary Developmental Biology. Edited by Edited by Hall BK, Olson WM. Harvard: Harvard University Press:190-201.

Wang, W., Grimmer, J.F., Van de Water, T.R., Lufkin, T. 2004. Hmx2 and Hmx3 homeobox genes direct development of the murine inner ear and hypothalamus and can be functionally replaced by Drosophila Hmx.Dev Cell , 7:439-453.

Warrens, A.N., Lombardi, G., Lechler, R.I.1994. Presentation and recognition of major and minor histocompatibility antigens. Transpl Immunol 2(2):103-107.

Webb, R. H.2011. If evolution is the answer, what is the question? J. Evol. Psychol. 9: 91-107.

Weber, M, Davies, J.J., Wittig, D., Oakeley, E.J., Haase, M., Lam, W.L., Schubeler, D. 2005.Chromosome-wide and promoter-specific analyses identify sites of differential DNA methylation in normal and transformed human cells. Nat Genet 37(8):853-862

Wells, J. C. K.2003. The thrifty phenotype hypothesis: thrifty offspring or thrifty mother? J. Theor. Biol. 221: 143-161.

Wilkins, A.S. 2002 The Evolution of Developmental Pathways. Sunderland: Sinauer Associates.

Williams, A., Peh, C.A., Elliott, T. 2002. The cell biology of MHC class I antigen presentation. Tissue Antigens 59(1):3-17.

Williams, G. C.1966 Adaptation and natural selection: a critique of some current evolutionary thought. Princeton, NJ: Princeton University Press.

Wilson, C.B, Rowell, E, Sekimata ,M.2009. Epigenetic control of T-helper-cell differentiation. Nat Rev Immunol 9(2):91-105.

Wray, G.A., Abouheif, E. 2003. When is homology not homology? Curr Opin Genet Dev 1998, 8:675-680.

Nielsen C, Martinez P. Patterns of gene expression: homology or homocracy. Dev Genes Evol , 213:149-154.

Youngson, N. A., Whitelaw, E.2008. Transgenerational epigenetic effects. Annu. Rev. Genom. Human Genet. 9: 233-257.

Zhang T.-Y., Meaney M. J.2010 Epigenetics and the environmental regulation of the genome and its function. Annu. Rev. Psychol. 61: 439-466.

Zufall, R.A. , Robinson, T. and Katz, L., A.. 2005. Evolution of developmentally regulated genome rearrangements in eukaryotes. J. Exp. Zool. B. Mol.Dev. Evol., 304: 448-455.

# CHAPTER 7

# Mechanism of Evolution

Genetic change by mutation, sexual recombination, gene flow, and genetic duplication causes evolution. Genetic change drives evolution causing **biological diversity**, natural selection does not but it acts on all populations. Natural-, sexual-, and artificial selections constrain the variation occurring through genetic change. Genetic drift occurs when genetic code is neither selected for, or against. Natural selection selects the favorable traits and do not select the negative traits for well-adaptedness of organisms to environment.

**Genetic drift** does not affect survival or sexual reproduction and their frequency in a population randomly change over time. Genetic drift has profound effect on small population. In sexual reproduction, (Shcherbakov, 2010) zygote is formed via fusion of maternal and paternal cells. A generation of sexual population becomes a self-reproducing unit called species. Granularity of selection increases. Random acts of nature affect evolution. No one doubted Weismann's idea that sexual reproduction creates genetic variability, material for natural selection and enhances species' evolutionary potential. Sexual

reproduction and genetic recombination are source of combinative variation in populations, but they do not produce new alleles, produce only new combinations of extant ones. If sexual reproduction accelerates evolution, than continual shuffling of genomes looks strange. A species with rapid evolution would not exist long. All organisms populating the earth today belong to species resistant enough to further evolution. Evolution is inevitable as systems created by evolution protects against evolution. Exchange of genetic information enables asexual populations to gain genetic diversity which result in outcomes like antibiotic resistance.

In a sexually reproducing population, genetic material exchange and recombination occur in haploid organisms like bacteria. Not all structural mutations change the function of gene and phenotype, though a considerable part of them change gene function. Clones of an asexual population could evolve independently. Sexual population evolves as a whole. Evolution of a population as a whole cannot be rapid, even from gene-centered view. Perhaps sexual reproduction succeeds as sexual populations are more resistant to evolution than asexual ones. An individual organism is a more organized system than a biological species. Biological species function to survive. Thousands of genes are involved in selection. Changing thousands of genes affect function of thousands of other traits. Changes in gene

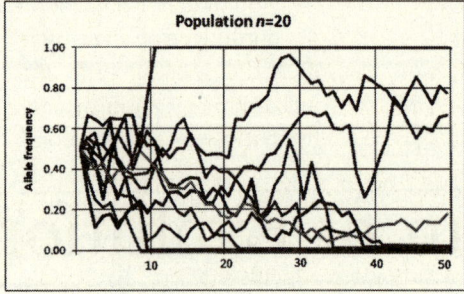

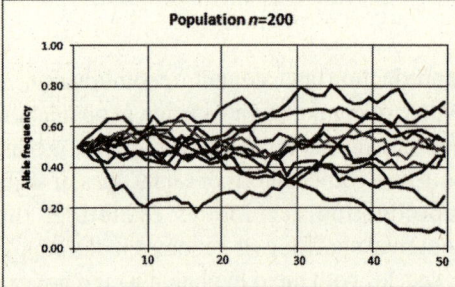

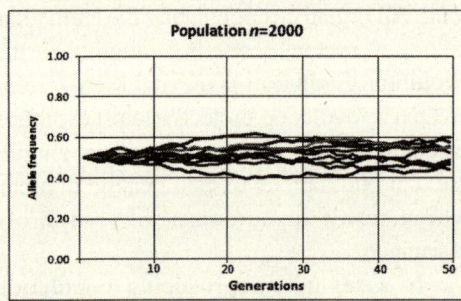

**Figure 7.1**  Ten Simulations of random genetic drift of a single given allele with an initial frequency distribution 0.5 measured over the course of 50 generations, repeated in three reproductively synchronous populations of different sizes. In these simulations, alleles drift to loss or fixation (frequency of 0.0 or 1.0) only in the smallest population.

frequencies (Figures 7.1, 7.2) in domestic bred are easy to observe. A breeder selects for a trait for changing gene frequencies of population. Monitoring gene frequencies are accurately performed by taking genetic samples from random sets of individuals and recording rate of occurrence of alleles in individuals in set. A change of gene frequency in population is detected when rate of occurrence of alleles is significantly different from one period of time to next. Change in gene frequency in wild populations is observed and verified in studies.

Microevolution, also called sequential evolution, involves continuous gradual change in interbreeding population, producing new subspecies and geographical races through changes in gene frequencies in population from one generation to next. Proponents of punctuated equilibrium hold view that speciation is analogous to mutation and replacement of one species by other is analogous to natural selection which is called

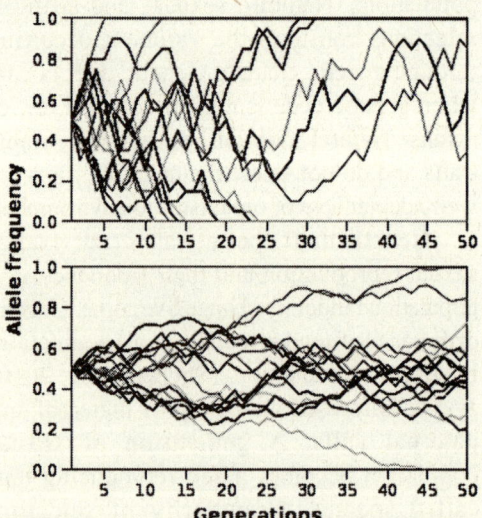

**Figure 7.2**  Simulations of genetic drift of 20 unlinked allele in populations of 10 (top) and 100 (bottom). Drift to fixation is more rapid in smallest population.

species selection. **Speciation** adds new species to species pool as mutation adds new alleles to gene pool. Species selection favors one species over another as natural selection favor one allele over other. Evolutionary trends within group may be result of selection among species. Natural selection at level of homeotic regulatory genes produce large changes in phenotypes in relatively short periods compared with kinds of changes. Light changes in homeotic genes can produce large changes in phenotypes in relatively short periods. Slight variations in homeotic genes become common in populations due to natural selection, producing relatively large changes in phenotypes.

Genome of eukaryotes is loaded with dead genes called pseudogenes which are copies of working genes that are inactivated by mutation. Most pseudogenes can not produce full protein. Hardy and others believed that evolution will not occur in a population if seven conditions like absence of mutation and natural selection are met .Hardy-Weinberg's Law states that relative frequencies of various genes in a large and randomly mating population tend to remain constant from generation to generation in absence of mutation and natural selection. This law describes a tendency of evolution to conserve gains of genetic changes and avoid frequent changes in genotype.

Allele frequencies remain unaltered unless mutation and natural selection change them. Previously it was thought that dominant alleles swamp recessive alleles out of existence over time. This theory called **genophagy** states that dominant alleles increase in frequency from generation to generation. Hardy and Weinberg showed that dominant alleles easily decrease in frequency and both allele and genotype frequencies in a population remain constant unless disturbed by non-random mating, selection, mutations, random genetic drift, gene flow and meiotic drive. Sewall Wright gave

the principles of gene frequency fluctuations in small populations. In small, non-randomly mating populations gene frequencies fluctuate purely by chance. In smaller population, fluctuation in gene frequency will be larger.

Law of large numbers predicts that in large population little change occurs over time due to genetic drift. When reproductive population is small, effects of sampling error alter allele frequencies significantly. Genetic drift is a consequential mechanism of evolutionary change primarily in small isolated populations. Although both processes affect evolution, genetic drift operates randomly while natural selection functions non-randomly. New organisms are not identical to their parent due to evolution. Changes in DNA express new characters. Natural selection does not drives evolution causing biological diversity, but it acts on all populations. Organisms primarily the prokaryotes reproduce by cloning. Perfect cloning stops evolution. Gene flow is the acquisition of new genes through insertion of foreign DNA, DNA swapping, and viral DNA exchanges.

Natural-, sexual-, and artificial selections constrain the variation occurring through genetic change. Genetic drift occurs when genetic code is neither selected for or against.

Sexual selection, a form of natural selection, occurs relative to mates. Traits that increase attractness to mates may arise via sexual selection like bright color body of fish, reptiles, and birds. Traits arise through sexual selection are not selected for survival benefit. They may have zero or negative fitness qualities and may not provide any survival benefit rather make survival difficult. Plumage of male peacock does not help in survival. Such traits develop when organisms overcome the negative attributes of traits through much overall fitness. This explains the commonness of sexually selected traits, as they signal to mates about fitness (Figure 7.3, 7.5).

Sexually selected traits identify viable potential mates for producing viable offspring. Sexually selective behavior keeps similar organisms together and creates mating boundaries. Even though specific trait is not being selected for it may be expressed similarly to a trait that is selected for, at least for some time. In small population few traits are critical; other traits are not selected for or against.

**Figure 7.5**   Illustration by Charles Darwin showing *Lophornis ornatus*; female on left, ornamented male on right.

## 7.1   Sexual Reproduction and Evolution

The Universe consists of elementary particles, planets, stars, galaxies, atoms, and molecules. Few configurations of matter are stable, intermediate ones are volatile. The Universe is structuralized and far from thermodynamic equilibrium. It contains information. The Universe exists based on mutual affinity of its constituents, their interaction with each other, and resisting the general aspiration for evenness. Natural selection acting by accumulation of small heritable changes thought to produce an even continuum of living beings was never corroborated. In biological world, the organisms form populations, which in turn create species. Species in combination with abiotic components form ecosystems. The intermediate configurations are volatile.

Stability of biotic entities depends on their physical durability and expedient behavior. They are organizations with survival function. The Universe evolves through interaction and

**Figure 7.3**   *Paradesia decora*; male above, female below.

**Figure 7.4**   Long-tailed widowbird (male)

cooperation of entities, which constitute the complexity and hierarchical structure of the Universe. Transitions in biological evolution (macromolecular replicator → prokaryotic cell → eukaryotic cell → multicellular organism → biological species) are the steps of cooperation (Maynard Smith and Szathmary, 1995). **Hierarchy** in biology doesn't mean complexity, or heterogeneity rather implies a functional predestination of their parts for sake of whole. Survival of parts depends on survival of whole. The operating principles of organization of higher rank do not necessarily related on properties of parts. The principles organizing an upper rank are novelties which are not necessarily predictable from rank below. The organizing restrictions of living entities cannot be deduced from any general principle, **but** can be understood in context of their history. Evolution of a higher entity cannot be presented as evolution of its constituents.

Several theoretical models explain widespread occurrence of sexual reproduction (Barton and Charlesworth, 1998, Otto et al., 2002). Weisman's idea that sex and recombination provide variation for natural selection is dominating. Natural selection selects the survived ones. Evolution and survival are not the same, rather they are opposite. To evolve means to change, but to survive means to persist. Persistence is the resistance to further change. We used to think that organisms die and species lost due to poor adaptation to environment, or due to competition with other organisms for substances and energy. Organisms die due to imperfection of their homeostatic mechanisms, for their limited ability to resist the increase of entropy, which is a driving force of evolution (Brooks and Wiley, 1986). Not only do those who die not survive but also those who evolve. The entities that change rapidly disappear rapidly and they are not observed among fossils and existing organisms.

## Sex as a solution to ecological problem

H.J. Muller pointed out that sexual recombination reduces fixation time of beneficial mutations that arise at different loci. Maynard Smith mentioned 2-fold advantage of asexual reproduction, as it would double the production rate of asexual females over sexual females. Such advantage would lead to swamping of populations with more prevalent asexual females and progressive elimination of sex. By 1970s, it was known that asexual reproduction was widely distributed among protists, animals, and plants. Sex had become crowned "the Queen" of problems by 1980s. Origin and maintenance of sex in context of ecological selection pressures were emphasized. Changing environments, competition between sibs, evolutionary arms races, parasite load, and genetic load were evolutionary and ecological mechanisms. Sex seemed to be ubiquitous, but without a suitable explanation based on ecological selection mechanisms. Since eukaryotic sex is complex and cumbersome, selection for its evolution must have been intense during its origin.

## Origin and maintenance of sex like genomic parasitism

Transposable elements could copy and spread in genomes due to selection on such DNAs to spread within and among genomes, not due to selection between organisms. Spread of such parasitic elements depends on sexual recombination and horizontal gene transfer. So long as cells and organism's DNA did not recombine (http://www.ncbi.nlm.nih.gov/pmc/articles/PMC2222615/) with each other, transposable elements would be selected to control their spread in genomes which led to proposal that sex originated as a device for **parasitic DNAs** to spread. Such parasitic elements could spread *de novo* in bacteria, making sense that ancestral unicellular eukaryotes have evolved sex by analogous means. Maintenance of sex was attacked from the point as sex is not beneficial to be maintained. When potentially asexual females suffer from continued fertilization by males, anisogamous sex could be maintained even if it was not evolutionarily beneficial. History of sex has been explained in terms of ability of new parthenogens to avoid sex with males.

### Sex as subject to natural selection at multiple levels

Accumulated experimental data suggests the occasional role of ecological factors in evolution of sex. Evolution of genetic systems now involves invocation of multiple levels of selection, between and within genomes, and molecular processes from transposition to unequal chromosome segregation during gametogenesis. Evolutionary ecology models for evolution of sex have been put in the proper perspective, as some mechanisms that contribute to scientific complexity of sex.

## 7.2 Survival by Means of Reproduction

Human body is the survival machine which houses DNA. Through sexual act, DNA replications machinery is transmitted to next generation .All things we do in our lives as steps toward this end. About each act of our daily lives can be viewed as either protection, or preservation of our DNA, or its transmission. Everything we learn serve in enhancement of our capacity to survive. Pleasure like sleep is necessary for us, when we need to recoup our energies. Combination of genetic predisposition and environmental influences varies. Organisms are complex highly organized systems far from maximum entropy. An organized system is a complex system, which

performs certain functions by virtue of its constituent parts (Denbigh, 1975). **Organized systems** are distinct from ordered ones. Both systems are not random. Ordered systems are generated as per a simple algorithm and lack complexity. Organized systems are assembled element-by-element following an external program or plan. Their structures are periodic, not random. Organization is complex, not random due to design or selection, but for necessity of crystallographic orders (Wicken, 1979).

Homeostatic mechanisms are not perfect. Individual can not avoid death but bypass thermodynamic limit by reproduction which is multiplication, renewal, replacement of the old bodies by new ones. Failure to retain their physical structures, organisms left information for their continuous regeneration. Maintaining the information rather than bodies is the distinct property of life. Reproduction based on genetic information is the contents of life. The copying genetic information should be precise. In practice, the copying is not errorless. Extra copying (multiplication), coupled with selection causes the possibility of evolution. In single-celled asexuals, reproduction occurs by doubling DNA and regeneration of other cellular components. Daughter cells inherit entire genome and entire equipment for reproduction of mature cell. It is not the separate genes nor does the whole genome represent the living entity. Biological form of existence starts with a cell. Only a cell, not any of its constituents, is "organization with function of survival" (Merser, 1981).

Majority of multicellular organisms pass through single-celled stage in their life cycle. In asexual multicellularity, little change occurs in information transmission between generations, except information itself contains a sophisticated ontogenetic plan. In sexual reproduction, the situation is itself sexually.

Zygote is formed via fusion of 2 different cells, maternal and paternal. A generation of sexual population becomes a self-reproducing unit called species. Daughter generation inherits the gene pool distributed among multiplicity of zygotes with their own developmental programs. Natural selection selects mainly and primarily the old-established things (Kimura, 1991). Selection works to check perfection of homeostatic mechanisms in conservative way. Without this conservatism, only chaos would result. Novelties get selected, if they improve, or do not worsen homeostasis. Adaptation to environment is important for homeostasis and chief vector of selection is the internal perfection and harmonization of system for its exact reproduction, which when achieved, entails stasis, i.e. a stop of evolution.

"Species are groups of interbreeding natural populations that are reproductively isolated from other such groups" (Mayr, 1996). Asexual organisms are the simple, while **obligatory sex** is a property of higher animals. Simple organisms exist continually due to high fidelity of individual genome replication. In organisms with large genome and complex development, fidelity of DNA replication is less, for exact reproduction of genome. Such species survive and remain unchanged in spite of continual changes of their genes. They would rapidly degrade and extinct. Result of transition to sexual reproduction is the creation of biological species.

Individual organisms forfeited their ability to autonomous reproduction to be forfeited their ability to evolution. They evolve as an entity of biological species. Renunciation of one's freedom and independence for reliability and well-being is common step in progressive evolution. Robust genomes created via sexual reproduction are not identical to each other. Organisms of same species are not identical genetically. Failure to retain precisely their genotypes, higher organisms retains information for continual regeneration of their phenotypes. To persist in time, single-cell organisms show homeogenomic reproduction where genomes of descendants and ancestors are identical.

Rate of spontaneous mutations in growing *Escherichia coli* is about 0.003 alterations in each genome per replication (Drake, 1991). Other microorganisms, including the eukaryotic possess similar, or higher fidelity of genome replication. Asexual lineages, being unicellular, or with small number of germ-line cell divisions, survive due to higher fidelity of DNA replication, which is enough for reliable self-reproduction. Replication with high fidelity is a costly process. The attained fidelity of DNA replication per base pair per cell division is almost close to maximum (Kondrashov, 1995). The per-generation rate of mutation in organisms with large genome and large number of germ-line cell divisions is very high, up to 3 orders of magnitude higher than in yeast (Lynch, 2009). Thus, genomes of higher organisms are not reproduced with high fidelity. In man, number of mutations per zygote is 60 or above (Lynch et al., 2003). Mutational deluge menaces to destroy both homeostatic mechanisms and evolutionary stability of species. Such species survive and remain unchanged in spite of continual changes in their genes. Andre Lwoff uttered," Problems do not exist in nature. Nature only knows solutions" (Shcherbakov, 2010).

### Apparent oddity of sex

Many biologists rarely ask about particular evolutionary process that produces such complex, troublesome, and risky way of multiplication. Asexual reproduction is simple, more effective in transmitting genes from ancestor to progeny. In asexual reproduction, each individual produces progeny; no problems arise for a mating partner. Each individual transmits all of its genes to offspring. Optimal genomes do not disperse in the next generation, but are transmitted to all offspring of a given individual.

High cost of sexual reproduction (Hastings, 1992) was realized first by Weismann (1889). Advantages of **asexuality** are contradictory to biological importance of sex. Some forms of genetic information shuffling occur in organisms. About all eukaryotic organisms are sexual (Schurko et al. 2008). In mammals, both male and female gametes are necessary for initiation of development (Marketrt, 1988). Reproductive behavior plays a dominant role in animal's life. Freud reduces the entire human psychology in expression of sexual instinct. Meeting of male and female chromosomes shuffle their information and create gametes with unique gene combinations. Such information exchange must outweigh all shortcomings and complications of sexual reproduction. Potentially lethal DNA damage (double-strand break) is used to start recombination in meiosis (Paques et al., 1999). A possibility of acceleration of evolution at amphimixis was substantiated by Fisher (1931) and H.J. Muller (1932). The idea of evolvability is still popular. It is often assumed that capability for diverse evolution is positive trait supported by natural selection, while shortage of evolutionary potential is fraught with extinction.

Evolvability as a selectable trait is in contradiction to known efforts of evolution aimed at creating genetic stability of organisms and lineages (Shcerbakov, 2010). Evolvability cannot be easily granted as a species homeostatic process. Direct selection for evolvability is theoretically impossible (Sniegowski and Murphy, 2006). So transition to sexuality needs explanation independent of evolvability. Alternative to evolution, or extinction leads to disappearance of species. If the first living cell was absolutely perfect and had infinite life without change, there

would be those cells only. There is little doubt that enhanced evolutionary potential could emerge as a by-product of development. Complexity of biological systems increases during evolution increases a range of evolutionary novelties and broadens opportunity for creation of more perfect organization that may resistant to enhanced entropic challenges. **Genetic polymorphism** increases morphogenetic homeostasis of a species (Severtsov, 1990). Genetic diversity of populations is valuable by providing an ecological plasticity and efficiency of interspecies interactions and cooperation (Ghiselin, 1974). It may play a crucial role in creating stable dynamics of population under varying environmental conditions (Robson et al., 1998).

During the last 3 decades, various hypotheses on origin and maintenance of sex and genetic recombination were recorded (Morran et al., 2009). Some works suggest that sexual reproduction accelerates evolution, whereas others remain conservative in stabilizing the function. Sexual reproduction is a powerful acquisition of evolution (Grantham, 2003), and its effects and consequences are found in different aspects of life. Selective advantages and shortcomings of sexual reproduction in comparison to asexual reproduction were observed. Results of few experimental works were diverse (Michod et al., 2008). In large-scale experimental evolution work on *Caenorhabditis elegans* out crossing favored populations under conditions of increased mutation rate and during adaptation to a novel environment.

Sexual and asexual individuals have different biological status. An asexual individual is a self-sufficient player on stage of life, whereas obligatory sexual individual is a biological species. There are 2 essential, though interdependent differences. First, a sexual individual is not a self-reproductive entity. Second, its behavior is aimed at the survival of species, not for its own survival and multiplication only. As a result, sexual and asexual populations must use different evolutionary strategies. Competition between individuals must have different biological consequences. Replacement of one lineage by other and working out of species organization, development of species robustness can successfully resist extinction and reach stasis through high fidelity of genome maintenance and by creation of diversity and genotypic plasticity. Adaptability and variability are instruments of evolution and its ontological matter is the creation of stable forms which are resistant to extinction and evolution.

## 7.3 Cohesion

Sexual populations remain in coherent systems. **Cohesion** is a very complex event. In asexual reproduction, a parent transmits all its genes to an offspring which are genetically identical to parent. All progeny of single individual produce clone of genetically identical individuals. Lineage of clone seems like a tree with single ancestor. If mutation occurs in individual, it transmits to its entire offspring, but not to other clones as genes are transmitted vertically. Population of different clones is genetically heterogeneous. Competition is reduced among the clones and those with selective advantages eliminate the rest. Reproductive isolation of asexual lineages is not complete and horizontal gene transfer plays role in evolution of prokaryotes and single-celled eukaryotes (Peeling, 2009).

Exchange of genetic information enables asexual populations to gain genetic diversity which result in outcomes like antibiotic resistance. In a sexually reproducing population, the genetic material exchange and recombination occur in haploid organisms like bacteria. Genuine sexual reproduction is the prerogative of eukaryotes in combination with alternation of generations. In most complete form as in vertebrates, haplophase reduces to gametes, oocytes (eggs) and spermatozoids. Life starts by fusion of male and female gametes and formation of zygote, in which 2 sets of genes of different origin combine. In meiosis, paternal and maternal chromosomes are randomly distributed in 2 haploid cells. They randomly exchange their genetic information. Thus, each generative cell gets unique genetic information of dual origin. Lineages of various individuals are interwoven into a multimeric network to form a shared population gene pool (Shcherbakov, 2010). Crucial feature of sexual reproduction is formation of genomes of individual by random picking them from continuously shuffled gene pool by intrinsic selection for perfection of organization, which starts before organism (Shcherbakov 2010) is tested by environment.

## 7.4 Levels of Selection

Operation of natural selection on genes, or individual organisms, or groups, or species is a problem in evolutionary thought, which occupies a great part of publications in last 50 years (Okasha, 2005). This incompatibility reflects difference between reductionistic and holistic philosophy. Multi-level selection is getting popular in comparison to gene-centered view. A species cannot stably exist unless it is provided with hierarchically structured system with survival suggesting constraints on behavior of system's constituents. Selection at lower level disrupts integration at higher level. (Szathmary and Maynard Smith, 1995). Hierarchy in biology implies a functional predestination of parts for sake of whole. There is no such thing as synchronic (Okasha, 2005) multi-level selection. Natural selection cannot select anything that is not a unit of reproduction.

In sexual organisms, minimal unit of

reproduction is population. "Selection at organismal level is Darwinism and is well understood only in asexual lineages with individual as a unit of reproduction. Sexual organisms have no power of self-reproduction. They are renewable temporary constituents of species. They provide genes into common gene pool with 100% **altruism**. One common feature to many transitions is that entities that were capable of independent replication before the transition can replicate only as part of a larger whole after it (Maynard Smith and Szathmary, 1995). Assuming the higher level selection does operate, lower levels selection if allowed can produce a parasitic DNA, or a malignant cell. **Ultra selfish genes** are factual parasites with net harmful effect on host which along with other parasites and harmful mutations represent the destructive force of nature. Evolution is unending struggle against the force.

### Intrinsic selection

Haploid phase in a life cycle of species is important for maintaining species robustness. In higher animals, this phase is generative cells. During haploid phase, new gene combinations are checked for vitality of basic cell functions. These new gene combinations including random exchanges of DNA sequences between homologous chromosomes and random segregation of homologs into daughter cells may lead to some interesting statistical result. If each homologous chromosome possesses 1 new mutation, crossing-over may create 2 daughter chromosomes, one with 2 mutations, and other with no mutation at all. Instead of 2 chromosomes with one error each, one chromosome without errors and one chromosome with 2 errors could be created. For any new inherited mutations, their recombination causes an increase of probability for zero class and class with a mutation number more than mean. Random segregation in

homologous chromosomes causes the same result at level of whole genome. Such events occur before formation of haploid gametes, where mutations are not protected by normal alleles. Gametes with multiple defects will most probably die, sacrificed for good ones. Cells with bad mitochondrion exhibit apoptosis during oogenesis (Krakauer and Mira, 1999). The purifying action of recombination may not be completely random. Biased gene conversion against common type of genetic damage can reduce mutational load (Bengtsson, 1985). One random recombination makes not random results of other stochastic phenomenon - distribution of mutations among chromosomes. Sexual reproduction enables a population to get rid of harmful mutations (Kondrashov, 1984).

Rate of recombination is different in different species. The less is the site separation and the more the recombining chromosomes are dissimilar, the higher is the specific rate of recombination. Due to high negative interference, probability for restoration of wild-type allele remains high even with multiple differences between recombining chromosomes. Multiplicity reactivation in bacteriophage T4 (Luria, 1947) is an example of such recombination effect, when viable phage particles are formed (Shcherbakov, 2010) after infection of bacterial cell with 2 or more, phage particles each bearing lethal damage. Effect may not be repaired by restoration of wild type sequence. Combination of different mutated sequences may produce functionally robust genomes. Recombination of related proteins preserves function with higher probability than random mutation (Drummand et al., 2005). It shows that intragenic recombination restores original DNA sequence and gene function on basis of a new DNA sequence. **Intrinsic selection** is the major one, which acts as a sole mode during fertilization and embryogenesis and is perhaps most important during maturation. It is difficult to interpret this selection as a competition between perfect and less perfect entities. Intrinsic selection is quantitatively most effectual and economical. Processes operating during gametogenesis may present a kind of intrinsic selection. Majority of generative cells and their predecessors show programmed cell death (Tilly, 2001). This mass suicide is partly aimed at selecting genetically robust germ cells.

## Multi-level Selection

Sober develops and explores the concept of multi-level selection of Lewontin. He offers with D.S. Wilson, an account to elucidate and defend a role for group selection. Sober considers himself to be a realist about level at which selection acts (Sober, 2011) in evolution of a trait. For Sober and Wilson, *"Objects at level X were units of selection in evolution of trait T if one of the factors that influenced T's evolution was that T conferred a benefit on objects at level X"* (Sober and Wilson, 1994). A honeybee's barbed stinger does not benefit the bee as using the stinger kills it. If it evolved to benefits the hive, then selection at level of hives was a factor in evolution of bee stingers. In cases where multiple levels benefit from an adaptation, what level selection operates on? Sober and Wilson answers this problem by concept of the the common fate. When parts of a whole share a common fate, they do not compete with each other, and selection works at higher level. If a whole benefits via a benefit to its part, but part is involved in competition with other parts, selection takes place at lower level. The idea that selection works where competition occurs can resolve the inadequately specified question of what entity is benefiting from a trait (Sober and Wilson, 1994). **Group selection** is the fitness differences between groups in a given population. Organism selection is the fitness

differences between organisms in a given group. Gene selection is the fitness differences between genes in a given organism (Sober, 2011) (Figures 7.6–7.10).

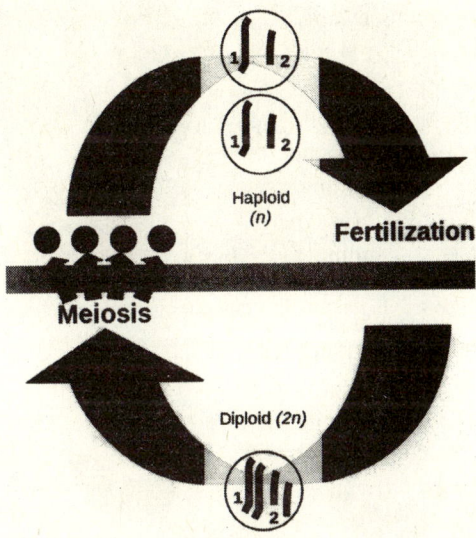

**Figure 7.6** In sexual reproduction, sex cells contain haplid number of chromosome, while the zygote is diploid

**Figure 7.7** Mating in Australian Emperor dragonfly

## 7.5 Muller's Ratchet

A gene consists of thousands, or millions of nucleotide pairs. Any change in nucleotide sequence like base insertion, substitution, or deletion of base pairs cause change in gene. Inversion is direct mutations of this gene at a

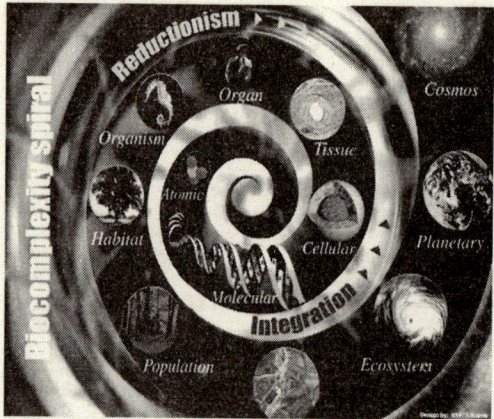

**Figure 7.8**  Figure illustrating biological complexity

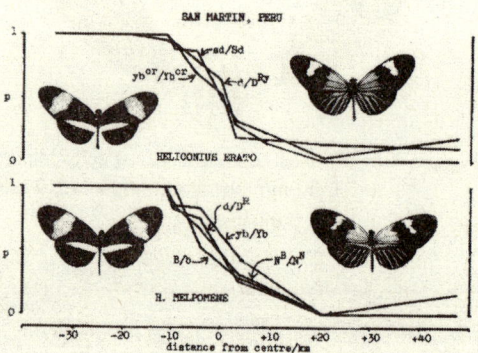

**Figure 7.9**    Frequency-dependent selection (intrinsic)

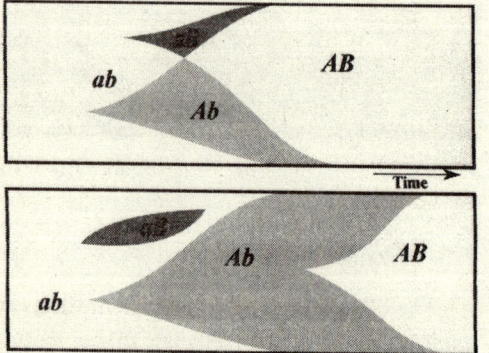

**Figure 7.10**    Rapid creation of novel genotypes when two adventageous alleles A and B occur  randomly.

structural level. Not all structural mutations change the function of gene and change the phenotype, though a considerable part of them can change gene function. Mutations are reversible at structural level. It can be assumed that direct and reversed mutations occur with equal probability on average. On functional level, a reversion of function is thousands times less probable than its damage. **Mutagenesis** on functional level is irreversible. Each mutant gene has a higher chance to get one more direct mutation than to reverse. Thus on functional level, mutagenesis is an entropic macro process with same formal description as molecular diffusion.

As mutations may be harmful, the outcome of this irreversibility must be fatal, which is saved by very high fidelity of gene reproduction to prevent mutations. A mutated gene will be eliminated, or inherited, but never returned to initial state as found in asexual population. Such shortcoming of asexual reproduction realized by G. Muller was presented as ratchet: *"an asexual population incorporates a kind of ratchet such as it can never contain, in any of its lines, a load of mutation small than already existing."* If genetic drift causes loss of the least-loaded line, ratchet makes a click and never rotates back. In a computer simulation study, reality of **Muller's ratchet** mechanism was confirmed (Felsenstein, 1974). A link between asexuality and mutation accumulation was obtained, which shows that mutational buildup could be rapid to contribute to short-term evolutionary mechanisms that favor sexual reproduction (Neiman et al., 2010). It is easy to imagine how an asexual population, being pushed by mutagenesis and attracted by adaptive advantages, in a blind alley without any return, if only to previous furcation. After each step, asexual population burns its bridges behind it. Sexual reproduction is not so careless. Since even mutations (Shcherbakov, 2010) in same gene, being functionally similar, are located in different sites, a scope to restore initial genotype by recombination always remains. For recombination, intra-species variations are reversible.

## 7.6    Tempo of Evolution

Clones of an asexual population can evolve independently. Sexual population evolves as a whole. Evolution of a population as a whole cannot be rapid, even from the gene-centered view. For becoming irreversible event, a new allele must be fixed if it is beneficial. Owner of a rare new allele will mate with partner lacking this allele, so that in the generation this useful allele will be in heterozygous state and not be expressed phenotypically avoiding positive selection. Multiplication of genotypes is a square function of their density in a population at sexual reproduction. Rare types cannot be multiplied even with very high fitness (Michod, 1999). Sexual reproduction and recombination cause genotypic diversity.

Positive meaning of diversity in accelerating evolution may not be applicable. Genotypic plasticity emerges during species ontogenesis to resist evolution (Ghiselin, 1974). Populations experiencing idiosyncratic risk have higher asymptotic growth rate than those experiencing aggregate uncertainty. Individuals of former type get a competitive advantage over individuals of the latter type. Genetic variation among offspring can reduce aggregate uncertainty, transforming it into more idiosyncratic (Robson , Bergstrom Pritchard, 1999) form of risk, which underlies the dynamics in several role models of out crossing in evolution of sex. At the idiosyncratic risk, the varying environment (Shcherbavov, 2010) favors some individuals, and combined variability restores the initial diversity of population. The species in all its complexity remains persist and remains unchanged despite individual's variability. Perhaps sexual reproduction succeeds as sexual populations are more resistant to evolution than asexual ones. Genotypes of outstanding individuals are not transmitted to next generation after mating with ordinary individuals, scatter and rearrange in new gene combinations (Shcherbavov, 2010), thus preventing the winner from exploiting a success. Recombination destroys favorable genotypes often than it creates them (Eshel, 1991).

Natural selection does not select outstanding individuals. Palaeontological evidence for evolutionary stasis is convincing (Erwin, 2000). Functional transfer of **mitochondrial genes** to nucleus is common in clonal plants than in out crossing plants that illustrates conservative function of sexual reproduction (Brandvain, 2007). This conservatism seems to contradict the data on evolutionary dynamics of biodiversity during Phanerozoic, which shows self-acceleration (Markov and Korotaev, 2007). Stability of established species implies its inability to speciate. The opposite is true: *"Sexual reproduction predominates among organisms mainly because most evolutionary change is concentrated in speciation events, and asexual species cannot speciate in the (Shcherbavov, 2010) normal sense"* (Stanley, 1975). The Robust, long lived parents produce many progeny.

## 7.7    Selfish Gene

Genes are a passive memory of cell. In prior-to-cell world, macromolecular replicators were evolutionary individuals. A gene is considered as an ideal replicator, but it is not. It is replicated. Genes are reproduced by cell, like RNAs, polypeptides, and organelles. A biological species is an organization of a higher rank as compared to an organism. An individual organism is a more organized system than a biological species. Biological species function to survive. Co-coordinated activity of constituent entities results in survival which is an emergent quality of entity that does not belong to any of its components. Genes have no freedom for security and reliability. The present-day genes care about welfare of cell, organism, and species. Frequency of harmful mutations exceeds that of beneficial ones by 4 to 5 (Taddei, 1997), or even 6 (Roth et al., 2003) orders of magnitude. Genes undergo mutation pressure and accumulate neutral changes and those that make them suitable for a genome, organism, and species. *"One gene - one trait - one selection vector"* is not found. One gene may affect several traits. Most traits depend on many genes. Phenotypic expression of an allele depends (Shcherbavov, 2010) on genetic context. Reversible replacement of white and black forms of peppered moths is rare event. Thousands of genes are involved in selection.

A good gene and good individual are not selfish, but those blending suitably with species gene pool to sustain their vitality. Genotypic plasticity is the important element of species robustness. For operation of selection, general population must contain more or less reproductively isolated groups (Lion and van vaalen, 2008). Such partial isolation must arise due to geographical and behavioral factors. Sexual and **filial imprinting** is common mechanism for limitation of panmixia, continuously creating local subpopulations (Shcherbavov, 2010). Various mechanisms preventing inbreeding result in genetic diversity of sub-population.

Phenomena like altruistic behavior (Figure 7.11) of individuals are difficult to explain from individual-centered position, but appear natural from the species-centered view. The known neodarwinistic explanations (Nowak, 2006) may be construed as group-selection. An individual organism works for survival of species by sacrificing its own

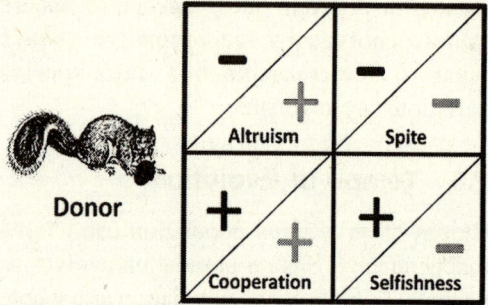

**Figure 7.11** Four social interactions including altruism

survival. Altruistic/egoistic phenotype is determined by many genes, and population is characterized by pure altruists to pure egoists. Owing to gene pool shuffling, it is totally transmitted to next generation, even if extreme altruists do not produce (Shcherbavov, 2010) their own offspring while extreme egoists exhibit less concern for offspring. The form of this distribution is optimized for species survival in evolution. One more phenomenon is phenoptosis, the programmed death of organisms (Longo et al., 2005) which occurs in expressive form in salmon: death of adult individuals after spawning. The phenomenon of aging is a slurred form of phenoptosis. Different longevity of the individual (Shcherbavov, 2010) life also exhibits the same phenomenon. (http://www.ncbi.nlm.nih.gov/pmc/articles/PMC3894905/).

General tendency in progressive evolution is to reduce fecundity. Relative fecundity reflects evolutionary success of an allele, or an individual organism, assuming that more the better. Extra copies are made to compensate for poor fidelity of reproduction and to compensate for random death of organisms. Competition between asexual lineages and between different species looks natural but competition between organisms of same species is akin to competition between, e.g., hepatocytes and neurons. Modern Synthesis assumes that heredity is hard as it is mediated by gene. **Mendelian heredity** was established through culmination of scientific debate (Mayr, E., 1982). Possibility of soft or Lamarckian inheritance, whereby acquired traits could be passed on to offspring was refuted. Structure of DNA discovered in 1953 was its death knell. Mendelian heredity possesses all the hallmarks of scientific revolution. Evidence points to inheritance mechanisms that operate beside Mendelian inheritance for inheritance of acquired traits. Nature of emerging pluralistic model of heredity (Mameli, 2005) recognizes diversity of inheritance mechanisms.

### Replicators and Interactors

Dawkin (1976) presented a controversial reconceptualization that ENS consists of differential replication. Genes are the paradigm replicators and for Dawkins evolution is always genetic .A meme is a cultural replicator which is copied, spreads, competes with rivals and is selected to continue in replicator pool. Mutation is a change in replicator structure, which is passed on by copying mechanism. Hull (1980, 1981) introduced the distinction between replicator and interactor which speaks about distinct role of phenotype and gene. Replicators are necessary for evolution (Godfrey-Smith, 2000). A general replicator based account of evolution is strongly reminiscent of Lewontin's (1970) view (Hull et. al., 2001). Selection was defined as *"repeated cycles of replication, variation, and environmental interaction (http:// www.cig.salk.edu/pics/gas-4.htm) so structured that environmental interaction causes replication to be differential"* and variation was conceived not in terms of differences existing within a population at a given time, but as new types that are continually introduced into population from one cycle of replication to next. Despite these differences, the 3 part structure of Lewontin conditions is included

in Hull's conception: there must be variation in population at any time; that variation must causally influence reproduction with some things replicate or reproduce more than others, and replication implies that copies have to be somewhat like the original.

Notion of replication is indispensable (Dawkins, 1976; Blackmore, 2000) which was supplemented with repeated cycles of variation (Hull et. al., 2001). Much of the debate around level of selection was that phenotype is visible to selection, and contrasted this with central role of genes in underwriting hereditary mechanism (Dawkins, 1982). Fisher does not incorporate it into his models. In cases where real biological entities reproduce asexually, abstracting away from the phenotype/genotype distinction is a natural move. So distinction is not even essential in biology. If we accept the evolution by natural selection at least in principle, it applies to non-biological domain. Since Hardy-Weinberg assumes diploid sex, it doesn't apply to bacteria, let alone **non-biological evolution**. The best corrective to domain-specific biases of classical evolutionary theory is to examine the evolution in non-biological domains also.

## 7.8 Evolutionary Forces

Evolutionary forces cause some traits to be widespread, or decline in population. A selection of trait causes its spread. If trait declines, then there is selection against it. This has not mentioned reproduction, replicatiors, genotypes, phenotypes, interactions, genes, novelty, variation, selection, or heredity. Lewontin and Fraccia (1999) draw a distinction between transformational evolution and variational evolution. Variational evolution is similar to ENS. Transformational evolution has rooted in individual's internal development. Individual changes over time following some internal organizing principle, or in response to the environment. Lewontin and Fraccia (1999) suggest that in variational evolution, individual elements may change during their lifetime in directions unrelated to dynamic of collection as a whole and on a much shorter time scale than evolutionary history of group. Reproduction, or replication has a central role in ENS involving competition. Darwinian fitness is the success in survival and reproduction. Sexual selection is a kind of natural selection. Interpretation regarding sexual selection is not contradicted to late 20th century approaches that treat it as separate from natural selection (Lande, 1980). Distinction between natural and sexual selection isolates selective factors that influence males differently than females. Sexual selection after Darwin generally not attracted much attention as an alternative to natural selection as a source of evolutionary change, until the last quarter of the 20th century. Knowledge on patterns of heredity came from acceptance of Weismann's doctrine of segregation of germ plasm. Darwinian evolutionary change other than selection was banished from theory supplied by Fisher, who argued against a significant role for mutation as a driver of evolutionary change.

Sober (1984) treat evolution as a theory of forces explicitly recognizing that evolution can arise from sources other than natural selection, including pleiotropy, drift, mutation, and migration. He was concerned with natural selection, and to a lesser extent drift, each of which serves as an evolutionary force. The evolution of a population is fixed by the combined effects of the various forces in transforming gene frequencies. **Evolutionary forces** like selection, drift, migration, or mutation may act independently to bias the population in one direction or another, favouring one allele and disfavouring its rival.

They claim that drift, as a process, acts to cause population to diverge from expectation, but that drift can also be used to refer to actual divergence from expectation of evolutionary change. Similarly, it is suggested, selection can be conceived either as a force in manner of Sober (1984) or as a product observable in output of an evolving system. Evolutionary change in gene frequencies that results from variation in fitness is this product of selection (Walsh, 2007). We can conceive the overall relation as follows: a population will have a probabilistic causal propensity to undergo a change in allele frequencies due to variation in fitness, identified as selection-the-process. It will have a further causal propensity to diverge from this expected change, which is function of size of population and is identified as drift. The actual change is selection, and divergence of that change from our expectations is drift. The position will advocate here could be seen as an attempt to split the difference between advocates of statistical and dynamical conceptions of evolution. Arguments of Walsh, Matthen, Ariew, and Lewens are most forceful and successful in attacking the idea that selection and drift should be conceived as causal processes. The fact that 2 central terms in evolutionary theory don't refer to evolutionary forces and does not show that evolution can't be understood as a theory of forces.

Neo-dynamical view can provide a coherent, general account of evolution by natural selection. In reforming the dynamical view; one challenge is a persistent ambiguity in relevant discussion. The ambiguity concerns the distinction between a force and a cause. Much of the discussion concerns broader issues of causation in evolutionary theory. The more specific question is whether selection and drift are **Newtonian forces**, in something like sense that Sober and Stephens variously defend. This is a separate issue from whether they are causal processes, or have a causal influence on evolution. The component of dynamical account is identified by Sober and his followers as **Hardy-Weinberg equilibrium**. By developing a more precise and formal account of forces and zero-force state, it could be shown that Hardy-Weinberg equilibrium is not zero-force state in evolutionary biology. Its misidentification as zero-force state has contributed to misidentification of drift as an evolutionary force. These considerations allow us to develop an improved account of evolution in terms of interaction of forces.

## Background Conditions and Drift

Hardy-Weinberg equilibrium does not even potentially represent a zero-force state. When a system in inertial motion is disturbed, once the force is removed the system is once again in zero-force state, albeit on a new trajectory. If system has discrete generations, it is expected to move to H-WE in next time-step; if reproduction is modelled in continuous time, it will take a little while. Is the process uncaused, since no forces are supposedly active? Naturally not, the processes responsible for restoration of H-WE are processes of random mating, random meiosis, random gamete sampling, and equal-fitness reproduction. If these causal processes are structured according to H-WE prerequisites, then population would find H-WE equilibrium again. Under most circumstances, the processes will lead to evolutionary change in population. H-WE just specifies situation where these neglected causal processes don't have any biasing effect - that is, those uncommon scenarios where this overlooked force of reproduction has zero magnitude. H-WE is a candidate zero-force state where central operating force is treated as background condition. Sober explicitly discusses this back ground of Mendelian processes, saying that a substantive Mendelian mechanism is assumed to be at working even when all evolutionary forces are absent. Assuming that Mendelian processes are the background conditions theory implicitly disqualifies them from affecting evolution. But we know that Mendelian processes do influence evolution, often dramatically. Sober together with Lewontin (1980) discusses how Mendelism can be a cause that acts alongside selection to change distribution of genotypes in population. Mendelism will act to change the population whenever its genotype frequencies are not those specified by H-WE - even if no other recognized evolutionary force is operating. So even when the drift is negligible, and there is no selection, and mutation and migration are not acting, Mendelism can act to change gene frequencies. It must be a force, on Sober's own account.

## Force of Drift

Consequences of failing, to distinguish a circumstance where there is a force operating with zero expected effect from a case where there is no force operating at all, can be seen in Sober's account of drift. Difficulties with identifying drift as a force are legion, and have occasioned much discussion. Sober even concedes that drift is a force of a different color (1984). Stephens calls it a different force but doesn't see the harm in treating it as a force otherwise. If we take HWE to be our zero-force state and treat Mendelian processes as background conditions, we are driven to position that sampling error is a Newtonian force (Sober, 1984). So in any finite population where none of other prerequisites of HWE are violated, drift must produce a rate of change, like any other Newtonian force. For Sober, drift is the stochastic element decomposed from an evolutionary process. He treats selection, mutation, and migration as deterministic forces, with drift being a separate in deterministic force statistically decomposed out of others to allow their deterministic treatment (1984). This decomposition is allegedly "*to facilitate comparison among populations*" (ibid). The idea is that, since 2 populations (http://doktorgee.worldzonepro.com/BlogFiles/NewPerspectives2007_2_41.pdf)might be undergoing selection of same strength, but be of different size, and expect different amounts of drift. This marks an important biological distinction. But this rationale does not warrant in treating drift as a separate force. It at most warrants treating population size as a causally relevant factor. One motivation for treating selection as a deterministic force and decomposing all randomness into a force of drift might be that is accords with tendency to compare strength of drift with selection in various contexts. This usage can be easily accommodated without mistakenly treating selection and drift as forces. Suppose, we don't create an artificial force that is sort-of-but-actually-not Newtonian by decomposing all the randomness out of every genuine evolutionary force. We would be left with probabilistic forces: that is, actual rate of change they produce could deviate from the expected value. Expected effect of this force is zero in conditions of H-WE. But since it is a stochastic force, it can cause a non-zero rate of change even when its expected effect is zero. This deviation from zero-force state shows existence of a net force, an impulse arising from Mendelian processes that contrary to our expectations bring about change. Decomposing the randomness from system and treating it formally as a separate cause of change might not be useful under some circumstances.

Dynamical level is defined in terms of some more basic inertial property, such as gene frequency. There is no need to limit our conception of evolution to gene frequencies. A population is defined by the individuals that constitute it. The effect of a **biological force** on a population must consist in an effect on particular individuals in population. This is the case, even if the force is causally sensitive

only to properties of groups of individuals. The changes brought about by such a process only count as changes in context of an evolutionary model in-so-far as they are changes in individuals, precisely because the evolution of population is defined in terms of individuals that make it up. This point implies nothing about what sort of individuals must be accounted for in evolutionary explanations, in particular, the individuals that define the population need not be organisms. Evolutionary change is something that happens to populations. A population is defined as a collection of things, which we call individuals. The individuals are defined by traits they possess which again, might be genes, or phenotypic traits. This definition intends to apply to evolution in non-biological contexts. In a certain sense, individuals are secondary, since what really counts as evolutionary change is change in distribution of traits in population. So just as building an account of forces impinging on a bungee jumper required us to isolate major distinct causal processes that could produce accelerations on jumper. Building an evolutionary account of forces requires to identify the major distinct causal processes that produce changes in our population, changes that might affect measurable quantities of evolutionary interest. A population of individuals can change in only 3 ways: individuals can leave the population, individuals can enter it, or individuals existing in population can change their traits. Distinct causal processes must suffice to do at least one of these 3 things to count as an evolutionary force. In identifying these causes, it is key to distinguish between potential forces and mere causal variables. A causal variable may influence the magnitude of a force, but it is insufficient to cause a rate of change in a conserved measurable quantity of a population. In biological context, to distinguish between biological causes and biological forces is important. This bias is certainly a cause of evolution in house mice, but it is not a force because it does not specify a rate of change. Relevant force is the set of processes that suffice to determine a rate of change in quantities of interest. Suppose the quantities of interest to us are the proportions of t-alleles and wild alleles in adult population. So at minimum, a force that incorporates causal effect of allele must be one that suffices to change the adult population. The most minimal way to do this is the model reproduction. Reproduction changes population via births, and can by itself, completely cause a rate of change in proportions of alleles in adult population. One causal variable that will be inevitably involved in modelling of reproduction will be the .95 bias towards t gametes. Full set of factors that we include in modelling reproductive change in population will be included in force of reproduction, including perhaps such factors as the infertility of t-homozygous males, and tendency of t allele to lower the viability of zygotes (compare Lewontin 1968, 1970).

### Contrasting accounts

Populations can change through few ways. Individuals can be subtracted or added to population. Reproduction and immigration correspond to such change. Individuals can change traits which correspond to mutation. Trait in population can have a systematic advantage on its rivals, with regard to any such population change, which can lead to evolution by natural selection. Trait can have a competitive advantage in regard to its fecundity, or its heritability, which can result in its increase in prevalence via natural selection. Such concept of selection is defined as broad trait fitness, tendency of traits to change in relative frequency. In classical view, selection involves only population change of narrowly specified kinds: individuals that leave

population via death, or enter it via reproduction. This concept is the narrow trait fitness which corresponds with population change Darwin identified with natural selection. However, the classical definition would be amended to consider the role of variation in heritability. The variance in individual fitnesses influences evolutionary outcomes, so we might expect one type to out compete another, even when their mean individual fitness is the same (Sober, 2006). Thus average propensity to survive and reproduce is not same as probability distribution of rates of survival and reproduction in population. Byerly and Michod (1991) make a cogent case for central role of F-fitness in evolution theorizing F-fitness as the trait fitness. Byerly and Michod do not adequately distinguish between narrow and broad varieties there of. **F-fitness**, the key concept in natural selection supports the idea that individual fitness is not the relevant fitness concept for classical evolutionary theorists.

## 7.9 Micro- Macro- & Mega- Evolution

Three stages in evolution based on degree of change and speed of evolution are:

1. Small evolutionary differences at sub specific level.
2. Large group of animals modify to produce species and genera by adaptive radiation.
3. Evolution of new types evolves due to large genetic changes.

**Microevolution**, also called sequential evolution, involves a continuous gradual change in interbreeding population, producing new subspecies and geographical races through changes in gene frequencies in population from one generation to next. This is produced by stabilizing or normalizing natural selections that operate in stable environment and in short time. Several lines of descent in sea urchin, *Micraster*, where gradual change in shape of test, structure of oral opening and form of ambulacra from *M. corbovis* to *M. coranguinum*, is noted.

## Link between macroevolution and Punctuated Equilibrium

Macroevolution is called adaptive radiation including evolutionary changes above species level that result in new adaptive types through genetic divergence. Changes are due to macro mutations that enter a new environment free of competition and result in establishment of new genera, families, and orders. Darwin called it orthogenesis. Evolution of horse is a perfect macroevolution, in which an increase in body size and legs and enlargement of teeth occurred. All body changes were related to life in open grasslands accompanied with fast running and feeding on harsh grasses causing new adaptive types. Adaptive radiation in Darwin's finches, divergence of reptiles, evolution of camel and elephant are also noteworthy. Modern evolutionary synthesis states the idea that evolutionary change at all levels can be explained by micro mutation, genetic drift, natural -, sexual selection, and reproductive isolation .In 1972, N. Eldredge and S.J. Gould (1972) proposed the theory of punctuated equilibrium (Figure 7.12) emphasizing macroevolution as different from microevolution. Features of punctuated equilibrium are as follow:

- Most speciation is the result of cladogenesis, rather than anagenesis.
- Pattern of descent with modification in an evolving line of organisms is a phylogeny.
- Cladogenesis is the splitting of one inter-related population of organisms into 2, or more reproductive isolated populations.

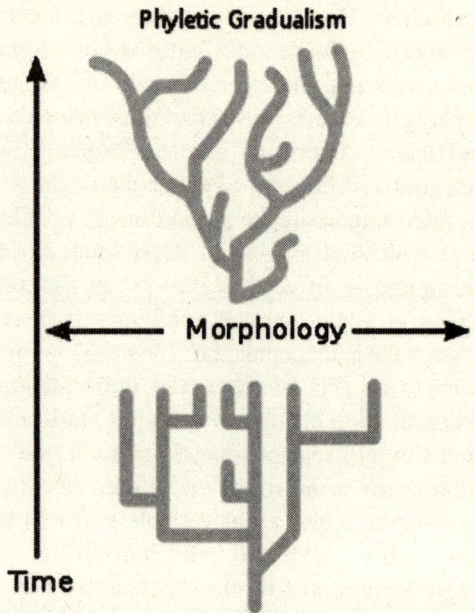

**Figure 7.12**   Punctuated Equlibrium

## Anagenesis

Anagenesis is the gradual change over time of a single phylogenetic line. Species remain unchanged for long time and then are quickly replaced by new species. If wide ranges are searched, transitional forms between 2 species are sometimes found in small, localized areas. In Jurassic brachiopods, *Kutchithyris acutiplicata* appears below *K. euryptycha*. Both were common and wide spread, but differ to extent that some argued to place them in different genus. In just one small locality, about 1.25 m sedimentary layer with these fossils is found. Both species are found along with transitional forms. in the narrow (10 cm) layer that separates 2 species, In other localities there is sharp transition. Gould proposed that major morphological change occurs quickly in small peripheral population during speciation. New forms enter the range of their ancestral species and at most locations, fossils are transitiory from one species to another. This abrupt change reflects migration, not evolution. To find transitional fossils, area of speciation must be found. Some accounts states that abrupt changes in fossil record are due to fast evolution. **Punctuated equilibrium** is presented as hierarchical theory of evolution. Proponents of punctuated equilibrium hold the view that speciation is analogous to mutation and replacement of one species by other is analogous to natural selection, which is called species selection. Speciation adds new species to species pool as mutation adds new alleles to gene pool. Species selection favors one species over another as natural selection favor one allele over other. Evolutionary trends within group may be result of selection among species.

Number of species produced over time is less than amount of different alleles that enter gene pools over time. Rates of evolution vary over time. Phylogenetic studies conflict the idea of clear link between speciation and morphological change. There are major polymorphisms within some species. Bluegill sunfish have 2 male morphs. One is a large, long-lived, mate-protecting male. The other is a small, short-lived male who sneaks mating

from females guarded by large males. Existence of within species polymorphisms shows that speciation is not a requirement for major morphological change. Trends in adaptation occur through mechanism of species selection. During rapid macroevolution, selection occurs at level of reproductively isolated groups, rather than at individual level. Punctuated equilibrium explains the fossil record patterns which include abrupt appearance of new species, relative stability of morphology in widespread species, and limited distribution of transitional fossils when those are found, differences in morphology between ancestral and daughter species, and species extinction pattern.

### Evo-Devo, the Engines of Speciation and Macroevolution

Gehring (1944) discovered the eyeless gene controlling formation of fruit fly eyes in 1994. Gehring put a mouse's eyeless gene into a *Drosophila*, resulting in normal *Drosophila* eyes. The same is applicable for homeobox genes for wings, heads, and legs showing that animals descended from common ancestor that passed along to them a set of homeobox genes (Figure 7.13), used to build several forms from

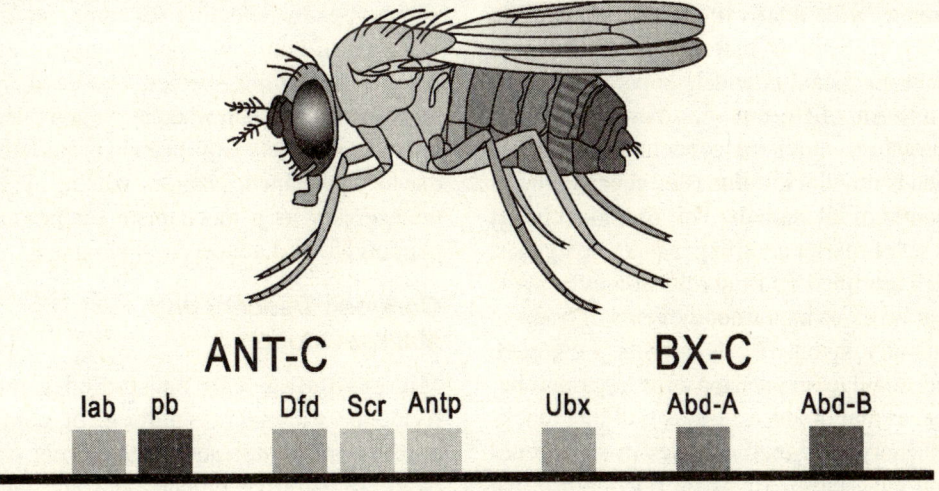

**Figure 7.13** Homeobox gene expression in *Drosphila melanogaster*

a few basic body plans. A simple, repeated pattern underlay the differing body structures of animals like segmented bodies of worms, in vertebrae of vertebrates, in head-thorax-abdomen construction of insects. In all these, modular units were arranged in a front-to-back line down the center of body. Development of body plans in animals is controlled by few virtually identical genes. In fruit flies, a malfunctioning body-plan gene produce a fly with legs sprouting from where antennae should be on head, or fly with extra set of wings . Genes providing instructions for body plan are called homeobox-containing genes. DNA code in homeobox genes directs cell to make chemical sequences, which regulate other genes that affect the positioning of cells in embryo. This simple mechanism could possibly cause enormous variation.

Evo-devo, a new concept for macroevolution explain the observed patterns of macro evolution as rapid, coordinated changes in major components of phenotypes following ecological disruption. Relatively rapid adaptive radiation of apparently genetically homogeneous types follows mass extinctions and/or invasion of new adaptive zones. Relatively large changes in phenotype correlate with relatively minor changes in genotype. Similar patterns of phenotypic change are found in widely separated animal groups in different environments. All observations point to concept of evo-devo which is the tool kit of master genes found in genomes of all animals. This tool kit consists of a set of master control genes, or hox genes, which are lined up in chromosomes in same linear order as major components of body of bilaterally symmetrical animals. Few such genes regulate expression of a large number of genes that produce various building blocks of phenotypes. Small changes in such genes could have large effects on phenotypes that genes regulate. Slight change in one homeotic gene in fruit fly can replace antennae by an entire leg. Genetic differences between finches with large and small beaks on Galapagos Islands has shown that such differences are the result of small genetic changes in single gene controlling development of jaw in vertebrates. Slight alterations of this homeotic gene cause extreme variations in jaw shape in African cichlid fishes. Light changes in homeotic genes can produce large changes in phenotypes in relatively short periods. Slight variations in homeotic genes become common in populations due to natural selection, producing relatively large changes in phenotypes.

Natural selection at level of homeotic regulatory genes produce large changes in phenotypes in relatively short periods compared with the kinds of changes. Evo-devo provides a new explanation of problems as evolutionary convergence, parallel evolution, and mimicry. Homeotic genes are highly conserved and widespread showing that all bilaterally symmetric animals have access to a similar tool kit during development. Rather than a particular phenotypic adaptation to evolve from scratch in each phylogenetic line of animals, it may be possible to produce similar phenotypes by selecting for same, relatively small set of homeotic gene arrangements. At level of diverging species, evo-devo might provide a mechanism whereby natural selection produces relatively large phenotypic differences due to small genetic changes, making allopatry unnecessary as a mechanism for producing phenotypic and genotypic divergence.

### Common Descent and Macroevolution

Microevolution can be studied directly. Evidence for macroevolution, or common ancestry and modification with descent, comes from comparative biochemical and genetic studies, comparative developmental biology, biogeography, morphology, anatomy and fossil record. Closely related species have similar gene sequences. Pattern of differences in closely related genomes is worth examining. All living organisms use DNA as their genetic material. Some viruses use RNA. DNA is composed of nucleotides. Four types of nucleotides are adenine (A), guanine (G), cytosine (C) and thymine (T). Genes are sequences of nucleotides that code for proteins. Within a gene, each block of 3 nucleotides is called a codon. Each codon designates an amino acid. Three letter codes are the same for all organisms. There are 64 **codons**, but only 20 amino acids to code for. So, most amino acids are coded for by several (http://www.talkorigins.org/faqs/faq-intro-to-biology.html)codons. Often the first 2 nucleotides in codon designate the amino acid. The 3rd position can have any of 4 nucleotides and not effect how the code is translated. A gene in use is transcribed into RNA. RNA transcribed from a gene is called **mRNA**. mRNA is then translated via cellular machinery called ribosome into string of amino acids - a protein. Some proteins function as enzymes (http://www.talkorigins.org/faqs/faq-intro-to-biology.html). Others are structural or involved in regulating development.

Gene sequences in closely related species are very similar. Often, the same codon specifies an amino acid in 2 related species; though alternate codons could serve functionally. Some differences exist in gene sequences especially in 3rd codon positions, where changes in DNA sequence would not disrupt sequence of protein. There are other sites in genome where nucleotide differences do not affect protein sequences. Genome of eukaryotes is loaded with dead genes called pseudo genes which are copies of **working genes** that are inactivated by mutation. Most **pseudo genes** can not produce full protein. They may be transcribed, but not translated.

Sometimes they may be translated with formation of truncated protein. Pseudo genes evolve faster than their working counterparts. Mutations in them do not get involved into proteins and they have no effect on fitness of an organism. Introns are sequences of DNA that interrupt a gene, but code for nothing. Coding portions of a gene are termed exons. Introns are spliced out of mRNA before translation, sometimes involved in regulation of gene but do not contribute information needed to make protein. Introns evolve faster than coding portions of gene (Figure 7.14).

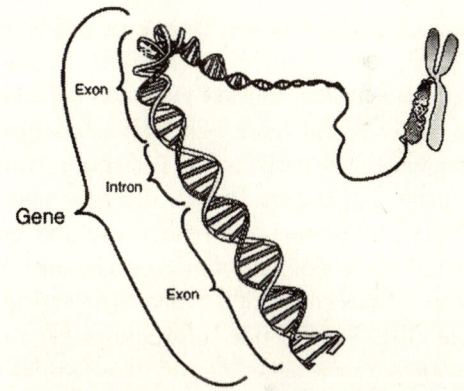

**Figure 7.14**   Intron and exon in a gene.

Changed nucleotide positions without changing sequence of a protein are called silent sites. Sites of changes resulting in an amino acid substitution are called replacement sites. Silent sites are expected to be polymorphic within population and show differences between populations. Both silent and replacement sites receive the same amount of mutations. Natural selection only infrequently allows changes at replacement sites. Silent sites are not as constrained, are more variable than coding sites and may not be entirely selectively neutral. Some DNA sequences are involved in gene regulation. Changes in these sites may be deleterious. Although several codons code for single amino acid, an organism preferred codon for each amino acid is called codon bias. If 2 species shared genetic information of the recent common ancestor, information like redundant nucleotides and the position of introns or pseudo genes are likely to be similar. Both species might inherit this information from the common ancestor.

Degree of similarity in nucleotide sequence is a function of divergence time. If 2 populations have recently separated, few differences would build up between them. If they separated long ago, each population (http://www.talkorigins.org/faqs/faq-intro-to-biology.html) may evolve many differences from common ancestor. Degree of similarity would be a function of silent versus replacement sites. Rates of evolution for silent vs. replacement rates were estimated from sequence comparisons of 30 genes from humans and rodents, which diverged about 80 mya. Silent sites evolve at an average rate of 4.61 nucleotide substitution per site per $10^9$ years. Replacement sites evolve at an average rate of 0.85 nucleotide substitutions per site per (http://www.talkorigins.org/faqs/faq-intro-to-biology.html) $10^9$ years. During evolution, organism's developmental pathway gets modified. An alteration near the end of developmental pathway is less likely to be deleterious than changes in early development.

Early changes may have cascading effect. Evolutionary changes in development are expected to take place at periphery of development, or in early development without later repercussions. For propagation of a change in early development, benefit of early alteration must outweigh the consequences to later development. Organisms go through the early stages of development that their ancestors did up to the point of divergence. The organism's development mimics its ancestors, although it fails to recreate it exactly. Development of flatfish, *Pleuronectes*, shows this. Early on,

*Pleuronectes* develops a tail that comes to a point. Top lobe of tail becomes larger than bottom lobe in the next developmental stage. When development is complete, upper and lower lobes are equal in size refelecting the evolutionary transitions.

Natural selection can modify any stage of life cycle and some differences are seen in early development. Evolution does not necessarily recapitulate ancestral forms. Butterflies have not evolved from ancestral caterpillars. Differences in appearance of early vertebrate embryos exist. Amphibians rapidly form a ball of cells in early development. Birds, reptiles, and mammals produce a disk. Shape of early embryo is the result of difference in yolk concentrations in the eggs. Birds' and reptiles' eggs are highly yolked. Their eggs develop similarly to amphibians except the yolk causing deformation of embryo's shape. The ball is stretched out and lying atop the yolk. Mammal's egg has no yolk, but they form a disk early. In all these vertebrates, the pattern of cell movements is similar despite superficial differences in appearance. All types quickly converge on a primitive, fish-like stage within few days. From there, development diverges.

Closely related organisms are usually found in close geographic proximity. Mammalian fauna of Australia is often cited as example. Marsupials fill most of equivalent niches that placentals fill in elsewhere. Species distribution would be a function of site of origin, if all organisms descended from a common ancestor. Physical separation of Australian mammals from sources of placentals means potential niches were filled by marsupial rather than the placental. Natural selection can mold available genetically based variation and provides no mechanism for advance planning. If selection can only tinker with available genetic variation, examples of jury-rigged design in living species could be seen. In lizards, *Cnemodophorus*, females reproduce parthenogenetically. Fertility in this lizard is increased when a female mounts another female and facilitates copulation. Such lizards evolved from sexual lizards whose hormones were aroused by sexual behavior. Now, the sexual reproduction has been lost, but means of getting aroused has been kept. Fossils having hard structures are less similar to modern organisms in progressively older rocks. Patterns of **biogeography** apply to fossils and extant organisms. In combination with plate tectonics, fossils give evidence of distributions and dispersals of ancient species. South America had a very distinct marsupial fauna until the land bridge formed between North and South America after which marsupial's begun to disappear and placentals took their place. This is commonly interpreted as placentals wiping out marsupials, but this may be over simplification.

### Linking the micro-and macroevolution

Life history theory does not adequately explain the diversity of life histories, partly for its inability to connect micro-and macroevolution. Huge difference in adult mortality rates in fruit fly experiment can be taken as an example. Large difference in selection pressure has not produced a large difference in life history traits. Because tradeoffs hold the life histories at equilibrium values not far from ancestral condition. Kinds of changes that microevolution can cause are not impressive when compared with what has actually evolved on the planet. Tradeoffs cause intrinsic stabilizing selection on characters. To bring about large change, the stabilizing selection converts into directional selection, at least for some characters. This requires that tradeoffs break down or transform other forms. Inability

of classical theory to explain complex life cycles becomes more apparent.

### Mega evolution

This includes formation of new groups, classes, or phyla from its predecessors by adaptation. In mega evolution, organisms of ancestral stalk enter a new environment where they face strong natural selection, and acquire pre-adaptations to survive in new zone. **Mega evolution** is caused by large genetic changes producing different types and disruptive or divergent natural selection causing population to occupy different environment. Amphibians were pre-adapted to live on land for short periods, since as fish they possessed lungs for air breathing and limbs to support body on land. Origin of birds from reptiles included development feathers and change in fore limb to wing to invade air and then developed beak, and sternal keel. During evolution of mammals, a false palate was formed, teeth became thecodont, and limbs moved under the body. Emergence of bats from primitive insectivores took place at the start of Coenozoic era. Skeletons of early Eocene bats show developed wings. Absence of **transitional forms** suggests that bats emerged by a mega evolutionary event.

### 7.10 Modern Theories of Evolution: Hardy-Weinberg Equilibrium

Evolution is a change in frequencies of alleles in gene pool of population. Suppose a trait is determined by inheritance of a gene with 2 alleles--C and c. If the parent generation has 91% C and 9% c and their offspring have 90% C and 10% c, evolution has occurred between generations. The entire population's gene pool evolved in direction of a higher frequency of the c allele. It was not those individuals who inherited the evolving c allele. This definition was formulated by Godfrey

Hardy and Wilhelm Weinberg who concluded in 1908 that gene pool frequencies are inherently stable and evolution is expected in all populations all the time. Hardy and others told that evolution will not occur in a population if seven conditions are met:

1. Mutation is absent
2. Natural selection is absent
3. The population is infinitely large
4. All members of the population breed
5. All mating is totally random
6. Everyone produces the same number of offspring
7. Absence of migration in or out of population

Since any of these 7 conditions, will not happen in real world, evolution is inevitable result. Hardy and Weinberg developed a simple equation called Hardy-Weinberg equilibrium equation to know possible genotype frequencies in population and to track their changes from one generation to the next. In equation ($p^2 + 2pq + q^2 = 1$), $p$ stands for frequency of dominant allele, q for frequency of recessive allele for trait controlled by a pair of alleles (A and a). p equals all alleles in homozygous dominant (AA) individuals and half of alleles in heterozygous people (Aa) for this trait in population. This is $p = $ AA + ½Aa .The q equals all of alleles in homozygous recessive (aa) individuals and the other half of the alleles in heterozygous (Aa) individuals.

$$q = aa + ½Aa$$

As there are only 2 alleles, frequency of one plus frequency of other should equal 100%, that is to say $p + q = 1$. The following is also correct: $p = 1 - q$. Chances of all possible combinations of alleles occurring randomly are $(p + q)^2 = 1$ or $p^2 + 2pq + q^2 = 1$. $p^2$ is the predicted frequency of homozygous dominant (AA) people , $2pq$ is the predicted frequency of heterozygous (Aa), and $q^2$ is the predicted

frequency of homozygous recessive (aa) ones. Frequency of homozygous recessive people, or $q^2$ in equation, is possible to know since they will not have the dominant trait from observations of phenotypes. Individuals expressing the trait in their phenotype could be homozygous dominant ($p^2$) or heterozygous ($2pq$). Since $p = 1 - q$ and $q$ is known, it is easy to calculate $p$. Knowing $p$ and $q$, it is a simple to place these values into equation ($p^2 + 2pq + q^2 = 1$) which provides the predicted frequencies of all 3 genotypes for selected trait.

By comparing genotype frequencies from next generation with those of current generation in population, it is easy to know that whether or not evolution has occurred. Alleles of a gene are expected to distribute in 3:1 ratio in F1 generation. It is now known that heterozygote appear in higher frequencies. Hardy-Weinberg's Law states that relative frequencies of various genes in a large and randomly mating population tend to remain constant from generation to generation in absence of mutation and natural selection. This law describes a tendency of evolution to conserve gains of genetic changes and avoid frequent changes in genotype.

Factors causing evolution are mutation, genetic drift, natural selection, non-random mating, and small population. Gene frequency is the proportion of an allele in gene pool as compared with other alleles at same locus. It can be calculated by dividing the number of a particular gene by total number of genes present in population. If a population consists of 100 individuals, 40 dominant MM, 40 heterozygous Mm and 20 recessive mm, frequency of dominant gene M (depicted by P), would be 40 + 20/100 = 0.6 and frequency of recessive gene m,( denoted by q), would be 20 + 20/100 = 0.4. When frequency of one gene increases, other must decrease. Genotype frequency is total number of one kind of individuals in a population exhibiting similar characters (genotype) in respect to locus. It is calculated by dividing the individual's number having one kind of genotype by total number of individuals in a population. In the example, the genotype frequency of MM is D/N = 40/100 = 0.4, that of Mm is H/N = 40/100 = 0.4 and that of mm is r/N = 20/100 = 0.2, where N denote total number of individuals and D,H and r denote respectively dominant, heterozygous and recessive. The next generation will have the following composition: 0.25 MM + 0.5 Mm + 0.25 mm, which can also be written as: 1 MM + 2 Mm + 1 mm or M2 + 2 Mm + m2 . If P denotes frequency of gene M , q denotes frequency of gene m, equation is $(P + q)^2 = P^2 + 2Pq + q^2$. This is called Hardy-Weinberg's equation, which is used to find out genetic composition of a population and to prove that gene frequencies remain unchanged from generation to generation. Let s consider a population of rats, with 60% brown (MM) and 40% white (mm) individuals. Gene frequencies will be P = 0.6 and q = 0.4 (P + q = 1.0). Both gene frequencies and genotype frequencies are same here. Now let us change the figures in Hardy-Weinberg's equation as follows:

$$P^2 + 2Pq + q^2 = (0.6)^2 + 2. (0.6) (0.4) + (0.4)^2$$
$$= 0.36 + 0.48 + 0.16$$

Genotype frequency in F1 generation will be 36% MM + 48% Mm + 16% mm but gene frequencies will l remain as P = 0.6 and q = 0.4. Here, dominant gene equals the recessive gene, but in nature the recessive gene is less in number. By knowing the number of homozygous recessive in a population, we can find out the entire gene pool of population or entire genotype frequency. If in a rat population 16% individuals are recessive white and rest are brown, then composition of population can be calculated. If white individuals are 16% (mm), then $m^2 = 16$ or $q^2 = 0.16$; q = square root of 0.16; q = 0.4. As we know the value

of q, the value of P is known by P = 1.0 – q; or 1.0 – 0.4 or P = 0.6.

Now to calculate the heterozygous Mm rats –2Pq = 2. (0.6)(0.4) = 0.48, which means 48% of rats were heterozygous brown (Mm), and as we know that 16% were white (mm), rest of population has to be MM or homozygous brown (100 – 48 + 16) = 16%. Final composition of population will be 36% MM + 48% Mm + 16% mm. Hardy-Weinberg equation is useful in finding out expressed, or hidden genotype frequencies in a population. Alleles the different forms of a gene produce different forms of a trait, code for phenotypes that change in time in evolving population. Changes in alleles in population produce changes in phenotypes of population. Thus, evolution is changes in allele frequency in population over time.

**Allele frequencies** remain unaltered unless mutation and natural selection change them. Previously it was thought that dominant alleles swamp recessive alleles out of existence over time. This wrong theory called genophagy states that dominant alleles increase in frequency from generation to generation. Hardy and Weinberg showed that dominant alleles easily decrease in frequency and that both allele and genotype frequencies in a population remain constant generation after generation. In real population, these disturbing influences are always in effect. Hardy- Weinberg equilibrium is an ideal state to provide a baseline against which change can be analyzed. Static allele frequencies in population across generations indicate no mutation, no migration or emigration, infinite population size, and without selective pressure for or against any genotypes.

### Derivation

A better, but equivalent, probabilistic description for HWP is that alleles for next generation for given individual are chosen randomly and independent. Consider 2 alleles, A and a, with frequencies p and q in population. Different ways to form new genotypes can be derived using a Punnett square, where fraction in each is equal to product of row and column probabilities. Sometimes; a population is created by bringing together males and females with different allele frequencies. In such case, assumption of single population is violated until after first generation, so first generation will not remain in Hardy-Weinberg equilibrium. Successive generations will have Hardy-Weinberg equilibrium.

## Random mating

HWP states the population will have the given genotypic frequencies after a single generation of random mating within population. When random mating assumption is violated, population will not remain in Hardy-Weinberg proportions. Inbreeding causes non-random mating, which increase in homozygosity of all genes. When a population disobeys one of the following 4 assumptions, population may continue to have Hardy- Weinberg proportions each generation, but allele frequencies will change over time. Selection changes allele frequencies often rapidly. While directional selection eventually leads to loss of alleles except the favored one, some forms of selection, such as balancing selection reach equilibrium without allele loss. Mutation effect allele frequencies at a very subtle. Mutation rates are of the order $10^{-4}$ to $10^{-8}$, and change in allele frequency will is, at most, the same order. Recurrent mutation maintains alleles in population, if even strong selection against them exists. Migration genetically links 2 or more populations. Allele frequencies become more homogeneous in populations. Some models for migration inherently include nonrandom mating. For those models, Hardy-Weinberg proportions stand generally invalid. Small population size can cause a random change in allele frequencies. This event due to sampling effect is called genetic drift. Sampling effects are most important when the allele is present in a small number of copies.

## Sex linkage

Where the A gene is sex linked, heterogametic sex have only one copy of gene, while homogametic sex have 2 copies. Genotype frequencies at equilibrium are $p$ and $q$ for heterogametic sex but $p^2$, $2pq$ and $q^2$ for the homogametic sex. In human, red-green color blindness is an X-linked recessive trait. Trait affects about 1 in 12, ($q = 0.083$) and about 1 in 200 females (0.005, compared to $q^2 = 0.007$), very close to Hardy-Weinberg proportions in western European males. If a population is brought together with males and females with a different allele frequency in each subpopulation, allele frequency of male population in next generation will follow that of female population because each son receives its X chromosome from the mother. The population reaches equilibrium very quickly.

## Generalizations

Simple derivation mentioned above can be generalized for more than 2 alleles and polyploidy.

## Generalization for more than 2 alleles

Consider an extra allele frequency, r. The 2-allele case is the binomial expansion of $(p + q) 2$, and the 3-allele case is the trinomial expansion of $(p + q + r) 2$. Consider the alleles A1. Al given by allele frequencies p1 to pi; giving for all homozygotes and for all heterozygote. Hardy-Weinberg principle may be generalized to polyploid systems, for organisms that have more than 2 copies of each chromosome. Consider again only 2 alleles. Diploid case is the binomial expansion of and the polyploid case is the polynomial expansion no migration, or selection of where c is the ploidy, for example with **tetraploid** (c = 4). Depending on whether the organism is a true tetraploid, or an **amphidiploid** will determine how long it will take for population to reach Hardy-Weinberg equilibrium. Hardy-Weinberg principle may be applied in 2 ways, when a population is assumed to be in Hardy-Weinberg proportions, in which genotype frequencies can be calculated, or when genotype frequencies of al 3 genotypes are known, they can be tested for statistical significant deviations.

### Application to complete dominance

Suppose that phenotypes of AA and Aa are indistinguishable due to complete dominance. Assuming that Hardy-Weinberg principle applies to population, then can still be calculated from f(aa). Thus an estimate of f(AA) and f(Aa) derived from respectively. Such a population cannot be tested for equilibrium using significance tests because it is assumed a priori.

### Significance tests for deviation

Testing deviation from HWP is performed using Pearson's chi-squared test, using observed genotype frequencies obtained from data and expected genotype frequencies obtained using HWP. For systems where there are many alleles, this may result in many empty possible genotypes and low genotype, because there are often not enough individuals present in sample to adequately represent all genotype classes. If this is the case, then asymptotic assumption of chi-squared distribution will not hold, and it is necessary to use a form of Fisher's exact test, which requires a computer to solve. Recently several MCMC methods of testing for deviations from HWP have been proposed

### Fisher's exact test

Fisher's exact test can be applied to testing for Hardy-Weinberg proportions. Since the test is conditional on allele frequencies p and q, problem can be viewed as testing for proper number of heterozygotes. Thus Hardy-Weinberg proportions are rejected, if number of heterozygote is too large or too small. The conditional probabilities for heterozygote, given the allele frequencies are given in Emigh as where n11, n12, n22 are the observed numbers of 3 genotypes, AA, Aa, and aa, respectively, and n1 is the number of A alleles, where An example Using one of the examples from Emigh, we can consider the case where n = 100, and p= 0.34.

### Inbreeding coefficient

Inbreeding coefficient, F (see also F-statistics), is one minus the observed frequency of heterozygotes over that expected from Hardy-Weinberg principle. Where expected value from H.W. equilibrium is given by for Ford's data above; fr 2 alleles, the chi-squared goodness of fit test for Hardy-Weinberg proportions is equivalent to test for inbreeding, F = 0. **Inbreeding coefficient** is unstable as expected value approaches zero, thus not useful for rare and very common alleles. For: E = 0, O > 0, F = −∞ and E = 0, O = 0, F is undefined. W.E. Castle showed that genotype frequencies would remain stable in absence of selection. K. Pearson found one equilibrium position with values of p = q = 0.5. R. Punnett, unable to counter Yule's point, introduced the problem to G. H. Hardy's view of biologists' use of mathematics which was published in his 1908 paper where he describes this as very simple. Take Aa as a pair of Mendelian characters, A as dominant, and that in generation the number of pure dominants (AA), heterozygotes (Aa) and pure recessives (aa) are as p: 2q: r. Assume that numbers are large,mating is random, sexes are distributed evenly in 3 varieties, and that all are equally fertile. In next generation, numbers will be as (p+q) 2:2(p+q) (q+r) : (q+r)2, or as p1:2q1:r1, say. It is easy to see that condition for this is q2 = pr. And since q12 = p1r1, whatever values of p, q and r, the distribution continues unchanged after 2nd generation.

## 7.11 Causes and Mode of Evolution

Jr B S Phalange revolutionized the theoretical population genetics begun by Hardy and others. He showed that genetic mutation could provide raw material for Darwinian natural selection and such mutations could do this even when their frequency in population was initially invisible to statistical analysis. Natural selection causes dominance in populations, even when the original expression of an allele at the start was recessive. Haldane observed that natural selection should result in pure selfishness on part of individuals, and no one should be willing to risk his own life to save another. Saving 2 brothers, or 4 cousins would result in same genetic contribution to next generation as that represented by one's own genome. This is now known as kin selection, in which natural selection act at gene level, rather than individuals.

## 7.12 Genetic Drift (Sewall Wright effect)

Sewall Wright gave the principles of gene frequency fluctuations in small populations. In small, non-randomly mating populations gene frequencies fluctuate purely by chance. Smaller the population, larger will be fluctuation in gene frequency. Genetic drift works on principle of tossing a coin. When a coin is tossed for number of times, then chances of getting heads and tail is equal and that standard error is low. When the coin is tossed few times, then standard error is high. Standard Error = square root of Pxq/n, where P = frequency of dominant gene, q = frequency of recessive gene and n = number of individuals

in a population. Let us imagine one pair black hamsters; one MM and other mutant Mm. If they produce only 2 offspring, then chance for first offspring with MM is 0.5 and that second offspring with MM is also 0.5. Chance for both offspring with MM is reduced to 0.5 × 0.5 = 0.25. In such case mutation m will be lost forever. By chance, both offspring can be Mm, where mutation m has a chance to express in next generation. Thus the population drifts for losing or fixing a mutation by chance.

Sewall Wright called this theory the shifting balance theory of evolution, as it emphasizes shift from one adaptive peak to another by random genetic drift. A population of moths that are adapted to forage for nectar in trumpet flowers is a good example. These moths need long tongues to reach nectar in flowers. A population of moths with tongues (average 3 cm in length) reaches nectar in short-necked trumpet flowers, but cannot reach nectar in long flowers. It would be better for moths to have long tongues. Selection should tend to cause the mean value for tongue length to about 3 cm. There is another theoretical peak at 5 cm, but selection would not allow for an increase in tongue length. If population of moths with 3 cm tongues can drift across the valley between adaptive peaks at 3 cm and 5 cm, selection can begin to shift moths with long tongues toward higher adaptive peak. In this way, selection and drift working together can result in shift in adaptive ness that could not have occurred by selection alone.

Gene frequency continues to fluctuate until one allele is lost and other fixed. Allele frequencies change due to chance. This is termed **genetic drift**. Drift is represented as binomial sampling error of gene pool. Alleles that form next generation's gene pool are a sample of alleles from current generation. When sampled from a population, frequency of alleles differs slightly due to chance. Alleles increase, or decrease in frequency due to drift.

Average expected change in allele frequency is zero, as increasing, or decreasing in frequency would equally result. Small percentage of alleles may change frequency in single direction for several generations. Very few new mutant alleles drift to fixation in this manner. In small population, variance in rate of change of allele frequencies is greater than in large population. The overall rate of genetic drift is independent of the population size. With the constant mutation rate, small and large populations lose alleles to drift at same rate as large populations have more alleles in gene pool, but they lose them more slowly. Smaller populations have fewer alleles, but these quickly cycle through. Thus mutation is constantly adding new alleles to gene pool. Selection is not operating on any such alleles. Sharp drops in population size change allele frequencies substantially. When a population crash, alleles in surviving sample may not be representative of pre crash gene pool. This change in gene pool is called founder effect. Small populations of organisms that invade a new territory are subject to this. Genetic changes brought about by founder effects may contribute to isolate populations developing reproductive isolation from their parent populations. In sufficiently small populations, (http://www.talkorigins.org/faqs/faq-intro-to-biology.html)genetic drift counteracts selection. Mildly deleterious alleles may drift to fixation.

Fisher thought populations were sufficiently large that drift could be neglected. As per Wright, population is often divided into subpopulations. Drift could cause allele (http://www.talkorigins.org/faqs/faq-intro-to-biology.html)frequency differences between subpopulations if gene flow was small enough. If a subpopulation was small enough, population could even drift through fitness valleys in adaptive landscape. Subpopulation could attain a larger fitness. Gene flow out of this subpopulation could contribute as a whole

to adapt the population. This is called Wright's Shifting Balance theory. Both natural selection and genetic drift decrease genetic variation. Populations would become homogeneous. Evolution would be impossible if they happen to be the only mechanisms. There are mechanisms that replace variation depleted by selection and drift. Decrease in number of certain species is due to genetic drift. When lethal mutations get fixed, as happened in case of passenger pigeon and cheetah. Carnivores have small populations and are affected by genetic drift. Marriage within their own communities in human tribes faces genetic drift, and accumulates lethal mutations.

## Fisher's Fundamental Theorem

For Fisher, struggle for existence is a kind of distraction from key element of natural selection, which is the Malthusian parameter which measures rate of increase of reproductive value of population from tables of births and deaths. **Reproductive value** is an abstract. Malthusian parameter would include demographic information, since demography influences rate of population growth. Rate of increase in reproductive value of population, m, is identified as fitness. Fisher's fundamental theorem of natural selection states that rate of change in m is equal to genetic variance in m at that time. This theorem identifies natural selection as a continual source of improvement, m, and fitness of population. Fitness always increases over time via natural selection. Introduction of novel variation lowers fitness. Fisher argues for detrimental mutational effects using an interesting geometrical argument. Fundamentalness of Fisher's theory is partly lies in its explanation for organic adaptation (Gardner, 2009). After all, organisms are marvelously and intricately adapted (Fisher, 1968) to conditions of their life. Fisher imagines adaptation as closeness to some target point in a multidimensional phenotype space. Target point represents ideal phenotype given the environmental circumstances. Selection tends to move the average values of population toward point, and reducing the variance around it (Gardner, 2009). Fisher's definition of 'environmental' force, includes all evolutionary factors that are not changing due to additive genetic variance in the population at a time. This broad and its significance has been questioned (Okasha, 2008). Individual organisms have expected and actual reproductive outputs, but these are not fitnesses.

Random drift by means of sampling error came to be known as Sewal Wright effect. Wright referred to all changes in allele frequency as either steady drift or random drift. Wright considered the random genetic drift by means of sampling error equivalent to that by means of inbreeding. But later work has shown them as distinct. Wright focused on effects of inbreeding on small relatively isolated populations and introduced adaptive landscape concept in which cross breeding and genetic drift in small populations could push them away from adaptive peaks, which allow natural selection to push them towards new adaptive peak. Wright thought smaller populations were more suited for natural selection, because inbreeding was sufficiently intense to create new interaction systems through random drift, but not intense enough to cause random no adaptive fixation of genes. Ronald Fisher conceded genetic drift played some role in evolution, but an insignificant one. To Fisher, viewing the evolution as a long, steady, adaptive progression is the only way to explain the ever-increasing complexity from simpler forms. In 1968, M. Kimura proposed neutral theory of molecular evolution, which states that most genetic changes are caused by genetic drift acting on neutral mutations.

## Analogy with marbles in a jar

Genetic drift can be illustrated using 20 marbles in a jar to represent 20 organisms in a population. Consider this jar of marbles as starting population. Half of the marbles in jar are red and half blue. Both colors correspond to 2 different alleles of one gene in population. In each generation the organisms reproduce randomly. To illustrate this reproduction, select a marble randomly from the original jar and deposit a new marble of same color as its parent into new jar. This process may be repeated until there are 20 new marbles in second jar. The second jar then contains a second generation offspring, consisting of 20 marbles of various colors. The second jar contains 10 red and 10 blue marbles, a random shift occurs in allele frequencies. Repeat this process several times, randomly reproducing each generation of marbles to form the next. Numbers of red and blue marbles fluctuate in each generation. Sometimes more red, sometimes less red. This fluctuation is genetic drift resulting from random variation in distribution of alleles from one generation to next. It is possible that in any one generation no marbles of a particular color are chosen, because they have no offspring. If no red marbles are selected, the jar contains new generation blue offspring. If this happens, red allele has been lost permanently in population, while remaining blue allele has become fixed. All future generations are entirely blue. Fixation can occur in just few generations in small populations. In this simulation, there is fixation in blue allele within 5 generations (Figure 7.15).

## Probability and allele frequency

Mechanisms of genetic drift can be illustrated with a simplified example. Consider a very large colony of bacteria isolated in a drop of solution. bacteria are genetically identical except for single gene with 2 alleles labeled

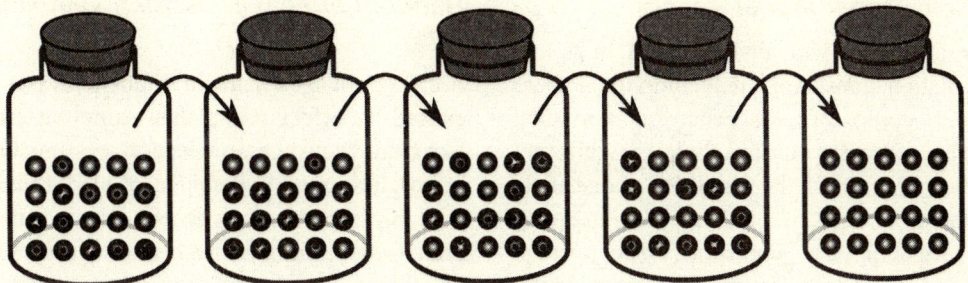

**Figure 7.15**  In this simulation, there is fixation in the blue "allele" within five generations

A and B. Half the bacteria have allele A and other half have allele B. Both A and B have allele frequency 1/2. A and B are neutral alleles. They do not affect the bacteria's ability to survive and reproduce. All bacteria in this colony are likely to survive and reproduce. The drop of solution then shrinks until it has only enough food to sustain 4 bacteria. All others die without reproducing. Among the 4 who survive, there are 16 possible combinations for the A and B alleles:

(A-A-A-A), (B-A-A-A), (A-B-A-A), (B-B-A-A), (A-A-B-A), (B-A-B-A), (A-B-B-A), (B-B-B-A), (A-A-A-B), (B-A-A-B), (A-B-A-B), (B-B-A-B), (A-A-B-B), (http://en.wikipedia.org/wiki/Genetic_drift )(B-A-B-B), (A-B-B-B), (B-B-B-B).If each combination with same number of A and B respectively is counted, we get the following table. Probabilities are calculated with slightly faulty premise that peak population size was infinite.

| A Combinations | B | Probability |
|---|---|---|
| 4 0 | 1 | 1/16 |
| 3 1 | 4 | 4/16 |
| 2 2 | 6 | 6/16 |
| 1 3 | 4 | 4/16 |
| 0 4 | 1 | 1/16 |

Probability of any one possible combination is where 1/2 is multiplied 4 times. The total number of possible combinations

to have an equal (conserved) number of A and B alleles is 6, and its probability is 6/16.Total number of possible alternative combinations is 10. The probability of unequal number of A and B alleles is 10/16. Total number of possible combinations can be represented as binomial coefficients. They can be derived from Pascal's triangle. Probability for any one of possible combinations can be calculated with formula where N is number of bacteria and k is number of A (or B) alleles in combination. Function '()' signifies the binomial coefficient and can be expressed as N choose k. Using formula to calculate probability that between them the surviving 4 bacteria have 2 A alleles and 2 B alleles. Genetic drift occurs when a population's allele frequencies change due to random events. In this example the population contracted (http://en.wikipedia.org/wiki/Genetic_drift) to 4 random survivors, a phenomenon called population bottleneck. Original colony began with an equal distribution of A and B alleles but chances are that remaining population of 4 members has an unequal distribution. Probability that this surviving population will undergo drift (10/16) is higher than the probability that it will remain same. Mathematical models of genetic drift can be designed using either branching processes or a diffusion equation describing changes in allele frequency in an idealized population.

### Wright-Fisher model

Consider a gene with alleles, A or B. In diploid population consisting of N individual there are 2N copies of each gene. An individual can have 2 copies of same allele, or 2 different alleles. We can call the frequency of one allele p and frequency of other q. Wright-Fisher model assumes that generations do not overlap. Annual plants have exactly one generation per year. Each copy of the gene found in new generation is drawn independently at random from all copies of gene in old generation.

### Moran model

This model assumes overlapping generations. At each time, one individual is selected to reproduce and one individual is selected to die. Thus number of copies of a given allele can go up by one, go down by one, or can stay the same. Thus transition matrix is tridiagonal and mathematical solutions are easier for Moran model than for Wright-Fisher model. Computer simulations are easier to perform using the Wright-Fisher model, because fewer time steps need to be calculated. In Moran model, where N is the effective population size, it takes N time steps to get through one generation. In Wright-Fisher model, it takes just one. Moran model and **Wright-Fisher model** produce qualitatively similar results, but genetic drift occurss twice as fast in Moran model.

### Other models of drift

If the variance in number of obbspring is greater than given by binomial distribution assumed by Wright-Fisher model, then same overall speed of genetic drift continues. Genetic drift is a less powerful force than selection. Even for the same variance, if higher moments of offspring number exceed those of binomial distribution, then again the force of genetic drift is weakened.

### Random effects other than sampling error

Random changes in allele frequencies can be caused by effects other than sampling error like random changes in selection pressure. One important alternative source of stochasticity probably important than genetic drift, is genetic draft. Genetic draft, the mathematical properties of which are different from genetic drift is the effect on a locus by selection on linked loci. Direction of the random change in allele frequency is auto correlated across generations.

## 7.13   Drift and Fixation

Hardy-Weinberg principle states that within sufficiently large populations, allele frequencies remain constant from one generation to next unless equilibrium is disturbed by migration, genetic mutation, or selection. Populations do not gain new alleles from random sampling of alleles passed to next generation. Sampling can cause an existing allele to disappear. Random sampling can remove, but not replace an allele. Random declines or increases in allele frequency influence allele distributions in next generation. Genetic drift drives a population to the genetic uniformity over time. When an allele attains a frequency of 1 (100%) it is fixed in population, and when frequency is 0 (0%) it is lost. After fixing an allele, genetic drift halt, and allele frequency cannot change without introduction of new allele in population via mutation or gene flow. Genetic drift is random, directionless process. It acts to eliminate genetic variation over time. If we assume genetic drift is the sole evolutionary force acting on an allele, after t generations in replicated populations, starting with allele frequencies of p and q, variance in allele frequency across those populations is the time to fixation or loss. At any given time probability that an allele finally become fixed in population is simply its frequency in population. If the frequency p for allele A is 75% and the frequency q for allele B is 25%, then given unlimited time the probability A will become fixed in population is 75% and probability that B will fixed is 25%. Expected number of generations required for fixation is proportional to population size. Fixation is predicted to occur rapidly in smaller populations. Effective population size, which is smaller than total population, is used to detect these probabilities. Effective population (Ne) considers factors like **inbreeding level**, life cycle stage in which the population is the smallest, and that some neutral genes are genetically linked to others that are under selection. Effective population size may not be the same for every gene in same population. One formula used for approximating the expected time before a neutral allele becomes fixed through genetic drift, according to Wright-Fisher model, is 10 simulations of random genetic drift of a single given allele with an initial frequency distribution 0.5 and is measured over 50 generations, repeated in 3 reproductively synchronous populations of different sizes. In such simulations, alleles drift to loss, or fixation (frequency 0.0 or 1.0) only in smallest population. Expected time for neutral allele to be lost through genetic drift can be calculated as when a mutation appears only once in a population large enough for initial frequency to be negligible. Formulas can be simplified to for average number of generations expected before fixation of a neutral mutation, and for average number of generations expected before loss of a neutral mutation.

## Time to loss with drift and mutation

If an allele is lost more often by mutation and gained by mutation, then mutation, and drift, may influence the time to loss. If the allele prone to mutational loss starts as fixed in population, and is lost by mutation at rate m per replication, expected time in generations until its loss in haploid population is given by Euler's constant. First approximation represents the waiting time until first mutant destined for loss, with loss then occurring rapidly by genetic drift, taking time Ne<<1/m. Second approximation represents the time needed for deterministic loss by mutation accumulation. In both, time to fixation is dominated by mutation via term 1/m, and is less affected by effective population size.

## Genetic drift versus natural selection

Law of large numbers predicts little change over time due to genetic drift when population is large. When reproductive population is small, effects of sampling error can alter allele frequencies significantly. **Genetic drift** is considered to be a consequential mechanism of evolutionary change primarily within small, isolated populations. Although both processes affect evolution, genetic drift operates randomly while natural selection functions non-randomly. **Natural selection** with direction guides evolution towards heritable adaptations to current environment. Genetic drift without direction is guided only by mathematics of chance. Consequently, drift acts on genotypic frequencies within a population without regard to their phenotypic effects. In contrast, selection favors spread of alleles whose phenotypic effects increase survival and/or reproduction of their carriers, lowers the frequencies of alleles that cause unfavorable traits, and ignores neutral. In natural populations, genetic drift and natural selection do not act in isolation. Both forces are always at play. Degree to which alleles are affected by drift or selection varies according to population size. Magnitude of drift on allele frequencies per generation is larger in small populations to overwhelm selection when selection coefficient is less than 1 divided by effective population size. As a result, drift affects frequency of more alleles in small populations than in large ones.

When allele frequency is very small, drift can overpower selection-even in large populations. While disadvantageous mutations are usually eliminated quickly in large populations, new advantageous mutations are almost as vulnerable to loss through genetic drift as are **neutral mutations**. The mathematics of genetic drift depends on effective population size. It is not clear how this is related to actual number of individuals in population. Genetic linkage to other genes that are under selection can reduce effective population size experienced by neutral allele. With a higher recombination rate, linkage decreases and local effect on effective population size occurs. This effect is visible in molecular data as correlation between local recombination rate and genetic diversity, and negative correlation between gene density and diversity at noncoding sites. Stochasticity associated with linkage to other genes that are under selection is not the same as sampling error, and is sometimes known as genetic draft in order to distinguish it from genetic drift.

## Adaptation versus exaptation

A feature may confer high fitness in a given environment, but may evolve initially for other (Losos, 2011) reason. As a result, 2 species may exhibit similar phenotypes in similar selective environment, even if both did not evolve the feature as an adaptation for environment. This is the distinction between adaptation and exaptation (Losos,2011).An adaptation is a feature that evolved in response to natural selection in environment in which it now occurs, whereas an **exaptation** provides enhanced fitness in an environment, but did not originally evolve in response to natural selection in that environment. In convergent exaptations, natural selection is still involved in favoring the feature in its current selective environment, even if it did not build feature

in (Losos, 2011) response to those selective conditions.

## Correlated response to selection on another character

If taxa share similar constraints, then selection on similar traits in multiple taxa may lead to similar correlated responses. Occupation of a particular habitat may lead to evolution of small size. If small size is achieved through paedomorphosis, paedomorphic features (Losos, 2011) like reduction in digit number may convergently evolve in same habitat. Although natural selection would have (Losos, 2011) driven this evolution, identifying target of selection as a trait other than body size is wrong. Interacting constraints and natural selection can produce phenotype-environment correlation. Convergent evolution can occur even in the absence of natural selection favoring convergent trait. Such convergence is not expected to be associated with function of selective environment. Even though natural selection usually drives such correlations, for several reasons it can do so without directly favoring convergent trait in question in that environment.

## Measurement of Natural Selection

One possibility to measure selection in natural populations is seeing if trait is favored and whether given trait is the actual target of selection (Lande and Arnold, 1983). This approach tests that whether a trait is beneficial in its present environment, in which case it would be an aptation (Gould and Vrba, 1982). Endler (1980) noted that guppies in parts of streams lacking predators are more colorful than those that occur in presence of predators. Predator-free populations rapidly evolve to be more colorful (Reznick and Ghalambor, 2005). Studies of a similar nature are common in laboratory microbial and viral evolution experiments.

## Functional Consequences of Trait Evolution

Selection does not operate on traits, but on their functional consequences (Arnold, 1983). One variant will be favored over another, if it confers increased ability that in turn leads to greater evolutionary fitness by increasing survival, mating success, or fecundity. Wing shape affects aerodynamic performance, enhance clinging ability topads, and cuticular secretions inhibit dehydration are questionable. In many cases, hypothesized functional relationships exist. Long legs allow lizards to run faster. Wing shape influence speed in insects, birds, bats. Leaf shape affects $CO_2$ uptake and water loss in plants (Reich et al., 1997). Functional studies do not support an adaptive hypothesis. Taxa that are phenotypically similar differ in functional capacities. Tall plants that branch well with single thickened trunk have evolved many times in plant history. This morphology is achieved in various ways in different lineages (Donoghue, 2005). Functional analysis of these various morphologies shows that not all these forms could have existed in the way we think of trees today. Prehistoric lycophyte trees lacked secondary phloem and a bifacial cambium and would have had difficulty in maintaining a tall structure for very long. Tree-like morphology of such plants would arise late in life cycle during reproduction, like a plant inflorescence (Donoghue, 2005). Webbed feed in salamanders do not always provide functional gain. Some species cling to smooth, wet surfaces. But many webbed salamander do not (Jaekel and Wake, 2007). In both cases, convergent phenotypes do not result in convergent functional capabilities and may not represent adaptations to the same selective environment. If it provides increased functional abilities, trait may not be favored by natural selection. Enhanced capabilities must lead to an increase in reproductive or survival success for providing a fitness advantage. Increased functional abilities are not relevant to some organism. *Anolis* lizards rarely jump .An increase in jumping ability seems to provide little benefit (Irschick and Losos, 1998). Even when natural selection is the primary force for evolution, comparative studies may not identify adaptations. Species adapt to same selective pressure in various ways, or they evolve the same phenotype to adapt to various selective pressures (Ridley, 1983).Recent conceptual developments, have added knowledge of extra ways in which natural selection in similar environments lead to adaptation.

## 7.14 Tenets of Modern Evolutionary Synthesis

Much of evolution is selectively neutral, especially at the level of nucleotide sequences in DNA. Genome of most organisms is chaotic, with bits and pieces of gene sequences being added, removed, and rearranged in nearly random patterns, often by **parasitic DNA**. Genes and traits are only weakly associated. Much of genome of most organisms, especially eukaryotes like us, is stuffed with redundant sequences, modular gene segments, and regulatory regions which do not code for polypeptides at all. Some are nevertheless essential to phenotypic expression. Phenotypic and protein evolution is largely decoupled from evolution of much of the DNA sequence in most organisms, especially eukaryotes. Mutation is immensely important, not only as a source of evolutionary novelty but also as a source of overall genetic change, especially in short-lived organisms such as bacteria.

## R. A. Fisher and Theorem of Natural Selection

Fisher provided mathematical models which undermine the Mendelian geneticists' theory of evolution via macro mutation showing that continuous variation could provide the basis for natural selection based on work of Hardy, Weinberg, and Castle. Fisher showed that trait characterized by continuous variation were common and could provide raw material for natural selection. Such traits are continuous in populations, but do not blend from parents to offspring. They are produced by unblending particles of inheritance. Mendelian inheritance conserves, rather than finally destroying the genetic variation that exists in natural populations. Fisher theorem states that rate of increase in fitness of any organism at any time is equal to its genetic variance in fitness at that time. Degree of change result from natural selection depends mainly on amount of genetic variation in population undergoing selection. With very little variation, natural selection cannot change characters of members of population much. Much genetic variation can bring considerable evolutionary change by natural selection. As per Fisher most variation required for natural selection to occur is hidden in heterozygote in populations. He believed that selection against dominant alleles should remove dominant alleles from populations almost immediately. Selection against recessive alleles should reduce their frequency until they are hidden among **heterozygote**. If selection shifts and recessive allele becomes favorable, it can increase in frequency quickly. The more hidden recessive alleles in a population, can bring selection more rapidly them forward if environment shifts and makes them beneficial. In larger population, the more recessive alleles can be hidden in it .Thus, natural selection works faster and effectively in large populations in which heterozygosit hidies recessive alleles.

## Evolution within a Lineage

Evolution, a change in gene pool of a population involves several factors. Mutation,

recombination and gene flow add new alleles to gene pool. Genetic drift and natural selection remove alleles from gene pool. Drift removes alleles randomly. Selection removes deleterious alleles. Amount of genetic variation found in a population is the balance between actions of these mechanisms. Natural selection increases frequency of an allele. Selection removing harmful alleles is called negative selection. Selection increasing frequency of helpful alleles is termed positive or positive **Darwinian selection**. A new allele can drift to high frequency. As change in allele frequency in each generation is random, nobody predicts positive or negative drift. Usually of high gene flow, new alleles enter gene pool as a single copy. Most new alleles added to gene pool are lost almost immediately due to drift or selection. Only a small percent ever reach a high frequency in population. Most moderately beneficial alleles are lost due to drift. A mutation can reappear numerous times. Fate of any new allele depends to some extend on organism alleles near it for many generations. A mutant allele can increase in frequency simply because it is linked to a beneficial allele at a nearby locus even if mutant allele is deleterious, although it should not be so deleterious to stop the benefit of other allele. A potentially beneficial new allele can be eliminated from gene pool because it was linked to deleterious alleles when it first arose. An allele riding on tails of a beneficial allele is called **hitchhiker**. Recombination will bring the 2 loci to linkage equilibrium. The more closely linked 2 alleles are, the hitchhiking will last longer. Action of selection and drift are combined. Drift is increased as selection pressures increase as increased selection reduces size of population and individuals passing alleles to next generation. Cumulative natural selection cause adaptation, and natural selection sifts mutation. Small changes, favoured by natural selection, could be the stepping-stone to changes. Sum total of large numbers of such changes is **macroevolution**.

## Life history

Life history evolution is part of evolutionary ecology, which attempts to explain phenotypic evolution. Developmental part of explanation includes evolutionary developmental genetics and morphology, now referred to as **Evo-Devo**. Classical life history theory based on optimization models aims to explain variation in size at birth, growth rates, age, and size at maturity, reproductive investment, clutch size, mortality and lifespan. By the 1990s the field had achieved consensus explanation of evolution of life history traits. Life histories are shaped by interaction of extrinsic and intrinsic factors. Extrinsic factors are ecological impacts on survival and reproduction. Intrinsic factors are tradeoffs among life history traits and lineage-specific constraints on expression of genetic variation. It is required to understand 2 things to understand the evolution of life histories. External one is the way the environment affects survival and re- production of organisms of different ages, stages, or sizes. The internal one is the way traits are connected to each other and constraints on how traits can vary. All modern theory builds on focus achieved by this simplifying claim. For example, the manner in which limited resources are acquired and allocated to survival and reproduction varies with type of organism. There is an intrinsic phylogenetic or developmental component to way the extrinsic factors interact with the organism, and which separate the extrinsic from the intrinsic.

## Evolution after Darwin

Darwin didn't explicitly write about purpose in Origin of Species. He constructed his argument in a way that purpose was not necessary, and was not mentioned. Since it isn't necessary for a scientific explanation of phenomena under investigation, then it is not included in such an explanation. This leaves open the question of whether purpose exists in nature and scientist's description of complex natural processes without reference to purpose. It's time to examine both of these questions in more detail. This does not undermine importance of natural selection in evolution. Following John Endler's lead, biologists no longer consider natural selection itself to be a cause of anything. It is an effect, caused by other processes like genetic variation, genetic inheritance, population expansion and contraction, and differential survival and reproduction. It is 4 underlying conditions working together to produce changes in genotype and phenotype that we call natural selection.

## Red Queen Hypothesis

The Red Queen Hypothesis first suggested by Leigh Van Valen states that probability of organisms becoming extinct bears no relationship to how long they already may have survived. Struggle for existence never gets any easier. Well adapted animal has the same chance of extinction as a newly formed species. It is genetics arms race in which an animal constantly run genetic gauntlet of being able to chase its prey, to make predators fool, and to resist infection. Organisms have to run fast-just to stay still. They constantly have to run to try to improve. Now Red Queen Hypothesis seems to be favorite of evolutionists worldwide to explain the reason as to why of sex.

# SUMMARY

1. Genetic change by mutation, sexual recombination, gene flow, and genetic duplication causes evolution. Evolution is inevitable as systems created by .evolution protects against evolution. Genetic change drives evolution causing biological diversity, natural selection does not but it acts on all populations. Natural -, sexual -, and artificial selections constrain the variation occurring through genetic change .Genetic drift does not affect survival or sexual reproduction but has profound effect on small population .The frequency of genetic drift in a population randomly changes over time. Change in gene frequency in wild populations is observed and verified in studies .A generation of sexual population becomes a self-reproducing unit called species. Sexual population evolves as a whole. Thousands of genes are involved in selection.

2. Microevolution, also called sequential evolution, involves continuous gradual change in interbreeding population, producing new subspecies and geographical races through changes in gene frequencies in population from one generation to next. Speciation adds new species to species pool as mutation adds new alleles to gene pool. Species selection favors one species over another as natural selection favors one allele over other .Natural selection at level of homeotic regulatory genes produce large changes in phenotypes in relatively short periods compared with kinds of changes. Hardy and Weinberg showed that dominant alleles easily decrease in frequency and both allele and genotype frequencies in a population remain constant unless disturbed by non-random mating, selection, mutations, random genetic drift, gene flow and meiotic drive. Sewall Wright gave the principles of gene frequency fluctuations in small populations. In small, non-randomly mating populations gene frequencies fluctuate purely by chance. In smaller population, fluctuation in gene frequency will be larger.

3. Law of large numbers predicts that in large population little change over time is due to genetic drift. When reproductive population is small, effects of sampling error alter allele frequencies significantly. Genetic drift is a consequential mechanism of evolutionary change primarily in small isolated populations. Although both processes affect evolution, genetic drift operates randomly; while natural selection functions non-randomly. Sexual selection, a form of natural selection, occurs relative to mates. Sexual populations remain in coherent systems .In sexual organisms; minimal unit of reproduction is population. Group selection is the fitness differences between groups in a given population. Organism selection is the fitness differences between organisms in a given group. Gene selection is the fitness differences between genes in a given organism (Sober, 2011).

4. Perhaps sexual reproduction succeeds as sexual populations are more resistant to evolution than asexual ones. "One gene - one trait - one selection vector" is not found. One gene may affect several traits. Most traits depend on many genes. Evolutionary change in gene frequencies that results from variation in fitness is the product of selection (Walsh, 2007). A population of individuals can change in only 3 ways: individuals can leave the population, individuals can enter it, or individuals existing in population can change their traits. Reproduction changes population via births, and can by itself, completely cause a rate of change in proportions of alleles in adult population.

5.  Individuals can change traits which correspond to mutation. Trait in population can have a systematic advantage or its rivals with regard to any such population change, which can lead to evolution by natural selection. In classical view, selection involves only population change of narrowly specified kinds: individuals that leave population via death or enter it via reproduction. This concept is the narrow trait fitness which corresponds with population change .Microevolution, also called sequential evolution, involves a continuous gradual change in interbreeding population, producing new subspecies and geographical races through changes in gene frequencies in population from one generation to next.

6.  Mega evolution is caused by large genetic changes producing different types and disruptive or divergent natural selection causing population to occupy different environment. Factors causing evolution are mutation, genetic drift, natural selection, non-random mating, and small population. Gene frequency is the proportion of an allele in gene pool as compared with other alleles at same locus. Alleles, the different forms of a gene produce different forms of a trait, code for phenotypes that change in time in evolving population. Changes in alleles in population produce changes in phenotypes of population. Thus evolution is changes in allele frequency in population over time.

7.  Recurrent mutation maintains alleles in population, if even strong selection against them exists. Migration genetically links 2 or more populations. Allele frequencies become more homogeneous in populations. In small, non-randomly mating populations gene frequencies fluctuate purely by chance. Smaller the population, larger will be fluctuation in gene frequency .With the constant mutation rate, small and large populations lose alleles to drift at same rate as large populations have more alleles in gene pool, but they lose them more slowly. Smaller populations have fewer alleles, but these quickly cycle through. Thus mutation is constantly adding new alleles to gene pool and selection is not operating on any such alleles. Sharp drops in population size change allele frequencies substantially.

8.  Populations do not gain new alleles from random sampling of alleles passed to next generation. Sampling can cause an existing allele to disappear. Random sampling can remove, but not replace an allele. Random declines or increases in allele frequency influence allele distributions in next generation. Genetic drift drives a population to the genetic uniformity over time. When an allele attains a frequency of 1 (100%) it is fixed in population and when frequency is 0 (0%) it is lost. After fixing an allele, genetic drift halt, and allele frequency cannot change without introduction of new allele in population via mutation or gene flow. Genetic drift is random, directionless process, but it eliminates genetic variation over time. At any given time probability that an allele finally become fixed in population is simply its frequency in population. In natural populations, genetic drift and natural selection do not act in isolation; both forces are always at play. Degree to which alleles are affected by drift or selection varies according to population size. Magnitude of drift on allele frequencies per generation is larger in small populations to overwhelm selection when selection coefficient is less than 1 divided by effective population size. As a result, drift affects frequency of more alleles in small populations than in large ones.

9. When allele frequency is very small, drift can overpower selection even in large populations Evolution, a change in gene pool of a population involves several factors. Mutation, recombination and gene flow add new alleles to gene pool. Genetic drift and natural selection remove alleles from gene pool. Drift removes alleles randomly. Selection removes deleterious alleles. Amount of genetic variation found in a population is the balance between actions of these mechanisms. Natural selection increases frequency of an allele. Selection removing harmful alleles is called negative selection. Selection increasing frequency of helpful alleles is termed positive or positive Darwinian selection. A new allele can drift to high frequency. As change in allele frequency in each generation is random, nobody predicts positive or negative drift. Usually of high gene flow, new alleles enter gene pool as a single copy.

# REVIEW QUESTIONS

## Short Answer Questions

1. What is horizontal gene transfer?
2. What is DNA?
3. What do you understand by homeostatic mechanism?
4. What do you understand by the evolability?
5. What is intrinsic selection?

## Long Answer Questions

1. Describe the selfish gene concept in your own words.
2. Write a note on various levels of selection.
3. Write a note on the evolutionary forces.
4. Describe the Hardy-Weinberg equilibrium with examples.

# REFERENCES

Ariew, Andre, and. Lewontin. R. C 2004. "The Confusions of Fitness." The british Journal for the Philosophy of Science, 55 : 347-363.

Barton, N.H., Charlesworth, B. 1998. Why sex and recombination? Science. ,281:1986-1990.

Dawkins R. 1976.The Selfish Gene. Oxford: Oxford University Press.

Denbigh, K. 1975. In: Entropy and Information in Science and Philosophy. Kubat L, Wicken, J.S. 1979. The generation of complexity in evolution: A thermodynamic and information-theoretical discussion. J Theor Biol, 77:349-365.

Donoghue, M. J. 2005. Key innovations, convergence, and success: macroevolutionary lessons from plant phylogeny. Paleobiology, 31:77-93.

Drummond, D.A., Silberg, J.J., Meyer, M.M., Wilke, C.O., Arnold, F.H.2005. On the conservative nature of intragenic recombination. Proc Natl Acad Sci USA. ,102:5380-5385.

Erwin, D.H. 2000.Macroevolution is more than repeated rounds of microevolution. Evol Dev., 2:78-84.

Eshel, I. In: Game equilibrium models I. Selton R, editor. Springer; 1991. Game theory and population dynamics in complex genetical systems: the role of sex in short term and in long term evolution; pp. 6-28.

Felsenstein, J. 1974.The evolutionary advantage of recombination. Genetics, 78:737-775.

Fisher, R.A.1931. The genetical theory of natural selection. Oxford: Oxford University Press.

Frank, S.A. 1995.George Price's contributions to evolutionary genetics. J Theor Biol., 175:373-388.

Ghiselin, M.T. 1980.In: The Evolutionary Synthesis: Perspectives on the Unification of Biology. Mayr E, Provine WB, editor. Cambridge, MA: Harward University Press. The failure of morphology to assimilate Darwinism; pp. 180-193.

Ghiselin, M.T. 1974. The Economy of Nature and the Evolution of Sex. Berkeley: Univ of California Press

Ghiselin, M.T. 1997.Metaphysics and the Origin of Species. Albany: State University of New York Press.

Ghiselin, M.T.1974. A radical solution of the species problem. Syst Zool. ,23:536-544.

Gould, S.J. 1990. Wonderful life. New York: W.W. Norton.

Gould, S. J. 2002. The structure of evolutionary theory. Belknap Press of Harvard Univ. Press, Cambridge, MA .

Gould, S. J., and Vrba. E.S. 1982. Exaptation-a missing term in the science of form. Paleobiology, 8:4-15.

Gould, S.J., Eldredge, N. 1977.Punctuated equilibria: the tempo and mode of evolution reconsidered. Palaeobiology. , 3:115-151.

Gould, S.J., Eldredge, N. 1993.Punctuated equilibrium comes of age. Nature, 366:223-227.

Hamilton, W.D. 1963.The evolution of altruistic behaviour. Am Nat., 97:354-356.

Hamilton, W.D.1975. Gamblers since life began: barnacles, aphids, elms. Q Rev Biol., 50:175-180.

Hull, D.L.1976. Are species really individuals? Syst Zool., 25:174-191.

Hull, D.L.1980. Individuality and selection. Annu Rev Ecol Syst. ,11:311-332.

Hull D.1981. In: The Philosophy of Evolution. Jensen U.J, Harre' R, editor. Brighton: Harvester Press. Units of evolution: a metaphysical essay; pp. 23-44.

Jackson, J.B., Cheetham ,A.H.1999. Tempo and mode of speciation in the sea. Trends Ecol Evol. ,14:72-77.

Jenkin, Fleeming. "Review of the Origin of Species." The North British Review, 46 (1867): 277-318.

Kimura, M.1991.Recent development of the neutral theory viewed from the Wrightian tradition of theoretical population genetics. Proc Natl Acad Sci USA.,88:5969-5973.

Kondrashov, A.S. 1984.Deleterious mutations as an evolutionary factor. I. The advantage of recombination. Genet Res Camb. ,44:199-217.

Kondrashov, A.S. 1993.A classification of hypothesis on the advantage of amphimixis. J Hered, 84:372-387.

Kondrashov, A.S. 1995.Modifiers of mutation-selection balance: general approach and the evolution of mutation rates. Genet Res. ,66:53-69.

Krakauer, D.C., Mira, A.1990. Mitochondria and germ-cell death. Nature, 400:125-126.
Bengtsson, B.O. 1985.Biased conversion as the primary function of recombination. Genet Res Camb. 47:77-80.

Lande, R. (1980) "Sexual dimorphism, sexual selection, and adaptation in polygenic characters." Evolution 34 : 294-305.

Lewontin, R.C.1970. The units of selection. Annu Rev Ecol Syst. ,1:1-18.

Lewontin, Richard C 1968. "The Effect of Differential Viability on the Population Dynamics of t Alleles in the House Mouse ." Evolution, 22 : 262-273.

Lion, S., van Baalen, M. 2008.Self-structuring in spatial evolutionary ecology. Ecology Letters. 11:277-295.

Longo, V.D., Mitteldorf, J., Skulachev, V.P.2005. Programmed and altruistic ageing. Nat Rev Genet. 6:866-872.

Luria, S.E.1974. Reactivation of irradiated bacteriophage by transfer of self-reproducing units. Proc Natl Acad Sci USA. ,33:253-264.

Lynch, M. 2008.The cellular, developmental and population-genetic determinants of mutation-rate evolution. Genetics., 180:933-943.

Malthus, T.R.1798. An essay on the principle of population. Oxford World's Classics reprint .

Markert, C.L.1988. Imprinting of genome precludes parthenogenesis, but uniparental embryos can be rescued to reproduce. Ann NY Acad Sci. ,541:633-638.

Maynard Smith ,J. 1978. The evolution of sex. Cambridge: Cambridge University Press.
Kondrashov, A.S.1988. Deleterious mutations and the evolution of sexual reproduction. Nature., 336:435-440.

Maynard Smith, J., Szathmary, E. 1995.The Major Transitions in Evolution. Oxford: Oxford University Press.

Mayr E. 1942.Systematics and the origin of species. New York: Columbia University Press.

Mayr, E. 1963. Animal species and evolution. Belknap Press of Harvard Univ. Press, Cambridge , MA .

Mayr, E. 1970.Populations, species and evolution. Cambridge, Massachusetts: The Belknap Press of Harvard University Press.

Mayr, E. 1996. What is a species, and what is not? Philosophy of Science. ,63:262-277.
Hastings, I.M. 1992. Population genetic aspects of deleterious cytoplasmic genomes and their effect on the evolution of sexual reproduction. Genet Res. ,59:215-225

Merser, E.H.1981. The Foundations of Biological Theory New York. Wiley-Interscience.

Michod, R.E. 1999. Darwinian dynamics. Evolutionary transitions in fitness and individuality. New Jersey: Princeton University Press.

Morgan, Lewis H. Ancient Society. New York: Holt, 1877.

Muller ,H.J.1964. The relation of recombination to mutational advance. Mutat Res. ;1:2-9.

Neiman M, Hehman G, Miller JT, Logsdon JM Jr, Taylor DR. Accelerated mutation accumulation in asexual Lineages of a freshwater snail. Mol Biol Evol. 2010;27:954-63.

Eldredge, N., Gould, S.J.1977. On punctuated equilibria. Science., 276:338-341.

Nowak, M.A.2006. Five Rules for the Evolution of Cooperation. Science. 314:1560-1563.

Okasha, S. 2006. "The two faces of fitness." In In Conceptual Issues in Evolutionary Biology, Third Edition, edited by Elliott Sober, 25-38. Cambridge MA: The MIT Press.

Okasha, S.2005. Maynard Smith on the levels of selection question. Biol Philos.; 20:989-1010.

Okasha, S.2011 "Realism, Conventionalism, and Causal Decomposition in Units of Selection: Reflections on Samir Okasha's Evolution and the Levels of Selection." Philosophy and Phenomenological Research 82, no. 1 : 221-231.

Okasha, Samir. 2006. Evolution and the Levels of Selection. Oxford: Clarendon Press.

Otto, S.P., Lenormand ,T.2002. Resolving the paradox of sex and recombination. Nat Rev Genet., 3:252-261.

Paques, F., Haber, J.E. 1999.Multiple pathways of recombination induced by double-strand breaks in Saccharomyces cerevisiae. Microbiol Mol Biol Rev. ,63:349-404.

Peeling ,P.J. 2009.Functional and ecological impacts of horizontal gene transfer in eukaryotes. Curr Opin Genet Dev.,19:613-619.

Price, G.R. 1972.Extension of covariance selection mathematics. Ann Hum Genet, 35:485-490.

Robson, A.J., Bergstrom, C.T., Pritchard, J.K. 1999. Risky business: Sexual and asexual reproduction in variable environments. J Theor Biol. ,197:541-556.

Severtsov, A.S. 1990 .Intraspecies diversity as a cause of the evolutionary stability. Zh Obshch Biol., 51:579-589.

Shcherbakov, V.c. 2010. Biological species is the only possible form of existence for higher organisms: the evolutionary meaning of sexual reproduction. Biol. direct Doi:10.1186/1745-6150-5-14

Shcherbakov, V.P. 2010. Stasis is inevitable consequence of every successful evolution. Biol Philos. in press .

Sniegowski, P.D., Gerrish, P.J., Johnson, T., Shaver, A. 2000.The evolution of mutation rates: separating causes from consequences. Bioassays, 22:1057-1066.

Sniegowski, P.D., Murphy, H.A. 2006 .Evolvability. Curr Biol. ,16:R831-R834.

Sober E, Wilson D S. 1989. Unto Others: the Evolution of Psychology of Unselfish Behavior. Cambridge, MA: Harward University Press

Sober, E., Wilson, D S. 1988.Unto Others: the Evolution and Psychology of Unselfish Behavior. Cambridge, MA: Harvard University Press.

Spencer, H. 1967."First Principles, 2nd ed." The Online Library of Liberty. 1867.

Stanley, S.M. 1975.A theory of evolution above the species level. Proc Natl Acad Sci USA, 72:646-650.

Stanley, S.M.1975. Clades versus clones in evolution: why we have sex. Science, 190:382-383.

Szathmary E, Maynard Smith J.1995. The major evolutionary transitions. Nature, 374:227-231.

Taddei, F., Radman, M., Maynard-Smith, J., Toupance ,B., Gouyon, P.H., Godelle, B. 1997. Role of Mutator Alleles in Adaptive Evolution. Nature, 387:700-702.

Tilly, J.L.2001. Commuting the death sentence: How oocytes strive to survive. Nat Rev Mol Cell Biol., 2:838-848.

Wade, Michael J. 1979."Sexual Selection and Variance in Reproductive Success." The American Naturalist 114, no. 5: 742-747.

Wake, D.B., Roth, G., Wake, M.H. 1983.On the problem of stasis in organismal evolution. J Theor Biol., 101:211-224.

Wake, D. B. 1991. Homoplasy-the result of natural selection, or evidence of design limitations. Am. Nat., 138:543-567.

Wake, D. B., and A. Larson. 1987. Multidimensional analysis of an evolving lineage. Science, 238:42-48.

Wake, D. B., Wake, M.H. and C. D. Specht. 2011. Homoplasy: from detecting pattern to determining process and mechanism of evolution. Science, 331:1032-1035.

Wake, D. B.. 1999. Homoplasy, homology and the problem of 'sameness' in biology. Pp. 24-46 in G. R.Bock and G.Cardew, eds. Homology. Novartis Foundation Symposium 222. John Wiley & Sons, New York .

Walsh, Denis M., Tim Lewens, and Andre Ariew. 2002."The Trials of Life: Natural selection and random drift." Philosophy of Science 69, 452-473.

Weismann ,A.1889. Essays upon heredity and kindred biological problems. Oxford: Clarendon Press.

West, A., Griffin, A.S. and Gardner A. (2006) "Social Semantics: altruism, cooperation, mutualism, strong reciprocity and group selection." Journal of Evolutionary Biology 20: 415-432.

Wilson, D.S., Sober, E. 1974.Re-introducing group selection to the human behavioral sciences. Behavioral and Brain Sciences. ,17:585-654.

Wilson, D.S., Sober, E. 1989.Reviving the superorganism. J Theor Biol. ,136:337-356.

# CHAPTER 8

# Molecular Basis of Evolution

Evolution occurs in every population. Mutations occur often without producing changes. Mutations provide raw material for engine of evolution. Gene flow is often restricted in fungi and plants that cannot move around. In them, genetic material gets moved from place to place. In animals, gene flow is a significant cause of deviations from previous allele frequencies. Effects of **gene flow** and its restriction are not simple. Most actual breeding populations are not large enough to ensure unchanged allele frequencies due to purely random accidents. Due to formulation and widespread acceptance of Hardy-Weinberg Equilibrium Law, natural selection was once again proclaimed the primary engine of evolution. Sexual selection is also an important engine of evolution in animals.

## 8.1 Heredity, Mutation, and Genes

Introduction of Mendelian genetics based on plant hybridization (Figure 8.1) into conception of evolution by natural selection transforms the very concept of evolution forwarded by Fisher. Genes are inherited. Evolution becomes the change in gene frequencies of population. Major concern for Fisher is the question of whether a novel mutant with higher fitness can establish itself in population. Mutation will tend to lower systematically the fitness of population, but this effect is small compared with force of natural selection in raising it. Overall fitness of population must remain constant, otherwise populations would be continually growing. For Darwin, the raw material for selection was provided by mysterious laws of variation, for Fisher mutation provided it. Evolution for Fisher is the genetic evolution of a population, in contrast to phenotypic evolution that concerned Darwin.

## 8.2 Lewontin Definition of Evolution by Natural Selection

Lewontin (1970) develops an account of selection, the essentials of which are still widely accepted. For Lewontin, **ENS** consists of 3 components which are:

1. Different individuals in a population have different morphologies, physiologies, and behaviours (phenotypic variation).

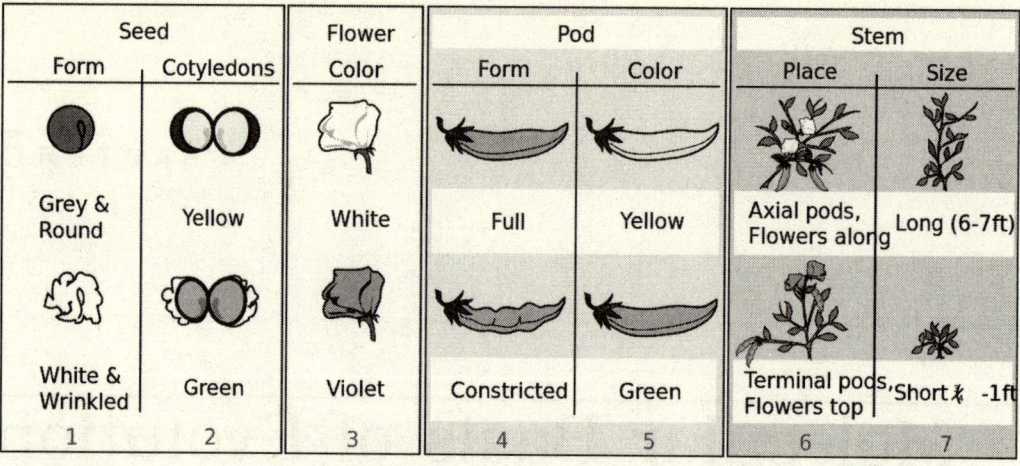

**Figure 8.1** Plant hybridization

2. Different phenotypes have different rates of survival and reproduction in different environments (differential fitness).

3. There is a correlation between parents and offspring in contribution of each to future generations (fitness is heritable).

If these 3 conditions are in place, a population undergoes evolutionary change (Lewontin, 1970). Mutation and migration potentially introduce alternate alleles into population, which then may spread if they are selectively advantaged (Lewontin, 1970). This conception is enlarged and fortified by Lewontin's development of idea of multi-level selection. Multi-level selection explains all evolutionary change in terms of selection and nothing else. Lewontin does not rule out the possibility of other sources of predictable evolutionary change. In Lewontin's discussion, evolution and natural selection are nearly identified, while mutation and migration are recognized by Lewontin as sources to provide new alleles for a population, which serves as raw material for ENS to occur. While it is accepted that mutation or migration could act to transform the population, the possibility is neglected in practice.

## 8.3 Selection vs. Mutation

R.A. Fisher first integrated the concept of natural selection with Mendelian genetics partly as a response to challenge posed to Darwinism by F. Jenkin (Jenkin, 1867). This synthesis overcame the many objections to Darwin's natural selection by integrating it with Mendelian genetics and statistical techniques. "Natural Selection is not Evolution" (Fisher 1968b), highlights a theme of central importance. Fisher advocates that particulate inheritance is more compatible with observed variation than blending inheritance, and on assumption of particulate inheritance, the effect of natural selection should greatly outweigh the evolutionary effect of mutation. Fisher considers 4 versions of such theories, variations of which were advocated by contemporary Mendelian biologists like Hugo de Vries:

1. Desires of organism influence the direction of mutation,

2. Use and disuse induce corresponding heritable mutations,

3. Environment causes mutations and so directs evolution,

4. Mutation represents the unfolding of an inner tendency or urge.

According to Fisher, inheritance is not blending and effect of natural selection dominate the mutation. His dismissal of directional mutation based on empirical evidence supports particulate inheritance. Mathematical consequences are based on particulate inheritance combined with empirical observations of pure strain beans and mutation rates in *Drosophila*. Genetic evidence later establishes that mutation is the sole driver of genetic evolution at many loci (Kimura, 1968). As per Fisher, while natural selection is not the same as evolution, it is the cause of evolution. No other factors are significant by comparison.

## 8.4 Random Mutation

One concept that has to be examined is the random mutation of genes. Mendel had not foreseen genetic mutation, but variation via interplay of dominant and recessive genes. Hugo de Vries in 1902 observed that gene could occasionally undergo radical changes. H. J. Muller in 1927 identified some agents of mutations. X-rays, UV light, and certain chemicals are **mutagens**. Taken individually, the impact of genetic mutations on species is minimal. In every human being 2 new mutations occur in the midst of its 100,000 genes. Over time, a few mutations might establish, but most will not. Mutations themselves do not determine the direction of evolution. Random mutations mean the genetic mutation that does occur to increase the organism's overall chances of success. Mutations may be beneficial to organism, or harmful, or neutral; viable and capable of being passed onto next generations, or not. Genetic mutation is not programmed into organism, with quasi-purpose of improving its chances for living. The term random, spontaneous or chance mutation is not intended to a notion of natural spontaneity. Mutations are considered cause, in deterministic sense of causation. Radiation or chemicals cause genes to mutate. Mutation is a physical reaction of gene. It is not akin to a response to stimulus or to make the organism more resistant to dangerous radiation or chemicals. Coincidence refers to 2 or more chains of events coming together at a certain time and place. They may all be determined, and yet their meeting is a matter of chance. An organism may stray into polluted place, which is not its natural habitat. Here, chance could refer to unpredictability of purely physical events in practice, but they are difficult to understand in theory; or broadly to possibilities of volition. We are now able to produce intentionally mutations by radiation or chemicals. Enlarging the concept of mutation include both artificial and natural mutation. The term random mutation applies only to latter, and not to mutations due to volitional intervention.

## 8.5 Founder Mutation

A founder mutation appears in DNA of one or more founder individuals of a distinct population. **Founder mutations** start with changes in DNA rather on single chromosome and pass to other generations. Original haplotype is the whole chromosome. Proportion of **haplotype** common to all carriers of mutation is reduced as generation's progress, which allows rough estimation of age of mutation. Founder effects, a special case of genetic drift, occur when a small group in a population split of from original population and forms a new one. New colony may have less genetic variation than original population, and through random sampling of alleles during reproduction of subsequent generations, continue rapidly towards fixation. Such inbreeding makes colony more vulnerable to extinction. When a newly formed colony is small, its founders strongly affect population's genetic makeup far into future. In humans, which have a slow reproduction rate, population remain small for generations, amplifying the drift effect until population reaches a certain size. Alleles present in relatively rare amount in original population move to 1 of 2 extremes. Thus, allele may soon lose altogether, or the allele survives and within a few generations become dispersed in population. New colony faces an increase in frequency of recessive alleles, and as a result, in an increased number which are homozygous for certain recessive traits.

Variation in gene frequency between original population and colony may also trigger the 2 groups to diverge significantly over many generations. As variance, or **genetic distance**, increases the 2 separated populations may become distinctively different, both genetically and phenotypically, although genetic drift, natural selection, gene flow and mutation contribute to divergence. This capacity for rapid changes in colony's gene frequency led to consider the founder effect a significant driving force in evolution of new species. Sewal Wright first attached this significance to random drift and newly isolated small populations with shifting balance theory. Ernst Mayr created many persuasive models to show that decline in genetic variation and small population size accompanying the founder effect are important for new species to develop.

Speciation by genetic drift is a specific case of **peripatric speciation**, which occurs rarely when random change in genetic frequency of population favours survival of few organisms of species with rare genes causing **reproductive mutation**. The surviving organisms then breed over long period of time to create whole new species whose reproductive systems or behaviors are no more compatible with the original population.

Changes in base pair sequences in DNA or RNA and changes in amino acid sequences and their molecular configuration in different proteins, from generation to generation are called **molecular evolution**. It is possible to measure differences between these molecules obtained from different organisms on a unit scale of amino acids or nucleotides and demonstrate their relationships. As molecular sequences are heritable, their variations produce molecular records that have been transferred from generation to generation during evolution. A triplet made of 3 pairs of nucleotides is called a codon which changes, if 1 of the 3 bases changes, it may or may not end a change in amino acid synthesized by it. Majority of changes are small and consequential but accumulate over long periods to bring about large alterations in gene frequencies in populations. Two kinds of such changes are possible:

**Silent site substitution:** These are the changes in DNA sequences which do not cause any change in amino acid synthesis and composition of proteins. They are usually changes in last base pair of codon. In mRNA strand, GCA codes for alanine and if guanine replaced adenine, the resulting GCG still code for same amino acid alanine. Silent site substitutions do not bring about any phenotypic changes.

**Replacement substitution:** They are the changes in bases of codons that result in synthesis of new amino acids and alter the structure of proteins and changes the phenotype. **Silent site substitutions** have higher rate of change as compared to replacement substitutions, since the former do not produce changes that can be exposed to natural selection but the latter do. For same reason, genes which are less vital to cell can undergo rapid changes by replacement substitution without showing harmful effects.

**Pseudogenes**, the duplicated sequences of bases do not code for proteins and are not exposed to natural selection, but undergo higher rate of evolutionary changes.

**Sequencing amino acids:** Comparing amino acid sequences in a protein in different species is the most popular methods to determine phylogeny. In haemoglobin, 2 pairs of alpha and beta sequences of polypeptide chains form a tetramer that is distinguished by different amino acid sequences in different species. In vertebrates, different types of globin chains appeared during evolution and in each species they followed their own evolutionary path by changes in amino acid sequences. They are all variations of a single globin ancestor that is controlled by similar globin genes which are believed to have originated by gene duplication of original type (Figure 8.2).

## 8.6 Mutation and Migration

Selection typically eliminates variation from the populations. If selection removes variation, there would be no variation for selection to work, and evolution will grind to a halt. However mutation restores genetic variation eliminated by selection and is the fundamental raw material of evolution. Simply, mutation changes allele frequencies and genotype frequencies. Forward mutation changes A allele to allele at a rate (u); backward mutation changes a to A at a rate (v). We can express the frequency (p) of A allele in next generation (pt+1) in terms of these opposing forward and reverse mutations, like forward and reverse chemical equations: $(p_{t+1}) = p_t(1 - u) + q_t(v)$ (<http://biomed.brown.edu/Courses/BIO48/7.mutation>). The 1st part on right is accounts for alleles not mutated (1-u), and 2nd part accounts for increase in p due to mutation from a to A (frequency of a times the mutation

rate to A). We say $\Delta p = (p_{t+1}) - (p_t)$. $\Delta p$ as: change in allele frequency between generations. This is useful as it calculates a theoretical equilibrium frequency, at which there is no change in allele frequencies, $\Delta p = 0$ which is when $(p_{t+1}) = (p_t)$; from above: $p_t (1 - u) + (1-p) t(v) = pt$ [remember, q=(1-p)]. Now solve for p and think equilibrium frequency $= p = v/(u+v)$. Similarly, equilibrium frequency of $q = u/ (u+v)$. The term **migration** describes gene flow, is the movement of alleles from one area to another. Gene flow assumes some form of dispersal or migration but dispersal is not gene flow. Consider x and y populations with frequencies of A allele of px and py. Assume that some individuals from y migrate into x. Proportion of y individuals that become parents in population x in next generation = m. After migration, x population could consist of migrant individuals (proportion m) and non-migrant individuals (proportion [1-m]). Thus frequency of A allele in population x in next generation (px t+1) is the frequency in non-migrant portion (= px [1-m]) plus frequency in migrant portion (py m). Thus: px t+1 = px t [1-m] + py m. (<http://biomed.brown.edu/Courses/BIO48/7.mutation>).

Change in Allele frequency due to gene flow is $\Delta p = (px t+1) - pxt$ which is just [pxt [1 - m] + pym] - pxt. Multiplying through and canceling terms leaves us with:

$\Delta p = -m(px t - pyt)$. Change in p depends on migration rate and difference in p between 2 populations. If we take a grid or array of populations and focus on one of those populations as recipient population with all other populations contributing equally to it, then py would be replaced by average p for all other populations. Many scenarios are possible. (<http://biomed.brown.edu/Courses/BIO48/7.mutation>)

# ACAGATATA - The Mutation Game

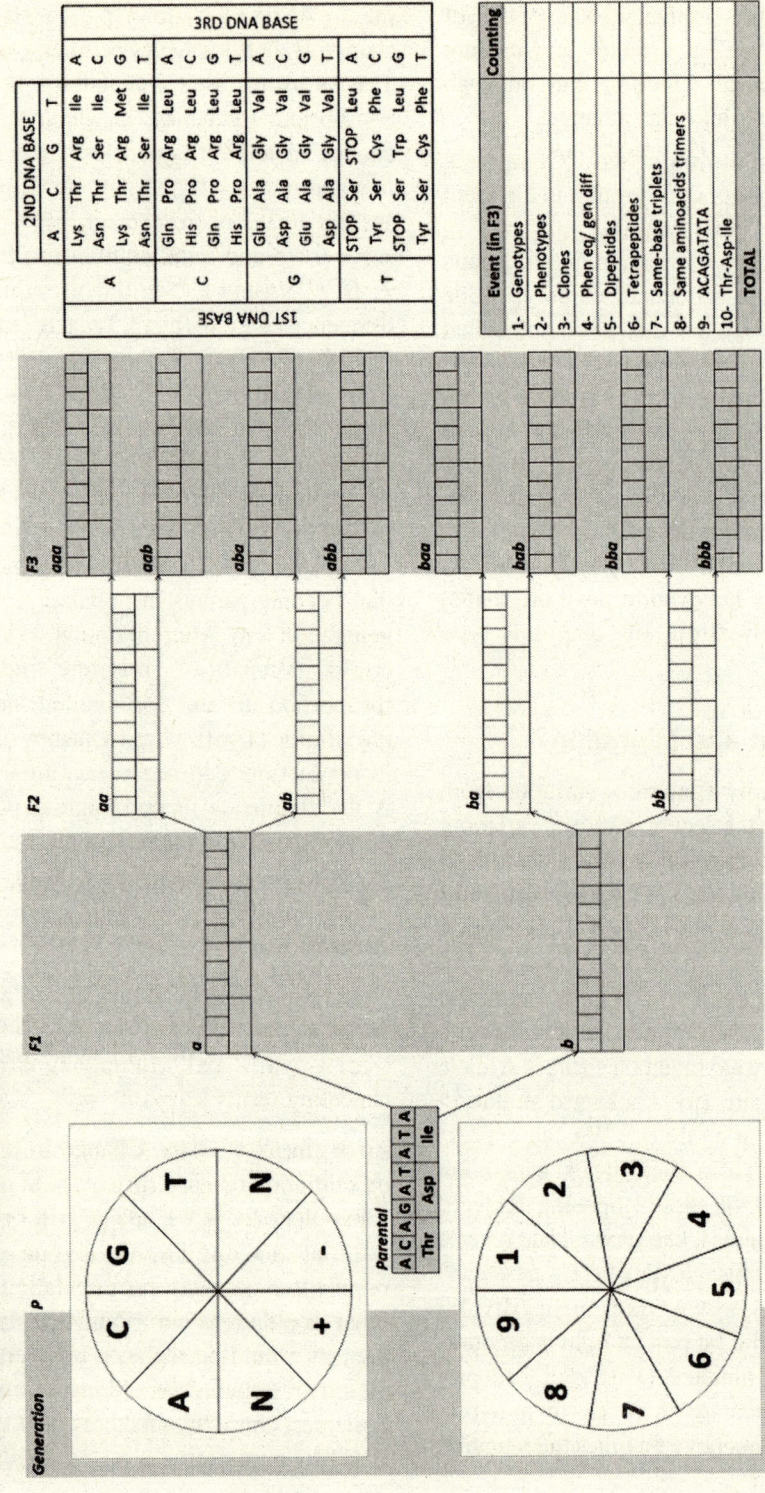

**Figure 8.2** The game board. It shows a phylogenetic tree to be constructed, with the parental DNA sequence ACAGATATA (center left, generation "P") branching in two offsprings (in F1), then four (F2) and, finally, eight "descendant" sequences (F3). Also printed in the board are two special roulettes to "induce" copy errors ("Mutation Wheel", left top, and "Position Wheel", left bottom), an adapted, unconventional Genetic Code table (right top), and a score table (right bottom).

## 8.7 Mutation and Selection Balance

In practice no specific evolutionary forces act alone. Some other force might counteract a specific force of interest. Detecting such opposing evolutionary forces depends on relative strengths of 2 (or more) forces. Lets consider a case where mutation introduces a deleterious allele into population and selection tries to remove it. As above take the mutation rate (u) as mutation rate to a allele. This tends to increase frequency of a (q will increase). The q increases at a rate of u (1-q); remember, (1-q) = p, or frequency of A allele. This mutation pressure increases the number of alleles which selection can act against. To select against an allele, first assume complete dominance, i.e., that deleterious effects of allele are only observed in aa homozygote. Frequency of "a" (q) decreases by selection at a rate of -sq2 (1-q), where s= the selection coefficient under these conditions. Amount of change created by this selection is a function of frequency of aa homozygote (q2) and frequency of A allele (1-q). Amount of change is proportional to amount of genetic variation. If these terms for mutation and selection are placed together, amount of change in a allele is :

$$\Delta q = u\ (1-q) - sq2\ (1-q)$$

if the fight between selection and mutation is a draw, then there is no change in allele frequency since mutation is increasing just as fast as selection is reducing it. Here Δq = 0, and equilibrium allele frequency = q-hat, has been reached (in formal notation q-hat is q with a circumflex over it). Rearranging the above formula gives: u (1-q) = sq2 (1-q). We solve this for q to give the equilibrium allele frequency, q-hat: q = sqrt (u/s) (sqrt stands for square root).Most mutation rates are small (about $10^{-6}$). So this equation

suggests that deleterious alleles are maintained in mutation selection balance at low frequencies.

## 8.8 Genetic Variation

Evolution requires genetic variation. If there were no dark moths, population could not evolve from light . For evolution, mechanisms exist to increase or create/decrease genetic variation. Mutation is a change in a gene. These changes (<http://www.talkorigins.org/faqs/faq-intro-to-biology.html>) are the source of genetic variation. Natural selection operates on variation. Two components of genetic variation are allelic diversity and non- random associations of alleles. Alleles are the different versions of same gene. A, B or O alleles in human determine their blood group. Most animals including humans are diploid containing 2 alleles for each gene at each locus, 1 inherited from mother and 1 inherited from father. Locus is the site of a gene on chromosome. AA, AB, AO, BB, BO or OO blood group are found in human. When the 2 alleles at a locus are the same, they are called **homozygous**. Individual with 2 different alleles at a locus (an AB individual) is called **heterozygous** (<http://www.talkorigins.org/faqs/faq-intro-to-biology.html>). At any locus different alleles may present in a population. At 45 % loci in plants more than one allele in gene pool is found. A plant is likely to be heterozygous about 15% of its loci. In animals, levels of genetic variation range from about 15% of loci with more than 1 allele in birds, to over 50% of loci being **polymorphic** in insects. Mammals and reptiles are polymorphic in about 20% loci. Amphibians and fish are polymorphic in about 30% loci. Linkage disequilibrium is a measure of association between alleles of 2 different genes. If 2 alleles are found together often than would be expected, alleles are in linkage disequilibrium.

If there are 2 loci in an organism (A and B) and 2 alleles at each of these loci (A1, A2, B1 and B2) **linkage disequilibrium** (D) is measured as D = f(A1B1) * f(A2B2) - f(A1B2) * f(A2B1) (where f(X) is the frequency of X in population). D varies between -1/4 and 1/4; the greater the deviation from zero, the greater the linkage (<http://www.talkorigins.org/faqs/faq-intro-to-biology.html>).

Linkage disequilibrium results from physical proximity of genes is maintained by natural selection, if some combinations of alleles work better. Natural selection maintains the linkage disequilibrium between color and pattern alleles in *Papilio memnon* in which a gene determines wing morphology. One allele at this locus leads to a moth with a tail; the other allele codes for tailless moth. Another gene determines wing colour. Four possible types of moths are brightly colored moths with, and without tails, and dark moths with, and without tails which are produced when moths are bred in the lab. Brightly colored moths with tails and darkly colored moths without tails are found in wild. Natural selection maintains non-random association. Bright, tailed moths mimic the pattern of unpalatable species. Dark morph is cryptic. Other 2 combinations are neither mimetic nor cryptic and are eaten by birds. Assortative mating causes a non-random distribution of alleles at a single locus. If there are 2 alleles (A and a) at a locus with frequencies p and q, frequency of 3 possible genotypes (AA, Aa and aa) will be p2, 2pq and q2, respectively. If the frequency of A is 0.9 and frequency of a is 0.1, frequencies of AA, Aa and aa individuals are: 0.81, 0.18 and 0.01. This distribution is called the Hardy-Weinberg equilibrium (<http://www.talkorigins.org/faqs/faq-intro-to-biology.html>). Random mating results in deviation from Hardy-Weinberg distribution.

We are likely to mate with someone of own race. In populations that mate in such way, few heterozygotes are found than is predicted in random mating. A decrease in heterozygote can be the result of mate choice, or simply result of population subdivision. Most organisms have limited dispersal capability, so their mate is chosen from local population. (<http://www.talkorigins.org/faqs/faq-intro-to-biology.html>).

## 8.9  Gene Flow

Gene flow is the transfer of alleles or genes from one population to another. Migration into or out of population causes a marked change in allele frequencies. **Immigration** may result in addition of new genetic variants to established gene pool of a particular species or population. Several factors affect rate of gene flow between different populations. Greater mobility of an individual provides great migratory potential. Animals are more mobile than plants. However, pollen and seeds may be carried considerable distances by animals or wind. Gene flow is the transfer of alleles (Figure 8.3) from a population to another population through immigration. Maintained gene flows between 2 populations lead to combination of 2 gene pools, reducing the genetic variation between 2 groups. Gene flow strongly works against speciation by recombining the gene pools of groups, and thus, repairing the developing differences in genetic variation that would have led to speciation. If a species of grass grows on both sides of a highway, pollen is likely to be carried from one side to other and vice versa. If this pollen is able to fertilize the plant where it ends and produce viable offspring, then alleles in pollen have been able to move from population on one side of highway to other.

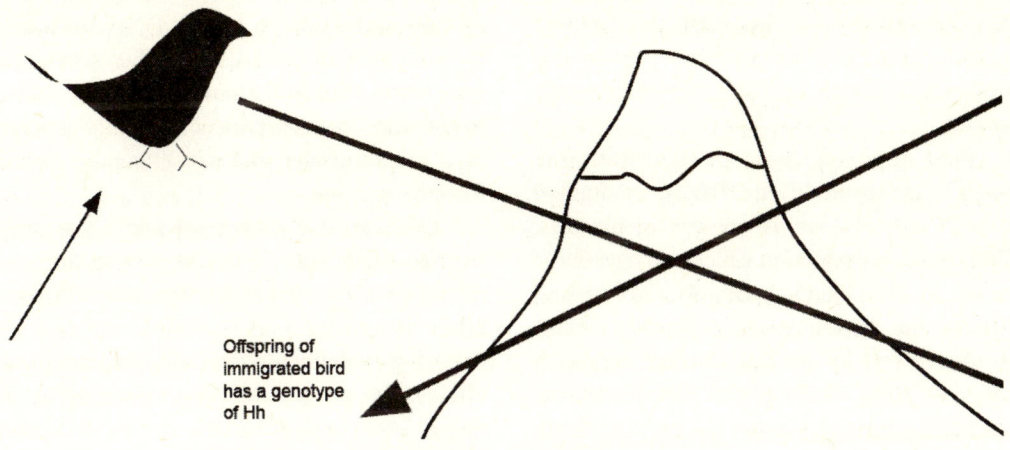

Offspring of
immigrated bird
has a genotype
of Hh

**Figure 8.3** Gene flow

Physical barriers to gene flow are, but not always natural including high mountain ranges, oceans, or deserts. They can be artificial man-made barriers, like the Great Wal of China, which hinders gene flow of native plants. One native plant, *Ulmus pumila*, showed a low prevalence of genetic differentiation than plants *Vitex negundo*, *Ziziphus jujuba*, *Heteropappus hispidus*, and *Prunus armeniaca* which inhabitated opposite side of Great Wal. *Ulmus pumila* has wind-pollination as primary means of propagation and the latter-plants carry out pollination through insects. Samples of same species which grow on either side have developed genetic differences, as there is little to no gene flow for recombination of gene pools. Barriers to gene flow need not be physical barrier alone. Species can live in same environment, yet show very limited gene flow due to limited hybridization or hybridization producing unfit hybrids. Female choice plays a role in hindering gene flow. Asymmetric recognition of local and non-local songs is found between 2 populations of black-throated blue warblers in the US, one in northern United States and other in southern United States. Males in northern population respond strongly to local male songs but relatively weakly to non-local songs of southern males. Southern males respond to local and non-local song. Northern male exhibiting differential recognition shows that northern female tend not to mate with hetero-specific males from south. Thus it is not necessary for northern males to respond to song from a southern chal enger. A barrier to gene flow exists from South to North as a result of female preference. Models of gene flow can be drawn from **population genetics**.

## 8.10 Gene Flow Mitigation

When cultivating genetically modified (GM) plants or livestock, it is necessary to prevent genetic pollution i.e. their genetic modification from reaching other hybridized or wild native plant and animals. Reasons to limit gene flow may include **biosafety** or agricultural co-existence, in which GM and non-GM cropping systems work side-by-side. Programmes of limiting gene flow in plants are Transcontainer, which investigates methods for bio-containment, SIGMEA, which focuses on biosafety of GM plants, and Co-Extra, which studies co-existence of GM and non-GM product chains. Three approaches to gene flow mitigation are keeping the genetic modification out of pollen, preventing formation of pollen, and keeping pollen inside flower.

First approach requires transplastomic plants in which modified DNA is not situated in cell's nucleus but is present in plastids. Chloroplasts are plastid in which photosynthesis occurs. In some plants, pollen does not contain plastids and modification in plastids cannot be transmitted by pollen. Second approach relies on male sterile plants which can not produce functioning flowers and cannot release viable pollen. Cytoplasmic male sterile plants produce higher yields. Researchers are trying to introduce this trait to GM crops. Third approach works by preventing the flowers from opening. This trait is called cleistogamy and occurs naturally in some plants. **Cleistogamous** plants produce flowers which either open only partly or not at al. It is unclear how reliable cleistogamy is for gene flow mitigation.

## 8.11 Gene Flow and Selection

New organisms may enter a population by migration from other population. If they mate, they can bring new alleles to local gene pool. In some closely related species, fertile hybrids can result from inter-specific mating. These hybrids can vector genes from species to species. Gene flows between distantly related species occur infrequently. This is called horizontal transfer. One interesting case of this involves genetic elements called P elements. Margaret Kidwell found that P elements are transferred from *Drosophila willistoni* group to *D. melanogaster*. These 2 species of flies are distantly related and hybrids do not form. Their ranges do overlap. **P elements** are vectored into *D. melanogaster*

by a parasitic mite that targets both these species. The mite punctures the exoskeleton of flies and feeds on juices. DNA from one fly can be transferred to another when the mite feeds. Since P elements actively move in genome, one incorporated itself into genome of a melanogaster and subsequently spread through species.

Consider that some weak allele is waiting over to other side of tracks, where they do not survive. An evolutionary pressure changes allele frequencies in one direction and an opposing evolutionary force eliminatintes those alleles. Depending on relative strengths of these 2 opposing forces, an equilibrium condition can arise. Let's consider the movement of "a" allele, and assume that it is completely recessive in its phenotype of death-by-sewage. Change in allele frequency from the $\Delta q = -m (qx\ t - qy\ t)$. Change in allele frequency due to selection against this allele is $-sq2\ (1-q)$ . Putting these 2 pieces together, we can write the expression for change in allele frequency $\Delta q = -m (qx\ t - qy\ t) - sqx2\ (1-qx)$, that is due to both gene flow and selection. When the fight between gene flow and selection is a draw, the system will be in equilibrium and there will be no change in q, and $-m(qx\ t - qy\ t) = sq2(1-q)$.

## 8.12 Gene Flow in Humans

In the US, gene flow between white European population and black West African population are recently brought together. In West Africa, **Duffy antigen** provides some resistance to malaria, and this allele is present in nearly all of West African population. Europeans possess the allele Fya or Fyb, as malaria is almost non-existent. The allele frequencies became mixed in each population because of movement of individuals as found by measuring the frequencies of alleles in the West African and European groups. This gene flow between European and West African groups is greater in Northern U.S. than in South.

Gene flow occurs between species through hybridization or gene transfer from bacteria or virus to new hosts. Gene transfer, defined as the movement of genetic material across species boundaries includes horizontal gene transfer, antigenic shift. Re-assortment is sometimes an important source of genetic variation. Viruses can transfer genes between species. Bacteria incorporate genes from other dead bacteria, exchange genes with living bacteria, and exchange plasmids across species boundaries. Sequence comparisons suggest recent horizontal transfer of genes among diverse species including across the boundaries of phylogenetic domains. Determining the phylogenetic history of species cannot be done by determining evolutionary trees for single genes. Gogarten suggests the original metaphor of a tree no longer fits the data from recent genome research. Biologists use the metaphor of a mosaic to describe various histories combined in individual genomes and use metaphor of an intertwined net to visualize rich exchange and cooperative effects of horizontal gene transfer. Using single genes as phylogenetic markers, tracing the organismal phylogeny in presence of HGT (horizontal gene transfer) is difficult. Combining simple coalescence model of cladogenesis with rare HGT events suggest absence of single last common ancestor containing all genes ancestral to those shared in 3 domains of life. Each contemporary molecule has its own history and traces back to an individual molecule ancestor. These molecular ancestors were present probably in different organisms at different times.

## 8.13 Gene Families

**Duplicate genes** are produced by irregular or unequal crossing over and produce similar phenotypic expression, but without cumulative effect. When such genes diverge slightly in their function, they form gene families that become sources of variations. Genes of a family share the following characteristics:

They originate by duplication of existing gene due to unequal crossing over, show structural homology with one another and produce distinct effect, but it is related to ancestral gene. Human haemogobin gene cluster called gene family or **multigene family** consists of 2 alpha duplicate tandem genes and 7 beta genes. Globin genes in mammals are an excellent example of gene families. Immunoglobin gene family is a branching lineage of duplicated genes and T-cell receptor genes that produce specific reaction against vast diversity of viruses and bacteria that invade our body. Ribosomal rRNA, tRNA and histone controlling genes are examples of gene families, but their coding sequences are identical and they produce same effect as their product is required in large quantities in cell. Genes coding for histones exist in tandem clusters of over 100 copies each in humans and up to 1000 copies in sea urchins. They all line up in sequence along chromosome and form gene families. The gene ancestral to modern haemoglobin genes are believed to have duplicated about 350 mya. Unequal crossing over is probably the primary source of gene families and they are known to evolve together in concert. Gene conversion is other source of gene families which occurs between homologous chromatids when cross-over products are being repaired.

## 8.14 Genetic Changes Over Generations

Environments change in subtle and complex ways. When changes are beyond tolerance of organisms, widespread death occurs. Natural populations have genetic diversity. Individuals which survive an environmental crisis reproduce. Their traits become more common in next generation. This natural selection resulting in evolution is found in laboratory in bacteria. When a lethal dose of antibiotic is added, a mass die-off results. Few bacteria usually are immune and survive. Next generation is mostly immune as they inherit immunity from the survivors.

### Evolution of antibiotic resistant bacteria

This same phenomenon of bacteria evolution occur when an antibiotic drug fail to remove completely bacterial infection. Antibiotic resistance demonstrates how mutation and fast reproductive rates of microorganisms can outpace modern medical breakthroughs. Species that mature and reproduce large numbers in short time have potential for fast evolutionary changes. Insects and microorganisms often evolve at such rapid rates that to combat them quickly lose their effectiveness. We are developing new pesticides, antibiotics, and other measures in an escalating biological arm race with these.

## 8.15 Genetic Pollution

Naturally evolved, region-specific species can be threatened with extinction through genetic pollution, causing hybridization, introgression and genetic swamping. Such processes cause homogenization or replacement of local genotypes due to numerical and/or fitness advantage of introduced plant or animal. Non-native species threaten native plants and animals with extinction by hybridization and **introgression** or by habitat modification. These phenomena can be detrimental for rare species coming into contact with abundant ones. **Interbreeding** between the species causes a swamping of rarer species gene pool. The extent of this event is not always apparent from outward appearance. While some degree of gene flow occurs during normal evolution, hybridization may threaten a rare species existence. For example, the Mallard is an abundant species of duck that interbreeds readily with a variety of other ducks and poses a threat to integrity of species.

## 8.16 Genetic Recombination

Each chromosome in sperm or egg cells is mixture of genes from our mother and father. Recombination could be considered as process of gene shuffling. Most organisms have linear chromosomes and their genes lie at specific location along them. Bacteria have circular chromosomes. In most sexually reproducing organisms, 2 of each chromosome type are found in every cell. In humans, every chromosome is paired, 1 inherited from mother, other inherited from father. When an organism produces gametes, the gametes end with 1 of each chromosome per cell. Haploid gametes are produced from diploid cells by meiosis. In meiosis, homologous chromosomes line up. DNA of chromosome is broken on both chromosomes in several places and rejoined with other strand. Later, the 2 homologous chromosomes are split into 2 separate cells that divide and become gametes. Because of recombination, both chromosomes are a mix of alleles from mother and father. Recombination creates new combinations of alleles. Alleles that arose at different times and various places can be brought together. Recombination could occur between genes, and within genes. Recombination within a gene forms a new allele. Recombination adds new alleles and combinations of alleles to gene pool. Genetic recombination is the production of new combinations of alleles, encoding a novel set of genetic information, e.g., by pairing of homologous chromosomes in meiosis, or by breaking and rejoining of DNA strands, which forms new molecules of **DNA**. This last type of recombination occurs between similar molecules of DNA, as in homologous recombination of chromosomal crossover, or dissimilar molecules, as in non-homologous end joining. V(D)J recombination in organisms with adaptive immune system is a type of genetic recombination that helps immune cells to rapidly diversify to recognize and adapt to new pathogens. Recombination is a common method of **DNA repair** in bacteria and eukaryotes. Recombination involving breakage and rejoining of DNA strands producing novel sequences that are different from parental strands. Genetic recombination could occur without breaking and rejoining of DNA strands in meiosis, with random pairing of homologous chromosomes (synapsis) and is catalyzed by many enzymes, called recombinases. RecA, the chief recombinase found in *Escherichia coli*, repair the DNA double strand breaks (DSBs). In yeast and other eukaryotic organisms 2 recombinases are required for repairing DSBs. The RAD51 protein is required for mitotic and meiotic recombination, whereas DMC1 protein is specific to meiotic recombination (Figure 8.4).

## Chromosomal crossover

In eukaryotes, recombinations in meiosis simplifies chromosomal crossover. Crossover process leads to offspring's with different combinations of genes from their parents, and occasionally produce new chimeric alleles. Shuffling of genes brought about by genetic recombination have many advantages, as it is a major engine of genetic variation and allows sexually reproducing organisms to avoid **Muller's ratchet**. Chromosomal crossover refers to recombination between paired

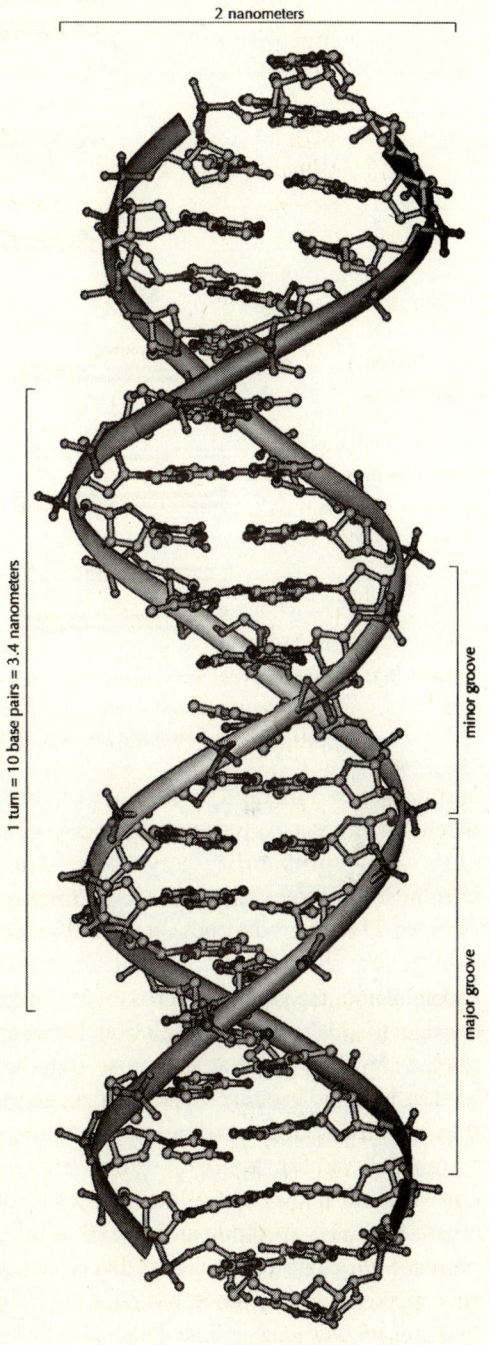

Figure 8.4a   DNA

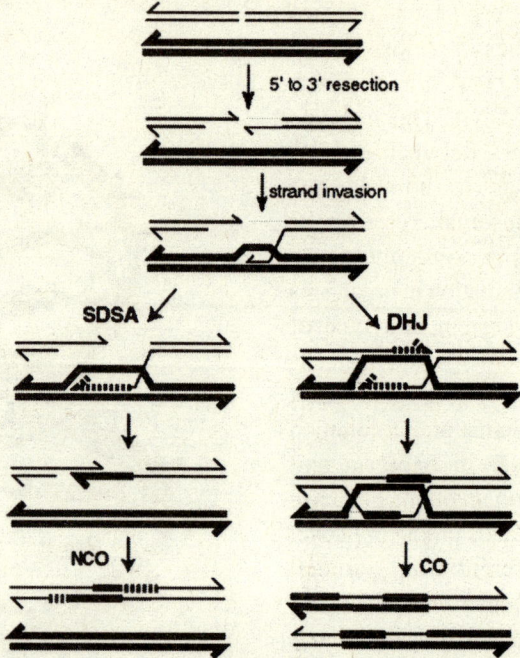

**Figure 8.4b** Meiotic recombination started by a double strand break, followed by pairiry with an homologom chromosome and straed intasion to repair. NCO = non-cronor, CO = croesoes DHJ = Double Holiday Jurction SDSA = Synthesis dependent stored anneating recombination.

chromosomes inherited from each of one's parents in meiosis. As recombination can occur with small probability anywhere along chromosome, frequency of recombination between 2 locations depends on their distance. For genes sufficiently distant on same chromosome, the amount of crossover is high enough to destroy the correlation between alleles. As 2 genes that are close together are less likely to become separated than genes that are further apart, biologists can deduce roughly how far apart 2 genes are on chromosome if they know the frequency of crossovers. This method can be used to infer presence of certain genes. Genes that typically stay together during recombination are said to be linked. One gene in linked pair sometimes is used as a marker to deduce presence of another gene. This is used to detect the presence of a disease-causing gene.

## Gene conversion

In gene conversion, a section of genetic material is copied from 1 chromosome to another, without donating chromosome being changed. **Gene conversion** occurs at high frequency during meiosis and at low frequency in mitosis. It is a process by which a DNA sequence is copied from one DNA helix to another DNA helix, whose sequence is altered. Gene conversion leading to non-Mendelian inheritance is recorded in fungal crosses.

## 8.17 Founder's Effect & Bottleneck Effect

When a small population migrates to new area, frequency of genes is determined by genetic drift. American Indians have no B blood group. In Asia, the ancestral home of American Indians, B group is prevalent.

Ancestral mongoloids that migrated across Bering strait to North America might have been small with all types of blood groups, but due to genetic drift, O group became fixed and B group removed by chance. **Bottleneck effect** is found in animal that exhibits seasonal cycle of dormancy. During breeding season, their population size are large but in adverse climate majority of individuals are killed, few find shelters and undergo diapauses. This small population produces next generation by genetic drift. When a population contracts to significant small size in a brief period due to some random environmental change it is called population bottleneck, which result in radical changes in allele frequencies. Impact of population bottleneck can be sustained, even if bottleneck is caused by event like natural catastrophe. After a bottleneck inbreeding increases causing inbreeding depression. Following a population bottleneck rapid and radical decline in population size reduce the population's genetic variation. These mutations are selected against, causing loss of genetical linked alleles through background selection. This reduces genetic diversity. Sustained reduction in population size increases the allele fluctuations from drift in generations to come. A population's genetic variation could be reduced by bottleneck, and even beneficial adaptations may be permanently removed. Loss of variation leaves the surviving population vulnerable to new selection pressures like disease, climate change, shifts in available food source, as adaptation to environmental changes requires sufficient genetic variation in population for natural selection to take place.

There were many recorded cases of population bottleneck in recent past. Before arrival of Europeans, North American prairies were habitat for greater prairie chickens. In Illinois, numbers plummeted from about 100 million in 1900 to about 50 birds in 1990s.

Population declined from hunting and habitat destruction with loss of genetic diversity. DNA analysis comparing birds from mid century to birds in 1990s documents a steep decline in genetic variation in latter few decades. Now greater prairie chicken faces low reproductive success. Over-hunting caused a severe population bottleneck in northern elephant seal in 19th century. Their decline in genetic variation can be deduced by comparing it to that of southern elephant seal, which were not aggressively hunted.

**Founder effect**, a special case of population bottleneck, occurs when a small group in population splinters of from original population and forms a new one. Random sample of alleles in new colony is expected to misrepresent grossly the original population in some respects. It is even possible that numbers of alleles for some genes in original population is larger than number of gene copies in founders, making complete representation impossible. When a newly formed colony is small, its founders strongly affect population's genetic make-up far into future. An example is found in Amish migration to Pennsylvania in 1744. Two members of new colony shared. When very few members of a population migrate to form a separate new population, the founder effect occurs. For a period after foundation, small population experiences intensive drift. This results in fixation of the red allele. Members of colony and their descendants tend to be religious isolates and remain relatively insular. As a result of many generations of inbreeding, Elis-van Creveld syndrome is now much more prevalent among Amish than in general population.

The loss of genetic variation due to founder effect was first outlined by Mayr in 1942, using existing theoretical work of Wright. As a result of loss of genetic variation, new population may be clearly different, both genetically and phonotypicaly, from parent population (Figure 8.5). In extreme cases, founder effect is thought to lead to speciation and evolution of new species. A population bottleneck may cause a founder effect even though it is not strictly a new population. New population is often a very small population and exhibits increased sensitivity to genetic drift, an increase in inbreeding, and relatively low genetic variation. This could be observed in limited gene pool of Iceland–Faroe Islands, Easter Islanders and those native to Pitcairn Island. The legendarily high deaf population of Martha's Vineyard is another example which resulted in development of Martha's Vineyard Sign Language.

## Serial founder effect

**Serial founder effects** occur when populations migrate over long distances typically involving relatively rapid movements followed by periods of settlement. Populations in each migration carry a subset of genetic diversity from previous migrations. Consequently genetic differentiation tends to increase regarding geographic distance. Migration of human out of Africa is featured by series of founder effects. Africa has the highest degree of genetic diversity, in consistence with an African origin of modern humans. After initial migration from Africa, the Indian subcontinent was the first major settling point for modern humans. India has the second highest genetic diversity in world. Genetic diversity of the Indian continent is a subset of Africa, and genetic diversity outside Africa is a subset of India (Figure 8.6).

## Founder effects in island ecology and in human populations

Due to various migrations throughout human history, founder effects are common in humans in various times and places. The French Canadians of Quebec exhibits a classic example of founder population. Over 150 years of

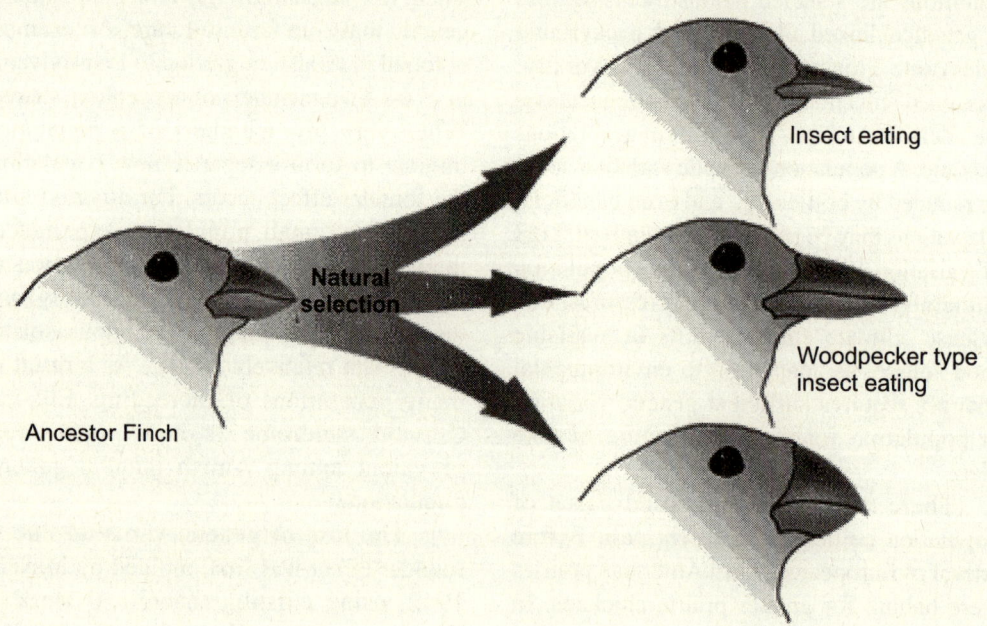

Insect eating

Natural selection

Woodpecker type insect eating

Ancestor Finch

**Figure 8.5** Variation best adapted to their surroundings survive

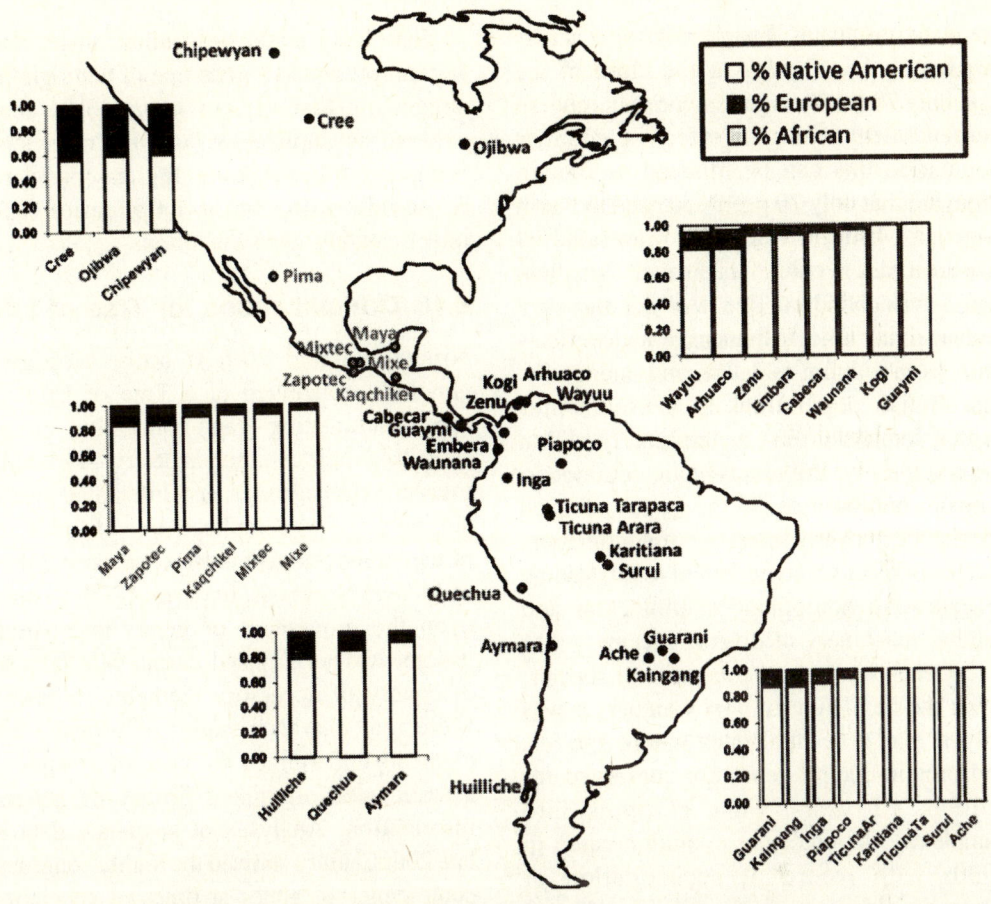

Map showing locations and African, European, and Native American ancestry proportions for 29 Native American populations.

**Figure 8.6** Latest admixture causing spvrions serial boarder effect (dienekes.blogspit.in/2011/09/labrot)

French colonization, it was estimated that 8,500 pioneer's married and left at least one descendent on territory. Following the takeover of colony by British crown, descendents of French settlers continued to grow in number because of high fertility rate. Intermarriage occurred mostly with deported Acadians and migrants from the British Isles. Since the 20th century, immigration in Quebec and mixing of French Canadians involve people from all over world. It is estimated that French Canadians of Quebec have ancestors of 6, 75 different origins on average. Genetic contribution of French founders is predominant; explaining ~90% of regional gene pools, while Acadians explain 4%, British 2% and Native American and other group contributed less. Founder effects occur naturally as competing genetic lines die out. An effective founder population consists only those whose genetic print is identifiable in subsequent populations. Because in sexual reproduction, genetic recombination ensures that with each generation, only half the genetic material of a parent is found in offspring, some genetic lines may die out entirely, even though there are many progenies. People migrating across the Bering land bridge at the close of ice age, only 70 left their genetic print in modern descendants, a minute effective founder population-this can be misread as though implying that only 70 people crossed to North America. Misinterpretations of Mitochondrial Eve are a case in point. It is difficult to explain that a mitochondrial Eve was not the only woman of her time. In humans, founder effects arise from cultural isolation and endogamy. The Amish populations in United States exhibit founder effects as they marry within the community. **Polydactyly** are common in Amish communities. If the genes of each species are perfectly tuned to present function, mechanistic convergence should result, leading to extensive homoplasy in molecular and cellular machinery of diverse species.

Genes were viewed as traits of species. From the beginning of 21st Century, it was known that gene duplicates within species, and protein-coding genes are conserved for millions of years. First glimpses into complexity of molecular evolution came in 1960s with record of accumulation of sequences of proteins from various organisms. Knowledge of sequences of proteins added a new hierarchical level for study. Proteins were then viewed not simply as characters of species. Amino acids could be treated as constituent characters of evolving proteins. Amino acid sequences were used to trace species history. Proteins also had their evolutionary histories - sometimes duplicating separately within a species with both copies persisting indefinitely. **Hemoglobin** was a key molecule for the discovery of history and complexity of protein evolution. Along with cytochrome C, hemoglobin was the first protein with amino acid sequence information from multiple species. These proteins showed deep homology across taxa separated by million years of evolution, showing that phylogenetic history

at gene level could be studied on its own. Ingram presented a gene tree of hemoglobin, suggesting that various hemoglobin chain evolved by duplication, and that myoglobin is paralog of hemoglobin. Ingram considered it novel, with major implications for understanding gene evolution.

## 8.18 Complications for Tree of Life

Nineteenth and 20th Century biologists generally conceived of a Tree of Life - a mostly bifurcating graph connecting species reflecting their common ancestry. Horizontal transfer, symbiogenesis, and differential lineage sorting of genes complicate tree of life. Each of these processes is fundamental assumptions of Modern Synthesis. In the mid 20th Century even the movement of genes in a single genome was not accepted, despite McClintock's discovery of accessory elements in maize. Molecular characterization of transposable elements undermined the view of genome as a static well-organized library of genetic information. Analyses of sequence data in late 20th Century showed the highly congruent coalescence of genes at times of speciation events. Coalescence times of alleles among species are highly variable. Species tree and gene tree cannot be equated. This phenomenon complicates the tree of life. **Tree of life** itself is hierarchical. A universal tree of species is largely a human-imposed ideal as components of any species have evolutionary histories that are not congruent with one another.

## 8.19 Genomic Elements of New Biology

The complete sequences of genomes that were first made available make interdependence of biological disciplines clear. Genomes rarely are tidy libraries of biochemical instructions for making cells, nor are abstract assemblages of numerous alleles of small effect. Genomes show the imprint of accidents in evolutionary history, biochemical constraints and selection laden with mechanistic and historical detail. Genomes are not universally elegant in their construction. Their elaborate detail implicates biochemical, organismal, ecological, cellular, and evolutionary machinery simultaneously.

New genomic foundations of biology are not nearly as convenient as Modern Synthesis:

- Genomes can have abundant DNA sequences without apparent functional benefit to organism.
- Much genomic DNA arises from proliferation of DNA sequences that evolves to proliferate within genomes, not benefit organisms.
- Protein-coding DNA sequences are often phylogenetically ancient, of far greater age than species that bear them.
- Genomes change rapidly due to selection operating on multiple levels simultaneously, and processes of mutation, recombination and transposition.
- Because genome is a complex and shifting patchwork subject to many evolutionary and biochemical constraints and pressures, simple models of cellular or organismal function will often fail.

## Genomic tools of new biology

Biologists discovered the lacunae of Modern Synthesis through employing same tools that were employed to build 21st Century biology. We start with obvious genomic tools and proceed to those that have received less attention.

- Rapid DNA sequencing is the key technology that undermined Modern Synthesis by revealing complexity and variety of genomes.
- Massive parallel assays of gene expression, from mRNA production to

protein level, have revealed interconnected gene networks on which cellular and organismal functions are based. These data undermined 20th Century notion of simple pathways of gene-enzyme determination for most biological processes, favoring network concept of biological machinery.

- Phylogenetic bioinformatics infer the sequence changes of nucleic acids and proteins with proper statistical validity, disclosing unity of biochemical machinery of life and speed at which that machinery evolves.
- **Quantitative-genetics** and genomic mapping are combining to build a genetics that move from organism to organism with greater speed and power than old model organism and single mutant genetics of 20th Century.
- **Molecular ecology** put DNA sequence variation and ecological processes together to increase the power of ecological research, and have revealed high levels of complexity and species diversity, especially microbial, underlying ecological phenomena.
- Large-scale mutagenesis, RNAi and other gene expression modifications, and experimental evolution complement genomic mapping in unraveling of gene networks, particularly by probing biological systems for their causal controls.

## 8.20 Study of Ageing

Around 1900, Metchnikoff tried to explain aging in terms of **autotoxification** effects of interaction between intestinal bacteria and immune system which was refuted by persistence of aging in bacteria-free rodents. Bidder tried to explain aging in terms of determinate adult body size, as opposed to absence of aging that he assumed in fish with indeterminate, indefinite, adult body growth. Comfort refuted this theory. Explaining aging in terms of general molecular mechanisms is well-established. Aging is ubiquitous among species, organs and tissues within individual animals, particularly the well-studied mammalians, like humans, rodents, and dogs. Szilard proposed that aging results from an accumulation of mutations in somatic cells. Orgel hypothesized that aging arises from positive feedback of errors in translation machinery, whereby errors in synthesis of components of machinery, like amino acyl tRNA synthetases, lead to further errors in synthesis of same components. These ingenious cell-level theories were complemented by Hayflick's finding of finite cell replication in vertebrate cells which allowed to divide without limit in vitro. Some metazoan cells, like sea anemones and some *Hydra*, show no universal tendency to age. Thus molecular mechanism which limit in vitro cell proliferation in humans often enhance organismal survival and function by impeding the establishment and spread of malignant tumors. Despite considerable effort devoted to find evidence for somatic mutations and error catastrophe mechanisms of cell aging, it was found that cellular translation machinery maintain its accuracy and somatic mutations are relatively limited in their damaging effects, the most significant debilitating effects of somatic mutation which lead to proliferate malignancies in vertebrates.

W.D. Hamilton supplied the analysis of age-dependent weakening of forces of Natural Selection which was placed on foundations by Charlesworth. Absence of aging in fissile organisms could be explained in absence of decreases in forces of natural selection that arise when reproduction proceeds by symmetrical fission. Much of comparative biology of aging fits within framework supplied by theory of Hamilton and Charles worth (http://www.biologydirect.com/content/2/1/30). Demonstrable mechanisms of aging in *Drosophila* are different from molecular mechanisms, revolving around stress resistance, investment in reproduction, and metabolic reserves. In 1992, experiments with dipteran cohorts showed that mortality rates did not indefinitely accelerate upward with adult age. Experiments with *Drosophila* proved that aging ceases with respect to both survival and fecundity, with sustained late-life plateau of stable health that can be manipulated in experimental evolution. At the start of 21st Century, the foundations of research on aging are different from those assumed by dominant paradigm of previous century (http://www.ncbi.nlm.nih.gov/pmc/articles/PMC2222615/)

## 8.21 Recombination , a by-product of Normal DNA Repair

The important point brought to debate was the intimate relationship between molecular machinery of recombination and **DNA repair**. DNA repair is the most fundamental needs of all organisms. Side-effect of double-strand break repair in cells with homologous chromosomes might be recombination. Selection for maintenance of chromosome would suffice. As of the first years of 21st Century, evidence that chromosomal recombination is a by-product of DNA repair has continued to grow (Figure 8.7).

## 8.22 Homology and Evo-devo: Ancient Genetic Toolkits for Development

In 1980s and 1990s, deep homologies in body patterning genes like Hox genes were observed. Organismal structures like segmented bodies, eyes, limbs, and hearts, were evolved

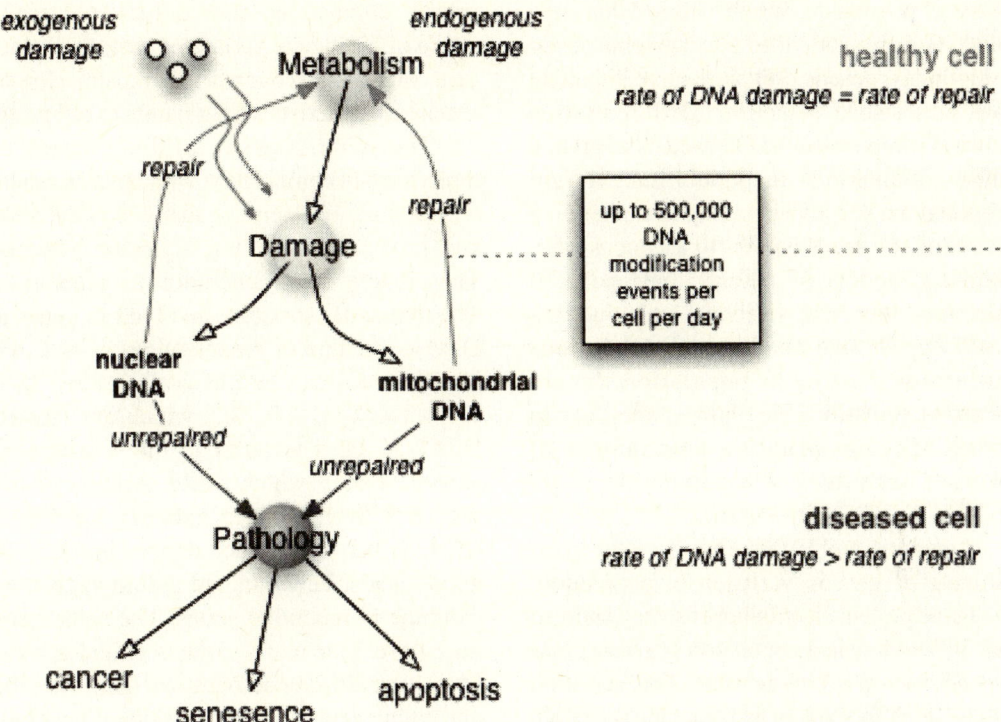

**Figure 8.7** DNA repair rate is an important determinant of cell pathology

de novo, multiple times, independently in various lineages in close conformity with need of function. The common idea was that organismal structures were simply traits of species. Species might have a specific morphological trait for particular function, or they might not have that trait if that function was not selectively favored. Morphological traits are complex patchwork of shared and derived elements, and are more analogous to baroque ornamentation. It is assumed that distinct developmental processes arose separately in different lineages especially when comparing different phyla with different body plans. However, the discovery of conserved developmental genetic processes for patterning the bodies of disparate organisms forced to consider common descent at deeper levels of biological organization.

Frequency of an allele would not change over time due to its rareness or commonness as shown by Hardy and Weinberg. They assume that reproduction of all alleles at same rate requires large population size and allele does not change in form. Mendel's laws explain continual traits when expression of such traits is caused by many genes. Fisher, Wright and Haldane founded population genetics to measure genetic variation in populations. Fisher studied the effect of natural selection in large population and showed that even small selective differences in alleles cause appreciable changes in allele frequencies over time. Rate of adaptive change in population is proportional to amount of available genetic variation. This is called Fisher's Fundamental Theorem of Natural Selection. The rate at which natural selection brings about adaptation depends on functioning of selection. Rarely, natural selection cause decline in mean relative fitness of population. Wright stressed that large population are subdivided into subpopulations. In his theory, genetic drift plays more important role rather than selection. Differentiation between sub-populations followed by migration causes adaptations in population. Wright proposed the adaptive landscape by extending results of one-locus models to a 2-locus case. Haldane models of natural and artificial selection show that selection and mutation could oppose one another and **deleterious mutations** remain in population due to recurrent mutation. He showed the cost to natural selection, placing a limit number of adaptive substitutions a population could undergo in given time span.

Lewontin and Hubby first gave the good estimate of genetic variation in population. Using the protein electrophoresis, they showed that 30% of loci in a population of *Drosophila pseudoobscura* is polymorphic. This variation was too high for explain by balancing selection. Motoo Kimura advocated that most variation in populations is selectively equivalent. Multiple alleles at a locus varied in sequence, but their finesses are the same. Kimura's **neutral theory** described rates of evolution and levels of polymorphism (http://en.mimi.hu/biology/multiple_alleles.html) in terms of mutation and genetic drift. Theory accepts that natural selection act on natural populations; but most natural variation is transient polymorphisms of neutral alleles. Selection does not influence rates of evolution or levels of polymorphism. Variety of observations seemed to be consistent with neutral theory. Several lines of evidence speak about less variation in natural populations than neutral theory predicts. There is much variance in rates of substitutions in various lineages to be explained by mutation and drift alone. Selection has impact on levels of nucleotide variation. Mathematical theory of evolution for exact prediction of rates of evolution and levels of heterozygosity in natural populations does not exist. Darwin spoke of biological variations, with indistinct idea. Mendel discovered the transmission of variant characteristics from parent to offspring.

The gene consists of long chain of molecules. In combination with environmental conditions they determine many physical traits and processes including behavior patterns. They inherit from generation to generation. The molecular structure involved in genes is DNA consisting of 4 nucleotides -- A, C, G, T. The latter can be physically paired in 4 ways TA, CG, AT, GC, which are labeled U, C, A, G. The latter is the 4 letters of genetic coding which could combine in 64 sets of 3 letters (called codons), i.e. UUU, UCA, GAU, CGA. Such triplets produce 20 amino acids. Few hundred amino acids may combine in repetitive series. The same gene may have 2 or more variants called alleles. Molecular difference between them may be one amino acid in a chain of several hundred. There is one allele for purple flowers and another for white. When these genes are brought together in reproduction, they cause many variations between individuals. Gene is stable and reproduces predictably. Mutation occurs in genes rarely. Genetic mutation may consist of substitution, deletion or addition of a single letter of genetic code. Mutation changes the nature and effect of gene. The genetic code is universal. Rare exceptions to the standard codon dictionery are bound. AUV, AUC, AVA and AUG are intiating codons. If mutation occurs in sex cells, mutant gene is transmitted to offspring.

If original gene is O and mutant gene is M, mutation is expressed in a proposition of form "gene O becomes gene M". If we take O and M to be 'the same' gene K, in sense that both refer to genetic key to specific biological trait or process, without specifying precise variation, change from O to M could be explained as change of K. **Genetic mutation** is presumed to imply the form of logical mutation, such that an individual gene (O) has itself physically become another individual gene (M). Mutant gene M might then go on and reproduce faithful copies of it. It would be conceivable for genes concerned (O and M) to be different individuals, former causing latter by faulty duplication process and 2 coexisting for sometime. Thus genetic mutation would not imply logical mutation, but form of reproduction not implying physical continuity between parent gene and its immediate offspring.

## 8.23 Explaining Variation

Molecular biology deals with mechanisms common to all living things. Cellular machinery that copies DNA make mistakes which alter gene sequence is called mutation. There are many types of mutations. In point mutation, one letter of genetic code is changed to another. Lengths of DNA can be deleted or inserted in a gene- these are also mutations. Genes or parts of genes can become inverted or duplicated. Typical rates of mutation are between $10^{-10}$ and $10^{-12}$ mutations per base pair of DNA per generation. Most mutations are neutral regarding fitness (http://www.scribd.com/doc/14970536/Introduction-to-Evolutionary-Biology).Only a small portion of genome of eukaryotes contains coding segments. Although some non-coding DNA is involved in gene regulation or other cellular functions, most base changes would have no fitness consequence. Mutations with phenotypic effect are deleterious. Mutations that result in amino acid substitutions could change the shape of protein. Changing or eliminating protein function lead to inadequacies in biochemical pathways or interfere with process of development. Very small percentage of mutations is beneficial. Ratio of neutral to deleterious to beneficial mutations is unknown.

Mutation limits the rate of evolution which could be expressed in terms of nucleotide substitutions in a lineage per generation. Substitution is the replacement of an allele by another in a population (http://reformation.edu/scripture-science-stott/evolution/pages/005-introduction-evolutionary-biology.htm). This is a 2 step process. First a mutation occurs in an individual, creating new allele which increases in frequency to fixation in population. Rate of evolution is $k = 2Nvu$ (in diploids) where k= nucleotide substitutions, N = effective population size, v = rate of mutation and u = proportion of mutants that eventually fix in population. A change in environment cause previously neutral alleles to acquire selective values. In the short term evolution run on stored variation and is independent of mutation rate. Other mechanism also contributes in selectable variation. Recombination produces new combination of alleles by joining sequence with separate micro evolutionary history in a population. Gene flow also provides the gene pool with variants. The final source of such variants is the mutation.

## 8.24 Development of Novel Traits

Novel traits, the result of mutations to DNA that affect individual characteristics were absent in a lineage of individuals in past. Development of such traits has observed by animal and plant breeders. Whether a trait is new in wild is difficult to determine. Novel traits could be studied in Cave populations. Many novel traits are identified in plants and animals which are essential for domestic breeding. Breeders specifically selected fancy pigeons which are decedents of common pigeon for their radical physical features. Breeders identify new traits in breeding populations, showing the present existence of novel traits.

## Mutant alleles

Mutation produces new alleles which enter gene pool as single copy among many. Most are lost from gene pool. Organism carrying mutation fails to reproduce, or reproduces but does not transmit the specific allele. Fate of mutant is shared with genetic background where it appears. New allele initially links to other loci in its genetic background, or even loci on other chromosome. If allele frequency increases in population, initially it pairs with other alleles at that locus. New allele is primarily carried in individuals which are **heterozygous** for that locus. Chance of pairing with itself is low till it attains intermediate frequency. In recessive allele, its effect is not found until a **homozygous** individual is formed. The ultimate fate of allele depends on whether it is deleterious, beneficial or neutral.

## Neutral alleles

Most neutral alleles disappear soon after their appearance. Mean time until loss of a neutral allele is estimated 2(Ne/N) 1n (2N) where Ne = effective population size, N= total population size. Small percentage of alleles fix. Fixation is the process of an increasing allele frequency at, or near one. Probability of fixation of neutral allele in a population is equal to its frequency. For a new mutant in diploid population, this frequency is 1/2N. When mutations are neutral in respect to fitness, substitution rate (k) is equal to mutation rate (v). It does not imply every new mutant finally reaches fixation. Alleles are included in gene pool by mutation at same rate they are lost to drift. **Neutral alleles** fix in an average of 4N generations. There are multiple alleles segregating in the population at equilibrium. Few mutations emerge each generation in small population. The ones that fix do so rapidly relative to large populations.

In large population, many mutants emerge over generations. The ones that fix take longer for fixation. Rate of neutral evolution does not dependent on population size. Rate of mutation rate determines heterozygosity level at a locus as per neutral theory. **Heterozygosity** is the proportion of population that is heterozygous. Equilibrium heterozygosity is given as H = 4Nv/(www.talkorigins.org/faqs/faq-intro-to-biology.html). H varies from very small number to about one. H is small in small populations. Heterozygosity approaches one in large population. Direct testing this model is difficult as N and v can be estimated for most natural populations. Heterozygosities are too low to be described by strictly neutral model. Neutralists solved this discrepancy hypothesizing that natural population may not be at equilibrium. There should be few alleles at equilibrium at intermediate frequency and many at very low frequency. This is called Ewens Watterson distribution. New alleles enter population each generation, most exist at low frequency until they are lost. Few drift to intermediate frequencies, very few drift all the way to fixation (www.talkorigins.org/faqs/faq-intro-to-biology.html). In *Drosophila pseudoobscura*, the protein Xanthine dehydrogenase (Xdh) has many variants (Pitts and Zwiebel, 2001). In a single population, Keith et. al., found that 59 of 96 proteins belong to one type, 2 others were represented 10 and 9 times and 9 other types were present singly or in low numbers (www.talkorigins.org/faqs/faq-intro-to-biology.html.)

## Deleterious alleles

**Deleterious mutant** is selected against but exist at low frequency in gene pool. Deleterious recessive mutant increase in frequency due to drift in diploid organisms. Selection cannot observe it when a dominant allele masks it.

Alleles causing diseases exist at low frequency. People who are carriers do not suffer from the negative effect of allele. Until they mate with another carrier, allele may continue to be passed on. Deleterious alleles exist in populations at low frequency for the balance between recurrent mutation and selection. This is termed mutation load.

## Beneficial alleles

Most new mutants even beneficial ones are lost. Probability of fixation of a **beneficial allele** is 2s as calculated by Wright. Allele that conferred 1% increase in fitness only has a 2 % chance of fixation. Probability of fixation of beneficial but recurrent mutation boosts up mutant. **Beneficial mutant** would lose several times, but it arises finally and sticks in a population. Directional selection reduces genetic variation at selected locus when the fitter allele sweeps for fixation. Sequence linked to selected allele increases in frequency for **hitchhiking**. The lower rate of recombination causes larger window of sequence that hitchhikes. Level of nucleotide polymorphism within and between species with recombination rate at a locus was compared. Low level of **nucleotide polymorphism** in species coincides with low recombination rates. This could be explained by molecular mechanisms if recombination was mutagenic. Here, recombination will be correlated with nucleotide divergence between species. No correlation was found between level of sequence divergence and recombination rate. They ended selection as cause. Correlation between nucleotide polymorphism and recombination ends that selective sweeps occur often to leave an imprint at level of genetic variation in population. Beneficial mutation is found in *Culex pipiens* in which a gene involved with breaking down organophosphates became duplicated. Progeny of mutated organism rapidly swept across worldwide mosquito population. Many examples of insects developing resistance to chemical, especially DDT which once heavily used in this country are recorded. Although good mutations occur less than bad one, organisms with good mutation thrive but organisms with bad mutation die out. If beneficial mutant arises infrequently, fitness difference in a population results due to new deleterious mutants and deleterious recessives. Selection simply weeds out unfit variants. Occasionally a beneficial allele sweeps through a population. Lack of large fitness differences segregating in population argues that beneficial mutant arise infrequently. Effect of a beneficial mutant on level of variation at a locus could be large. Many generations are required for a locus to regain considerable levels of heterozygosity following selective sweep.

## 8.25 Neutral Theory of Molecular Evolution

M. Kimura proposed that most base substitutions preserved in population is neutral in regard to natural selection. Positive substitutions appears to be rare. Negative changes are removed rapidly by natural selection. Natural selection, likely favour neutral changes, which determines the overall rate of sequential evolution. Pseudogenes exhibit highest substitution rate, but changes are absolutely neutral regarding selection. Kimura's theory contrasts classical Darwinism but fails to explain fixation of various types of alleles in various sizes of population. It advocates that rate of fixation of neutral mutations is not dependent on population size but genes are eliminated, or fixed by genetic drift. Theoretical framework for testing and predicting **molecular evolution** in absence of positive selection are given by the neutral theory (Figures 8.8-8.11).

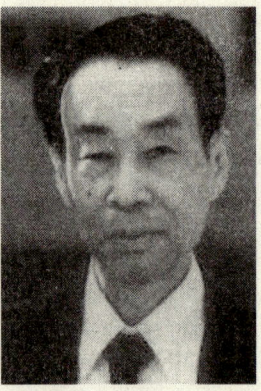

**Figure 8.8** Motto Kimura.

**Motoo Kimura:**
- troubled by cost Haldane's dilemma:
  - 1 substitution every 300 generations
- troubled by Zukerkandi and Pauling's (1965) molecular clock:
  - 2 substitution every 2 year

Published a model of neutral evolution in 1968

King and Jukes:
Independentfy arrived at same conclusion as Kimura
Published (1969) under the "Non-Darwinian evolution"

**Figure 8.9** Neutral theory of molecular evolution

*k* = rate of nucleotide subsitution at a site per generation [year]

*k* = new mutations × probability of fixation

Number of new mulations = $\mu \times 2N_o$

Prebability of fixation = $\frac{1}{2}N_o$

Hence, the neutral rate is:

$$k = \mu \times 2No \times \tfrac{1}{2}N_o$$

$$k = \mu$$

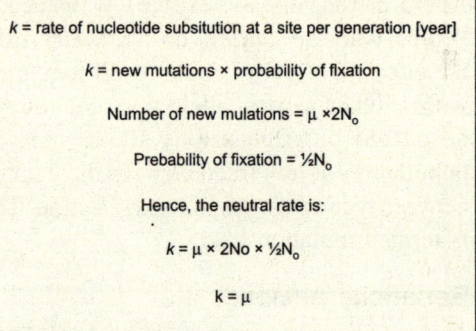

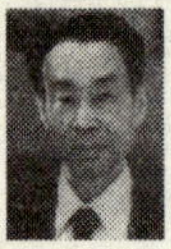

**Figure 8.10** Neutral theory elegant simplicity

## 8.26 Molecular Clock

Zuckerkandl and Pauling observed that divergence rate in amino acid sequences of haemoglobin and cytochrome c in various species pairs in mammals always remain same. Changes in base sequences of DNA and substitution of resultant amino acid accumulates in clock-like regularity in population over time branching evolution. This is called molecular clock, which before using

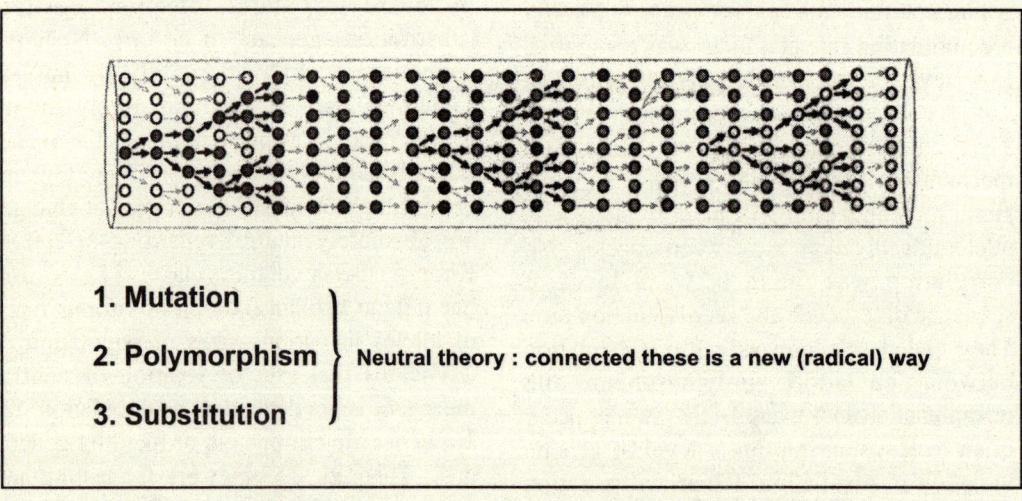

**Figure 8.11** Diagramnetic view of Neutral theory
(awarnach.mathstat.dal.ca/~job/3046/PDFs/slides/slides_neutral_T2.pdf.)

require calibration with matching genetic divergence of living population with absolute time of divergence as found in fossil records. Differentiation in haemolymph proteins of Hawaiian drosophilids shows split of phyletic lines and colonization of Hawaiian Islands about 40 mya from North America by these flies. Using this method Sarich & Wilson estimated the divergence of hominids from apes by calibrating the amount of molecular differentiation achieved between 2 groups in relation to time by taking example of divergence of Old World and New World monkeys (http://www.ncbi.nlm.nih.gov/books/NBK21122/). Measurement of divergence in albumin gave the time of split of hominids and apes at 5 million years from present, which is supported by other evidences. The following 3 types of changes are considered for molecular clock.

The base sequences in which substitution occur in 3rd position of codon, which are expected to be neutral. As changes in **pseudogenes** are not subjected to natural selection they are likely to give good result. When natural selection is very strong, neutral substitutions are likely to be fixed only in population. Linear and constant changes in mtDNA are commonly used in molecular clock. The divergence is about 2% per million years in mtDNA of mammals. In sea urchins, this rate varies between 1.8-2.2%, which is similar to mammals. Sharks have reliable molecular clock and rate of change is 8 times slower than in mammals. Universal molecular clock even for mtDNA of animals does not exist .Substitution rates varies in genes in various species due to differences in mutation rate and generation time. The same molecule is known to evolve at various rates in various evolutionary lines. Introns, **transposons**, regulatory genes, and gene families may show considerable deviations in their divergence rates. The clock might work in a specific lineage but not work in others, or might work at different rates in various lineages.

## SUMMARY

1. Mendel's laws explain continual traits if expression of such traits are caused by many genes. Fisher, Wright and Haldane founded population genetics to measure and explain the levels of genetic variation in populations. Fisher studied the effect of natural selection on large population and showed that even very small selective differences among alleles cause appreciable changes in allele frequencies over time. Rate of adaptive change in population is proportional to amount of available genetic variation.

2. Kimura's neutral theory described rates of evolution and levels of polymorphism solely in terms of mutation and genetic drift. The theory accepts that natural selection act on natural populations; but most natural variation is transient polymorphisms of neutral alleles. Selection does not influence rates of evolution or levels of polymorphism.

3. Selection has impact on levels of nucleotide variation. Mathematical theory of evolution for exact prediction of the rates of evolution and levels of heterozygosity in natural populations does not exist. Darwin spoke of biological variations, with indistinct idea. Gene is stable and reproduces predictably. Mutation occurs in genes rarely. Genetic mutation may consist of substitution, deletion or addition of a single letter of genetic code. However, mutation changes the nature and effect of the gene.

4. The cellular machinery that copies DNA sometimes make mistakes which alter the gene sequence is called mutation. There are many kinds of mutations. In point mutation, one letter of the genetic code is changed to another. Lengths of DNA can be deleted or inserted in a gene; these are also mutations. Typical rates of mutation vary. Most mutations are neutral regarding fitness (http://mt1.cyclingforums.com/showthread.php?p=877852)

5. A new allele will initially be linked to other loci in its genetic background, even loci on other chromosomes. If the allele increases in frequency in population, initially it will be paired with other alleles at that locus — the new allele will primarily be carried in individuals heterozygous for that locus.

6. Deleterious alleles also remain in populations at a low frequency due to a balance between recurrent mutation and selection. This is called the mutation load.

7. Directional selection depletes genetic variation at the selected locus as the fitter allele sweeps to fixation. Sequences linked to the selected allele also increase in frequency due to hitchhiking. Founder effects are a special case of genetic drift, occurring when a small group in a population splinter of from the original population and forms a new one. New colony may have less genetic variation than original population, and through random sampling of alleles during reproduction of subsequent generations continue rapidly towards fixation.

8. The variation in gene frequency between the original population and colony may also trigger the 2 groups to diverge significantly over many generations. As the variance, or genetic distance, increases, the 2 separated populations may become distinctively different, both genetical, and phenotypically, although not only genetic drift but also natural selection, gene flow and mutation will contributes to this divergence.

9. Founder effect is a significant driving force in the evolution of new species. Human haemogobin gene cluster is called gene family or multigene family that consists of 2 alpha duplicate tandem genes and 7 beta genes. Globin genes in mammals form an excellent example of gene families.

10. Immunoglobin gene family is a branching lineage of duplicated genes and T-cell receptor genes produce specific reaction against huge diversity of viruses and bacteria that invade our body. Most actual breeding populations of organisms are not large enough to ensure that there will be no changes in allele frequencies as result of purely random accidents.

11. A form of evolution, called genetic drift occurs whenever populations are small enough for random accidents to cause changes in allele frequencies. This process called the Sewall Wright effect. Survival is virtually never random. Non-random survival is just another name for natural selection, Darwin's original engine of evolutionary change.

12. Sewall Wright discovered the random genetic drift and proposed that in small population of organisms, random sampling errors could cause significant changes in allele frequencies in those populations. He showed mathematically that in smaller population the greater the effect of such random events on allele frequencies. Evolution could proceed by mechanisms of Darwinian natural selection and random genetic drift.

13. Wright then went onto develop a formal model for genetic evolution in which allele frequencies present in population were visualized as an adaptive landscape. Mutation is a change in a gene. These changes are the source of new genetic variation. Natural selection operates on this variation. Genetic variation has 2 components allelic diversity and non- random associations of alleles. Alleles are different versions of the same gene.

14. Gene flow is the transfer of alleles from one population to another population through immigration of individuals. Maintained gene flows between 2 populations can lead to a combination of the 2 gene pools, reducing the genetic variation between the 2 groups. Gene flow can occur between species, either through hybridization or gene transfer from bacteria or virus to new hosts.

15. Gene transfer, defined as the movement of genetic material across species boundaries includes horizontal gene transfer, antigenic shift, and reassortment is sometimes an important source of genetic variation.

16. Viruses can transfer genes between species. Bacteria can incorporate genes from other dead bacteria, exchange genes with living bacteria, and can exchange plasmids across species boundaries. Sequence comparisons suggest recent horizontal transfer of many genes among diverse species including across the boundaries of phylogenetic domains. Determining the phylogenetic history of a species cannot be done conclusively by determining evolutionary trees for single genes

## REVIEW QUESTIONS

### Short Answer Questions

1. What do you understand by random mutation?
2. Define founder mutation.
3. What is linkage disequilibrium?
4. Define gene flow.
5. What is genetic pollution?

### Long Answer Questions

1. Describe the process of genetic recombination with suitable illustrations.
2. Discuss the importance of founder effect in light of modern study.
3. Write a note on molecular clock.

## REFERENCES

Jenkin, Fleeming.1867.Review of the Origin of Species.The North British Review 46 : 277-318.

Kimura, Motoo.1968 Evolutionary rate at the Molecular level. Nature 217 : 624-626.

Lewontin, Richard C. 1970. The Units of Selection. Annual Review of Ecology and Systematics 1 : 1-18.

Pitts, R.J./ and Zwiebel, L.J. 2001.Isolation and characterization of the Xanthine Dehydrogenase gene of the Mediterranean fruit fly, Ceratitis capitata. Genetics, 158: 1645-1655

http://biomed.brown.edu/Courses/BIO48/7.mutation

http://www.talkorigins.org/faqs/faq-intro-to-biology.html

http://www.biologydirect.com/content/2/1/30

http://www.ncbi.nlm.nih.gov/pmc/articles/PMC2222615/

http://en.mimi.hu/biology/multiple_alleles.html

http://www.scribd.com/doc/14970536/Introduction-to-Evolutionary-Biology.

http://www.ncbi.nlm.nih.gov/books/NBK21122/

http://mt1.cyclingforums.com/showthread.php?p=877852

# CHAPTER 9

# Evolution of Primates and Human

Primates, a group of mammals include monkeys, apes, and humans. Their generalized form is not specialized for specific adaptive strategy. **Primates** are found in environmental niche, from the tropics to artic. Three main evolutionary trends of primates are found in their limbs and locomotion, dentition and diet besides cranio-neurology and behavior. They are pentadactyl; possess nails, prehensile hands and feet, and clavicle. They maintain erect posture. They are omnivorous and their generalized dental pattern is designed for varied diet. Their snout is reduced and olfactory senses are ill developed. **Stereoscopic vision** is their marked character besides increased complexity of brain. They exhibit increased parental investment, higher dependency on learned behavior and more group cohesion. Adult males permanently associate with group. Their adaptation to an arboreal niche is mainly responsible for most of their unique evolutionary trends. Dentition, diet, locomotion and limb architecture are due to adaptation of life in trees. Increased dependence on vision,

stereoscopic ability, reduction of snout and ill developed senses are due to arboreal adaptations. Omnivorous diet takes the benefit of various food found in forest. Dental generalization takes advantage of varied dietary possibilities found in trees. Matt Cartmill has proposed the **visual predation hypothesis**, and considers that primates unique adaptations are due to use of low forest canopy niche and reliance on insects and small fruits . This hypothesis fails to explain the development of larger primates, which could not subsist on small insects and fruits.

Modern humans have been originated at Sub-Saharan Africa at around 200,000 years ago. The divergence of *Homo sapiens* and *H. neanderthalensis* from a common ancestor probably occurred 400,000 years ago. The average life span of mammal species is about 10 million years. Hominids diverged from the ape lineage about 7 mya. Archaic *Homo sapiens* appeared about 400,000 years ago. The earliest modern humans date back 170,000 years. Humans and chimpanzees were diverged

7 mya. Our own lineage is bushy with different species at a time. The ancestors of modern humans evolved in Africa and remained there until 1.5 mya ago. Then *Homo erectus* populations left Africa and moved across Europe and Asia. In the late Miocene, the ancestors of the African apes migrated from Eurasia for adaptive radiation into Africa.

Evolutionary history of primates is interesting as it provides the context for human evolution. Comparative study of primates is a well-established discipline. Except tree shrews - commonly classified as primates earlier, but now placed in order Scandentia - about 375 modern primate species are recognized (Groves, 2005).Partly due to geographical distribution, the extant primates belong to 6 natural groups: Madagascar lemurs, lorisiforms, tarsiers, New World monkeys, Old World monkeys, apes and humans(Martin, 1990). The last 3 groups (monkeys, apes and humans) the higher primates Figure 9.1 are distinguished from primitive prosimians (lemurs, lorisiforms and tarsiers). The earliest fossil primates date back to beginning of

Eocene, found in North America, Europe and Asia. The first substantial primate fossil from the earliest Eocene is a about complete skull of *Teilhardina asiatica* known from China (Ni et al., 2004). Fossil mammals identified as archaic primates, known clearly from the Palaeocene, have only a tenuous connection with evolution of "primates of modern aspect" **(euprimates)**. Plesiadapiformes is believed to be branched away before the common ancestor of euprimates. Fossil euprimates can be crudely divided into early Tertiary forms and other fossil species. The latter occur from the early Miocene upward, although exceptional cases dated back to middle Eocene. Most early Tertiary primates are subdivided into lemur-like **Adapiformes** and tarsier-like Omomyiformes (Martin, 1990). Adapiformes are linked to strepsirrhine primates while **Omomyiformes** are linked to halothanes (Fleagle, 1999). Alternatively Adapiformes and Omomyiformes together constitute a separate radiation of early primates with no direct connection to radiation that led to modern primates (Ross, 2003. About 400 fossil

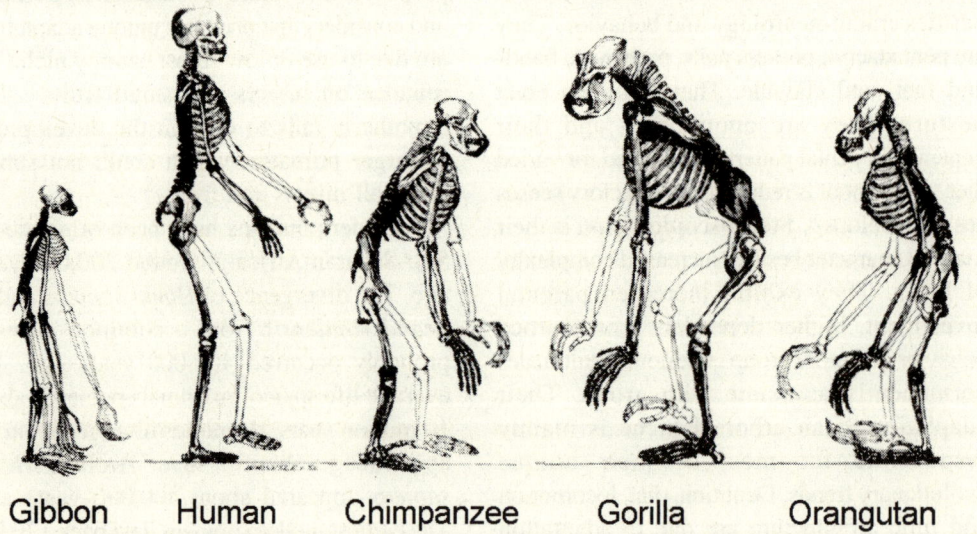

    Gibbon    Human    Chimpanzee    Gorilla    Orangutan

**Figure 9.1**  Skeletons of Gibbon, Human, Chimpanzee, Gorilla, Orangutan.

euprimate species were recognized (Soligo et al., 2006), with various stages of primate evolution over the past 55 my. Broad comparisons including morphology of extant and fossil forms along with chromosomal and molecular evidence for living species have led to consensus to broad outlines of phylogenetic tree of primates. The mainstream palaeontological interpretation suggests that euprimates originated not long before the earliest known fossil forms and not earlier than 65 mya (see Kay et al., 1997). A direct reading

of known euprimate fossil record suggests the site of origin in the northern continents.

## 9.1 Interpretation of Fossil Record

Fossil record (Figure 9.2, 9.3) is important in understanding of phylogenetic history of living organisms. Hypothetical reconstructions of relationships between species can be generated through analysis of characters of living forms. Analyses of molecular data serve to show patterns of branching among lineages and

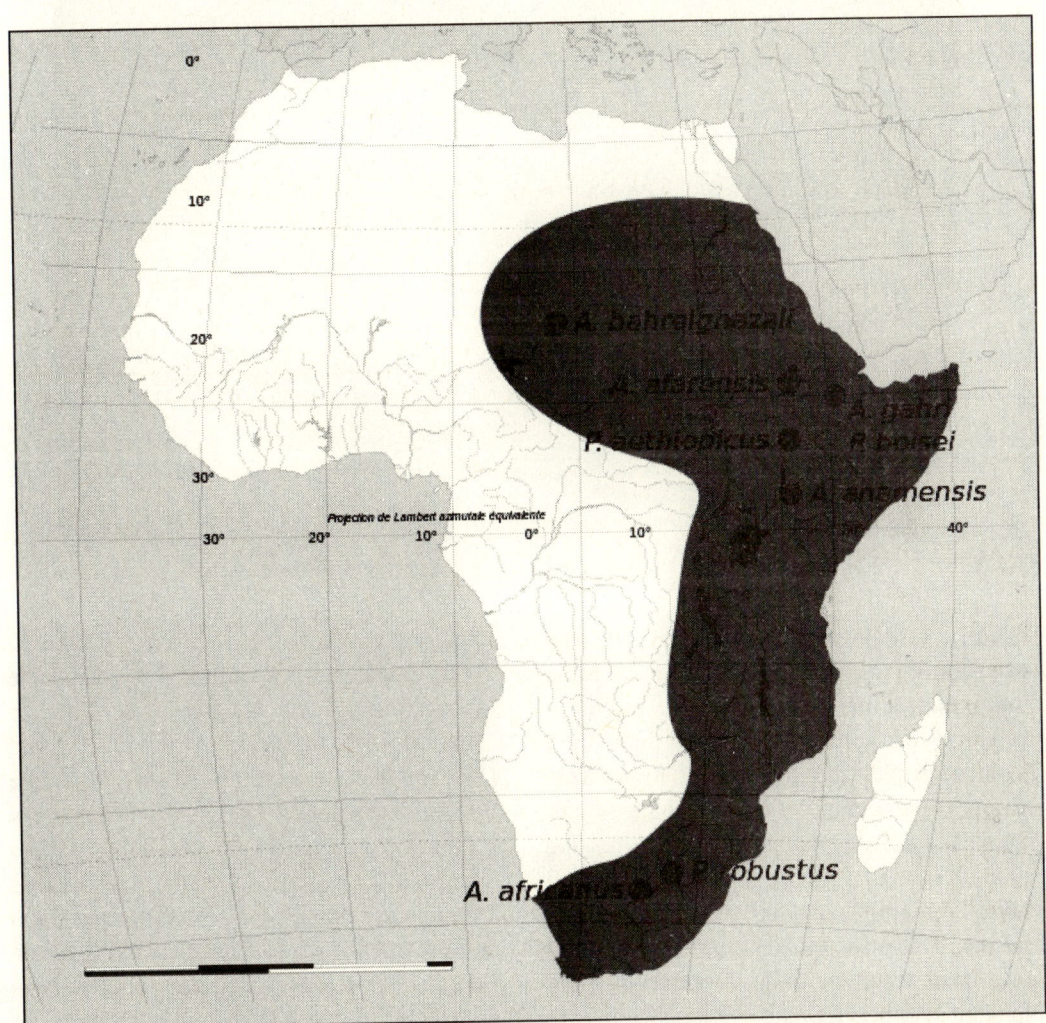

**Figure 9.2** Map of the fossil sites of the early hominids

**Figure 9.3**   Fossil Hominids.

usually tell little about characters and functioning of past organisms. Reliable interpretation of important fossil record is subject to many problems. Many of these result from incompleteness of record, ranging from partial preservation of individual specimens through regional variation in fossilization, and onto existence of major gaps in the **Tree of Life**. Problem of incompleteness afflicts the primate fossil record. On strepsirrhine side of primate tree, few direct fossil relatives of modern members (lemurs and lorisiforms) are known. The known fossil record for undoubted lorisiforms (with a modern 28 species; Groves, 2005) extended back only to early Miocene (about 20 mya) and was limited to 4 genera and 8 species found in East Africa and Pakistan (Phillips & Walker, 2002). Discovery, in late Eocene deposits in Egypt, of fragmentary remains of a relative of bush babies (Saharagalago) and a relative of lorises (Karanisia) doubled the recorded geological age of **Loris** form primates about 40 **mya** (Martin, 2003). New fossil discoveries can have a dramatic impact. Fossil record for Madagascar **lemurs** is less satisfactory. Not a single fossil lemur has been discovered in Madagascar. The existence of the lorisiforms - the sister group of lemurs - by around 40 mya shows that lemurs must have been in existence since at least that time.

An Oligocene primate Bugtilemur, discovered in deposits in Pakistan dating back to about 30 mya, has been linked to dwarf lemur family Cheirogaleidae (Marivaux et al., 2001). The haplorhine side of primate tree is better documented with substantial gaps. Direct relatives of modern tarsiers are limited to one early Miocene species from Thailand, an Oligocene species from Egypt and 2 middle( Martin, 2009) Eocene species from China (Gunnell & Rose, 2002). Higher primates are documented by the late Eocene in Africa, about 40 mya, and questionably by middle Eocene in China and South-East Asia (Miller et al., 2005). Modern higher primates are documented only from the latest Oligocene (for New World monkeys) or from early Miocene. Large gaps remain within these groups. Among the New World monkeys, we do not find fossil relatives of small bodied callitrichids (marmosets, tamarins and Goeldi's monkey), which could account for 43 of the 128 extant species. Among Old World apes not a single direct fossil relative of 14 species of lesser apes (gibbons) has yet been recorded, despite probable separate existence of this lineage for a comparable period of time.

Existence of extensive **gaps in fossil record** of primates gives rise to major problems in interpretation, with respect to determining the time of origin of any given lineage and to infer its geographical site of origin. Because the earliest known primates of modern aspect (euprimates) date back to basal Eocene (about 55 mya), the common practice leads to an inferred origin for euprimates somewhere in Palaeocene, around 60 mya. The earliest known primates always date back to no more than

55 mya, and yet they possess key features of their respective orders.

Extinction of dinosaurs and several groups of organisms was precipitated - or at least highly influenced -by a giant meteorite impact at **K/T boundary** about 65 mya. Many view the adaptive radiation of modern mammals as a sequel to major extinctions at the K/T boundary. Direct inference of times of origin from fossil record is subject to 2 problems :(1) If there are substantial gaps in record, the first known fossil specimen is likely to be younger than first occurrence in phylogenetic tree. (2) Biases of various kinds in fossil record influencing preservation and discovery may entail more error. However, it is important to recognize that any time of origin inferred directly from a first appearance in fossil record must be a minimum date and that there is no equivalent direct indicator of a maximum date. It should be intuitively obvious that degree of underestimation of a time of origin based on time of first appearance in fossil record must increase as patchiness of that record increases. It is expected that a direct reading of that record might lead to serious underestimation of time of origin of primates and times of divergence within primate tree (Martin, 1990). Very simple quantitative approach to incompleteness in primate fossil record (Martin, 1993) was made by assuming a linear increase in number of species over time and a species survival time of 1 my, equivalent to average. Calculation based on this shows that only 3 % of extinct euprimate species have so far been recorded, with about 97 % remain to be discovered. Trees in which only 3 % of extinct species are known revealed that dating of time of origin from the first known fossil form would lead to serious underestimation of actual time of origin (Figure 9.4).

All fossil euprimates species are lumped together in a single figure which is very

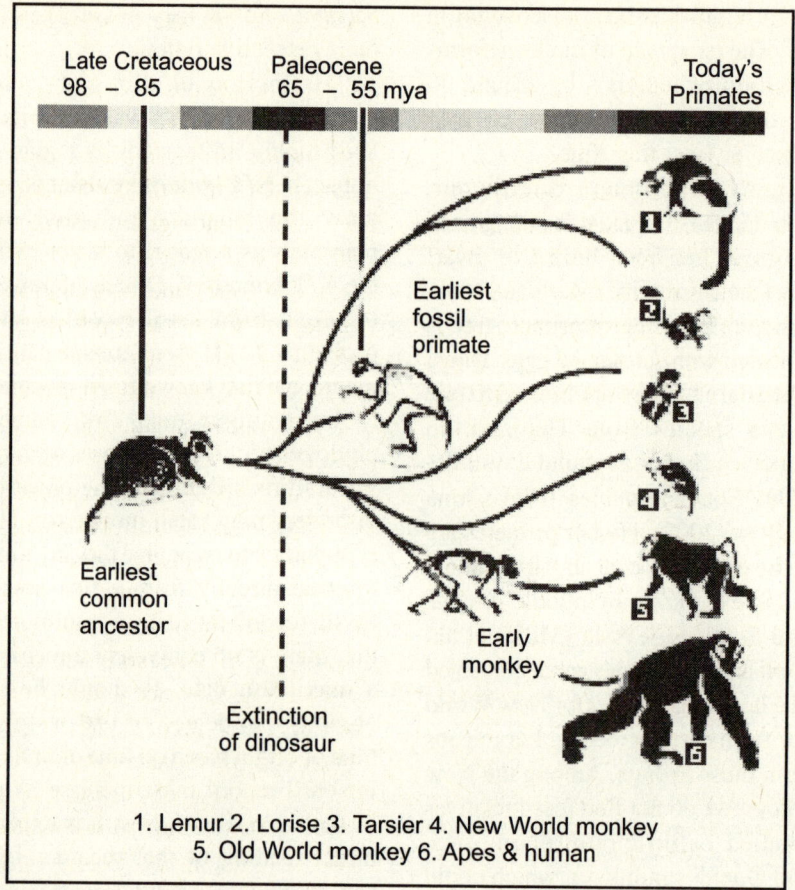

Late Cretaceous    Paleocene                              Today's
98 – 85            65 – 55 mya                            Primates

Earliest
fossil
primate

Earliest
common
ancestor

Early
monkey

Extinction
of dinosaur

1. Lemur 2. Lorise 3. Tarsier 4. New World monkey
5. Old World monkey 6. Apes & human

**Figure 9.4**  New evolutionary tree of primates (Modified after Martin, 2005)

misleading if there are major fluctuations in sampling levels over time. This is best illustrated by considering (Martin, 2009) a period of 6 my in middle of the Oligocene (26-32 mya) for which not a single fossil primate species is known. A direct reading of the fossil record requires a precipitous decline in numbers of species about 32 mya, followed by a rapid re-expansion from 26 mya onwards. There has been particularly poor preservation and/or discovery of primates in the middle Oligocene. The same kind of probability calculation as that used by Gingerich & Uhen (1994) can be applied to assess the likelihood that primates existed during the period 26-32 mya. Instead of simple linear expansion in number of species over time, a more biologically realistic logistic model was taken in which 90 % of modern number of primate species was attained by 49 mya. Effective evaluation of data required subdivision of primate fossil record into narrow time intervals which was done following standard geological subdivisions, with an average interval length of about (Martin, 2009) 4 my. (Tavaré et al., 2002).

It is necessary to distinguish between date of initial divergence of target group from its sister group and subsequent date of initial diversification of extant members of target group from the common ancestor. Either of these dates could be taken as the "time of origin" of a group. Key point is that diversification of extant members of a target group may take place some time after founding lineage diverged from the sister group. It might be argued that primate stem lineage indeed diverged from other placental mammals back in the Cretaceous, but that date of initial diversification of modern primates did not occur until after the K/T boundary. Early relatives of primates might have existed during the latter part of Cretaceous, but they were not recognizable as such because defining features of primates were not present until common ancestor of modern primates emerged at a later stage. This argument simply does not apply to the date of 81.5 mya inferred by Tavaré et al. (2002), because the method estimated the date of initial diversification of extant primates and not the date of initial divergence from other placental mammals.

## 9.2  Recent Findings from Molecular Comparisons

Molecular evidence has added a valuable new dimension to **phylogenetic reconstruction**. More easily quantifiable information on likely relationships between groups of organisms, such evidence has opened new possibilities for the inference of divergence times through application of the concept of the **molecular clock** (Bromham & Penn, 2003). Although it is now evident that rates (Martin, 2009) of molecular evolution can in fact vary markedly between lineages, such that it is necessary to think in terms of **local clocks**. The approximate regularity of molecular (Martin, 2009) change is sufficient to permit crude application of the clock concept. It should not be overlooked that a molecular clock must be calibrated and that the only method currently available requires use of information from the fossil

record to infer the age of at least one node in the tree. If gaps in the fossil record (Martin, 2009) lead to serious underestimation of any divergence date because of direct reliance on first appearance of a fossil relative, that underestimation will have ramifications (Martin, 2009) throughout any molecular tree calibrated with that date. If it is indeed true that the first appearance of a euprimates fossil in the record is considerably younger than the actual age of the last common ancestor of **euprimates** (55 mya versus 81.5 mya or more), and if such underestimation applies throughout the primate record, it follows that calibration of molecular trees using any dates derived from fossil primates is severely misleading. Among other things, this includes inferences regarding the timing of the divergence between African great apes (chimpanzees and gorillas) and humans (see Arnason et al., 2000). Accumulating evidence over the past decade from several studies of DNA sequences using several calibration dates external to primates has, confirmed an early date for the initial divergence between primates and other groups of placental mammals (Soligo et al., 2006). In one of the first such studies (Martin, 2009; Janke et al., 1994) a comparative analysis was conducted using sequence data for a marsupial and several placentals.

After testing revealed that rates of evolution for 8 mitochondrial genes were compatible with a **molecular clock** model, a conservative calibration date of 130 mya for the divergence between marsupials and placentals was applied. This yielded a date of about 93 mya for the divergence between primates and a cluster containing artiodactyls, cetaceans and carnivores. A subsequent study (Martin, 2009) analyzed DNA sequence information for both mammals and birds, taking a large sample of 48 nuclear genes with relatively constant rates of change (Hedges et al., 1996). A very early calibration date of

310 mya was used, derived from the well-documented separation between diapsid reptiles (which led to modern reptiles and birds) and synapsid reptiles. Divergence times estimated on this basis were greater than 90 mya for the separation between primates and 2 other groups of placental mammals (rodents and artiodactyls).In a later publication by the same research team including more species and nuclear DNA sequences (Kumar and Hedges, 1998), it was reported that vertebrate divergence times calibrated in the same way fitted well with most early (Palaeozoic)and late (Tertiary) dates derived from the fossil record, but that considerable gaps were revealed for the Mesozoic (Triassic, Jurassic and Cretaceous). It was inferred that at least 5 modern lineages of placental mammals diverged more than 100 mya and most orders had diverged by the end of the Cretaceous. In a different approach, a combined analysis of DNA sequences from 3 mitochondrial genes and 2 nuclear genes (Springer et al., 1997) showed that a group of endemic African mammals(hyraxes, elephants, sea-cows, golden moles, elephant shrews, and aardvarks) descended from a specific ancestral stock during the adaptive radiation of the placentals. Using a panel of 9 different calibration dates (including a marsupial/placental split at 130 mya and a ruminant/cetacean split at 60 mya), the mean divergence time between this African group of mammals (Afrotheria) and other orders of placental mammals (including primates) was estimated to be about 90 mya.

In another key study (Arnason et al., 1996), divergences between various placental (Martin, 2009) mammals, including 7 primate species, were reconstructed using data for complete **mitochondrial DNA** sequences. The resulting tree was calibrated with a (Martin, 2009) date of 55 mya for the minimum age of the cetacean lineage, yielding an inferred divergence of primates (Martin, 2009) from other orders of placental mammals about 90 mya. A double calibration based on the fossil record for hoofed mammals was then applied to an expanded dataset (Arnason et al., 1998), taking 60 mya for the divergence (Martin, 2009) between artiodactyls and cetaceans and 50 mya for the divergence within perissodactyls between horses and rhinoceroses. The outcome was an estimate of 95 mya for (Martin, 2009) the time of divergence between primates and hoofed mammals. Given the broad array of DNA sequences and calibration dates used, it is striking that all these studies consistently show that primates diverged from other placental mammals about 90 mya. Regardless of the reliability of the molecular clock, the ages of some first known representatives of other mammalian groups are simply incompatible with the interpretation that primates diverged only 60-65 mya (Soligo et al., 2006). One clear illustration of this is provided by studies of the relationships of cetaceans (whales and dolphins). It has long been accepted that cetaceans have a sister-group relationship to artiodactyls, but recent molecular evidence has consistently showed that cetaceans are actually nested within artiodactyls as relatives of hippopotamuses. This conclusion initially showed by immunological data (Sarich, 1993), is now supported by nuclear gene sequences (Gatesy et al., 1999), by insertions of interspersed elements (retroposons) in nuclear genome(Nikaido et al., 1999) and by complete mitochondrial DNA sequences (Ursing& Arnason, 1998).

The combined evidence supports the following sequence of divergences, from most ancient to most recent, during the evolution of hoofed mammals (ungulates): between perissodactyls and artiodactyls; within artiodactyls between camels+pigs and ruminants+hippos+cetaceans; between ruminants and hippos+cetaceans; between hippos and cetaceans. Given that the first known fossil representative (Martin, 2009) of the cetaceans is dated at 54 mya, it follows that the first of these 4 divergences in ungulate evolution must have occurred at a relatively early date and that the separation between ungulates and primates must have taken place even earlier. A date of only 60-65 mya for the divergence (Martin, 2009) of primates from other placental mammals hence seems inherently unlikely. In fact, calibration of a molecular tree with a date of 56.5 mya for the divergence between hippos and cetaceans, while allowing for variation in rates of (Martin, 2009) evolution, yielded a date of 97.6 mya for the divergence between primates and a cluster containing artiodactyls, perissodactyls and carnivores (Huelsenbeck et al., 2000). Here, it should be emphasized that all the molecular studies cited have focused primarily, or exclusively on the time of separation between primates and other groups of placental mammals. Although the molecular evidence, following calibration with various fossils (Martin, 2009) dates outside the primate tree, consistently shows that the lineage leading to living primates diverged from other placental mammal lineages about 90 mya. It could be imagined that morphologically recognizable primates did not emerge until 60-65 years ago.

Few molecular studies have addressed the question of the age of the last common ancestor of living primates. It is obvious from the short genetic distances involved in that part of the tree that the divergence between strepsirrhine and haplorhine primates must have occurred relatively soon after the primates diverged from other placental mammals. Even if marked variations in rates of molecular evolution can occur, it is highly improbable that the molecular data would be compatible with a divergence between primates and other placental mammals about 90 mya followed by a period of 25-30 my before the common ancestor of euprimates

emerged. In one of the few studies that has provided information directly relating to this issue, Arnason et al. (1998)indicated that the split between strepsirrhines and higher primates occurred about 80 mya, some 10-15 my after the primate lineage diverged from other placental mammals. An inferred age of about 80 mya for the initial time of diversification of modern primates fits remarkably well with the age of 81.5 mya estimated by statistical evaluation of the euprimate fossil record allowing for gaps (Soligo et al., 2006) (Figure 9.5).

Yoder & Yang (2004) estimated primate divergence dates by applying a Bayesian method permitting variation in rates of molecular evolution to DNA sequences from 4 unlinked genetic loci. Although that study focussed primarily on lemur evolution, several non-primates out groups were included. Fossil evidence was used to calibrate 8 nodes (4 for primates and 4 for non-primates) with upper and lower bounds, including a range of 63-90 mya for the last (Martin, 2009) common ancestor of strepsirrhines and anthropoids (following Tavare et al., 2002). Taking all genetic data together (Martin, 2009), the age of the ancestral primate node was estimated at 84.9 mya. Two points are of particular interest. First, when the tree was calibrated with a fossil date for just one node, markedly younger divergence dates were obtained. Second, the study by Yoder & Yang (2004) included the early divergence time of 38-42 mya between lorisids and galagids indicated by new fossil evidence (Seiffert et al., 2003). A previous analysis conducted before that information had yielded a strikingly concordant divergence time of about 40.5 mya (Yang & Yoder, 2003).

Substantial **molecular datasets** have been used to generate overall phylogenetic trees for mammals generally, to clarify likely relationships among different orders of

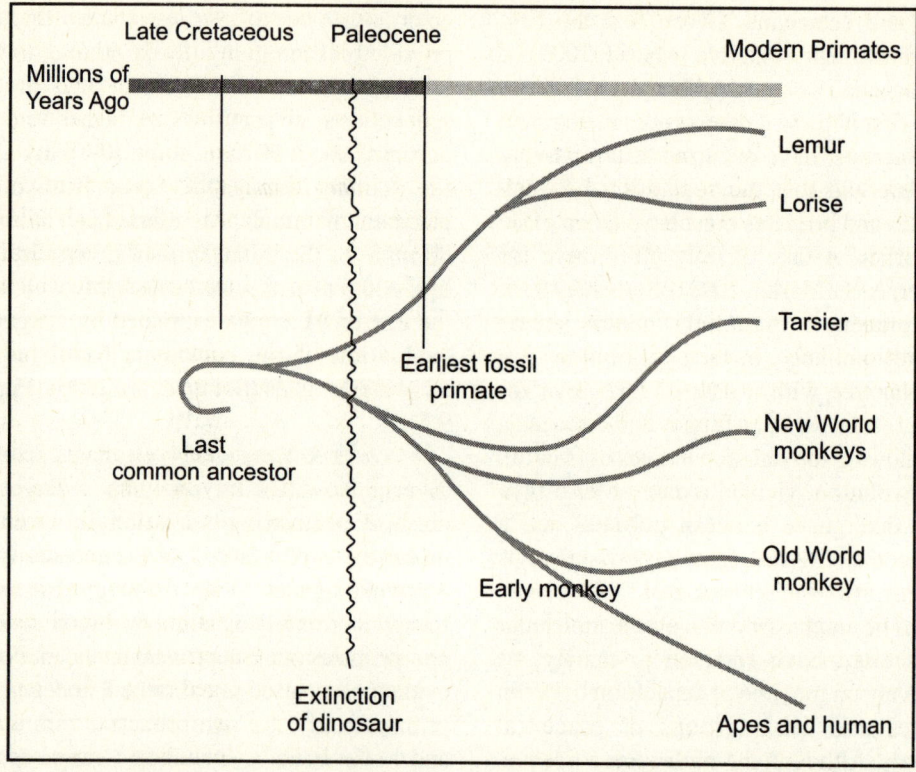

**Figure 9.5**    Simplified phylogenetic tree for primates. The last common ancestor is set at about
85mya, 20 mya before the extinction of dinosaur (Modified after Martin, 2005)

placental mammals (Madsen et al., 2001).
Various attempts to reconstruct higher-level
relationships among mammalian orders using
classical morphological evidence (e.g. Novacek
& Wyss, 1998), is absent. One major finding
from the new molecular studies is clear
confirmation of the existence of the endemic
group of African mammals Afrotheria
identified in previous analyses of DNA
sequences (Waddell et. al., 1999) and recently
supported by analysis of retroposons in the
nuclear genome (Nikaido et al., 2003). This
is one of 4 super groups of placental mammals
that can now be identified with some
confidence: **Afrotheria**, Euarchontoglires,
Laurasiatheria and Xenarthra. Afrotheria, as

now recognized, contains the tenrecs of
Madagascar besides golden moles, elephant
shrews, hyraxes, elephants, sirenians and
ardvarks. Although a potential link between
hyraxes and elephants had long been suspected,
links with other orders in Afrotheria had not
been previously showed by morphological
evidence. The inclusion of golden moles and
tenrecs splits the long-accepted mammalian
order Insectivora. The super group
**Euarchontoglires** includes primates, colugos
and tree shrews (archontans) along with rodents
and lagomorphs. Laurasiatheria combines
artiodactyls, cetaceans, perissodactyls,
carnivores, pangolins, bats and certain
insectivores. The fourth group **Xenarthra** is

a relatively small assemblage restricted to toothless (edentate) mammals now restricted to South America: anteaters, armadillos, and sloths. Springer et al. (2003) used a large molecular dataset for representatives of all extant orders of placental mammals, with sequences from 19 nuclear and 3 mitochondrial genes, to estimate basal divergence times. Their method permitted variation in rates of molecular evolution while applying 9 calibration dates based on first known occurrences in fossil record. All results showed that divergences between placental orders took place in the Cretaceous; whereas diversification within orders took place mainly after the K/T boundary. However, 4 placental orders (Eulipotyphla, Primates, Rodentia, Xenarthra) showed diversification prior to the K/T boundary, the earliest being the initial divergence in primate evolution at 77 mya.

## 9.3   Broader Context of Mammalian Evolution

The first appearance of mammals in fossil record coincides about with boundary between Triassic and Jurassic, about 200 mya. They are descendants of mammal-like reptiles (synapsids), which diverged at least 310 mya from diapsid reptiles that gave rise to modern reptiles and birds. For about two thirds of their evolutionary history, between their first appearance near Triassic/Jurassic boundary (200 mya), and the K/T boundary (65 mya), the mammals are poorly recorded in the fossil record, particularly for the southern hemisphere. This reflects major gaps in fossil record, as advanced mammal-like reptiles and first mammals are particularly well documented in southern continents. Initial stages of mammalian evolution were well under way in the south, but that later developments shifted abruptly and predominantly to north. Jurassic/ Cretaceous fossil record of mammals in

southern continents was disappointingly limited to a toothless mandible of *Brancatherulum* from late Jurassic of Tanzania and Argentinian footprints of similar age attributed to a mammal. Although dentition of *Brancatherulum* is unknown, new research on deposits from Tanzania has revealed dental remains of 3 other early mammals, a triconodontid, a eupantothere and a haramyid (Heinrich, 1999).

The southern continental record has been expanded by discoveries of fragments of early Cretaceous mammals in Cameroon and Morocco (Sigogneau-Russell, 1995), thus confirming that early mammals were present in southern continents. Fossil evidence remains exceedingly fragmentary and enormous gaps in record remain. Despite inadequacies of fossil record, it now seems highly probable that 2 main groups of modern mammals, marsupials and placentals, diverged early in the Cretaceous at least 125 mya, as fossil mammals of that age identified as a placental (*Eomaia*) and as a marsupial (*Sinodelphys*) have been reported from China (Luo et al., 2003). This means that placentals and marsupials must have existed for at least 60 my before the K/T boundary but are very poorly known from fossil record, particularly for southern continents. Although **adaptive radiation** of modern placental mammals is comparatively well documented above K/T boundary, there are still major gaps for primates. There is a systemic problem with respect to origins of modern placentals in that placental mammals known from the Palaeocene generally belong to archaic groups with no clear connection to modern orders. Just as Palaeocene **archaic primates** have at the most only a remote connection to euprimates, various other groups of archaic placentals are of dubious affinities. Mesonychians which were traditionally linked directly to cetaceans but have been sidelined by new fossil discoveries of early terrestrial relatives of cetaceans (Thewissen et al., 2001). A major turnover of terrestrial mammalian fauna occurred close to the Palaeocene/Eocene boundary, coinciding with an episode of global warming (Berggren et al., 1998).

The warming process at the Paleocene-Eocene boundary, which was evident at high latitudes, took about 2 my. During late Paleocene and early Eocene, Europe, North America and Asia together constituted a single Holarctic biogeographical province with common floristic and faunal elements. Deciduous vegetation that characterized this Holarctic province during Paleocene was gradually replaced during early Eocene by markedly different rainforest vegetation, which extended to latitudes as far as 60° north and south of equator (Wolfe, 1987). International recognition of Paleocene/Eocene boundary was initially linked to a abrupt major turn over in mammalian fauna characterized by disappearance of archaic groups and immigration by modern placental groups. In Europe, condylarths, adapisoriculid insectivores, multituberculates and plesiadapiforms declined and were replaced by artiodactyls, perissodactyls, rodents, bats, euprimates and certain other groups. In North America, transition was accompanied by a relatively rapid decline in mesonychians, not ungulates and plesiadapiforms, with replacement by artiodactyls, perissodactyls, rodents, insectivores and euprimates. With respect to new appearances of placental mammal groups in North America, Wing (1998) made the following observation: "*These groups do not appear to have mid-Paleocene ancestors in North America, implying that they arrived from a different continent. Several authors have suggested that the late Paleocene-early Eocene mammalian immigrants to North America originated at more tropical latitudes in the Americas, Asia, or Africa (Krause and Maas, 1990), and then migrated to middle and high latitudes as global climate warmed in late Paleocene and early Eocene epochs.*" As noted by Hooker (1998), the same applies to comparable faunal turnover that took place in Europe and Asia: "*There has been much speculation on centers of origin of main mammal groups that appeared synchronously in Europe, North America, and Asia, namely orders Artiodactyla, Perissodactyla, Primates, and Chiroptera, and families Hyaenodontidae (Creodonta) and herpetotheriine Didelphidae (Marsupialia). Krause and Maas (1990) has thoroughly investigated the problem of these origins in a temporally broader study of mammalian immigrants into North America. Low-latitude areas are involved, in view of absence of Paleocene representatives in any mid latitude faunas in northern hemisphere [...].*" In all cases, ancestors of these immigrating placental mammal groups at the Paleocene/Eocene boundary in the northern hemisphere have yet to be discovered. Inadequate documentation of early placental mammals is apparent throughout fossil record( Figure 9.6, 9.7).

## 9.4   Role of Continental Drift

If the initial divergence and subsequent diversification of various modern groups of placental mammals took place considerably earlier than has been traditionally supposed, namely during the latter half of the Cretaceous rather than during the Palaeocene. This introduces possibility that continental drift might have played a significant role (Martin, 1990). It also increases the likelihood that southern continents, which have a particularly poor fossil record for period between 90 and 65 mya, played an important part in early diversification of placental mammals. If the initial radiation of placental mammals began during the mid-Cretaceous, about 90mya, it would have coincided with a period of maximal

# MAMMALIA

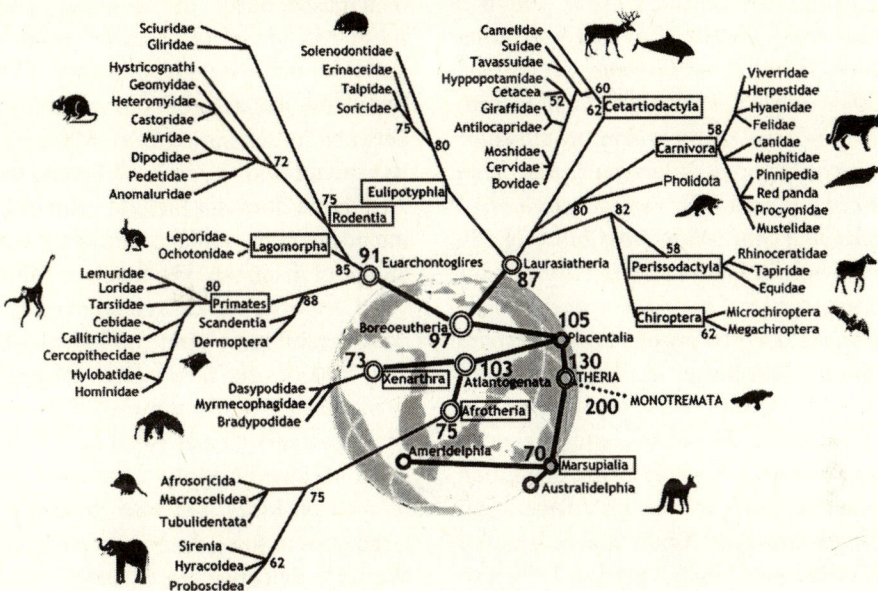

**Figure 9.6** Genome diversity and karyotype evolution of mammals

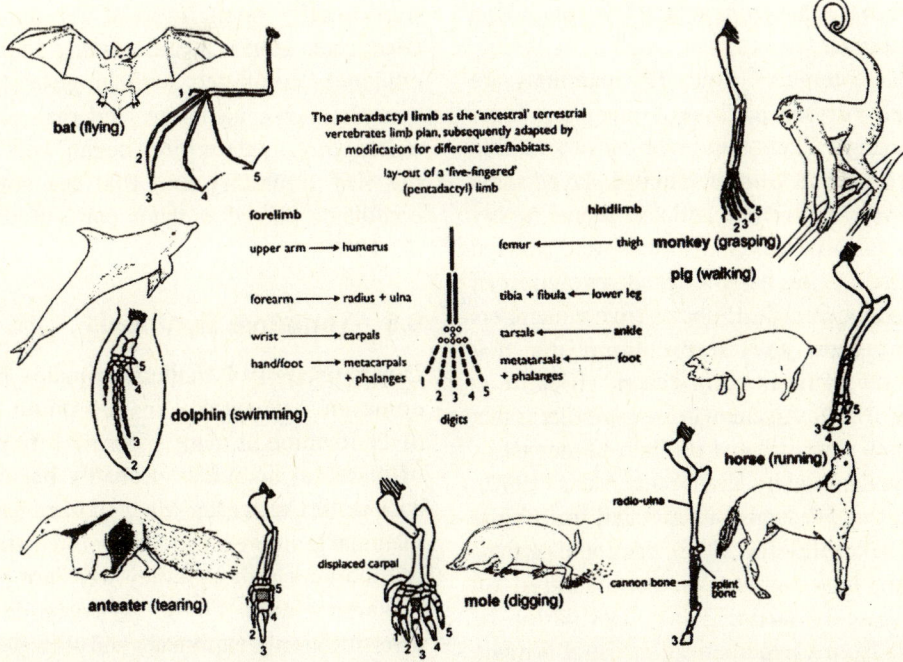

**Figure 9.7** Adaptive radiation of the forelimb of mammals illustrates the principles of homology. The basic pentadactyl pattern are modified for different useages

subdivision of land masses through a combination of continental drift and extensive formation of epicontinentalseas. Such a possibility was explicitly invoked by Kumar & Hedges (1998): "*For example, the sudden appearance (in the Early Tertiary fossil record) of mammalian and avian orders, which show large morphological differences, has been taken to imply rapid rates of morphological change at that time. Now, possibility of 20-70 Myr of prior evolutionary history relaxes that assumption and suggests a greater role for Earth history in evolution of terrestrial vertebrates.*" Murphy et al. (2001b) linked subdivision between their 4 super groups of placental mammals directly to continental drift. They suggested that Afrotheria (the first group to diverge) became isolated on Africa at an early stage and that Xenarthra originated through isolation in South America. They also proposed that **Laurasiatheria** and Euarchontoglires became isolated in Laurasia, although it is not evident how this conclusion was reached.

As divergence times for mammals are pushed further back into past, it seems increasingly likely that break-up of southern super continent **Gondwanaland** played some part in their diversification. It seems highly likely that afrotherian mammals were isolated on Africa at an early stage of evolution of placental mammals. Primates are not members of Afrotheria, so it seems improbable that ancestral primates occurred on Africa, (e.g. Martin, 1990). An alternative possibility is that primates were isolated on Indo-Madagascar, as hypothesized by Krause and Maas (1990). Given that Madagascar separated from India about 88 mya(Storey, 1995), this could perhaps explain how lemurs became isolated on Madagascar(Martin, 2000). Derivation of lemurs from a hypothetical ancestral primate stock in Africa was problematic as Madagascar separated from Africa at least 130 mya. Some

palaeontological evidence shows exchanges between India and other, supposedly isolated land masses during the Mesozoic. The known fossil records of India lacks peculiar fauna and flora that would be expected if India had been fully isolated for some 30 million years between its separation from Madagascar and its contact with Eurasia. Whereas the group Afrotheria does not include primates, it does include tenrecs, which is one of the 4 endemic groups of mammals now found on Madagascar.

Carnivores colonized Madagascar some considerable time after the lemurs (Yoder et al., 2003), so a single, all-embracing explanation of mammalian invasions of Madagascar is clearly ruled out. Many of the same arguments apply to adaptive radiation of modern birds. Because there are similar large gaps in fossil record, a direct reading of available evidence would seem to show that there was an explosive adaptive radiation of modern birds after K/T boundary, as has traditionally been inferred for mammals (Feduccia, 1995). Accumulating molecular evidence, combined with biogeographical considerations, has suggested that adaptive radiation of modern birds began well before the K/T boundary and that the southern continents played a major part (Waddell et al., 1999).

## 9.5    Tentative Synthesis

Diversification of modern primates from a common ancestor, in common with diversification of many other modern groups of placental mammals probably began well back in the Cretaceous. Many modern placental mammal groups existed long before extinction of dinosaurs, although they were not present in same places. Given an early date for diversification of euprimates and other placental mammals, it is highly likely that continental drift played a significant role. Mammalian super group including primates (Euarchontoglires), perhaps along with Laurasiatheria, might have been isolated on drifting landmass of Indo-Madagascar. A renewed search for the Cretaceous mammals in India might uncover new evidence. If the geographical distributions of euprimates over time are examined, it can be seen that earlier Tertiary representatives (32-55 mya) are largely restricted to northern continents. This contrasts starkly with distribution of extant primates, which are confined to southern continents. **Fossil primates** from later Tertiary deposits (0-26 mya) show an intermediate condition, occurring widely in both northern and southern continents. Primates originated in northern continents and progressively shifted to south. Alternative interpretation supported by evidence now available is that primates were prevalent in northern continents only while there was a period of higher world temperatures during the Eocene. Global cooling ensued around the Eocene-Oligocene boundary, and primates progressively disappeared from the northern continents.

## 9.6 Human Evolution

Modern man evolved about 100, 000 years ago. The almost complete skeleton of a woman found encased in thick limestone was about 28 million years old. The Calaveras skull, the complete mineralized human skull dated back to Pliocene. **Human footprints** in various sites in the US and Africa include Glen Rose tracks and Antelope Springs tracks. Human footprints, 15 and 21½ inches in length in early Cretaceous rock are called Glen Rose tracks. The sandaled human tracks stepping on trilobites in Cambrian strata in Utah is called Antelope springs tracks. The name hominids refer to supposed half-man or half-ape remains. Of 2 theories of human ancestry, first views that human and ape had the common ancestor about 20 mya.

Second views that the human descended from the ape. Since then, man switched over from physical evolution to cultural- and social evolution. Our DNA is different from that of monkeys and apes. Number of vertebrae is different in human and apes. Cranial capacity in human and great apes also differs. Although existing view suggests our evolution from the great apes several differences between man and ape are:

1. Birth weight as a percent of maternal weight in human is about twice that of the great apes (5.5 and 2.4), but about same as found in monkeys (5-10) and gibbons (7.5).
2. Teeth arrangement is the same in man and in the Old World monkeys, but different from the great apes.
3. Man and gibbon walk upright; the great ape does not.
4. The neck hinge is found at the back of human, but at the front on ape. Shape and arrangement of teeth are different in apes and man. Many male primates fight and defense with large canine teeth. Spaces exist in places where upper canines meet with lower jaw, between the opposing teeth. Canine diastema is notable in jaws of baboons, gorillas and monkeys. They are diagnostics in studying fossils but are absent in hominids. Primate jaws with canine diastema are perhaps linked to ape, or monkeys, but not to human family.

Human groups migrated to new areas and for a time in Stone Age cultures in New Guinea lived under trees. Men gathered worms, leaves, and fruit with few crude stone and bamboo tools. Many believe that those primitive Stone Age peoples are not the evidence of earlier human, but are tribes which slipped back from us. Some suggests that certain hunting tribes in South America, Africa, Central India and western Pacific are not relics of Stone Age but are the wreckage of developed societies forced through circumstances to lead less developed life. Some human group lived in caves before building homes in new land, wandered from warm climates to colder ones. Some primitive people are found among people living along the shores of the Red Sea in caves along with barbarian tribes. Both advanced civilizations and backward cave cultures thus existed at the sometime. In some places in Palestine, people resembling the Neanderthal race lived in caves while Jericho people dwelt in well-built, decorated houses.

The cavemen are known as the Neanderthals. In a cave in the Neander Valley, Germany pelvis, ribs, limb bones and a skull cap were found. R. Virchow said the bones were those of modern men affected with rickets and arthritis. Bowed legs due to rickets were caused due to lack of sunlight. In 1886, 2 similar skulls were found at Belgium. In early 1900s, several similar specimens in southern France were found. Boule said that they were ape-like creatures. Diet of Neanderthal man lacked Vitamin D. Neanderthal features include larger brow ridge and arthritis could make them more prominent. Deficiency of Vitamin D causes osteomalacia and rickets, increased the size of orbit, especially vertically. The femur was curved. Congenital syphilis could cause bone deformities in Neanderthal. The ice ages came due to worldwide volcanic dust pollution. The weather in Europe was cold enough that they stayed so much in their caves and deprived of sunlight due to overcast sky conditions. They perhaps lived longer than men today. Neanderthal existed for several hundred years. Facial bones exhibited growth throughout life. Neanderthal skeleton was found in Poland and Philippine Islands with larger cranium and brain than ours reflecting regression of our race from a long-lived, more intelligent race. Brain case of Neanderthal was more than 13% larger than of modern man on an average.

## 9.7  Miocene Apes

Apes evolved in Africa when it was a separate land mass. The best known early apes, the Proconsul retained monkey-like characters of backbone, pelvis and forelimbs. Lowered sea levels about 17 mya created a land bridge between Africa and Eurasia. Some early apes entered Eurasia, along with elephants, pigs and antelopes. By this time the apes developed a coating of enamel on teeth to eat harder foods. Within 1.5 million years of apes moving into **Eurasia** they diversified into at least 8 different forms. By 13 mya apes were found throughout Eurasia, including lineages of *Dryopithecus* and *Sivapithecus*. Both had similar anatomy to modern great apes. These 2 lineages survived major climate changes while many Eurasian ape species went extinct. They survived by migrating into Southeast Asia (*Sivapithecus*) and back into Africa (*Dryopithecus*). Living great apes are descended from these 2 lineages. Phylogenetic analyses show that *Sivapithecus* is the likely ancestor of orang-utan and *Dryopithecus* is probably forebear of African apes and humans.

## 9.8  The Earliest Hominids

Molecular biology suggests that humans and chimpanzees last shared a common ancestor about 8 mya. Descriptions of *Orrorin tugenensis* (in 2001) and *Sahelanthropus tchadensis* (in 2002) have provided more knowledge of this period. Comparing chimp and human genetic sequences and using the differences between them estimate the date of divergence of 2 species and suggest humans and chimps diverged no more than 6.3 mya (Pennisi, 2006). The data also raise possibility that 2 new species may have hybridized for some time after their initial separation (Patterson et al., 2006).

## Orrorin tugenensis

*Orrorin tugenensis* from Kenya is dated at 6 million years old. Its remains consist of limb bones, jaw material and few teeth. Its discoverers place it in hominid family tree as bipedal, suggesting that bipedalism in hominids evolved very early. Not all researchers agree that *Orrorin* is a hominid as its canine teeth are more ape-like. Its lower limb bones are those of bipedal organism. The neck of femur shows that bone is thickest on underside of neck, as seen in humans (Galik et al. 2004). In chimpanzees, the neck is uniform in thickness.

## Sahelanthropus tchadensis

*Sahelanthropus tchadensis* was described from a partial mandible and some teeth dated at 6-7 million years old. Michel Brunet's team describes *Sahelanthropus* as a hominid for its shape and angle of face and skull and dentition. Reconstruction of cranium places the foramen magnum well under skull suggesting *Sahelanthropus* was bipedal (Zollikofer et al., 2005). Comparisons of reconstructed cranium with those of modern apes and other fossil hominins show that it belongs on hominin lineage.

## Ardipithecus ramidus

*Ardepithecus ramidus* is a 3rd ancient hominid. Some place it in the genus *Australopithecus*. The oldest remains of this species, belong to subspecies *kadabba*, are known from Ethiopian rocks dated about 5.8 million years old. Fossil material of this species consists of skull fragments and some teeth and a toe bone. *Ardipithecus* lived in a forested environment.

## 9.9  The Australopithecines

This large group comprises both gracile and robust australopithecines. *Australopithecus* (Figure 9.8) include variety of ape bones in various African sites, including Sterkfontein, Swartkrans, KoobiFora, Olduvai, Hadar, and Orno River. They had a range of varieties. They are all apes. The most famous was named Lucy. Some believe that Australopithecines were descended from Ramapithecines. In the 1960s, Louis Leakey found some teeth and skull fragments at Olduvai and thought that they belong to human family, and name them *Homo*. That brain enlargement marked the beginning of man was long popular. The endocranial volume of australopithecine group was not larger than those of gorilla. Charles Oxnard and Sir Solly Zuckerman ended that australopithecines are not intermediate between man and living apes.

**Figure 9.8**  *Australopithecus africanus*

## Australopithecus anamensis

The earliest australopithecine is *Australopithecus anamensis*, which lived about 3.9 mya. This had teeth and jaws resembling those of older fossil apes. They were likely bipedal with human-like upper limbs. Anamensis may be the ancestral to all later hominids.

## *Australopithecus afarensis*

The best-known member of this species is Lucy, discovered in 1974 by Donald Johanson & Tom Gray and estimated to be around 3.2 million years old. The skeleton is remarkably complete for its age, providing data of size, gait and posture. Other finds the 13 individuals of the First Family. As per Johanson, Lucy had massive V-shaped jaws. In 1981, he said that Lucy was un-*Homo* like. Johansson's theory is based on assumption linking 2 fossils 1,000 miles apart. Although the Lucy fossils were initially dated at 3 million years, Johanson noted them as 3.5 million because the species was same as a skull found by Mary Leakey at Laetoli, Tanzania. By proposing Mary Leakey's finding as type specimen for *Australopithecus afarensis*, he identified Lucy with another half a million years older fossil. Johanson soon after changed his mind based on view of Tim White. They now describe the bones as too ape-like in jaws, teeth and skull to be considered *Homo*, yet sufficiently different from other, later **australopithecines** to warrant their own species.Afarensis was strongly sexually dimorphic, with males larger than females. The afarensis maximum height was 152 cm, and maximum cranial capacity was 550cc. Face and cranium of afarensis was ape-like with distinct brow ridge, low forehead, and the prognathous muzzle lacking chin. Teeth are intermediate between apes and human. Molars are large and canines though smaller than that of living ape are larger and pointed than that of human. Shape of dental arcade lies between human parabolic form and apes' rectangular shape and foramen magnum, while more forward than in apes, is not directly under cranium like human. Their postcranial skeleton is far closer to that of modern humans. The pelvic leg and foot bones show that this was bipedal, though not well suited for running.

The finger & toe bones were curved and longer than that of humans which speaks that afarensis spent time in trees, their hands are otherwise human-like. *Homo erectus* lived between 1.8 million and 300,000 years ago, and was probably first hominid species to move out of Africa and colonize Europe and Asia

## *Lucy's child*

The knowledge of *Australopithecus afarensis* is extended with description of juvenile afarensis (Alemseged et al., 2006). The complete skull and partial skeleton are probably like a female of about 3 years old. It yields information lacking in other afarensis remains: an endocranial cast of cranium; shoulder blades and collarbones - permitting an estimate of extent of arm movement in this species; and a hyoid bone showing voice box of ape-like structure. The form of shoulder blade and fact that arms could be swung above head and presence of long, curved fingers all suggest that species was at least partly aboreal.

## *Kenyanthropus platyops*

Until recently our knowledge of early hominids suggested that *A. afarensis* is the best example of a single early-middle Pliocene lineage. In 2001, Maeve Leakey's research team found well-preserved cranium that they placed in a new name: *Kenyanthropus platyops*. This fossil is a mosaic, with combination of ape-like and hominid features. On ape side, it has a small ear hole, thick enamel on molar teeth, cranial capacity of about 400cc, and flat nose. Its face has several novel features not found in other gracile australopithecines: a flat face, vertical cheek region, and small brow ridge and cheek teeth.

## *Australopithecus garhi*

*A. garhi* was described in 1999 from fossils found in Awash region of Ethiopia. The 2.5

million-year-old fossils found in association with animal bones had marks on them that appeared to be from stone tools. Very simple stone tools were found in nearby site of same age. Fossils of *A. garhi* comprise partial cranium, fragment of other partial skeletons, other postcranial bones and 2 mandibles. Many features of skull, including small cranial capacity of 450 cm$^3$, are similar to *A. afarensis*. There was a sagittal crest and an ape-like dental arcade. Molars and canines were large. Femur-to-humerus ratio was like that of modern humans, but ratio of forearm-to-humerus was like that of chimpanzee.

### Australopithecus africanus

Raymond Dart's Taung child, consisting of teeth, jaws, facial bones, and endocranial cast was the first australopithecine fossil to be found. All its milk teeth were present. Its cranial sutures suggest that this individual was 3 years old. Subsequent discovery of many adult africanus fossils confirmed his classification. Africanus lived between 3.3 and 2 mya. It showed strong sexual dimorphism, and bones of feet, legs, pelvis and spine show that it was bipedal. Both body size and cranial capacity (420 - 500cc) was slightly larger in africanus, which had larger molar, small canine, and full parabolic dental arcade. *Australopithecus anamensis*, *afarensis*, africanus, and *Kenyanthropus platyops* are collectively known as gracile australopithecines, for their light, slender build.

### Australopithecus robustus

*A. aethiopicus* and *boisei*, *robustus* are the robust australopithecines. All of them had heavily built skulls, large jaws, sagittal crests, and thick enamel on molar. Robustus lived between 2 and 1.5 mya. While its body size was like that of *A. africanus*, it had a large, more robust cranium and very large molars. The face was large, without forehead. Many specimens found to possess sagittal crests in addition to their big brow ridges. Combination of such ridges and crests with large jaws and molar suggest that this species ate coarse, tough food. While tool use was regarded for many years as characteristically human feature, robustus was probably the first hominids that used tools. This explanation was based on bones found with robustus fossils as worn ends of these bones suggest they may have been used for digging.

### Australopithecus aethiopicus

This is the oldest of 3 robust australopith species, living between 2.6 and 2.3 mya. There is one major fossil, the "Black Skull", so named as it had been stained by minerals in soil. Some researchers consider it an ancestor of both robustus and boisei. At 410cc, its cranial capacity is little more than that of a chimpanzee, and posterior parts of skull are like *A. afarensis*. But it has heavily built face and jaws of other robust species, and largest sagittal crest ever seen in hominid.

### Australopithecus boisei

*A. boisei* is the most robust of all robust australopithecines. Louis Leakey nicknamed it Nutcracker Man for its huge molar teeth. Its cranial capacity is same as robustus; about 530cc. Its face and jaws are more massively built. The "hyper-robust" nature of species suggests its high specialization to chew hard, low quality foods. Probably boisei became extinct as it was so highly adapted to a specific ecological niche and could not evolve fast enough to adapt when the environment changed.

## 9.10 Homo Species

### Homo habilis

Until Jane Goodall's pioneering studies of chimpanzees, most palaeoanthropologists believed that tool use was a hallmark of humanity. When Louis Leakey's team found very simple stone tools closely associated with hominid remains, in Olduvai Gorge, he named hominid *Homo habilis*. The associated tools are assigned to Oldowan tool culture. *Homo habilis* lived from 2.4 until 1.5 mya, and resembles the australopithecines. Recent study suggests that habilis would be classified in *Australopithecus*. The face still projects forwards but facial angle is less in comparison to *A. africanus*. Mean cranial capacity in habilis is about 650cc, and range is 500–800cc - this overlaps both australopithecines and *H. erectus*. Analysis of wear patterns on teeth suggest that habilis was adding meat to its diet - probably as a scavenger. Postcranial remains are fragmentary. Only one set of limb bones has been securely assigned to habilis. This broken skeletal material suggests that average height of *Homo habilis* was around 127cm. They were probably 45kg in weight. They were bipedal. Some argument exists about the taxonomic status of habilis specimens.

## Homo erectus, or Homo ergaster

*Homo erectus* lived between 1.8 million and 300,000 years ago. It was probably the first hominid to move out of Africa and colonize Europe and Asia. This event must have happened early in species' history. Some recognize 2 sister species: erectus is assigned to the Eurasian specimens while *Homo ergaster* is for African specimens. Ergaster has a smaller cranial capacity, and 2 differ in features of skull, like shape of brow ridges. Erectus had a long, low skull, with little forehead and cranial capacity up to 1225cc. The smaller brain sizes are associated with older specimens. Face was prognathous. Protruding jaws supported large molar but lacked chin. This had a projecting, rather than a flattened nose. The post-cranial skeleton of *Homo erectus* was robust. Members of this species in Africa were tall. Some place

them in upper quartile of height range for *H. sapiens*. Few remains from China (**Peking man**) are shorter individuals reflecting adaptation to local climate. Tall, slender individuals with a high surface area to volume ratio (SA:V) are well adapted to lose heat in hot weather. Those living in colder regions need to conserve heat. Heat loss is reduced in short, stocky individuals with a lower SA: V.Studies of erectus pelvic bones, particularly those of Turkana (Nariokotome) Boy, shows that members had a narrow pelvis and pelvic canal. This indicates that their babies were smaller-brained at birth. Erectus was probably efficient at walking than sapiens. Erectus was a competent toolmaker. Their tools are classified in the *Acheulean* tool culture used to butcher animal carcases. Meat made up a portion of erectus diet. They first use fire as evidenced from charcoal deposits from Choukoutien caves near Beijing, where fossils of Peking man were found.

## Homo floresiensis

*Homo floresiensis* found on Indonesian island of Flores was named on the basis of partial remains - including skull, jaw, and teeth of an adult female, and fragments from several individuals. It was only 18,000 years old and tiny. Adult female had cranial capacity of just 380cm3. Up till now, scientists believe of only 2 hominin species in Asia: *Homo sapiens*, and *H. erectus*. This is yet evidence for bushy nature of human family tree. The Flores skull shows a mixture of advanced and primitive characters like the low cranium, prominent brow ridges, along with small relatively flat small face. This hominin was bipedal. The pelvis is similar as that of australopithecines. Various features of skull, shape of brain suggest that floresiensis evolved from *H. erectus* was dwarfed. Floresiensis was found in association with bones of a dwarfed elephant species. *Homo floresiensis* was probably a toolmaker & tool-user. Several stone tools were found with remains. Since Flores is separated from larger island of Java by a deep ocean channel, it is likely that ancestors of the species have arrived by sea.

## Other fossil records

**Cro-Magnon man** (*Homo sapiens fossalis*): A cave at Les Eyzies was found to contain several skeletons which were believed as the missing link between man and ape. Cro magnon, means big hole, were truly human. Some were over 6 feet tall, with a cranial capacity than that of men today. They were normal people and do not provide evidence of transition from ape to man. Experts have worked with jaw fragments, broken skull pieces, and other bones. Full or half-complete skeleton, linking man with the other animals are yet to be found.

## Java Man (*Pithecanthropus erectus*)

**Java man** is a classic instance. Dubois found a skull cap, a femur and 3 teeth and assumed that all these bones were from the same individual. He found another 2 human skulls (known as *Wadjak* skulls), with a cranial capacity above modern man and admitted the *Wadjak* skull was an ape. Dubois promoted his discovery of Java man, *Pithecanthropus erectus*. Most evolutionists believe that his finding were bones from a modern human.

## Piltdown Man (*Eoanthropus dawsoni*)

In 1912, **Piltdown man** *Eoanthropus dawsoni* was found with one specimen. J. Weiner and K. Oakley conducted fluorine test to the bones and found that Piltdown man was a grand hoax. Someone took an ape jaw and put it with human skull, filed the teeth, and then carefully stained it so that bones looked both ancient and a matching set.

## Rhodesian Man

In 1921, Rhodesian man was discovered in a cave. Presence of dental caries which modern diets produce proves that Rhodesian man was not very ancient.

## Peking Man (*Sinanthropus pekinensis*)

**Peking man** emerged on the scene in the 1920s. Animal bones were found in a pit near Peking, with 14 human skulls in varying conditions, 11 jawbones, 147 teeth and some small arm bone and femur fragments but with small brain capacity, and with prominent brow ridges which are find in Neanderthals and *Australopithecus*. There are present day races with larger brow ridges than modern humans. Some Philippine women have brow ridges which men generally have. Braincase of 1,000cc is not sub-human. Present day people have braincase between 1,000 and 2,000cc, with rare low of 750cc. All the skulls disappeared during the World War II. Peking man was found mainly by skulls. The foramen magnums were widened and smashed.

## Nutcracker Man

Nutcracker Man was found in 1959 by Louis Leakey in Olduvai Gorge in East Africa. Louis Leakey called this *Zinjanthropus boisei*, but press called it Nutcracker Man as it had jaw much larger than skull. The skull was very ape like. Leakey conceded that Nutcracker Man was another ape skull as Dart's Taung Man.

## SKULL 1470

In 1972, Richard Leakey found a human-like fossil skull, and dated it 2.8 million years old. The official name of this skull is KNM-ER 1470, commonly called Skull 1470. It looks like a modern small-brained person. Skull 1470 is about midway in this category and not like that of humans. It has a long upper lip area like apes . Viewing 3 skulls from the rear Skull 1470 was found similar to that of *Australopithecus*.

### Comment

Sir Arthur Keith refused to accept the African australopithecine fossils as human ancestors as their brains were too small. Keith proclaimed that human qualities of mind can only appear when brain volume is at least 750 cubic centimeters; a point nick named Keith's rubicon. And at 450cc. *Australopithecus africanus* didn't qualify. In Keith's day, the *Homo erectus* skulls at 950cc. could comfortably treated as humans, since their range overlaps our own species (1,000cc.-2,000cc.). The *Homo habilis* skulls discovered later measured about 640cc. Just on other side of Rubicon, skulls of *Australopithecus* adults is about 500cc. which is larger than chimps, but small than *Homo habilis* (Figure 9.9–9.16) illustrates various aspects of human evolution.

## 9.11 Trends in Human Evolution

Trends like brain size, changes in posture, and facial angle are often misused to give the idea of linear manner of evolution, from primitive ancestor through series of descendants. The evolutionary history of humans like most organisms is best reconstructed as a bush, where several related species existed at a time. These trends do give a useful overview of the evolutionary changes that have occurred in our biological history.

## 9.12 Trends in Cranial Capacity

Gradual increase in **cranial capacity** over course of human evolution is noted. Thus *Sahelanthropus* and early australopithecines

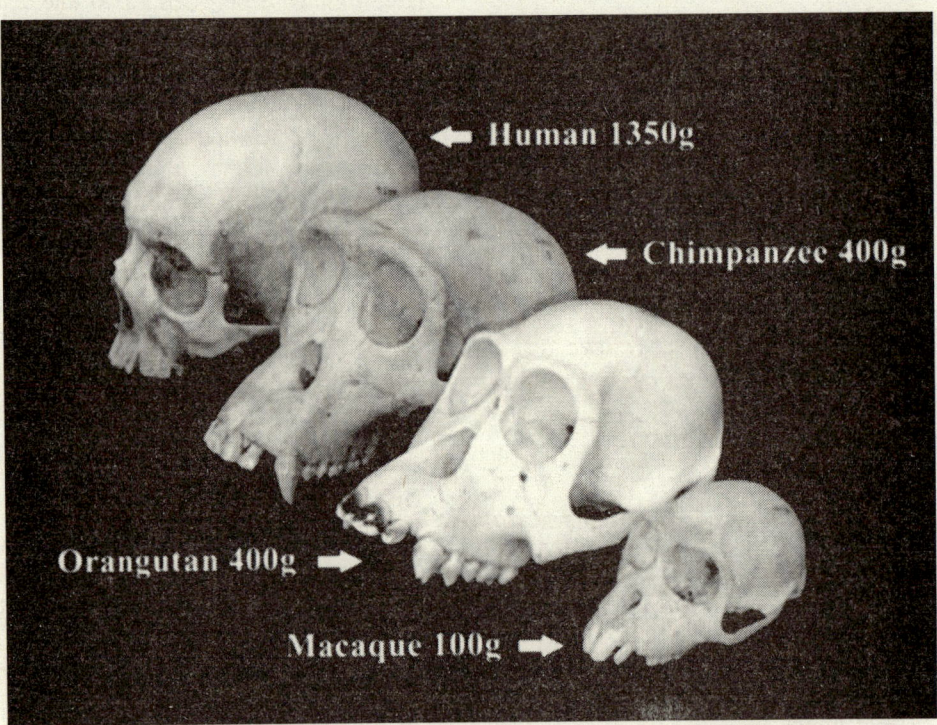

**Figure 9.9** Skull of human, Chimpanzee, Orangutan and Macaque

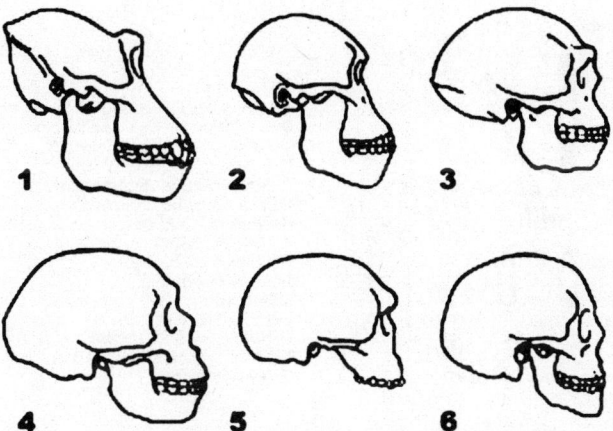

**Figure 9.10** Evolution of skull in human

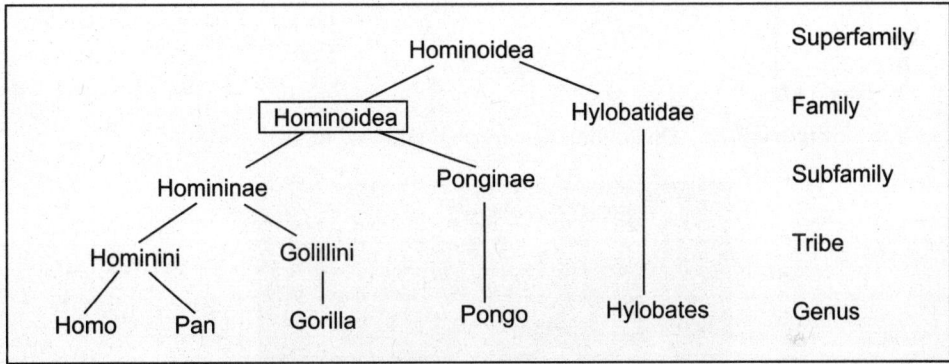

**Figure 9.11** In search of ancetsry of *Homo*

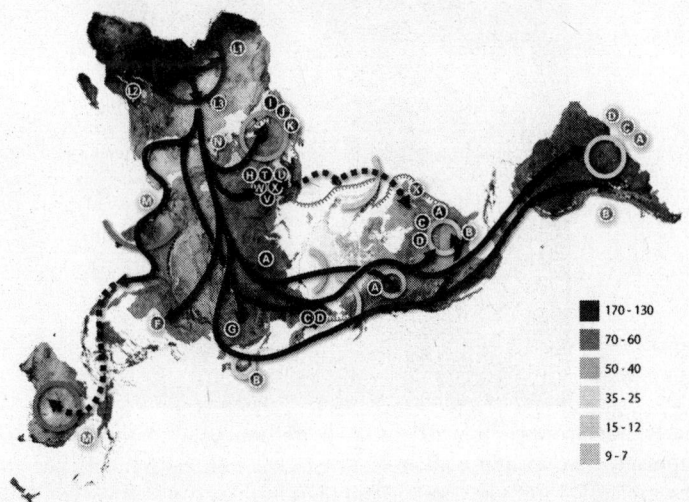

**Figure 9.12** Map showing human migration

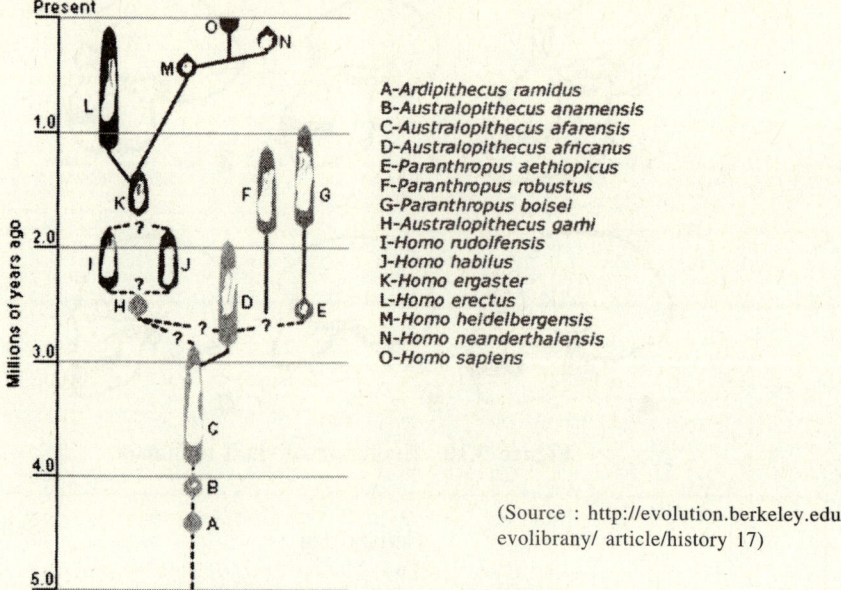

A-*Ardipithecus ramidus*
B-*Australopithecus anamensis*
C-*Australopithecus afarensis*
D-*Australopithecus africanus*
E-*Paranthropus aethiopicus*
F-*Paranthropus robustus*
G-*Paranthropus boisei*
H-*Australopithecus garhi*
I-*Homo rudolfensis*
J-*Homo habilis*
K-*Homo ergaster*
L-*Homo erectus*
M-*Homo heidelbergensis*
N-*Homo neanderthalensis*
O-*Homo sapiens*

(Source : http://evolution.berkeley.edu/
evolibrany/ article/history 17)

**Figure 9.13**   Diagrammatic representation of human evolution

**Figure 9.14**   Early human upright walking. A 6 million-yea-old thighbone, or femur (center), represents an upright walker of one of the earliest human ancestors (*Orrorin*), and resembles 2-3 million year old thighbones of australopiths (left, bottom). Thighbones of Homo (right) mark a transition toward a more modern gait about 2 million years ago. **Credit: artwork and composite by John Gurche, photograph by Brian Richmond**

**Figure 9.15**  Lucy

**Figure 9.16**  *Homo heidelbergensis*

had cranial capacities within range of modern chimpanzees; with later australopithecines reaching 550cc. Skull was about 510cc in early *Homo* followed by a marked increase in *Homo erectus* with brain size of up to 1225cc. Average cranial capacity of Neanderthals was larger than modern humans (1450cc and 1350cc respectively), but this may be due to the larger body mass of *neanderthalensis*. A strong positive correlation between body size and brain size within the species exist. Male have large body mass and large cranial capacity than female. The brain size of hominids, particularly of *Homo* species are greater.

## 9.13  Bipedalism

Bipedalism appeared very early in our evolutionary history. There was dispute regarding the posture of *Sahelanthropus*. A computer-assisted reconstruction of this fossil shows a foramen magnum beneath the cranium, and small nuchal crest show this species as bipedal (Zollikofer et at., 2005). This is not to say that their posture was fully erect. While the pelvic, leg, and foot bones of *Australopithecus afarensis* clearly show that this species was bipedal, it was not well adapted for running. The position of foramen magnum suggests that afarensis did not hold its head fully erect. *Homo erectus* was possibly an even more efficient biped than modern human, due to narrow pelvic outlet in *Homo erectus*. The wide pelvic outlet in sapiens permit the birth of large-brained infants, places hip joints further apart. Achilles tendon is another feature linked to bipedalism, which links calf muscles with tarsal bones of heel in *Homo*. The great apes lack this structure and calf muscle extends right down to tarsal bones. Presence of Achilles tendon in humans makes us relatively good endurance runners - a feature for active hunting on the open savannah. Tendon acts as a spring, storing energy when it is stretched and releasing it again as foot pushes off for the next stride (Bramble & Lieberman, 2004). This can save up to 50% of metabolic energy costs of rapid locomotion.

## 9.14   Trends in Morphology of Skull

**Dental arcades:**   In an ape, the teeth are arranged in rectangular dental arcade, where left and right cheek teeth are parallel. Dental arcades in *Australopithecus* are more rectangular than parabolic, while it is a full parabola in *Homo*, broader at the back than at front. There is strong trend in tooth size. The cheek teeth of modern human are small than those of australopithecines. Even within *H. sapiens*, there is a marked decrease in tooth size over the last 30,000 years.

**Crests and ridges:**   Both the great apes and early hominids have crests and ridges. The sagittal crests provide anchorage for large chewing muscles, and prominent in species which eat hard, tough material. Brow ridge development in hominids may be related to diet. The large brow ridges aid to redirect the stresses on skull by a diet of coarse vegetable matter. Presence of pronounced nuchal crest provides information about a species' posture, together with position of foramen magnum. Crest provides anchorage for neck muscles in those species where skull is not balanced vertically atop the spine. Modern humans have complete upright posture. Foramen magnum is centrally placed beneath skull. A nuchal ridge is absent. Early australopithecines such as Lucy (*A. afarensis*) have a foramen magnum position intermediate between the human and apes, and small nuchal ridge. This tells us that their posture was not completely upright.

## 9.15   Facial Angle

There is a general trend towards a flatter facial angle with the appearance of more recent hominids, culminating in the vertical (orthognathous) face of *Homo sapiens*. In spite of considerable variation even among the older members of our lineage, *Kenyanthropus platyops* is named for its relatively flat face. Its name means "**flat-faced ape-man** from Kenya".

## 9.16   Reduced Sexual Dimorphism

All hominoids show some differences in size between the sexes and in features as the shape of pelvis and in crests on skull. Thus male gorillas weigh perhaps twice as much as females. Size difference is much less in chimpanzees and even in modern human, where average male is 1.2 times as heavy as female. This trend towards a lesser degree of sexual dimorphism can be traced in hominin fossils. Skeletons of australopithecines show marked sexual dimorphism, which is less in early hominids.

## 9.17   Changes in Size of Ribcage

Lucy (*Australopithecus afarensis*) had the funnel-shaped ribcage while it is has a barrel-shaped in *Homo erectus*. Allan Walker (1996) relates these changes in ribcage to a change in diet. Gorillas have a ribcage similar to Lucy's which accommodates the extensive gut to process the gorillas' rough diet and extract sufficient nutrients. The barrel-shaped ribcage and the waist in erectus suggest the consumption of significant quantities of meat. That meat is higher in calories, easy to digest than vegetable suggests an omnivore, or a carnivore, has a short gut than the vegetarian. An increase in amount of meat could fuel the high energy demand of large brain. Average cranial capacity increases markedly in *Homo erectus*.

## 9.18  Human Cultural Evolution

Human cultural evolution include tool making, use of fire, manufacture of shelter and clothing, art and other non-utilitarian products, cooperative hunting behaviour, and domestication of wild plants and animals. These allowed humans to have greater control of their environment. The development of skill directly contributes to survival of individuals practicing these behaviours. Some aspects of cultural evolution are easy to trace. Stone tools made by hominin species are relatively common. Tools made of wood or bone is not well preserved in stratigraphic record. Changes in behaviour, such as development of cooperative hunting groups or changes in social structure leave no direct traces at all. Their presence must be inferred from other evidence. Evidence of developing human culture appears far back in time. The name *Homo habilis* signifies its association with crude cobble tools and probably with *Australopithecus garhi* and *A. robustus*. Manufacture and use of tools was generally viewed as sole human activity. The Oldowan tools are the earliest evidence of culture in our ancestors. Now it is known that various animals use tools and best exemplified in chimpanzees. Jane Goodall first documented this. Animals use rocks, twigs and vegetation as simple tools, and modified them by stripping a twig of leaves, and breaking it to right length, to fish for termites in insects tunnels. Young chimps acquire such skills by observing their elders. Chimpanzees from different areas have distinctly different tool-making cultures. One difference between chimp and human culture is that chimps seldom carry tools, or raw materials for tool making, for any distance. Chimps make tools only immediately before using them. Tools used by early human were worked and reworked at different locations. Art works like jewellery, carving and cave paintings, do not recorded until 30-40,000 years

ago which follows the development of *Aurignacian* tool kits associated with Cro-Magnon culture. Some suggest that use of highly sophisticated language accompanied this flowering of culture, and marked the appearance of significant capacity for abstract thought. **Cultural evolution** has occurred in different times in different places. This is a reflection both of time at which different regions of globe were settled, and nature of biology & geology of an area, which poses constraints on domestication of wild plants & animals.

### Tools and tool use

At present the earliest-known evidence of manufacture and use of tools is a 2.5 million-year-old site, probably associated with *Australopithecus garhi* containing primitive stone tools, but without hominin remains. Animal bones recovered together with garhi remains from a nearby site of same age show cut marks from the stone tools which are often called as cobble tools. Two main types of **cobble tools** are core tools and flake tools. **Core tools** are stones with one or more flakes. Flake tools were not modified to use. The Oldowan tool culture persisted for about million years in Africa. *Homo erectus* developed the sophisticated highly modified Acheulean toolkit with sharp and straight edges formed by removal of more and small flakes. So-called hand axe is perhaps the best-known *Acheulean* tool with a pointed end and sharp sides. Other *Acheulean* tools are hammers, cleavers, and flake-based tools like knives. Erectus was probably the first hominin to leave Africa. *Acheulean* tools (Figure 9.17) are found in Europe and Asia. The Chinese erectus populations were believed to use **Oldowan tools**. The 1.3 million years old *Acheulean* tools are found in China. Sophisticated tools changed the types of foods eaten. While australopithecines, and perhaps *H. habilis*, were

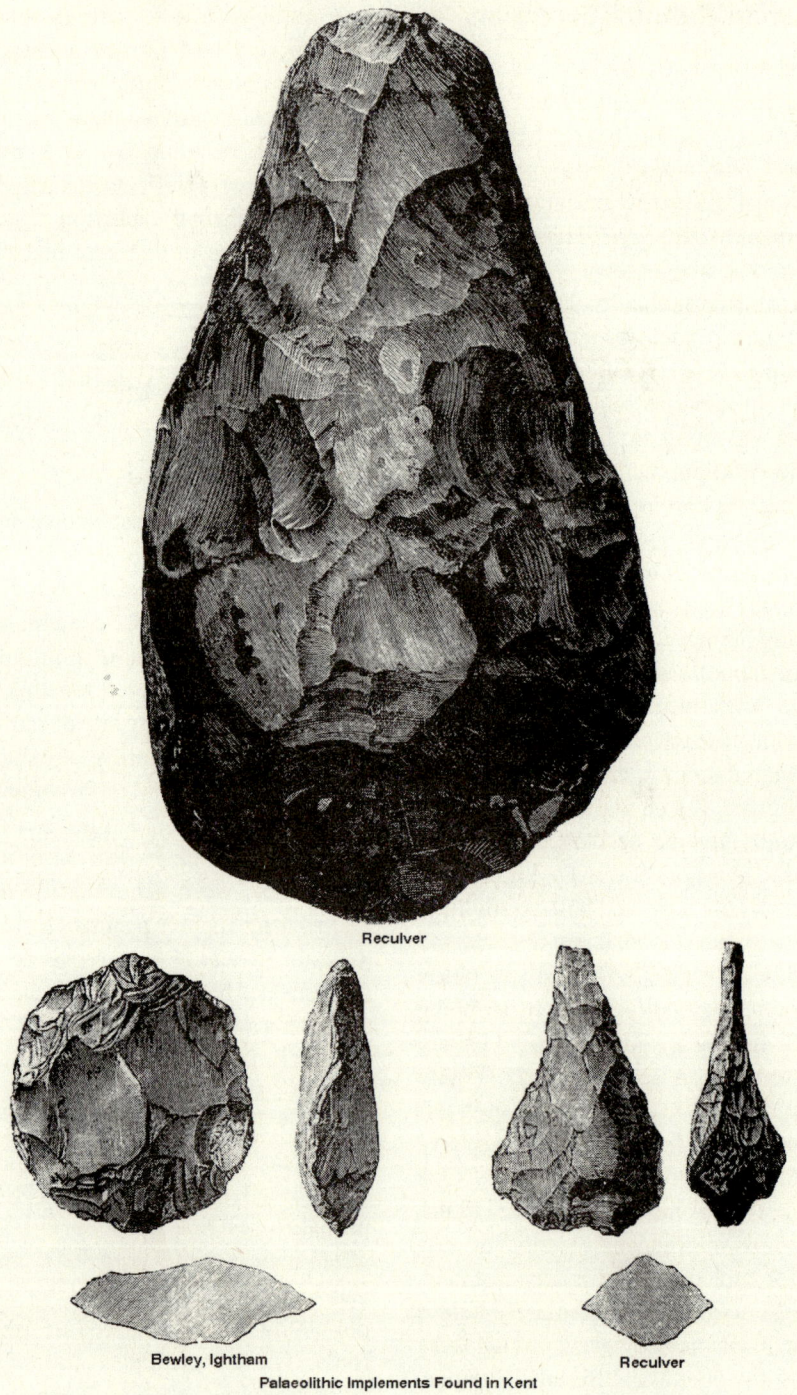

Reculver

Bewley, Ightham

Palaeolithic Implements Found in Kent

Reculver

**Figure 9.17**   Acheulean Handaxes

vegetarian, meat was included in erectus diet. Remains from many sites show large scale meat consumption.

## 9.19  Fire

Bones dating back perhaps 1.5 mya found in the Swartkrans Cave in South Africa provide some of the earliest evidence for use of fire. Analysis of bones showed that they had been heated to high temperatures. The earliest site with evidence of regular use of fire is the Zhoukoudian near Beijing. Here charred bones, charcoal, and rocks modified by exposure to fire are found. Many bones of game animals suggest organized hunts by local erectus population. Learning to use fire in a controlled manner was major step for our ancestors. It gave them greater control over their environment with potential to make available variety of foods. Fire protected from predators, and helped to survive in cold environments. The controlled use of fire is evidence of ability to plan ahead and would have aided social interactions as people gathered round the hearth.

## 9.20  The Intelligence

Theory tells that natural selection, random mutations made cross-species changes and produced life-forms to survive in the environment. Human brain does not fit into evolutionary theory. Man's mind is too advanced for survival needs. Darwin assumed that Europeans were highly intelligent. They competed with third-world natives who only had intelligence slightly above that of apes as per Darwin. Wallace lived with natives in primitive tropical lands and recorded their advanced minds like the Europeans. Their knowledge was different, but not their mental faculties. All humanity had intelligence more than any animal in world. Darwin ended that

natural selection makes an animal only as perfect as it needs for survival. Wallace reasoned, humans living as simple tribal hunter-gatherers would not need more intelligence than gorillas.

## 9.21  Monkey Talk

*Homo sapiens*, perhaps the most adaptable single species on the earth with learning and behavioral attitude has evolved in same environment following the same rules as all other organisms. Alexander noted the inheritance of genetic code for form and function in all organisms. **Sexual recombination** and random mutations produce endless variation in succeeding generations. Natural selection, the demands of environment select from this endless genetic variety those organisms' best suited to survive and reproduce. Alexander states that interaction of these phenomena and successions of environments in which organisms have lived account for traits and history of all forms of life. Natural selection with a directional component is the driving agent in evolution of traits which are adaptive to explicit circumstances and selected. There is an explanation for all existing traits and behaviors. If evolutionary theory is the truth, then species have a vast potential for material and mental progress. If natural selection shapes the evolution of brain, power of esthetic judgments must have arisen by same process. Brain promotes the survival and multiplication of genes that direct its assembly. Conrad Waddington compared the relationship of inherited and acquired behavior in humans to topography of a landscape. For physical traits like eye color topography is a single deep channel. Once egg and sperm meet, only one eye color is possible. The adaptive topography of human behavior is broader and complicated, but it is still topography. Even many behaviors are similar in form across culture, suggesting that behavior follows a defined path. Paul Ekman showed pictures of **facial expressions** to children of American and of Stone Age cultures; 80% of children in both recognized the emotion being expressed in face of other culture.

Topography of language is more complex. Humans have an innate drive to acquire speech. Unlike other primates they babble, invent words, and experiment with meaning. This first language is universal in humans across culture, and shows that "upper slope" of **topography of language** is well-defined; though the lower slope of "second language" is broad and ill-defined, shifting with culture. Primary mental abilities and perceptual and motor skills are most influenced by heredity, while personality traits are least influenced. It seems that ability needed to cope with relatively unvarying problems in physical environment develop along narrow channels, while qualities of personality, which represent adjustments to rapidly shifting social environment, are more malleable. Actions that must be induced rapidly are guided by emotion, for example childhood phobias, fears of snakes, spiders, rats, heights, dislike of different tastes. These were all potentially dangerous in ancient environment. In early human history, fears and phobias might have provided an extra margin of security needed to ensure survival.

If humans evolved according to evolutionary laws, then human propensity to create religious and secular ritual may be due to biological messages. The universal human tendency to classify into categories like members versus non-members, kin verses non-kin, makes sense in terms of genetic advantage. Individuals seek to perpetuate their own genes at expense of stranger. Intraspecific competition is a basic tenet of evolutionary theory. The key to many aspects of human culture is the hypertrophy, exaggerated growth of pre-existing structures. Basic social responses of hunter-gatherers have metamorphosed from relatively modest environmental adaptations into elaborate, sometimes monstrous forms in more advanced societies. Nationalism and racism are the culturally nurtured outgrowths of simple tribalism. Within Kungtribe in Africa, the Nyae Nyae Kung believed themselves as clean and perfect and other Kung people as alien murderers who use poisons. More developed civilizations have raised this mechanism of self aggrandizement to an art form, exalting them by divine allow and denigrating others with false histories. Often origins of these cultural elaborations are hidden. Marvin Harris has suggested that chronic meat shortage may be the underlying cause in evolution of certain religious beliefs. Large game was deficient in the Ancient Mexico. Collection of meat protein became difficult as Aztec civilization flourished in the Valley of Mexico. When in 1521 Cortez entered the Aztec capital of Tenochititlan, he observed thousand skulls. The Aztecs believed that human sacrifice was necessary to please the high gods.

## 9.22   The Languages

**Human language** is unique without significant analogue in animal world. This is different from any communication system in other animals. Non-human vocable is the interjection. They show the physical or emotional state of the individual. Chimpanzees talk in sign language, signal for things and get them. They don't argue and make moral decisions. They have no value system and know knowledge about death. Human language is closely associated with superior thinking ability. In communication ability, man differs from other animals than he does in his learning or thinking. Writing did not begin until after 2500 B.C. Ancient writings were more complicated than modern ones when some races alternately write backward and forward: one line from left to right, and next line from right to left. Complicated ancient languages shows that person back then had better mental capacities. The so-called **primitive languages** throw no light on language origins. Most of them are complicated in grammar than tongues spoken by civilized people.

Many languages of non-literate peoples are more complex than modern European ones. Evolution in language is the opposite of biological evolution. Languages have evolved from complex to simple. Many primitive languages are more complex and more efficient than the languages of the so-called higher civilizations. Elgin remarks, '*the most ancient languages for which we have written texts-Sanskrit, are often far more intricate and complicated in their grammatical forms than many contemporary languages.*" This clearly suggests that ancient men were more intelligent than those living on earth today. The oldest language can be reconstructed is sophisticated, already modern from Simpson's evolutionary point of view. Evolution of language is a story of progressive simplification.

## 9.23 Learning

Classical learning theory suggests that by trial and error, a given animal can learn to respond to wide variety of stimuli. This suggests certain plasticity to learning process which may often not exist. It has been shown for example that rat can easily learn to associate the taste of food to sensation of nausea and can quickly learn to associate food size to physical pain, but cannot make inverse associations, that is, fails to associate food taste to physical pain and food size with nausea. This is likely because an animal can get sick in nature from eating tainted food, while food size is rarely a threat. The rats learning are specific to its environment. **Insight learning** is the ability

to do a correct or appropriate behavior on the first attempt in a situation with which animal has no prior experience. If a chimpanzee is placed in an area with a banana hung too high above its head to be reached, and several boxes on floor, the chimp can reflect cognitively on the situation and then stack the boxes to allow it to reach food.

Insight is best developed in primates even in this group the number of insights often varies substantially from one situation or species to another. Consciousness can be thought of as an extension of insight learning. Consciousness is a system by which in some sense aware of us and our relationship to others and rest of the world. It has likely evolved in context of success in social matters. How others see us is crucial to social success. Consciousness may be a system for testing the question of how others see us and adjusting our image in our self-interest. A central aspect of consciousness is the ability to look ahead, the capability that we call foresight. It is the ability to plan, and in social terms the ability to outline scenario that is going to happen. Thus it is a system whereby we improve our chances of doing those things that will represent our own best interests. During using our consciousness and foresight to build scenarios and plan our social interactions, we visualize alternatives and test them one by one. We see these alternatives as available to us if we choose to use them. Our ability to visualize alternatives, particularly about social interactions, may be the basis for concept of free will. In this sense, free is not incompatible with idea of an evolved tendency to maximize inclusive fitness.

Plasticity of human behavior and the specie's unprecedented ability to learn new behaviors suggest we can somewhat choose modes of existence that are adaptive and functional in rapidly changing environments. This is certainly present human condition, evolving rapidly, as we have, from hunter gatherers to densely populated urban dwellers, from spear throwers to missile launchers. We are more controlled by our genes than we have realized. So it seems that the more we discover about the mechanisms of genetic control, more possible it will become to transcend these controls through awareness and conscious choice. In the words of A. Rosenfeld, "*for the first time in our history, work for ourselves instead of for our genes, exercise truly free will and free choice, give free rein to our minds and spirits, attain something close to our full human hood*."

A prehistoric human skull was found in a cave with holes in cranium. Holes matched perfectly with canines of a leopard, which had apparently dragged the body into cave before consuming it. Life has always been perilous for *Homo sapiens*. Theory of natural selection states that offspring are variable. All species have more offspring than can survive, and on average, those individuals that are best adapted to the current environment will endure. Our 46 human chromosomes come in pairs, one from mother and one from father. When sex cells are produced, each paired set sorts independently of others; the possible ways to recombine these 46 chromosomes totals 64 trillion. **Sexuality** and aggression are related, We are concerned in observing that people have a strong tendency to adopt religions beliefs all over the world. Chimpanzees are capable of insight learning: reviewing possibilities and scenarios inside the brain and then acting on them. Human were hunter gatherers for 98% of the 300,000 years that the species has existed.

## 9.24   Domestication of Plants and Animals

Evidence of the earliest crops and domesticated animals is always in the Near East, generally in plains below eastern Turkey where the Ararat Mountains are located. Using carbon-14 dating the earliest wheat cultivation originated in Palestine or Turkey about 7000 B.C. Maize and other plants were cultivated in Central America and Peru. The first-known dogs and sheep are found in the Near East from about the same time. Sheep found in Iran dated back to 6700 B.C. During the same time in Jericho, goats appeared. No evidence of evolution of dogs has been found. The earliest pigs were kept in Iran by 7000 B.C. The first cats are dated back to 3000 B.C. in Egypt. The earliest remains of cattle come from Greece and date to about 6500 B.C. Humped cattle of India first appeared in Mesopotamia about 3000 B.C. Domesticated cattle were in Egypt by 3700 B.C. The earliest use of elephant for transport comes from India. The first domesticated plants would have been annual plants with large seeds or fruits, like peas, legumes, and cereals. Apples and olives came later. Rice was domesticated in Asia.

## 9.25   Mitochondrial DNA

All of the genes in eukaryote cell are not found in chromosomes. Mitochondria of all eukaryote cells and chloroplasts of green plants and algae have DNA. In multicellular organism, the DNA of mitochondria (**mtDNA**) and chloroplasts is inherited about exclusively in the maternal line. Zygote inherits their organelles from ovum. mtDNA is used to determine phylogenetic relationships. Organisms whose DNA sequence for a particular gene differs by only a few bases are likely to be closely related. Different parts of genome mutate in different rates. Ribosomal DNA (**rDNA**) can be used to determine the relationships between species that last shared a common ancestor as it mutates relatively slowly. mtDNA accumulates changes in base sequence rapidly and it can be useful in studying the evolutionary history of species.

## mtDNA and the African Eve hypothesis

mtDNA is used to examine our own recent history. Studies have tried to put a date to the time when various human populations diverged and found that the mtDNA of humans from a range of different geographical origins is relatively uniform. This shows a very recent date for the populations to have separated. They have also shown that the mtDNA from African populations is more diverse than any other group, showing that these populations have relatively longer history. The common origin for all modern human populations lies in Africa - the so-called *African Eve hypothesis*. Dates for this origin range from around 60,000 to 400,000 years ago. More recent studies have placed the founding population at 200,000 years or less. These younger dates are supported by the discovery of modern human skulls dating back 160,000 years ago, from the Middle Awash of Ethiopia.

## mtDNA and history of Polynesian migrations

Studies of genomic DNA have shown that, in the NZ population as a whole, Maori and Polynesians have lower heterozygosity than the rest of the population. This fits with the idea that NZ was colonized by a series of "island-hopping" migrations, each of which would have led to a founder effect, with the result of inbreeding and decreased heterozygosity. More detailed information on the history of these migrations can be obtained by studying mitochondrial DNA (mtDNA). This is because mtDNA is inherited almost exclusively down maternal lines. Studies of mtDNA lineages from the NZ population have confirmed earlier findings that the ancestors of NZ's indigenous population followed the Austronesian migration route, from Taiwan, into the Philippines and Indonesia, and then across the Pacific and finally to New Zealand. And like the studies of genomic DNA, they show a decrease in genetic diversity along this migration route. The concept of a Taiwanese homeland is also supported by studies of genes coding for enzymes involved in the metabolism of alcohol in the liver. Many NZ Maori and Polynesians have a variant form of an enzyme that speeds this process up, a trait that they share with the indigenous tribes people of Taiwan. mtDNA data also make it possible to estimate the number of female colonists on the migration from Eastern Polynesia to New Zealand. There may have been more than 100 female settlers, which suggest that the migration was planned rather than the result of voyagers becoming lost between islands. Studies of **Y chromosome** DNA, which is passed exclusively from father to son show that genetic lineage in male have different origins from the lineages of female. This results from gender-biased migration following contact between Austronesian and Papuan groups.

## mtDNA and recent human evolution

Study of mtDNA has shown that gene frequencies have changed during the last 50,000 years. Human populations are still subject to evolution. Some mutations in mtDNA cause aerobic respiration less effective in which mitochondria produce more heat and be advantageous for humans living in the cold climates that prevailed during the Ice Age. The mtDNA from over 1,000 people has revealed that such a mutation is found in East Asians, Northern Europeans, and Amerindian populations. Of those live in Arctic regions, 75% had the mutation, which was found in 14% of sample living in temperate zones. Some ancestors would have lived in Siberia, and all would have faced the Ice Age's glaciations. The mutation is absent in people of African ancestry. Study concludes that correlation between habitat and presence of the beneficial mutation is evidence of positive selection for the changed gene sequence. That is, mutation was selected for because those people who had them were able to generate more body heat in an extremely cold climate.

## 9.26 Ongoing Human Evolution

Detecting the influence of evolution on the present-day human gene pool is easy as we can view the development of our species' family tree (Baulter, 2005). In the developed world, the combined effect of modern medicine,

**modern agriculture** and technological advances, and cultural alterations have reduced the effects of natural selection. In developing countries people are exposed to such selection pressures. The spread of some alleles giving resistance to malaria. In malaria endemic regions, anyone having genotype for resistance to malaria would be advantageous, because they would be likely to survive and reproduce, passing their genes to their children. The overlap between geographic distributions of **malaria** in Africa with sickle-cell allele is an example.

## SUMMARY

1. The earliest modern humans date back 170,000 years. Humans and chimpanzees were diverged 7 mya. The ancestors of modern humans evolved in Africa and remained there until 1.5 mya ago, when *Homo erectus* populations left Africa and moved across Europe and Asia.

2. The Calaveras skull dated back to the Pliocene. Human footprints in various sites in the US and Africa include Glen Rose tracks and Antelope Springs tracks.

3. Of 2 theories of human ancestry, first views that the human and ape had the common ancestor about 20 mya. Second views that the human descended from the ape. Man switched over from the physical evolution to cultural- and social evolution. Human groups for a time live in Stone Age cultures as in New Guinea and other areas lived under trees. Men gathered worms, leaves, and fruit with few crude stone and bamboo tools.

4. Phylogenetic analyses show that *Sivapithecus* is the likely ancestor of the orang-utan and *Dryopithecus* is probably the forebear of African apes and humans. The oldest remains of this species, belong to the subspecies *kadabba*, are known from Ethiopian rocks dated about 5.8 million years old. Fossil material of this species consists of skull fragments and some teeth and a toe bone.

5. *Ardipithecus* lived in a forested environment. In the 1960s, Louis Leakey found some teeth and skull fragments at Olduvai and name them *Homo*. That brain enlargement marked the beginning of man was long popular.

6. *Homo floresiensis* found on the Indonesian island of Flores was named on the basis of the partial remains - including the skull, jaw, and teeth of an adult female, and fragments from several other individuals.

7. There has been a gradual increase in cranial capacity in human evolution. Thus *Sahelanthropus* and the early australopithecines had cranial capacities within the range of modern chimpanzees. Skulls attributed to early *Homo* begin at around 510cc, and

there was a marked increase with *Homo erectus*, where later specimens had brain sizes of up to 1225cc, well within the modern range.

8. There is a strong positive correlation between body size and brain size; even within species. Male humans have larger body mass than females, and correspondingly larger cranial capacity. Equally importantly, the brain size in hominids, particularly *Homo* species, is greater than would be predicted for animals of their body mass.

9. There are many examples of human cultural evolution. They include: tool making, the controlled use of fire, manufacture of shelters and clothing, appearance of art and other non-utilitarian products, development of cooperative hunting behaviour, and domestication of plant and animal species. All of these features allowed humans to have greater control of their environment, rather than responsive to it.

10. The mind of man is the unanswerable to the concept of evolution. The theory tells that natural selection, random mutations, made cross-species changes and produced life-forms to survive in their environment. The human brain does not fit into evolutionary theory. Man's mind is too advanced for survival needs.

11. Human language is unique without significant analogue in the animal world. Human language is different from any communication system in other animals. Non-human vocables are the interjections.

12. Until recently biologists believed sex speeded up the rate of evolution, so that over long periods of time sexual species replaced asexual by virtue of their more rapid evolutionary response to environmental change.

13. The plasticity of human behavior and the specie's unprecedented ability to learn new behaviors suggest we have modes of existence that are adaptive and functional in rapidly changing environments. mtDNA accumulates changes in the base sequence relatively rapidly, up to 10 times as fast as nuclear DNA and can be useful in studying the evolutionary history of species that diverged only recently, or to look at populations of the same species.

## REVIEW QUESTIONS

### Short Answer Questions

1. Give two salent features of Primates.
2. Give four major features of Primate evolution.
3. Give the scientific name of Cro-magnon man.
4. Give the scientific name of Java man.
5. Give the scientific name of Piltdown man.

## Long Answer Questions

1. Discuss the important features of Primate evolution.

2. Discuss the important features of human evolution.

# REFERENCES

Arnason, U., Gullberg, A. & Janke, A. 1998. Molecular timing of primate divergences as estimatedby two nonprimate calibration points. Journal of Molecular Evolution, 47: 718-727.

Arnason, U., Gullberg, A., Burguete, A. S. & Janke , A.2000 .Molecular estimates of primate divergences and new hypotheses for primate dispersal and the origin of modern humans. Hereditas, 133: 217-228.

Berggren, W. A., Lucas, S. G. &. Aubry ,M.P.1998. Late Paleocene-Early Eocene climatic and biotic evolution: an overview. In: Aubry, M.-P., Lucas, S. G. & W. A. Berggren (eds.) Late Paleocene-Early Eocene Climatic and Biotic Events in the Marine and Terrestrial Records, New York: Columbia University Press, pp. 1-17.

Bramble, D.M. & Lieberman, D.E .2004. Endurance running and the evolution of *Homo*. Nature, 432: 345-352

Bromham, L. & Penn, D. 2003. The modern molecular clock. Nature Reviews Genetics, 4 :. 216-224.

Brown, P. et al.2004. A new small-bodied hominin from the Late Pleistocene of Flores, Indonesia. Nature 431, 1055 - 1061

David R. B .2003. Planet of the Apes .Scientific American ,289(2): 64-73

Diamond, J. 1997. Guns, germs and steel: a short history of everybody for the last 13,000 years pub. Vintage

Diamond, J. 2002. Evolution, consequences and future of plant and animal domestication. Nature, 418: 700-706

Fleagle, J.G.1999. Primate Adaptation and Evolution, 2nd edition, Academic Press, New York.

Foote, M. & Sepkoski, J. J. 1999. Absolute measures of the completeness of the fossil record.:Nature, 398 : 415-417.

Galik, K. et al.,2004. External and Internal Morphology of the BAR 1002'00 *Orrorin tugenensis Femur*. Science, 305 (5689):1450-1453

Gee , H.2001. Return to the planet of the apes .Nature ,412:131-132

Gibbons, A.2007. A new body of evidence fleshes out *Homo erectus*. Science, 317: 1664

Gingerich, P. D.1980. Evolutionary patterns in early Cenozoic mammals. Annual Review of Earth and Planetary Sciences ,8: 407-424.

Gingerich, P. D. &. Uhen, M. D 1994. Time of origin of primates. Journal of Human Evolution,27: 443-445.

Groves, C. P.2005. Order Primates. In: Wilson, D. E. & D. M. Reeder (eds.), Mammal Species of the World: A Taxonomic and Geographic Reference, Volume 1, Baltimore: Johns Hopkins University Press, pp. 111-184.

Hecht J .2004. Donkey domestication began in Africa New Scientist on-line: newscientist.com accessed 17 June 2004

Hedges, S. B., Parker, P. H., Sibley, C. G. & Kumar. S. 1996. Continental breakup and the ordinal diversification of birds and mammals. Nature, 381: 226-229.

Heinrich, W.D. 1998. Late Jurassic mammals from Tendaguru, Tanzania, East Africa. Journal of Mammalian Evolution ,5 : 269-290.

Heinrich, W.D.1999. First haramyid (Mammalia, Allotheria) from the Mesozoic of Gondwana. Mitteilungen des Museums für Naturkunde Berlin, Geowissenschaften ,2: 159-170.

Hooker, J. J.1998. Mammalian faunal change across the Paleocene-Eocene transition in Europe.In: Aubry, M.-P., Lucas, S. G. & W. A. Berggren (eds.), Late Paleocene-Early Eocene Climatic and Biotic Events in the Marine and Terrestrial Records, New York: Columbia University Press, 428-450.

http://www.becominghuman.org - an website with a wealth of information and links to other sites and sources

http://www.modernhumanorigins.com - this site provides good information on the workings of natural selection and genetic drift

http://www.rsnz.govt.nz/news/beinghuman/beinghuman.pdf - proceedings of a conference organised by RSNZ; contains an excellent article by Colin Groves on human physical & cultural evolution

http://www.stanford.edu/~harryg/protected/evolve6.htm - discusses changes in tool-making techniques, & tool-kits characteristic of various species

http://www.talkorigins.org/faqs/homs/garhi_cg.html - mentions various evolutionary trends in a discussion of the evolutionary relationships of Australopithecus garhi

http://www.talkorigins.org/faqs/homs/species.html

Huelsenbeck, J. P., Larget, B. & Swofford, D. 2000. A compound Poisson process for relaxing the molecular clock. Genetics, 154: 1879-1892.

Janke, A., Feldmaier-Fuchs, G., Thomas, W. K., von Haeseler, A. & Pääbo, S. 1994. The marsupial mitochondrial genome and the evolution of placental mammals. Genetics, 137: 243-256 .

Kay, R. F., Ross, C. & Williams, B. A. 1997. Anthropoid origins. Science, 275: 797-804.

Krause, D. W. & Maas, M. C. 1990. The biogeographic origins of late Paleocene-early Eocene mammalian immigrants to the Western Interior of North America. Geological Society of America Special Papers 243 , 71-105.

Kumar, S. & Hedges S. B. 1998. A molecular time scale for vertebrate evolution. Nature ,392: 917-920.

Leakey, M G. etal.2001. New hominin genus from eastern Africa shows diverse middle Pliocene lineages, Nature, 410: 433 - 440

Leakey M. G., Feibel, C. S., McDougall, I.,Walker, A.2002. New four-million-year-old hominid species from Kanapoi and Allia Bay, Kenya. Nature, 376: 565-571

Lemonick ,M., & Dorfman, A.1999. Up from the apes Time August 23: 36-44

Leslie C. Aiello & Mark Collard .2001. Our newest oldest ancestor? Nature, 410: 526-527

Lieberman D. E .2 001. Another face in our family tree. Nature, 410: 419-420

Lieberman D. E. 2007. Homing in on early *Homo*. Nature, 449: 291-292

Lordkipanidze, D et al. 2007. Post-cranial evidence from early *Homo* from Dmanisi, Georgia. Nature ,449: 305-310

Luo, Z.X., Ji, Q., Wible, J. R. &.Yuan C.X .2003. An early Cretaceous tribosphenic mammaland metatherian evolution. Science, 302 : 1934-1940.

Madsen, O., Scally, M., Douady, C. J., Kao, D. J., Debry, R. W., Adkins, R., Amrine, H. M.,Stanhope, M. J., de Jong, W. W. &. Springer, M. S. 2001. Parallel adaptive radiations in two major clades of placental mammals. Nature, 409:610-614.

Marivaux, L., Welcomme, J. L., Antoine, P.O., Métais, G., Baloch, I. M., Benammi, M.,Chaimanee, Y., Ducrocq, S. & Jaeger, J.J. 2001. A fossil lemur from the Oligocene of Pakistan. Science, 294: 587-591.

Martin, R. D. 1990. Primate Origins and Evolution: A Phylogenetic Reconstruction, New Jersey:Princeton University Press.

Martin, R. D. 1993. Primate origins: plugging the gaps. Nature, 363 : 223-234.

Martin, R.D. 1999. New Light on Primate Evolution http://bbaw.opus.kobv.de/volltexte/2009/1202/pdf/VI_03_Martin.pdf?origin=publication_detail

Martin, R. D. 2000. Origins, diversity and relationships of lemurs. International Journal of Primatology, 21 : 1021-1049.

Martin, R. D. 2003 Palaeontology: Combing the primate record. Nature, 422: 388.

McCrone ,J.2000. Fired up. New Scientist 20 May : 30-34

Michael Bulter.2005. Are humans still evolving? Science, 309: 234-237

Michel Brunet et al. .2002. A new hominid from the Upper Miocene of Chad, Central Africa. Nature, 418: 145-151

Murphy, W. J., Eizirik, E., Johnson, W. E., Zhang, Y.P., Ryder, O. A. & O'Brien S. J. 2001a. Molecular phylogenetics and the origins of placental mammals. Nature, 409: 614-618.

Murphy, W. J., Eizirik, E., O'Brien, S. J., Madsen, O., Scally, M., Douady, C. J., Teeling, E.,Ryder, A. O., Stanhope, M. J., de Jong, W. W. & Springer, M. S. 2001b. Resolution of the early placental mammal radiation using Bayesian phylogenetics. Science, 294: 2348-2351.

Ni, X., Wang, Y., Hu, Y. & Li. C. 2004. A euprimate skull from the early Eocene of China. Nature, 427: 65-68.

Nikaido, M., Nishihara, H., Hukumoto, Y. & Okada, N.2003. Ancient SINEs from African endemic mammals. Molecular Biology and Evolution ,20: 522-527.

Nikaido, M., Rooney, A. P. & Okada, N. 1999. Phylogenetic relationships among certartiodactyls based on insertions of short and long interspersed elements: Hippopotamuses are the 404 Robert D. Martinclosest extant relatives of whales. Proceedings of the National Academy of Sciences,USA 96: 10261-10266.

Novacek, M. J. & Wyss, A. R. 1998. Higher-level relationships of the recent eutherian orders: Morphological evidence. Cladistics, 2 : 257-287.

Phillips, E. M. & Walker, A. C.2002. Fossil lorisoids. In: Hartwig, W. C. (ed.), The Primate Fossil Record, Cambridge: Cambridge University Press, pp. 83-95.

Richmond, B. G. & Strait, D. S. 2002 Evidence that humans evolved from a knuckle-walking ancestor. Nature, 404: 382-385

Ross, C. F. 2003. Review of "The Primate Fossil Record": Edited by Walter C. Hartwig. Journal of Human Evolution, 45: 195-201.

Ruiz-Pesini, E., Mishmar,D.,Brandon, M., Procaccio, V.and Wallace, D. C.2004. Effects of Purifying and Adaptive Selection on Regional Variation in Human mtDNA. Science , 303(5655): 223 - 225

Seiffert, E. R., Simons, E. L. & Attia, Y. 2003. Fossil evidence for an ancient divergence of lorises and galagos. Nature, 422: 421-424.

Shadan, S. 2007. You are what you ate. Nature, 449: 155

Shreeve, J.1995. The Neandertal Enigma: solving the mystery of modern human origins, Morrow

Sigogneau-Russell, D. 1995 .Further data and reflexions on the tribosphenid mammals (Tribotheria)from the Early Cretaceous of Morocco. Bulletin du Muséum National d'Histoirenaturelle de Paris, 4. série (C) 16: 291-312.

Soligo, C., Will, O., Tavaré, S., Marshall, C. R. & Martin, R. D. 2006. New light on the dates of primate origins and divergence. In: Ravosa, M. J. & M. Dagosto (eds.), Primate Origins: Adaptations and Evolution, New York: Springer, pp. 29-49.

Springer, M. S., Cleven, G. C., Madsen, O., de Jong, W. W., Waddell, V. G., Amrine, H. M.&. Stanhope, M. J. 1997. Endemic African mammals shake the phylogenetic tree. Nature, 388: 61-64.

Storey, M. 1995. Timing of hot spot-related volcanism and the breakup of Madagascar and India. Science, 267: 852-855.

Stringer, C., & Gamble C. 1993. In Search of the Neanderthals: solving the puzzle of human origins. Thames & Hudson.

Tavare, S., Marshall, C. R., Will, O., Soligo, C. & Martin R. D. 2002. Using the fossil record to estimate the age of the last common ancestor of extant primates. Nature, 416: 726-729.

Thewissen, J. G. M., Williams, E. M., Roe, L. J. & Hussain , S. T.2001. Skeletons of terrestrial cetaceans and the relationship of whales to artiodactyls. Nature ,413: 277-281.

Ursing, B. M. & Arnason, U.1998. Analyses of mitochondrial genomes strongly support a hippopotamus-whale clade. Proceedings of the Royal Society, London B 265: 2251-2255.

Waddell, P. J., Cao, Y., Hasegawa, M. & Mindell, D. P. 1999. Assessing the Cretaceous super ordinal divergence times within birds and placental mammals by using whole mitochondrial protein sequences and an extended statistical framework. Systematic Biology ,48: 119-137.

Walker, A., & Shipman, P. 1996. The Wisdom of Bones: in search of human origins. Weidenfeld & Nicolson, UK

White, T., Asfaw, B., Degusta, D., Gilbert, H., Richards, G., Suwas, G. & Howell, F. C.2003. Pleistocene *Homo sapiens* from Middle Awash, Ethiopia. Nature ,423: 742-747

Wing, S. L.. 1998. Late Paleocene-Early Eocene floral and climatic change in the Bighorn Basin,Wyoming. In: Aubry, M.-P., Lucas, S. G. & W. A. Berggren (eds.), Late Paleocene-Early Eocene Climatic and Biotic Events in the Marine and Terrestrial Records, New York: Columbia University Press,. 380-400.

Wolfe, J. A. 1987. Distribution of major vegetational types during the Tertiary. Sundquist, E. T.& W. S. Broecker (eds.), The Carbon Cycle and Atmospheric $CO_2$: Natural Variations,Archean and Present, Washington, D. C.: American Geophysical Union, pp. 357-375.

Wood, B.2002. Hominid revelations from Chad. Nature, 418: 133-135

Yang, Z. H. & Yoder, A. D. 2003. Comparison of likelihood and Bayesian methods for estimating divergence times using multiple gene loci and calibration points, with application to aradiation of cute-looking mouse lemur species. Systematic Biology, 52: 705-716.

Yoder, A. D., Burns, M. M., Zehr, S., Delefosse, T., Veron, G., Goodman, S. M. & Flynn, J. J. 2003. Single origin of Malagasy Carnivora from an African ancestor. Nature, 421: 734-737.

Yoder, A. D. &. Yang ,Z. H.2004. Divergence dates for Malagasy lemurs estimated from multiplegene loci: geological and evolutionary context. Molecular Ecology, 13: 757-773.

Zollikofer, C., Ponce de Leon, M., Lieberman, D., Guy, F., Pilbean, D., Likius, A., Mackaye, H., Vignaud, P. and Brunet, M .2005. Virtual reconstruction of *Sahelanthropus tchadensis*. Nature, 434: 755-759.

# CHAPTER 10

# Coevolution and Mimicry

Co evolution is a process of reciprocal evolutionary change in interacting entities which could be species, nucleotides, amino acids, proteins, genomes, proteomes, organisms and perhaps even ecosystems. Most theories of antagonistic interactions concentrate on the evolution of 1 of the 2 species (Boots and Haraguchi, 1999). Evolution of one species affects selection on other. A fully co evolutionary framework is important to understand evolutionary outcome. The co evolutionary model of **host-parasite interactions** shows that co evolutionary outcomes cannot always be predicted from evolutionary outcomes single of species. **Parasite evolution** is highly sensitive than that of host, often resulting in evolution of virulent parasites. Differences in mutation rates lead to highly variable parasite virulence.

Branching in one species does not induce branching in other. Relative shapes of trade-off relationships are crucial in determining evolutionary outcome. These results are important to understand host-parasite interactions and co-evolutionary dynamics. Modern host-parasite co evolutionary game theory emphasizes the importance of dynamic evolutionary behaviors (Christiansen, 1991). Important asymmetries in co evolutionary outcomes are due to ecological characteristics and a fully co evolutionary model is the only way to predict them. We only fully understand the evolution of parasites in context of co evolution. Perhaps the level of **parasite virulence** is highly dependent on evolutionary dynamics of host. Evolution in host can induce selection for highly virulent parasites with higher transmission and shorter infectious periods than predicted by purely evolutionary models. Such fast, acute, and deadly parasites are a cause for great concern which may arise due to host evolution that maximizes resistance and are likely to evolve in polymorphic host populations, in which proportion of host individuals are infected easily and killed quickly. Any imposed changes in host resistance in agriculture (Gurr and Rushton, 2005), or medicine (de Clercq, 2004) will affect parasite evolution and risk selecting for high virulence. Understanding the processes that determine how deadly particular parasites are to their hosts is crucial to wildlife, agriculture, and human health. In the spillover parasites like Ebola in humans(Pedersen and Fenton, 2007) and squirrel pox in the Eurasian red squirrel (Tompkins et al. 2003), virulence is manifested in an evolutionary dead-end host from which there is little transmission.

Trade-offs with transmission explain parasites induced mortality as parasite optimizes R0 (Bremerman and Thieme, 1989), but host resistance can drive much higher virulence than would optimize the parasite R0 in the absence of host resistance. **Host heterogeneity** may lead to either reduced (Regoes et al., 2000) or increased parasite virulence (Ganusov et al., 2002), depending on nature of heterogeneity. Polymorphism in avoidance resistance, selection for very high virulence may lead to evolution of effectively obligate killers, like insect baculoviruses (Dwyer et al., 1997), in which challenged hosts do not become infectious, or are killed. Rates of mutation can alter co evolutionary systems (Marrow et al., 1996).

Sensitivity of parasite to host evolution leads to significant transient changes in its virulence, if it has a higher mutation rate. Co evolutionary processes explain some variation in parasite virulence in nature. Whenever a parasite has a higher mutation rate than its host, it undergoes rapid change because relatively slow changes in its host have large impacts on parasite virulence. Viruses and bacteria have much faster generation times than their vertebrate, or invertebrate hosts. This potential to evolve much faster than their hosts has been one justification for modeling parasite evolution alone. The slower host evolution leads to dramatic changes in parasite transmission and virulence. The parasite's evolutionary optimum is very sensitive to current level of host resistance. Virulence is a transient state brought on by evolution of new resistance mechanisms

in host until host has itself reached its evolutionary optimum. This has implications for emergence of disease and management of virulence (Ebert and Bull, 2003). If **parasite virulence** is a moving target, managing disease in host populations becomes much more difficult. Only one instance of evolutionary branching may occur and that co evolution cannot itself cause branching when **evolutionary dynamics** alone do not. In co evolutionary predator-prey studies, high levels of polymorphism can evolve through repeated branching from a single ancestor (Weitz et al., 2005). These studies link traits to functional forms leading to asymmetric competition. More complex trade-off relationships would produce similar results. There has been debate about the potential for evolutionary branching in nature (Butlin and Tregenza, 2005) since it appears to be much easier in models than in real systems. Models deal with evolution of only one species, whereas co evolution is the rule in real systems, may help to explain this discrepancy.

Total transmission is decomposed into host susceptibility and parasite transmission in a straightforward multiplicative fashion More complex interactions between susceptibility and transmission may lead to a greater chance of polymorphisms. Parasite cannot branch if density dependence acts on host birthrate but can if density dependence acts on host death rate (Pugliese, 2002). This remains true when parasite coevolves with its host. Convergence of parasite is found to be very sensitive to host strategy and requires the host to have evolved to its singularity before parasite branching. This sensitivity is accentuated further when both host and parasite could potentially branch, that is, when co evolutionary response of host may prevent parasite branching but as a consequence cause the evolution of a highly virulent parasite.

Red Queen's Hypothesis states that an evolutionary system continues development to maintain its fitness relative to systems. Antagonist interactions in evolution create the concept of antagonist co evolution. Branching strategies for asexual population dynamics in limited resource environments are modeled using **Lotka-Volterra equations**. Model based on adaptive dynamics and data of floral and proboscis lengths, nectar consumed and pollen deposited during pollination of long tubed irisy by long-proboscid fly shows co evolutionary dynamics, including 2 types of **Red Queen dynamics**, evolutionary branching and trap. Selective pressure between 2 species can include host-parasite co evolution. This antagonistic relationship leads to necessity for pathogen to have the best virulent alleles to infect organism and for host to have the best resistant alleles to survive parasitism. As a result, allele frequencies changes through time depending on size of virulent and resistant populations and generation time where some genotypes are preferentially selected. Genetic change accumulation in both populations explains a constant adaptation to have lower fitness costs and avoid extinction in accordance with Red Queen's hypothesis.

Co evolution sometimes results in astonishingly exaggerated traits. Very high level of tetrodotoxin in some newt species protects them from most predators (Brodie et al., 2002). Complex suites of chemical and anatomical defenses of some plants protect them from herbivores. **Co evolutionary arms races** between species can favor exaggeration of traits for attack and defense (Franceschi et al., 2005)). Huge antlers and long spines are the structures of co evolutionary puzzle. Interacting species co evolves in different ways, often producing a geographic mosaic of traits and counter-traits. **Co evolutionary mosaics** can arise from geographic differences in intensity of natural selection, or in specific traits that are the focus of selection.

## 10.1　Types of Co evolutionary Interactions

Intensity of selection on coevolving interactions varies among regions which seem initially counterintuitive (Toju and Sota, 2005). The work on Japanese camellias (*Camellia japonica*), which occur throughout Japan is mentionable. Plants protect their seeds in fruits consisting of heavily thick, woody pericarp that contains up to 10 seeds. However, *Camellia* weevils (*Curculio camelliae*) using an unusual long rostrum, penetrate the pericarp and chew down to the seed. These weevils are the major cause of seed death. Weevil larvae destroy more than half the seeds in some populations. Plants and weevils from Yakushima Island have shown that, the longer the beetle rostrum, the higher (Thompson, 2005) the probability of successful boring. Thus, there is a high potential for an escalating co evolutionary arms race. Natural selection on increasing thickness of plant's pericarp may favor increasing length of weevil's rostrum. Camellia pericarp thickness varies in populations from 6 to 20 mm, and weevil (Thompson, 2005) rostrum length varies from 9 to 20 mm (Toju and Sota, 2005) showing a geographic pattern. The pericarp thickness and rostrum length increase from the North to South and are correlated in their pattern of geographic variation. Change in traits from south to north differs between these 2 species, creating geographic mismatches in traits of 2 species and in ecological outcomes of interaction. Proportion of excavations is higher in northern populations and lower in southern populations. These results were used to construct interaction fitness functions to assess how different combinations of plant and weevil traits would affect the fitnesses of interacting individuals (Toju and Sota, 2005). The procedure requires knowledge of phenotypic distributions of traits under selection. Distributions are incorporated into a set of linear equations to describe pattern of covariance of traits and evaluate the direction and strength of selection on coevolving populations.

Analyses showed that though the weevils now have advantage in the north, these plant populations are under weaker (Thompson, 2005) selection for increased pericarp thickness than those in the south. The present distribution of plant and weevil traits in northern populations results in high levels of seed infestation that correlation between traits and fitness is lower than in the south. The relatively low correlation decreases potential strength of selection for increased pericarp thickness. Although plants in the south are already highly defended, a strong correlation between plant traits and fitness in these populations exist, which results in stronger current **phenotypic selection** than in the north. Direct and necessary relationship between ecological intensity of an interaction like proportion of plants, or seeds attacked and strength of natural selection on that interaction is absent. Strength of selection imposed by an enemy on a plant population depends on degree to which that enemy differentially reduces the fitness of some plant genotypes more than others. High level of random attack would impose no selection on a host, or prey population. Moderate level of differential attack could exert strong selection. When co evolution is focused on same, few traits in a pair of interacting species (Thompson, 2005), selection is likely to vary geographically. Some local interactions will be co evolutionary hotspots, exhibiting strong reciprocal selection on the interacting species. Other local interactions will be co evolutionary cold spots, with selection acting on one species, or neither

species. In many interactions, there will often be regions where one species occurs without the other. These different forms of **co evolutionary cold spots** will lead to relax selection on traits that are escalating to varying degrees in **hotspots**. Gene flow among these regions, random genetic drift, metapopulation dynamics (Thompson, 2005), and selection for novel traits rather than exaggeration of current traits can all further mitigate relentless escalation of coevolving traits across the geographic range of an antagonistic interaction (Thompson, 2005).

Geographic mosaics of antagonistic co evolution have now been demonstrated over the past decade in an increasingly wide range of interspecific interactions, including those between vertebrates (Thompson, 2005) and their prey (Brodie et al., 2002), insects and plants (Thompson and Cunningham, 2002), fungi and plants (Thrall et al., 2002), and slave-making ants and their victims (Foitzik et al., 2003). Toju and Sota's (2005) study is the first to suggest how steepness of geographic clines in coevolving traits may shape geographic patterns in strength of co evolutionary selection suggesting that co evolution is a pervasive process that continually reshapes inter-specific interactions across broad geographic areas. Diseases like malaria vary geographically both in parasite virulence and host resistance, potentially creating regions of co evolutionary hotspots and cold spots. Spread of introduced species seems to create new geographic mosaics of co evolution as some species become invasive and co evolve with native species in different ways in different regions, or drive rapid evolution in native species, sometimes in less than a hundred years, or (Thompson, 2005) so (Callaway et al., 2005). Results for Japanese camellia and camellia weevils reinforce the developing view that interaction co evolve as a geographic mosaic across landscapes.

Coevolving species, the part of a broader ecological community do not live in isolation. Community-level interactions influence the dynamics of reciprocal selection, other micro evolutionary aspects of co evolution (Dercole, 2010) co speciation and other macro evolutionary patterns . A well-known example of the influence of coevolution involves Red Crossbills (*Loxia curvirostra*) and Rocky Mountain lodge pole pines (*Pinus contorta*) .In the western United States, the crossbills are the dominant seed predator which select for large, thick-scaled cones that protect pine seeds. Increased cone size exerts reciprocal selection (Harbison and Clayton, 2011) on crossbills for increased beak size. In areas where red squirrels (*Tamiasciurus hudsonicus*) are the dominant seed predator and out-compete crossbills, birds adapt to average cone size (Harbison and Clayton, 2011) and have small beaks (Benkman, 2001). As populations of crossbills with different beak sizes tend not to interbreed crossbills with and without squirrels have undergone speciation (Smith and Benkman, 2007). Thus, squirrels influence the co evolutionary dynamics of crossbills and pines over both micro- and macro evolutionary time (Figures 10.1, 10.2).

Another example of community interactions on co evolutionary patterns involves fungus-growing attini ants, the fungi they cultivate, and parasitic (Harbison and Clayton, 2011) fungi of cultivars. Phylogenies of these 3 groups are broadly congruent, reflecting 50 million years of close association (Currie et al., 2003, Harbison and Clayton, 2011). Phylogenies do not mirror one another perfectly as periodic switching of fungi between unrelated host lineages. Parasitoid wasps, or mites play a role in dispersing fungi between host lineages (Caldera et. al., 2009). Phoresis (Houck and O'Conner, 1991), is another way in which members of a community can conceivably alter patterns of co evolutionary history (Balakrishnan and Sorenson, 2007) involving birds and feather-feeding lice (Phthiraptera: Ischnocera). **Lice** complete their entire life cycle on body of host and are transferred to new hosts during direct contact, such as that between parent birds and their (Harbison and Clyton, 2011) offspring in nest. Some groups of lice are also capable of transmission by hitchhiking rides on more mobile parasitic flies (Harbison et al., 2008). Lice feed on feathers and dead skin. Birds defend themselves against feather lice by destroying, or removing lice with their bill during frequent bouts of preening (Clayton, et al., 2005) two groups of feather lice, wing and body lice, found on pigeons and doves (Columbiformes) have similar natural histories and are considered ecological

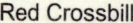
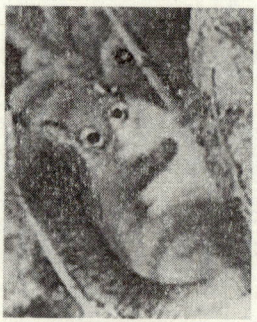

Red Crossbill                    Lodgepole pine                    Red Squirrel

**Figure 10.1** Coevolution between red crossbill, lodgepole pine and Red squirrel

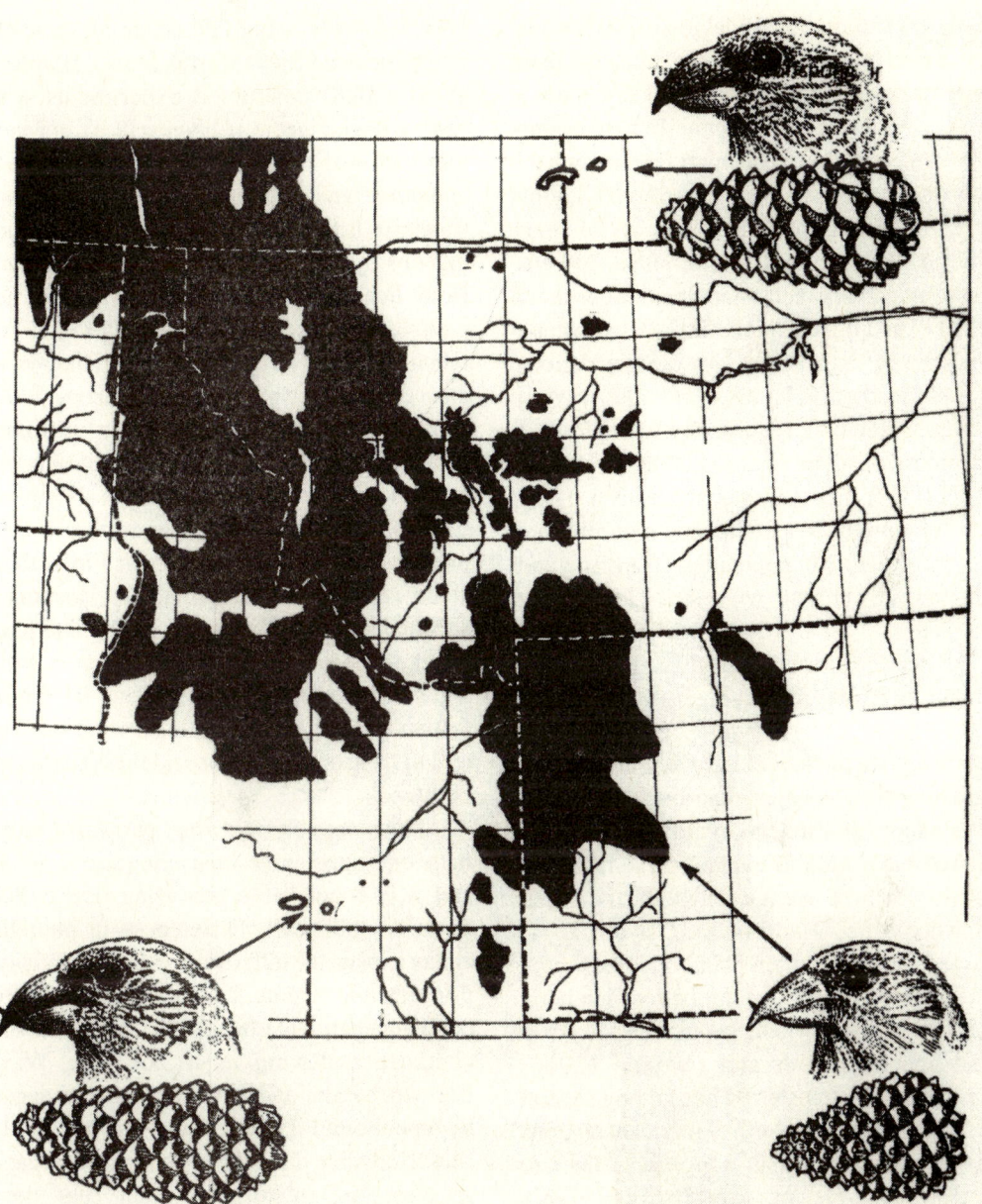

**Figure 10.2**  The distribution of lodgepole pine (black) (modified from Critchfield and Little 1966), the location of study sites, and representativecrossbills and cones in the Rocky Mountains, the Cypress Hills, the South Hills, and the Albion Mountains showing coevolution between crossbills and pine (modified after Benkmen, 199 American naturalist 153)

replicates (Johnson and Clayton , 2003). Wing lice spend most of their time and lay their eggs on large flight feathers. They have a long, slender shape that facilitates hiding between coarse barbs of flight feathers where they are protected from preening (Clayton et al., 1999). Body lice spend all of their time and lay their eggs on abdominal feathers. Their oval shape and short legs allow them (Harbison and Clayton, 2011) to escape preening by burrowing into downy regions of these feathers (Clayton et al., 1999). Wing and body lice both require downy regions of abdominal feathers for food. Wing and body lice differ in their patterns of host use. Body lice are more host-specific, and exhibit more genetic structure in population (Harbison and Clayton, 2011) than wing lice (Johnson et al., 2002).

Differences in specificity, population genetic structure, and phylogenetic (Harbison and Clayton, 2011) congruence demonstrate that wing lice are less closely associated with particular host lineages than body lice, both over micro- and **macro evolutionary time**. Studies have tried to explain these different patterns of host use. One hypothesis is that wing lice switch (Harbison and Clayton, 2011) host species more often than body lice which could result if wing lice are better than body lice at establishing viable populations on novel host species. Bush and Clayton (2006) experimentally transferred lice among several species of captive North American pigeons and doves showing that 2 groups of lice are equally capable of establishing viable populations on novel hosts. Differences in establishment ability could not explain different patterns of host use in wing and body lice. Another hypothesis is that wing and body lice differ in their ability to disperse to novel host species in the first place. Neither group is capable of independent (Harbison and Clayton, 2011) locomotion off the body of

host. However, wing lice attach phoretically to hippoboscid flies (Martin,1934 ). Harbison et al. (2008) conducted experiments with captive Rock Pigeons (*Columba livia*), in which they showed that wing lice (*Columbicola columbae* and *Columbicola tschulyschman*) transmit horizontally between individual pigeons by hitchhiking on hippoboscid flies. Body lice (*Campanulotes compar*) were not capable of phoresis because they could not remain attached to flies, even when they were placed on experimentally (Harbison, 2009, Harbison and Clayton, 2011). This functional constraint likely results from morphological adaptations of body lice for living in abdominal regions. They have short legs for burrowing in dense downy feathers, which limits their ability (Harbison and Clayton, 2011) to remain attached to flies (Harbison, 2009). The long outrigger legs of wing lice provide a wide stance suited to coarse surface of flight feathers. Using their long legs, wing lice are able to hang onto hipposboscid flies (Harbison, 2009).

Many hippoboscid flies parasitize more than one species of host (Bequaert, 1953) and it is conceivable that wing lice switch hosts via **phoresis**. Differences in phoretic ability explain the differences in host specificity, population genetic structure, and cophylogenetic (Harbison and Clayton, 2011) congruence of wing versus body lice. Wing lice switch hosts, and that switching is mediated by hippoboscid (Harbison and Clayton, 2011) flies. Body lice did not switch hosts or engage in phoresis, despite the fact that they outnumbered wing (Harbison and Clayton, 2011) lice on donor pigeons, and that flies spent more time in **microhabitats** (Harbison and Clayton, 2011) favored by body lice. Only 2 body lice were ever found on mourning doves-one in an experimental set and one in a control set-suggesting accidental dispersal on our clothing. The first step in phoresis-mediated host switching is that louse must be capable of locating a fly to which it can attach. One study is consistent with recent behavioral (Harbison and Clayton, 2011)work showing that wing lice orient to hippoboscid flies, whereas body lice show no response to same flies (Harbison, 2009). An underlying assumption of phoresis-mediated host-switching is that individual hippoboscid flies move between host species under natural conditions. Corbet (1956) showed that about 7% of flies in a marked population dispersed between host species over course of a week.

Flies were observed on (Harbinson and Clayton, 2011) both recipient pigeons and doves. Of 458 flies observed, 24% were on doves, despite the fact that *P. canariensis* (Harbinson and Clayton, 2011) is a parasite of Rock Pigeons. Both data from the field and experimental data from captive birds confirm that **hippoboscid flies** move between species of birds. Another assumption of phoresis-mediated host switching is that lice attach to flies. Three of 120 (2.5%) flies were observed with lice on them. Published studies of wild fly populations (Harbinson and Clayton, 2011) report lice on 8% to 43% of flies (Bequaert, 1953). Although a single gravid female may be capable of founding a new population, dispersal of several lice to a new host (Harbinson and Clayton, 2011) undoubtedly improves the probability of establishing a viable population on that host. In the analysis of over 400 published records of lice attached to flies, 60% of cases involved multiple lice on single flies, with up to 31 individual lice attached (Harbinson and Clayton, 2011) to a single fly (Peters, 1935). Recipient doves harbouring both adult and immature lice, suggest that lice are capable of reproducing on doves after they dispersed to them. It is possible that most of the immature lice on doves were themselves phoretic. Published records of immature lice attached to flies do exist (Walter, 1989) (Figures 10.3, 10.4).

**Figure 10.3**  Pigeon fly, *Pseudolynchia canariensis*

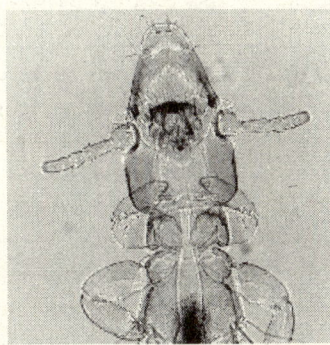

**Figure 10.4**  Narrow head pigeon louse
**Source:** www.people.uper.ca.sgreenhood/html/techinghtml.

Little is known about frequency of lice dispersing between wild birds, but several records involve lice from distantly related hosts. For example, a fly removed from a woodpecker was carrying four *Brueelia marginata*, a species of louse that normally infests songbirds (Ash, 1952). Similarly, the literature contains 2 records of flies removed from swifts with ischnoceran songbird lice attached (Bequaert, 1953). Swifts are normally host only to lice from suborder Amblycera, which do not engage in phoresis because their mouthparts do not allow them to hang onto flies (Price et al., 2003). The final step in phoresis-mediated host-switching is that lice must disembark onto a novel host species. In the study the percentage of wing lice found on doves was not significantly different from percentage of flies found on doves, suggesting that lice leave flies at the first opportunity. Flies sometimes knock lice off onto novel hosts (Harbison et al., 2009) showing that, in vitro, flies use grooming to dislodge lice that have been placed on them experimentally. Although flies presumably cannot groom when they are flying, newly attached lice can be groomed off once the fly has landed on a new host.

Differences in co evolutionary history of wing and body lice can be explained by differences in host-switching. **Nonphoretic** body lice track host lineages more closely than phoretic wing lice, leading to greater congruence between phylogenies of body lice and their hosts, compared with wing lice and the same hosts (Clayton and Johnson, 2003). Because body and wing lice are equally capable of establishing viable populations on novel hosts once they are reached (Bush and Clayton, 2006), difference in dispersal ability best explains the different patterns of association. Future work comparing the gene flow of phoretic and nonphoretic lice within and between host species could further clarify the impact of phoresis on processes such as local adaptation, specialization, and patterns of host-parasite co speciation. Columbiform wing and body lice are a powerful comparative system to study ecological factors influencing

macro evolutionary differences in host association. It is difficult to assess the relative influence of factors, like ongoing phoresis in bird-louse associations that lack this ecological replicate framework. Over one-third of all phoretic events found in survey literature involved *Brueelia* indicates that phoretic dispersal often occurs in this genus. Phoresis may govern differences in specificity of Brueelia from brood parasitic indigo birds (*Vidua* spp.), versus nonphoretic lice (*Myrsidea*) (Balakrishnan and Sorenson, 2007). Phoretic dispersal may explain how a single species of *Brueelia* infest every North American migrant thrush species (*Catharus* spp.) and maintain gene flow among these species (Bueter et al., 2009). Phylogeny for *Brueelia* shows no congruence with host phylogeny. This lack of congruence has been attributed to frequent host-switching mediated by phoresis (Johnson et al., 2002). Linking community interactions to co evolutionary dynamics is a stated goal of evolutionary ecologists (Johnson and Stinchcombe, 2007). Phoresis provides just one example of how broad community interactions can influence co evolutionary dynamics over both micro- and macro evolutionary time. Results highlight the importance of adopting a broad, community approach in studies of co evolution and also demonstrate the potential of host-parasite systems for unraveling connections between community ecology and co evolutionary biology.

**Co evolution** implies on 2 interacting species. Species interactions may be tight to diffuse based on frequency of interaction and impact of interaction on reproductive success. The more frequent and the stronger the fitness effects, the tighter are an interaction. Host-parasite interactions are often tight. **Predator-prey interactions** can be diffuse. The classical examples of mimicry illustrate different types of co-evolutionary interactions. Darwin described plant pollinator co evolution and emphasized the importance of mutual interactions of organisms and described the process of co evolution between bees and clover. Darwin wrote a book on **pollination biology** of orchids describing many specialized interactions between orchids and insects. The Star orchid (*Angraecum sesquipedale*) from Madagascar has 25 cm long flower spur which predict the existence of an insect with a matching long tongue. In 1903, the Hawk moth *Xanthopanmorganii praedicta* was described. T. Wasserthal could observe the predicted pollination mechanism suggesting an alternative hypothesis for the co evolutionary process. In the original hypothesis of orchid-moth co evolution, as proposed by Darwin and Wallace, the basic idea was that an orchid evolves a long spur to ensure that specialized pollinator can get access to nectar. A pollinator species evolves a long tongue and specialize on this plant. This pollinator always visits same flower species to pollinate it, effectively which is beneficial for orchid. Wasserthal suggested a different way of evolution of long spur and long tongue indirectly. He stated that there is first Predator-moth co-evolution which is followed by plant co adaptation. He suggests that moth evolved long tongues to increase distance-keeping and sideways oscillation hovering while nectar feeding, to avoid predation by bats and spiders. Plants then follow and evolve long spurs to bring their pollen closer to the moth.

**Sucking lice** belonging to the Order Anoplura are specialized obligate ectoparasites of eutherian mammals and complete their life cycle on the host (Marshall, 1981). Their modified mouthparts are adaptive to feed on host blood and host's blood differs in terms of providing nutrition to louse (Murray and Nicholls, 1965)). High host specificity in sucking lice could be reinforced by interactions with host's immune system, or due to a long

history of co speciation between hosts and their parasites. Co speciation is the speciation (cladogenesis) in a parasite lineage due to host cladogenesis (Brooks, 1979).Host-parasite associations can be the result of co speciation, or historical events like host switching, sorting events (extinction and lineage sorting), duplication events (parasites speciating on single host) and failure of parasite during **host speciation** (Paterson and Gray, 1997). By comparing the phylogenies of hosts and their parasites using reconciliation analysis, historical processes can be detected (Charleston and Page, 2003).Two genera of sucking lice parasitize human, one shared with chimpanzees (*Pan* spp.) and other with gorillas (*Gorilla gorilla*). Head and body lice of human and chimpanzee belong to the genus *Pediculus* (*Pediculus humanus* and *Pediculus schaeffi* respectively). *Pediculus* does not parasitize gorilla. Human public lice and gorilla lice belong to genus *Pthirus* (*Pthirus pubis* and *Pthirus gorillae*). *Pthirus* does not parasitize chimpanzee. *Pediculus* and *Pthirus* are sister taxa based on morphology and molecular data. Primate lice have cospeciated with their hosts for at least 25 million years (Reed et al., 2004). Divergence between *Pediculus humanus* and *Pediculus schaeffi* occurred at the same time as split between human and chimpanzee about 6 mya (Stauffer et al., 2001).The split between *Pediculus* and *Pthirus* occurred with split between gorillas and lineage leading to chimpanzees and humans about 7 mya (Reed et al, 2004). Such events were followed some time later by one host switch of a *Pthirus* species from gorillas to humans. **Host switching** in lice is very common in many groups of birds and mammals (Johnson et al., 2002). This recent host switch hypothesis predicts that the divergence of *Pthirus pubis* and *Pthirus gorillae* is more recent than chimpanzee / human split. Alternative hypothesis involves an ancient louse duplication event happened on the ancestor of chimpanzee, gorilla and human, which would have created the lineages leading to 2 extant genera, *Pediculus* and *Pthirus*. Timing of divergence between *Pthirus pubis* and *Pthirus goriillae* would correspond to their host about 7 mya. In this case, human would have retained both genera, but chimpanzee would have lost *Pthirus* and gorilla would have lost *Pediculus* species. *Pthirus pubis* has been suggested to be associated with humans for several million years and reached human via host switching from gorillas. Parasites switch from a given species to a predator of that species (Whiteman et al., 2004) and are sometimes found to switch to unrelated hosts in communally used areas like roosting or nesting sites (Jonson et al., 2002).

## 10.2   Evidence from Macroevolution

Macroevolution describes the pattern of evolution at above the species level. Identified taxonomic relationships give valuable hints that co evolution might have played role. If 2 lineages mutually influence each other's evolution, they might tend to change and speciate together .This might result in co phylogenies. Aphids and their bacterial endosymbionts seem to speciate together. They show, at least partly, mirror-image phylogenies although co phylogenies are not always proof of co speciation. Although there are co phylogenies, the time scale shows that the species did not split together. Viruses may have switched preferentially between closely related host species creating the mirror-image phylogeny. In pocket gophers and louse, mirror-image phylogenies coincide with **simultaneous speciation**, as supported by substitution rates of nucleotides .Together this gives good evidence of co evolution. Co phylogenies may even affect more than 2 interacting taxa, as in the ancient tripartite co evolution of leafcutter ants, their food fungus and a pathogenic fungus. The ant, the cultivated food fungus and the parasitic fungus show partly mirror-image phylogenies. A forth party involved a bacterium that produces antifungal agent.

## 10.3   Evidences from Microevolution

Microevolution, the process of evolution within populations acts on a rapid timescale. We can directly study genetic changes that show evolutionary change. Good **micro evolutionary evidence** for co evolution is still largely lacking. Co adaptational cycles of parasite and host allele frequencies are perhaps based on frequency-dependent selection. Dynamics of host-parasite co evolution have been described with the Red Queen hypotheses. Hosts and parasites never stand still. There is constant evolutionary change that is adaptation of one part, e.g. the parasite, almost inevitably lead to counter adaptation on the host, resulting in a never-ending arms race. For host-parasite co evolution, the moving environment is the evolving parasite. Parasites evolve more rapidly than their host, as they have a greater relative evolutionary potential. During co evolution, the interacting species with the shorter generation time, more genetic variation for the interaction trait and sexual reproduction evolve more rapidly. The studies on snails *Potamopyrgus antipodarum* and their trematode parasites *Microphallus* in New Zealand supports the Red Queen hypothesis. If parasites are ahead in the arms race, then **sympatric parasites** should be superior to allopatric ones. Such local adaptation of parasites was found. Under frequency-dependent selection an advantage of rare host clones is expected, which could also be demonstrated.

## 10.4 Dynamics of Co evolution

The co evolutionary interactions are very dynamic in time which may translate into spatial differences, in particular local adaptation. Certain host genotypes are particularly susceptible to certain parasite genotypes and vice versa. Such specific host parasite interactions have been observed. They deliver suggestive genetic evidence for dynamic specific interactions. Theory predicts that migration enhances local adaptation, as long as it does not completely homogenize populations. Experimental proof for this notion comes from **bacteriophage** parasite of *Pseudomonas fluorescens*.

## 10.5 Geographic Mosaic of Adaptation

Geographic mosaic theory of co evolution is based on geographic differences in species interactions due to variation. In natural selection a constantly changing genetic landscape shaped by gene flow and other forces. Reciprocal selection occurs in only some communities (**co evolutionary hotspots**). This heterogeneity could lead to the extreme case that species interaction can be mutalistic in some habitats, but commensal, or even antagonistic in others.

## 10.6 Evolution of Virulence

Evolution of virulence is best illustrated in the Myxoma virus that was introduced to Australia to reduce the rabbit, which had reached high densities and was regarded as a pest species. Within a few years, the virus reduced its virulence to an intermediate value. Rabbits evolved partial resistance. Although virulence must be seen as a trait that is shaped by host and parasite, the main clue to understanding virulence evolution is the fitness of the parasite, with the basic reproductive rate R0 as an important magnitude. Transmission is a function of parasite-induced host mortality which often indicates that parasite face a trade-off between within-host and between-host components of fitness, i.e. more growth or replication within a host is often associated with higher host mortality and thus reduced transmission to next host. Reduced transmission may result from host mortality, or morbidity. Vertically transmitted parasites harm themselves when they strongly reduce fitness of their host. They have lower virulence than parasites that are transmitted horizontally in population. Within-host competition might lead to higher virulence especially when competitors are unrelated.

## 10.7 Levels of Co evolution

An example is the co evolution in interacting biological macromolecules, like tRNAs and ribosomal proteins. The reciprocal evolutionary change in both parties led to refined interplay between those entities. Other well-known examples are the ancient endosymbioses between prokaryotes and host cells that led to mitochondria and chloroplasts. There have been further secondary and tertiary endosymbioses.

Co evolution of ligand-receptor pairs is an example where proteins are the coevolving entities. Luteinizing hormone (LH), human chorionic gonadotropin (hCG) and follicle-stimulating hormone (FSH) coevolves with their receptors. Advances in proteomics and bioinformatics enable large-scale studies on the co evolution of interacting proteins. A high-throughput screen of yeast (*Saccharomyces cerevisiae*) proteins shows that proteins that interact a lot evolve more slowly, i.e. connectivity of well-conserved proteins in the network is negatively correlated with their evolution rate as a greater proportion of the protein is directly involved in its function. Interacting proteins evolve at similar rates implying that sites are important for interaction between proteins. Evolutionary changes may occur largely by co evolution, in which substitutions in one protein result in selection pressure for reciprocal changes in interacting partners. Even though a completely different level was studied, these results are comparable with co phylogenies on the level of macroevolution.

## 10.8 Geographic Mosaics of Co Evolution

Species constantly engage in strong interactions with other species - parasites, predators, prey, and mutualists. As a result, their traits may co evolve and diversify in geographic mosaics. G. E. Hutchinson argued that evolution is a play taking place on an ecological stage (1965). His point was that ecological interaction between species and the living and non-living aspects of their environment are the drivers of **adaptive evolution**. His examples demonstrate the ubiquity of site-specific interactions that dominate the ecology of species. He showed that the foraging niches of organisms as diverse as rhinoceros, birds, cone shells, copepods, and rotifers seem to have diversified due to competition among similar species within those groups. Most species also interact with other species as a consumer or parasite, a food resource or prey item or both. Most species are engaged in several kinds of mutualisms with other species, in which both species typically benefit. Plants may engage simultaneously in mutually beneficial interactions with fruit-dispersing birds, pollinating insects, nitrogen-fixing bacteria, and mycorrhizal fungi that act like extended root systems in the soil. One dramatic result of such interactions can be coevolution, in which the traits of each of 2 species evolve in response to the traits of other, resulting in

relentlessly reciprocal evolutionary changes in traits of those interacting species. Darwin and Wallace both recognized the importance of co evolutionary process in evolutionary diversification, emphasizing the primacy of species interactions as drivers of adaptive evolutionary change in species. Darwin's book on orchids provided the catalyst for all subsequent studies of co evolution and evolution of extreme specialization (Thompson, 1994). Studies on co evolution have also shown that species interactions can drive rapid and sustained evolutionary change in species at multiple spatial and temporal scales, generating genetic diversity within populations, leading to adaptive differentiation among populations, and often leading to ecological speciation (Schluter, 2009). Much of the diversity on earth is a consequence of co evolutionary diversification in species interactions (Thompson, 2005). Studies of co evolution in species interactions can lend insight into the fundamental processes generating and maintaining biodiversity, including genetic and phenotypic diversity within, and between species. Co evolutionary selection between 2 species within a single local community has the potential to act as a potent evolutionary force on the traits of species by virtue of the feedback driven by its inherent reciprocality. Reciprocality of co evolutionary selection means that fitnesses of 2 interacting species depend not only on their own genotypes, but also on each other's genotypes. There is a genotype by genotype interaction on fitnesses of both species. As a result of this reciprocality, adaptive changes in traits of one species may trigger subsequent adaptive changes in traits of 2nd species, which in turn feed back to cause further adaptive evolution of first species, and so on. Relative fitness of genotypes in one species is context-dependent and context itself also evolves (Wade, 2003). Relative fitnesses of different genotypes of hummingbird and humming bird-pollinated plant may depend not only on hummingbird bill shape, or on flower morphology, but also on match between 2, generating a G x G interaction for both **humming bird** fitness and plant fitness. If hummingbird bill shape adapts due to selection driven by average flower morphology in plant population, nature of reciprocal selection on flower morphology changes.

Although local co evolutionary selection is an important building block of co evolutionary dynamics, studies on co evolution in a wide range of species interactions have shown that co evolution is inherently geographically structured. Co evolution between 2 species can progress along very different trajectories in different places, causing trait differences to evolve among different populations of same species. As most species are collections of genetically variable populations distributed among environments, and as environments often differ in abiotic factors and/or community composition, pattern and strength of natural selection imposed by species on each others' traits are expected to be highly variable among environments. That co evolution is an inherently geographically-structured process has been synthesized as **Geographic Mosaic Theory** of Co evolution (Thompson, 2005). GMTC proposes that co evolution exhibits 3 inherently geographic characteristics viz., co evolutionary hotspots and cold spots, geographic selection mosaics, and trait remixing beyond local co evolutionary selection. Co evolutionary hotspots are communities in which pair wise interaction between 2 species exhibits co evolutionary selection and these hotspots are perhaps embedded in a geographic matrix of cold spots. In cold spots, selection is not reciprocal, including communities in which only one of the species occurs. Geographic selection mosaics occur when a fitness function of co

evolutionary selection differs among environments. Trait remixing is the suite of processes that influence geographic distributions of alleles at gene loci underlying coevolving traits , including mutations, gene flow, genetic drift, and population extinction/recolonization dynamics.

Mathematical modeling studies of geographic mosaics of co evolution have shown the interaction of 3 key elements of GMTC to generate a variety of outcomes of co evolution including patterns of specificity, local adaptation/maladaptation, patterns of genetic variation within, and between populations (Nuismer, 2006). Studies of species interactions - conifers and crossbills (Benkman et al., 2001), snails and trematodes (Lively et al., 2004), camellias and weevils (Toju & Sota., 2006), garter snakes and newts (Brodie et al., 2002), and Greya moths that act as, both pollinators and seed parasites of *Lithophragma* plants (Thompson & Cunningham, 2002) have provided support for key elements of theory (Thompson, 2005). Johnson et al. (2010) studied interactions between big bluestem grass (*Andropogon gerardii*) and its arbuscular mycorrhizal (AM) fungal symbionts from 3 different populations. They estimated plant and fungal performance in all reciprocal combinations of plant populations (GP), whole AM fungal guilds of species (GF) and sterile soils (E) in a greenhouse experiment. They found a significant GP x GF x E interaction for a key trait of interaction: formation of arbuscules, which are the fungal structures in AM plant roots through which nutrients pass. Arbuscule formation was highest in local combinations of plants, fungi, and soils, suggesting that a selection mosaic has resulted in local adaptation of plants and fungi to each other in interaction. It remains challenging to perform such rigorous tests of key phenomena central to the GMTC

(Gomulkiewicz et al., 2007), since ideal tests require that all combinations of genotypes of all important interacting species are replicated and studied across multiple field sites or artificial gradients of key environmental variables (Nuismer & Gandon, 2008).

Experiments have not yet been performed to test the importance of ongoing co evolution and geographic mosaics of co evolution in dominant species interactions, like those between plants and mycorrhizal fungi (Hoeksema, 2010). **Non-reciprocal adaptation** of species to biotic and abiotic environmental factors has been repeatedly showed to be an important force driving diversification in species (Saenz-Romero et al., 2006). The relative roles of non-reciprocal selection from biotic and abiotic factors versus the 3 elements of the GMTC (selection mosaics, hotspots and cold spots, and trait remixing) in driving trait diversification of species needs to be addressed. At least 3 hypothetical answers for this question for a particular species interaction (Hoeksema, 2010) exist. One hypothesis states that all 3 GMTC elements - hostpots/coldspots, **selection mosaics** and trait remixing-are important and interacts to play prominent roles in driving diversification within, and across communities in which 2 species occur. A second hypothesis states that co evolutionary selection is relatively important but is consistent among populations and hotspots/cold spots are relatively unimportant or uncommon. A third hypothesis states that local co evolutionary selection is consistently less important compared to non-reciprocal selection on species by abiotic environmental factors, or by one partner species on the other. Importance of host-parasite interactions has led to theory on evolution of parasite virulence (May and Anderson, 1983) and host resistance (Boots and Haraguchi, 1999). Interactions between hosts and their parasites are likely to depend on interplay of both species' evolutionary characteristics: the co evolutionary dynamics. Studies on evolution of parasites, in which virulence is assumed to be traded off against transmission, show that selection often leads to intermediate levels of virulence, as parasites seek to maximize the epidemiological $R0$ (May and Anderson, 1983). Provided that hosts are only infected by one parasite at any one time, a competitive exclusion principle dictates that evolution favors the parasite strain with maximal $R0$ (Nowak and May, 1994). When density dependence acts on host mortality, then disruptive selection may allow evolution of coexisting parasite strains through evolutionary branching (Pugliese, 2002).

Recognition that evolution of host defense mechanisms may itself affect host-parasite interaction has led to development of models focused on trade-offs between host resistance and other life-history traits (Miller et al., 2007). Coexistence of host strains is possible for defense mechanisms like avoiding infection (avoidance), increased clearance, and reducing the within-host parasite growth rate (Miller et al., 2005). Tolerance of the damaging effects of parasite growth does not lead to the coexistence of host strains (Roy and Kirchner, 2000). These evolutionary models give insight into how ecological feedbacks shape selective pressures on the host or parasite, but in nature host and parasite often evolve together. Models that examine the co evolution of quantitative traits in hosts and parasites are relatively few (Bonds, 2006) which search for co evolutionary stable states (CoESSs), which once attained cannot be invaded by other strategies/strains. A single CoESS exists if host defense is through avoidance (Restif and Koella, 2003), whereas with resistance through recovery there may be bistability such that resistance and virulence are either both low, or both high (van Vaalen, 1998). This approach examines only the evolutionary stability (ES) of outcomes, that is, their ability to be invaded once reached. The other important component of evolutionary process is **convergence stability** (Christiansen, 1991). It determines whether through evolutionary time the population will move toward, or away from a CoESS and whether evolutionary branching will occur. Impact of co evolution on ES and CS is now well understood (Marrow et al., 1996). A model of predator-prey co evolution focus on the impact of tradeoff shapes on co evolutionary dynamics (Kisdi, 2006). Given that most natural systems will be co evolutionary, there is a clear need to assess whether the predicted dynamics from models in which only one of the 2 species evolves accurately match the outcomes of fully co evolutionary systems. There is variation within both host defense and parasite transmission in natural systems and it is important to understand whether this variation has co evolutionary consequences that may be missed by purely evolutionary studies.

## 10.9 Mimicry

Co evolution, the change of organism triggered by the change of a related object occurs at various levels. Microscopic co evolution is correlated mutations between amino acids in protein and co varying traits between different species. Co evolution is exemplified in evolution of host and its parasites, in relationship between Bumble bees and flowers they pollinate, predator and prey, host and symbiont, but many cases are not distinct. A species may evolve in response to several other species; to a set of species which is referred as diffuse co evolution as found at level of populations and species. Viruses and their hosts co evolve in various scenarios. Each party in a co evolutionary relationship exerts selective pressures on other affecting each other's evolution. Use of the word mimicry dates

back to year 1637 which is derived from the Greek term imetikos, imitative, in turn from the mimetos, to imitate. It was used to describe people and applied to other forms of life after Mimicry is the similarity of one species to another to protect from predator, or exhibiting an apparent aggressiveness which is not at all real. Similarity can be in appearance, behaviour, sound, scent, and location. Mimicry occurs when mimics evolve to share common perceived characters with models. Models are sometimes difficult to define, for example eye spots may not bear similarity to any specific organism's eyes. Model is another species except in cases of automimicry. Evolution driven by selective action of a signal-receiver may be intermediate organism like common predator of 2 species, or model itself, like a moth resembling its spider predator. Birds use sight to identify palatable insects and avoid the noxious models. This is called **mimicry complex**. Mimicry is advantageous to mimic and harmful to receiver. It may increase, reduce, or has no effect on model's fitness. In camouflage, a species resembles its surroundings is a type of visual mimicry. **Camouflage** often cannot be attributed to particular model. In mimesis, the mimic takes on properties of a particular object or organism, but one to which the dupe is not responsive. Lack of clear distinction between 2 phenomena is found in animals that resemble bark, leaves, twigs, flowers, and are often classified as camouflaged, but sometimes as mimics.

**Crypsis** is a broad concept encompassing all forms of avoiding detection like mimicry, camouflage, and hiding. Visual mimicry is clear to us along with olfaction or hearing. Sometimes more than one type of signal may be employed. Mimicry involves morphology, behavior and other properties. Signal deceives the receiver by preventing it from identifying the mimic. This is a form of co-evolution involving an evolutionary arm race and should not be confused with convergent evolution, which occurs when species resemble one another by adapting similar lifestyles. Mimics may have various models for different life cycle stages. They may be polymorphic with different individuals imitating different models. **Models** may have 2 or more mimic. The frequency dependent selection favours mimicry where models outnumber mimics. Models tend to be closely linked organisms. Mimicry of many species is known. Most mimics are insects; though include mammals, plants and fungi.

Butterflies in the Amazon rainforest are pioneer example in the field. **Mimics** are less likely to be found in low proportion to their model, a phenomenon called negative frequency dependent selection which applies to most other forms of mimicry. Ash Borer (*Podosesia syringae*), a moth of family Sesiidae, is a **Batesian mimic** of common wasp as it looks like the wasp, but unable to sting. A predator when learns to avoid wasp similarly avoids Ash Borer. Plain Tiger, *Danaus chrysippus*, an unpalatable model has several mimics. Several hoverflies, mimic stinging wasp species. Common crow, *Euploea core* an unpalatable model has several mimics. *Consul fabius* and *Eresia eunice* imitate unpalatable *Heliconius ismenius*. Several palatable butterflies resemble highly noxious *Battus* species. Several palatable moths produce ultrasonic click calls to mimic unpalatable tiger moths. The snake *Malpolon moilensis* is mildly venomous but harmless colubrid snake which mimics the hood of an Indian cobra's threat display. The Snake *Heterodon platirhinos* mimics the threat display of poisonous snakes. The milk snake resembles the deadly coral snake. Vespid wasps bear resemblance with moths, beetles and hoverflies. The Octopus, *Thaumoctopus* alter their body shape and color to resemble

dangerous sea snakes, or lionfish. Many types of mimicry often based on function with respect to the mimic are described. Some may belong to 2 or more class. **Auto mimicry** and aggressive mimicry are not mutually exclusive, as described in species relationship between model and mimic, while the other describes the function for mimic.

### Defensive Mimicry

Defensive or protective mimicry occurs when organisms can avoid harmful encounters by deceiving enemies by treating them something else. The first 3 such cases entail mimicry of organisms protected by warning colouration. In **Batesian mimicry**, a harmless mimic poses as harmful. In Müllerian mimicry, 2 or more harmful species mutually advertise themselves as harmful. In **Mertensian mimicry**, a deadly mimic resembles a less harmful but lesson-teaching model. In Vavilovian mimicry, weeds resemble crops, is important. Human are the agent of selection of Macroxiphus. Katydid mimics an ant. In Batesian mimicry, the mimic shares signals similar to the model, without attribute that makes it unprofitable to predators. A Batesian mimic is thus a sheep in wolf's clothing. It is named after H. W. Bates who studied Müllerian mimicry

In **Müllerian mimicry** 2 or more species have similar warning, or aposematic signals and both share genuine anti-predation attributes. F. Müller put forward the first explanation for this phenomenon. If 2 species are confused with one another by common predator; individuals in both are likely to survive. Such mimicry is unique. Here mimic and model benefit from interaction, which could be classified as mutualism. The signal receiver is advantaged despite being deceived regarding species identity, as it avoids potentially harmful encounters. The clear identity of mimic and model are blurred. In cases where one species is scarce and another abundant, rare species is the mimic. When both are present in similar numbers, each one is treated as co mimics than distinct mimic and model, as their warning signals converge toward something intermediate between the two. The 2 species may exist on a continuum from harmless to highly noxious. Examples: Lepidoptera

Monarch Butterfly (*Danaus plexippus*), a member of Müllerian complex with Viceroy Butterfly (*Limenitis archippus*) shares color patterns and display behavior. Viceroy has subspecies with different coloration each one matching local *Danaus* species. In Florida, the pairing of Viceroy and Queen some hawk-cuckoos resembles hawks like Shikra. *Heliconius* from tropics of Western Hemisphere are the classical model of Müllerian mimicry. In Mexico, the Viceroy resembles the Soldier Butterfly. Viceroy, a single species is involved in 3 different Müllerian pairs. While *L. archippus* is bad-tasting, *Danaus* tends to be toxic rather than just repugnant, due to their different food plants. Unpalatable *Euploea* species look very similar. Morpho is palatable but some species are very strong fliers. Birds which are specialized for catching butterflies on wing find it difficult to catch them. The conspicuous blue colour shared by most Morpho species may be Müllerian mimicry. Orange complex of species includes heliconiines *Dryadula phaetusa*, *Agraulis vanillae*, and *Dryas iulia* which all taste bad. Tiger moths make ultrasonic clicking calls to warn bats that they are unpalatable. A bat may learn to avoid any signalling moths, which make this an example of Müllerian mimicry. Various bees and numerous vespid and sphecoid wasps are examples of Müllerian mimics with the aposematic yellow and black stripes. Females of such species are potentially harmful to predators, fulfilling the 2nd requirement of Müllerian mimicry. In all such species, males are harmless, and can be auto mimics of their conspecific females. Many genera in such groups where females are not capable of stinging, and still possess **aposematic coloration**, are Batesian mimics.

### Emsleyan/Mertensian Mimicry

In Emsleyan, or Mertensian mimicry first proposed by Emsley for coral snake mimicry in New World, where deadly prey mimics a less dangerous species. It was elaborated on by Wickler who named it after Mertens. In other mimicry, the most harmful species is the model. If a predator dies, it cannot recognize the warning signal, for example bright colors in some pattern. There is no advantage in being aposematic for an organism that is likely to kill any predator by poisoning. Such animal profit from being camouflaged, to avoid attacks altogether. If there is some other species that is harmful but not deadly and aposematic, predator recognizes its particular warning colours and avoids such animals. A deadly species profits by mimicking the less dangerous aposematic organism, if this result in less attack than camouflage. Ignoring any chance of animal learning by watching a conspecific die is the possibility of not learning that it is harmful in first place as an exception: instinctive genetic programming to be wary of certain signals. Here, other organisms benefit from this programming, and Batesian or Müllerian mimics of it evolve. Some species have innate recognition of some aposematic warnings. Hand-reared Turquoise-browed Motmots (*Eumomota superciliosa*), avian predators, avoid snakes with red and yellow rings instinctively. Other colors with same pattern, and even red and yellow stripes with same width as rings, were tolerated. Models with red and yellow rings were feared, with birds flying away and giving alarm calls sometimes. This provides one alternative explanation to Mertensian mimicry. Some Milk Snake

(*Lampropeltis triangulum*) are moderately toxic. False coral snakes (*Erythrolamprus*), and deadly Coral Snakes all have a red background color with black and white or yellow rings. Here the milk snakes and deadly coral snakes are the mimics, whereas false coral snakes are the model (Figures 10.5, 10.6, 10.7).

### Wasmannian Mimicry

This mimicry refers to cases where mimic resembles a model along with which it lives in a nest, or colony. Most models here are social insects such as termites, wasps, and bees.

### Vavilovian Mimicry

This mimicry is illustrated in weeds which share characters with a domesticated plant through artificial selection. It is named after N. Vavilov. Selection against weed occurs by killing the weed manually, or through separation of its seeds from those of crop. The latter process called winnowing can be done manually, or by machine. Vavilovian mimicry is an unintentional selection by man. While some cases of artificial rye are secondary crop, originally selection goes in the desired direction, such as selective breeding, this being a mimetic weed of wheat. Weeders do not want to select weeds that look more like the cultivated plant. **Vavilovian mimics** may domesticate themselves. Vavilov called these weed-crops as secondary crops. This is classified as defensive mimicry as weed mimics the protected species. This bears affinity to Batesian mimicry as weed does not share the properties to provide protection to model, and both the model and dupe are harmed by its presence. In Batesian mimicry, model and signal receiver are enemies, but here the crop and its human growers are in a mutualistic relationship. The crop gain advantage from being dispersed and protected by people, despite being eaten by them. The crop's only protection relevant here is its use by humans. Secondly, the weed is not consumed, but destroyed. Motivation for killing the weed is its effect on crop production. Such mimicry does not occur in ecosystems unaltered by humans. In case of *Echinochloa oryzoides*, a species of grass is found as a weed in rice (*Oryza sativa*) field. The plant is similar to rice and its seeds are often so mixed in rice that it is difficult to separate. This close similarity was increased by weeding process

Texus coral snake

Mexican milk snake

**Figure 10.5**  Mertenisan mimicry

**Figure 10.6** Several species including several hoverflies mimic stinging wasp

**Figure 10.7** Batesian mimicry between *Dismorphia* species (top row, third row) and various 1 thomini (second row, bottom row)

which is a selective force that increases the similarity of the weed in each subsequent generation.

### *Gilbertian Mimicry*
### *Protective Egg Decoys*

This mimicry involves only 2 species. The potential host, or prey drives away its parasite, or predator through mimicking it. Pasteur named for such rare mimicry systems after L. E.Gilbert. This protective mimicry occurs in Passiflora. Leaves of this plant contain toxins to deter herbivorous animals. Some *Heliconius* butterfly larvae evolve enzymes to break down such toxins, allowing them to adapt on this genus. This creates more selection pressure on host plants, which evolve stipules that mimic *Heliconius* eggs. *Heliconius* tend to avoid laying eggs near existing ones to avoid intra-specific competition between caterpillars. Those lay on vacant leaves provide

their offspring with a greater chance of survival. Most *Heliconius* larvae are cannibalistic. Those leaves with older eggs hatch first and eat the new arrivals. Such plants evolve egg dummies due to grazing herbivore enemies. Decoy eggs are nectaries though attract predators of caterpillars like wasps and ants. Effectiveness of their mimetic function is difficult to assess. Use of eggs is not essential in this system, only the species composition and protective function are essentic. Mimicry also involves eggs. Cuckoo eggs mimick their host, or spreading of plants seeds by ants, who treat them as their own eggs are examples. **Protective mimicry** within species the Browerian mimicry, named after L. P. Brower and J. V. Z. Brower, is a form of auto mimicry where model belongs to same species as mimic. This is analogous to Batesian mimicry within single species and occurs when palatability spectrum within a population exists. Monarch Butterflies (*Danaus plexippus*) feed on milkweed species of varied toxicity and stores toxins from its host plant even in the adult form. As the levels of toxin vary depending on diet during the larval stage, some individuals are more toxic than others. The less palatable organisms are the mimics of the more dangerous individual, with their already perfected likeness. This is not the case in sexually dimorphic species. One sex may face more a threat than other species, which mimic the protected sex.

## Aggressive Mimicry

Here predators share same characters as harmless to avoid detection by their prey. Mimic may resemble prey, or host, or other organism which is neutral or beneficial to signal receiver. Model may be affected negatively, positively or not at all. **Host-parasite mimicry** is treated as subclass of aggressive mimicry. Mimic may have particular significance for duped prey. **Aggressive mimicry** is common in spiders in both luring prey and stealthily approaching predators. *Nephila clavipes* spins a golden colored web in well-lit areas. Bees associate the webs with danger when yellow pigment is not present, as occurs in less well-lit areas where web is difficult to see. Other colors were learned and avoided, Bees seemed least able to associate effectively with yellow pigmented webs with danger. Yellow colour is found in many nectar bearing flowers. So perhaps avoiding yellow is not worthwhile. Mimicry is based on pattern. *Argiope argentata* employ prominent patterns in middle of their webs such as zigzags. These reflect ultraviolet light, and mimic the pattern seen in many flowers called nectar guides. Spiders change their web day wise, which is explained by bee's ability to memorize web patterns. Bees link some pattern with spatial location. The spider spins a new pattern, or suffers from scarcity of prey. In other case

males are lured to sexually receptive female. Lloyd's investigation of female *Photuris* showed that they emit the same light signals that males of the *Photinus* use as a mating signal. Male fireflies from different genera are attracted to these signals, and are captured and eaten. Female signals are based on that received from male. Each female having a repertoire of signals, matching delay and duration of female of corresponding species. This mimicry may evolve from non-mating signals that are modified for predation.

The katydid *Chlorobalius leucoviridis* of inland Australia attracts male cicadas by imitating species-specific reply clicks of sexually receptive female cicadas. This acoustic aggressive mimicry is similar to *Photuris*. *C. leucoviridis* attracts males of many cicada species including *Cicadettine cicadas*, though mating signals of cicada show species-specificity. Increasing the rate of prey capture through mimicry is found in carnivorous plants. Luring is not a necessary condition, as predator has a significant advantage by simply not being identified. They resemble a mutualistic symbiont, or a species of little relevance to prey. The latter situation is found in cleaner fish and its mimic, though the model is disadvantaged by presence of mimic. Cleaner fish, the allies of many other species allow them to eat parasites and dead skin. The Blue streak cleaner wrasse (*Labroides dimidiatus*) is the model of a mimetic species, *Aspidontus taeniatus*, and the Sabre-toothed blenny. This wrasse, cleans the Spotted Predatory Katydid (*Chlorobalius leucoviridis*) is an acoustic aggressive mimic of cicadas. Two Blue streak cleaner wrasse cleaning *Epinephelus tukula*, and *Epinephelus sp. resides* in coral reefs and is recognized by other fish who allow it to clean them. A species of blenny looks like it in size and coloration, but even mimics the cleaner's dance. Having fooled its prey it bites the prey, tear a piece of its fin before fleeing.

Fish grazed on in this way learn to distinguish mimicking from model, but because the similarity is close between the 2, they become more cautious of the model, such that both are affected. Victims' ability to discriminate between foe and helper cause the blennies to evolve close similarity, right down to the regional level. The Zone-tailed Hawk, which resembles Turkey Vulture, does not involve any luring. It flies among the vultures; suddenly break from the formation and ambushing its prey. The hawk's presence is not significant to the vultures. When Parasites are aggressive mimics, the situation is different from those outlined above.

Some predators draw prey and parasites mimic their host's natural prey and are being eaten themselves, a pathway into their host. The flatworm, *Leucochloridium* become adult in the digestive system of songbird. Their eggs then pass out of the bird via feces. They are taken up by a snail *Succinea*. The eggs develop in this intermediate host and find a suitable bird to mature. As the host birds do not eat snails, so sporocyst adopts strategy to reach the host's intestine. Brightly colored sporocysts move in a pulsating fashion. Sporocyst-sac pulsates in eye stalks of snail's to be an irresistible meal for songbird. It can bridge the gap between hosts, allowing it to complete its life cycle. A nematode (*Myrmeconema neotropicum*) changes the colour of abdomen of workers of canopy ant *Cephalotes atratus* to appear like ripe fruits of *Hyeronima alchorneoides*. It changes the **behaviour** of ant so that gaster is held raised. This increases the chances of ant being eaten by birds. The other ants collect the droppings of birds to feed to their brood, helping to spread the nematode. Planidium larvae of some beetles of *Meloe* form a group and produce a pheromone that mimics the sex attractant of its host bee species. When male bee arrives

and attempts to mate with larvae, they climb onto his abdomen, and from there transfer to female bee, and from there finally to bee nest for parasitizing the bee larvae. In 2 species system of host-parasite mimicry, a parasite mimics its host. Cuckoos are an example of brood parasitism, where mother raises its offspring by other unwitting organism, reducing mother's parental investment. The laid eggs which mimic host eggs are the vital adaptation. Adaptation for various hosts is inherited through female line in gentes. Intraspecific **brood parasitism**, where a female lays in conspecific's nest, is found in Golden eye duck (*Bucephala clangula*), do not represent a case of mimicry. Reproductive mimicry occurs when actions of dupe directly aid in mimic's reproduction. This is common in plants, which may have deceptive flowers that do not provide the reward. Other forms of mimicry have a reproductive component, such as Vavilovian mimicry involving seeds, and brood parasitism, which also involves aggressive mimicry.

### Bakerian Mimicry

This mimicry named after H. G. Baker is a form of auto mimicry where female flowers mimic male flowers of their own species, cheating pollinators out of a reward. In this mimicry, members of the same species may still exhibit some degree of sexual dimorphism. It is common in species of Caricaceae. Like Batesian mimicry, **Dodsonian mimicry** (refers to C. H. Dodson) is a form of reproductive floral mimicry, but model belongs to different species than mimic. By providing similar sensory signals as model flower, it lures its pollinators. Like Bakerian mimics, no nectar is given. *Epidendrum ibaguense* resembles flowers of *Lantana camara* and *Asclepias curassavica*, and pollinates by hummingbirds and Monarch

Butterflies. Similar cases are seen in other species of same family. Mimetic species may pollinators of its own. For example a lamellicorn beetle which pollinates correspondingly colored Cistus flower is known to aid in pollination of *Ophrys* species that are normally pollinated by bees.

### Pseudo Copulation

**Pseudo copulation** occurs when a flower mimics a female of a certain insect species, the males of which try to copulate with it. This is like the aggressive mimicry in fireflies, but with more benign outcome for pollinator. This mimicry is common in orchids which mimic female Hymenoptera, and account for about 60% of pollinations called Pouyannian mimicry, after Pouyanne who first described it. A pollen sac called pollinia is attached to the abdomen, or head of the male. This is then carried to stigma of next flower. Male tries to inseminate causing pollination. Visual mimicry is the clear sign of this deception, but the visual aspect is the minor. Senses of touch and olfaction are most important.

### Inter-sexual Mimicry

Inter-sexual mimicry occurs when individuals of one sex in a species mimic members of other sex, as found in male forms of marine isopod, *Paracerceis sculpta*. Alpha males are the largest and guard the harem of females. Beta males manage to enter harem without being detected by alpha males to mate. Gamma males, the smallest males mimic juveniles and mate with the females without alpha males detecting them.

### Auto Mimicry

Auto mimicry or intraspecific mimicry occurs within single species where one part of an organism's body resembles another part. In some snakes, tail looks like head and show behavior like moving backwards to confuse predators. Insects and fishes with eyespots on their hind ends look like head. Mimic imitates other morphs of same species in auto mimicry. In **sexual mimicry**, males mimic the females or vice versa. Many insects have filamentous tails at the ends of their wings which combines with patterns of markings on the wings themselves to create false head which misdirects predators. Several pygmy owls bear false eyes on back of their head to fool predators into believing the owl is alert to their presence (Figures 10.8, 10.9, 10.10).

**Figure 10.8** Vavilovian mimicry.

**Figure 10.9**   Male *Chlorobalius* showing aggresive mimicry

**Figure 10.10**   Northern pigmy owl

## 10.10   Bat and Moths

Bats have evolved sonar to detect and catch prey. Moths evolve to sensitive high-pitched calls of bat and to avoid being caught. Bats evolve for catching the moths over water. Rough-skinned newt produces tetrodotoxin as a defence mechanism. Toxin binds reversibly to sodium channels in nerve cells and influences flow of sodium ions in and out of cell inducing death and paralysis. The common garter snake exhibit resistance to **tetrodotoxin** produced in skin of newts. Toxin binds to protein that acts as sodium channel in snake's nerve cells. A genetic disposition in several snake populations is found where protein is configured to prevent binding of toxin. In each population, snakes exhibit resistance to toxin and prey on the newts. Mutations in snake's genes conferred resistance to toxin resulted in a selective pressure that favors newts to produce more potent toxin. Increases in toxicity in newt are the result of selective pressure favoring snakes with mutations conferring greater resistance. This **evolutionary arms race** results in newts producing levels of toxin far in excess which is needed to kill any conceivable predator. Toxin resistant garter snakes are the animals today that can eat a rough-skinned newt and survive.

## 10.11   Hummingbirds and Ornithophilous Flowers

Hummingbirds and ornithophilous flowers evolve as mutualistic. Nectar in flowers is suited to birds' diet. Color of flowers attracts the birds' vision and their shape fits with birds' beak. The flowering times coincide with hummingbirds' breeding seasons. Flowers converge to get benefit from similar birds. Flowers compete for pollinators. Adaptations minimize the unfavorable effects of this competition. Bird-pollinated flowers with sufficient nectar and high sugar production than those pollinated by insects. This meets the birds' high energy requirements, a factor of their flower choice. Following their respective breeding seasons, several hummingbird species occur at same locations in North America and several hummingbird flowers bloom simultaneously in such habitats. Different lengths and curvatures of corolla tubes of flowers affect the extraction efficiency in hummingbird in relation to bill shape. Tubular flowers influence the beak in a way when probing flower, especially when bill and corolla are curved. This causes the plant to place pollen on part of bird's body opening the door form morphological co-adaptations. Birds have great spectral sensitivity and fine hue discrimination at the red end of visual spectrum. Hummingbirds see ultraviolet colors. Prevalence of ultraviolet patterns and nectar guides in nectar-poor entomophilous flowers warns the bird to avoid these flowers. Hummingbirds form family Trochilidae, have 2 subfamilies, the haethornithinae (hermits) and Trochilinae. Each subfamily has evolved in conjunction with a set of flowers. Most Phaethornithinae species are associated with large monocotyledonous herbs, while Trochilinae prefer dicotyledonous plant species.

## 10.12 Old World Swallowtail and Fringed Rue

The old world swallowtail caterpillar lives on fringed rue plant. Rue produces etheric oils which repel plant-eating insects. Caterpillar developed resistance to these poisonous substances, thus reducing competition with other plant-eating insects. Co evolution of predator and prey species is illustrated by the Rough-skinned newt and common garter snake. Newts produce a potent neurotoxin that concentrates in their skin. Garter snakes

have evolved resistance to this toxin through a series of genetic Old world swallowtail caterpillar on fringed rue mutations, and prey on the newts. Relationship between these animals results in evolutionary arms race that has driven toxin levels in newt to extreme levels. Here, differential survival caused each organism to change in response to changes in the other.

### California buckeye and pollinators

When beehives are populated with bee species that have not coevolved with California buckeye, sensitivity to aesculin, a **neurotoxin** present in its nectar, may be noticed. This sensitivity is thought to be present in honeybees and other insects that did not co evolve with *A. californica*.

### Acacia ant and bullhorn acacia tree

Acacia ant protects the bullhorn acacia from preying insects and from other plants competing for sunlight. The tree provides nourishment and shelter for ant and its larvae. Some ant species exploit trees without reciprocating are called cheaters. Cheater ants damage reproductive organs of trees, but their net effect on host fitness is difficult to forecast and not necessarily negative.

### Yucca Moth and the yucca plant

In mutualistic symbiotic relationship, the yucca plant is pollinated exclusively by *Tegeticula maculate* which relies on yucca for survival. Yucca moths tend to visit flowers of only one yucca species. In flowers, moth eats the seeds of plant, gather pollen on special mouth parts. Pollen is very sticky, and remains on mouth parts when moth moves to next flower. Yucca plant provides a place for moth to lay its eggs, deep within flower where they are protected from any potential predators. Adaptations that both species exhibit characterize co evolution as species have evolved to become dependent on each other. Spread of avian influenza was a bad news. WHO told about a new mutation in H5N1 viruses in Turkish birds which results in substitution in a virus protein to help the **flu virus** to bind to host cell's receptors. It increases affinity of the virus for human receptors over poultry ones and increase affinity for receptors in nose and throat rather than lower respiratory tract which increases risk of human to-human transmission. Avian influenza is at this time was a bird disease. Co evolution happens when 2 (or more) species influence each other's evolution.

## SUMMARY

1. Antagonistic interactions concentrate on the evolution of 1 of the 2 species. Evolution of one species affects selection on other. Co evolutionary model of host-parasite interactions shows that co evolutionary outcomes cannot always be predicted from the single-species evolutionary outcomes.

2. In addition, parasite evolution is highly sensitive than that of host, often resulting in evolution of virulent parasites. Host-parasite co evolutionary game theory emphasizes importance of dynamic evolutionary behaviors. Perhaps the level of parasite virulence is highly dependent on evolutionary dynamics of host. Evolution in host can induce selection for highly virulent parasites with higher transmission and shorter infection periods than predicted by purely evolutionary models.

3. Any imposed changes in host resistance in agriculture and medicine affect parasite evolution and risk selecting for high virulence. Trade-offs with transmission explain parasite induced mortality as parasite optimizes R0 , but host resistance drive the higher virulence than would optimize the parasite R0 in absence of host resistance. Host heterogeneity may lead to reduced, or increased parasite virulence depending on the type of heterogeneity. Rates of mutation alter the co evolutionary systems.

4. Sensitivity of parasite to host evolution leads to transient changes in its virulence if it has a higher mutation rate. Co evolutionary processes thus explain some variation in parasite virulence. When a parasite has a higher mutation rate than its host, it undergoes rapid change because relatively slow changes in its host have large impacts on parasite virulence.

5. Viruses and bacteria have faster generation times than their hosts. This potential to evolve faster than their hosts is the justification for modeling parasite evolution alone. The slower host evolution dramatically changes parasite transmission and virulence. The parasite's evolutionary optimum is very sensitive to the current level of host resistance. Virulence is a transient state brought out by new resistance mechanisms in the host until the host itself reached its evolutionary optimum with implications for the emergence of disease and the management of virulence.

6. If parasite virulence is a moving target, managing disease in host populations becomes more difficult. The parasite cannot branch if density dependence acts on host birth rate but density dependence on host mortality remain true when parasite coevolves with its host.

7.  Antagonist interactions in evolution have created the concept of antagonist co evolution. Host-parasite interactions are often tight. Predator-prey interactions can be diffuse. Mimicry illustrates nicely different types of co-evolutionary interactions. Microevolution provides good tests of co evolution. We might expect co adaptational cycles of parasite and host allele frequencies that are based on frequency-dependent selection.

8.  Dynamics of host-parasite co evolution have been described with Red Queen hypotheses. Hosts and parasites never stand still. For host-parasite co evolution, the moving environment is the evolving parasite. Parasites evolve rapidly than their host, because they have a greater relative evolutionary potential. During co evolution, the interacting species with shorter generation time, more genetic variation for interaction trait, and sexual reproduction evolve rapidly.

9.  Species constantly engage in strong interactions with other species - parasites, predators, prey, and mutualists. Most species also interact with other species as a consumer or parasite, a food resource or prey item, or both. One potentially dramatic consequence of these interactions can be co evolution, in which traits of each of 2 species evolve in response to traits of other, resulting in relentlessly reciprocal evolutionary changes in traits of those interacting species. Studies on co evolution have shown that species interactions drive rapid and sustained evolutionary change in species at multiple spatial and temporal scales, generating genetic diversity within populations, leading to adaptive differentiation among populations, often leading to ecological speciation and can lend insight into fundamental processes generating and maintaining biodiversity, including genetic and phenotypic diversity within, and between species.

10. Co evolutionary selection between 2 species within a single local community has the potential to act as a potent evolutionary force on traits of species by virtue of feedback driven by its inherent reciprocality. Reciprocality of co evolutionary selection means that fitnesses of 2 interacting species depend on their own genotypes and on each other's genotypes. There is a genotype by genotype interaction on fitnesses of both species. As a result, adaptive changes in traits of one species may trigger adaptive changes in traits of 2nd species, which in turn cause further adaptive evolution of first species, and so on.

11. Co evolution is inherently geographically structured. Co evolution between 2 species can progress along very different trajectories in different places, causing trait differences to evolve among different populations of same species.

12. Coexistence of host strains is possible for avoiding infection (avoidance), increased clearance, and reducing the within-host parasite growth rate. Microscopic co evolution is correlated with mutations between amino acids in protein and co varying traits between different species. Co evolution is exemplified in evolution of host and its parasites, in relationship between Bumble bees and flowers, predator and prey, host and symbiont, but many cases are not distinct.

13. A species may evolve due to several other species; to a set of species which is referred as diffuse co evolution as found at level of populations and species. In Batesian mimicry, a harmless mimic poses as harmful. In Müllerian mimicry, 2 or more harmful species

mutually advertise themselves as harmful. In Mertensian mimicry, a deadly mimic resembles a less harmful but lesson-teaching model. In Vavilovian mimicry, weeds resembling crops is important. Katydid mimics an ant. In Batesian mimicry the mimic shares signals similar to model, without attribute that makes it unprofitable to predators.

## Review Questions

1. Define mimicry.
2. Define auto mimicry.
3. Define Batesian mimicry.
4. Define Mullerian mimicry.
5. Define Bakerian mimicry.
6. Define inter-sexual mimicry.
7. Define co evolution.
8. What do you understand by parasite's virulence?

## Long Answer Questions

1. Describe co evolution with two examples
2. Discuss about the Batesian and Mullerian mimicry with suitable examples

# REFERENCES

Aguirre, U.L.A., Lozoya, S.A.1991. Mallophaga of domestic birds in southeastern Coahuila, Mexico. Folia Entomol Mexicana, 82:93-105.

Ash, J.1952. Records of Hippoboscidae (Dipt.) from Berkshire and Co., Durham, in 1950, with notes on their bionomics. Entomol Mon Mag, 88:25-30.

Balakrishnan, C.N., Sorenson, M.D.2007. Dispersal ecology versus host specialization as determinants of ectoparasite distribution in brood parasitic indigo birds and their estrildid finch hosts. Mol Ecol, 16:217-229.

Baum, V.H.1968. Biology and ecology of the feather lice of blackbirds. (Translated from German) Angew Parasitol, 9:129-176.

Benkman, C.W., Holimon, W.C., Smith, J.W.2001. The influence of a competitor on the geographic mosaic of co evolution between crossbills and lodgepole pine. Evolution, 55:282-294.

Benkman, C.W., Parchman, T.L.,Favis, A., Siepielski, A.M.2003. Reciprocal selection causes a coevolutionary arms race between crossbills and lodgepole pine. Am Nat, 162:182-194.

Bennett, G.F.1961. On three species of Hippoboscidae (Diptera) on birds in Ontario. Can J Zool 39:379-406.

Bequaert, J.1953. The Hippoboscidae or louse-flies (Diptera) of mammals and birds. Part I. Structure, physiology and natural history. Entomol Am, 32:1-209.

Bonds, M. H. 2006. Host life-history strategy explains pathogen induced sterility. American Naturalist, 168:281-293.

Boots, M., and Haraguchi, Y. 1999. The evolution of costly resistance in host-parasite systems. American Naturalist, 153:359-370.

Bremermann, H. J., and Thieme, H.1989. A competitive exclusion principle for pathogen virulence. Journal of Mathematical Biology,27:179-190.

Brodie, E. D. Jr. et al. 2002.The evolutionary response of predators to dangerous prey: Hotspots and coldspots in the geographic mosaic of coevolution between newts and snakes. Evolution, 56: 2067-2082 .

Brooks, D.R. 1979. Testing the context and extent of host-parasite coevolution. Syst Zoolog , 28:299-307.

Brown, N.S.1971. A survey of the Arthropod parasites of Pigeons (*Columba livia*) in Boston. J Parasitol ,6:1379-1380.

Bueter, C., Weckstein, J., Johnson, K.P., Bates, J.M., Gordon, C.E.2009. Comparative phylogenetic histories of two louse genera found on Catharus thrushes and other birds. J Parasitol ,95:295-307.

Bush, S.E., Clayton D.H.2006. The role of body size in host specificity: Reciprocal transfer experiments with feather lice. Evolution, 60:2158-2167.

Butlin, R. K., and Tregenza, T. 2005. The way the world might be.Journal of Evolutionary Biology, 18:1205-1208.

Caldera, E.J., Poulsen, M., Suen, G., Currie C.R.2009. Insect symbioses: A case study of past, present, and future fungus-growing ant research. Environ. Entomol, 38(1):78-92.

Campbell, M. A., Fitzgerald, H. A. and Ronald, P. C 2002. Engineering pathogen resistance in crop plants. Transgenic Research, 11:599-613.

Charleston, M .A ., Page, R .D .M. 2002. TreeMap. v. 2.0.2. Software distributed by authors .

Christiansen, F. B. 1991. On conditions for evolutionary stability for a continuously varying character. American Naturalist, 138:37-50.

Clayton, D .H ., Bush, S .E ., Johnson, K .P.2004. Ecology of congruence: Past meets present. Syst Biol ,53(1):165-173.

Clayton, D .H ., Drown, D.M.2001. Critical evaluation of five methods for quantifying chewing lice (Insecta: Phthiraptera) J. Parasitol. 87:1291-1300.

Clayton, DH, et al.2005. Adaptive significance of avian beak morphology for ectoparasite control. Proc. Biol. Sci, 272:811-817.

Clayton, D. H., Johnson, K .P.2003 Linking coevolutionary history to ecological process: Doves and lice. Evolution, 57:2335-2341.

Clayton, D .H., Lee, P .L .M ., Tompkins, D.M., Brodie, E .D III.1999. Reciprocal natural selection on host-parasite phenotypes. Am. Nat. 154:261-270.

Corbet, GB.1956. The life-history and host-relations of a hippoboscid fly *Ornithomyia fringillina* Curtis. J. Anim. Ecol., 25:403-420.

Corbet, GB.1956. The phoresy of Mallophaga on a population of *Ornithomyia fringillina* Curtis (Dipt., Hippoboscidae) Entomol. Mon. Mag., 92:207-211.

Currie, C R. 2001. Prevalence and impact of a virulent parasite on tripartite mutualism. Oecologia, 128(1):99-106.

Currie, CR, et al.2003. Ancient tripartite co evolution in the attine ant-microbe symbiosis. Science, 299:386-388.

de Clercq, E. 2004. Antivirals and antiviral strategies. Nature Reviews Microbiology, 2:704-720.

Dercole, F., Ferriere, R. , Rinaldi, S.2010. Chaotic Red Queen coevolution in three-species food chains. Proc. Biol. Sci., 277:2321-2330.

Dranzoa C,Ocaido M, Katete P.1999. The ecto-, gastro-intestinal and haemoparasites of live pigeons (*Columba livia*) in Kampala, Uganda. Avian Pathol. 28:119-124.

Dwyer, G., J. Elkinton, and Buonaccorsi. J. 1997. Host heterogeneity in susceptibility and disease dynamics: tests of a mathematical model. American Naturalist 150:685-707.

Ebert, D., and Bull, J. J. 2003. Challenging the trade-off model for the evolution of virulence: is virulence management feasible? Trends in Microbiology, 11:15-20.

Edwards, R .1952. Flat flies taken in the laboratory during 1951. Fair Isle Bird Obs Bull, 6:37-38.

Fukatsu, T., et al..2007. Bacterial endosymbiont of the slender pigeon louse, *Columbicola columbae*, allied to endosymbionts of grain weevils and tsetse flies. Appl. Environ. Microbiol, 73:6660-6668.

Gomulkiewicz, R. et al. 2000.Hot spots, cold spots, and the geographic mosaic theory of co evolution. The American Naturalist, 156: 156-174.

Gurr, S. J., and Rushton, P. J. 2005. Engineering plants with increased disease resistance: what are we going to express? Trends in Biotechnology,23:275-282.

Haloin, J.R., Strauss, S.Y.2008. Interplay between ecological communities and evolution: Review of feedbacks from micro evolutionary to macro evolutionary scales. Ann. N Y Acad. Sci, 1133:87-125.

Harbison, C.W., Bush, S.E., Malenke, J.R., Clayton, D.H.2008. Comparative transmission dynamics of competing parasite species. Ecology, 89:3186-3194.

Harbison, C.W., Jacobsen, M .V., Clayton, D.H.2009. A hitchhiker's guide to parasite transmission: The phoretic behaviour of feather lice. Int. J. Parasitol., 39:569-575.

Hoeksema, J. Tansley, D.2010. Ongoing coevolution in mycorrhizal interactions. New Phytologist, 187: 286-300

Houck, M .A., O'Conner ,B.M.1991. Ecological and evolutionary significance of phoresy in the Astigmata. Annu. Rev. Entomol, 36:611-636.

Iannacone, J.A.1992. Registering a case of phoresis: *Columbicola columbae* (L) (Phthiraptera: Insecta) on *Psuedolynchia canariensis* (Diptera: Insecta) in the area of Lima, Peru. (Translated from Spanish) Boletin de Lima, 84:17-18.

Johnson, K.P ., Adams, R .J ., Clayton, D .H.2002. The phylogeny of the louse genus Brueelia does not reflect host phylogeny. Biol. J. Linn. Soc. Lond, 77:233-247.

Johnson, K.P., Clayton, D.H.2003. In Tangled Trees: Phylogenies, Co speciation, and Co evolution, Co evolutionary history of ecological replicates: Comparing phylogenies of wing and body lice to Columbiform hosts, ed Page RDM (The University of Chicago Press, Chicago), pp 262-285.

Johnson, K.P., Williams, B.L., Drown, D.M., Adams, R.J., Clayton, D.H.2002. The population genetics of host specificity: Genetic differentiation in dove lice (Insecta: Phthiraptera) Mol. Ecol, 11(1):25-38.

Johnson, M.T.J., Stinchcombe, J.R .2007. An emerging synthesis between community ecology and evolutionary biology. Trends Ecol. Evol., 22:250-257.

Johnson, N. C. et al.2010 Resource limitation is a driver of local adaptation in mycorrhizal symbioses. Proceedings of the National Academy of Sciences of the United States of America, 107: 2093-2098.

Kisdi, E. 1999. Evolutionary branching under asymmetric competition. Journal of Theoretical Biology, 197:149-162.

Kisdi, E. 2006. Trade-off geometries and the adaptive dynamics of two co-evolving species. Evolutionary Ecology Research, 8:959-973.

Klei, T.R., DeGiusti, D.L.1975. Seasonal occurrence of *Haemoproteus columbae* Kruse and its vector *Pseudolynchia canariensis* Bequaert. J. Wildl. Dis, 11:130-135.

Macchioni , F, Magi ,M., Mancianti, F., Perrucci, S.2005. Phoretic association of mites and mallophaga with the pigeon flies *Pseudolynchia canariensis*. Parasite, 12:277-279.

Markov, G.S.1938. The presence of phoresy in Mallophaga. Zool Zh, 17:634-636.

Marrow, P., Dieckmann, U. and Law, R. 1996. Evolutionary dynamics of predator-prey systems: an ecological perspective. Journal of Mathematical Biology, 34:556-578.

Marshall, A.G.1965. The Ecology of Ectoparasitic Insects. London: Academic Press.

Murray, M.D., Nicholls, D.G. 1981. Studies on the ectoparasites of seals and penguins: I. The ecology of the louse *Lepidophthirus macrorhini* Enderlein on the southern elephant seal, Mirounga leonina (L.). Aus. J. Zoolog , 13:437-454

Martin, M.1934. Life history and habits of the pigeon louse (*Columbicola columbae* [Linnaeus]) Can Entomol ,66(1):6-16.

May, R. M., and Anderson, R. M. 1983. Epidemiology and geneticsin the coevolution of parasites and hosts. Proceedings of the Royal Society B: Biological Sciences, 219:281-313.

Miller, M. R., White, A.and Boots, M. 2005. The evolution of hos tresistance: tolerance and control as distinct strategies. Journal of Theoretical Biology, 236:198-207.

Nowak, M. A., and May, R. M. 1994. Superinfection and the evolution of parasite virulence. Proceedings of the Royal Society B: BiologicalSciences, 255:81-89.

Nuismer, S. L..2006. Parasite local adaptation in a geographic mosaic. Evolution, 60: 24-30.

Nuismer, S. L. & Gandon, S.2008. Moving beyond common-garden and transplant designs: Insight into the causes of local adaptation in species interactions. The American Naturalist, 171: 658-668.

Paterson, A M, Gray, R.D. 1997. Host-parasite cospeciation, host switching and missing the boat. In Host-Parasite Evolution: General Principles and Avian Models. Edited by Clayton DH, Moore J. Oxford: Oxford Univ. Press; 236-250.

Pavlovic, I, et al.1995. Significance of Arthropoda in health problems of pheasants which are bred artificially. Veterinarski Glasnik, 49:745-749.

Pedersen, A. B., and A. Fenton. 2007. Emphasizing the ecology inparasite community ecology. Trends in Ecology & Evolution, 22:133-139.

Peters, H.S.1935. Mallophaga carried by hippoboscids. Ann. Carnegie Mus., 24:57-58.

Price, R.D., Hellenthal, R.A., Palma, R.L., Johnson, K.P., Clayton, D.H.2003. The Chewing Lice World Checklist and Biological Overview (Illinois Natural History Survey, Special Publication). Proceedings of the National Academy of Sciences: www.pnas.org> vol. 108 no. 23

Pugliese, A. 2002. On the evolutionary coexistence of parasite strains. Mathematical Biosciences, 177:355-375.

Reed, D.L., Smith, V.S., Hammond, S.L., Rogers, A.R., Clayton, D.H. 2004. Genetic analysis of lice supports direct contact between modern and archaic humans. PLoS Biol 2004, 2(11):e304.

Regoes, R., Nowak, M., and Bonhoeffer. S. 2000. Evolution of virulence in a heterogeneous host population. Evolution, 54:64-71.

Restif, O., and Koella. J. C. 2003. Shared control of epidemiological traits in a co evolutionary model of host-parasite interactions.American Naturalist, 161:827-836.

Roy, B. A., and Kirchner, J. W. 2000. Evolutionary dynamics of pathogen resistance and tolerance. Evolution, 54:51-63.

Saenz-Romero, C. et al. 2006.Altitudinal genetic variation among *Pinus oocarpa* populations in Michoacan, Mexico: Implications for seed zoning, conservation, tree breeding and global warming. Forest Ecology and Management, 229: 340-350.

Schluter, D. 2009.Evidence for ecological speciation and its alternative. Science, 323: 737-741 .

Smith, J.W., Benkman, C.W. 2007. A coevolutionary arms race causes ecological speciation in crossbills. Am. Nat, 169:455-465.

Sol, D., Jovani, R., Torres, J. 2000. Geographical variation in blood parasites in feral pigeons: The role of vectors. Ecography, 23:307-314.

Stauffer, R.L., Walker, A., Ryder, O.A., Lyons-Weiler, M., Hedges, S.B. 2001. Human and ape molecular clocks and constraints on paleontological hypotheses. J. Hered , 92(6):469-474.

Strauss, S.Y., Sahli, H., Conner, J.K. 2005. Toward a more trait-centered approach to diffuse (co)evolution. New Phytol, 165(1):81-89.

Teel, P.D., Fleetwood, S.C., Hopkins, S.W., Cruz, D.1988. Ectoparasites of Eastern and Western meadowlarks from the Rio Grande plains of south Texas. J. Med. Entomol, 25(1):32-38.

Thompson, G.B .1947. Association of Hippoboscidae and Mallophaga: Further notes and records. Ent. Mon. Mag, 83:212-214.

Thompson, J. N. 1994. The Co evolutionary Process. Chicago, IL: University of Chicago Press.

Thompson, J. N. 2005.The Geographic Mosaic of Co evolution. Chicago, IL: University of Chicago Press, 2005.

Toju, H. & Sota, T. 2006. Imbalance of predator and prey armament: Geographic clines in phenotypic interface and natural selection. The American Naturalist 167, 105-117.

Tompkins, D. M., A. White, and M. Boots. 2003. Ecological replacement of native red squirrels by invasive greys driven by disease. Ecology Letters 6:189-196.

Urban, MC, et al.2008. The evolutionary ecology of metacommunities. Trends Ecol Evol, 23:311-317.

van Vaalen, M. 1998. Co evolution of hosts and parasites. Proceedings of the Royal Society B: Biological Sciences 265:317-325.

Wade, M. J. 2003.Community genetics and species interactions. Ecology. 84: 583-585 .

Walter, VG.1989. Phoresy and hyperparasitism in *Ornithomyia* (Diptera, Hippoboscidae) in the Federal Republic of Germany. (Translated from German) Angew Parasitol 30(1):43-46.

Ward, RA.1953. Additional record of phoresy of Mallophaga on Hippoboscidae. Bull Brooklyn Entomol. Soc. 48:128.

Weitz, J. S., H. Hartman, and S. A. Levin. 2005. Co evolutionary arms races between bacteria and bacteriophage. Proceedings of the National Academy of Sciences of the USA, 102:9535-9540.

Whiteman, NK, Santiago-Alarcon D, Johnson KP, Parker PG. 2004. Differences in straggling rates between two genera of dove lice (Insecta: Phthiraptera) reinforce population genetic and cophylogenetic patterns. Int. J. Parasitol., 34:1113-1119.

# CHAPTER 11

# Evolution Beyond Biology

Modern evolution started as a biological concept with the introduction of natural selection. Evolution was recognized to have wider application since the time of Darwin. Evolutionary force help adapts evolutionary thinking in non-biological contexts. The process of drift operating from simplicity results in diversity and complexity over time. The diversity arises from the continuous introduction of novelty by random process. In fact diversity and complexity arise due to 'drunkards walk' away from simplicity (Gould, 1996; McShea and Brandon, 2010). That large-scale evolutionary pattern passes through stages in a predictable order is linked with social progress, moral progress, and scientific progress .In the social sciences, the evolution is a predictable change, often through well defined stages (Sanderson, 2007). L. H. Morgan's evolutionary anthropological theory posited a progression of human culture from savagery, to barbarism, to civilization through the continuing process of technological progress (Morgan, 1877).

**Life**, a complex and undirected system is based on biochemical processes. Cell collects organic molecules for producing DNA; make their structures for copying DNA from the organic molecules .The plants produce organic molecules from the solar energy. Non-photosynthetic organisms collect the products from the plants. Non-photosynthetic organisms have evolved to kill other organisms. Acquiring organic molecules requires metabolic activity. Animals can not exist without this theft. Organism, a package of assembly of organic molecules, mates to produce offspring and to pass their traits to next generation. The evolutionary tree of life previously thought as a simple ladder. Now it is seen as extremely complex with many side branches that went extinct.

The DNA changes through various ways and passed onto organisms through reproduction, cell to cell transfers, and acquisition of floating DNA in environment. **Evolutionary biology**, an empirical science, the principles of which are supported by observational evidence. The empirical method with 2 sharp cutting edges cuts through confusion and misunderstanding. Some argues about philosophical implications of current understanding of evolutionary biology, implications that contradicts our cherished beliefs. The evolution, the most significant revolutions in human thought solves problems and makes predictions.

**Genetic code** controlling a majority of form and function of about all organisms has evolved instructions as specific adaptations to specific environmental problems and potentialities through time. *Homo sapiens* with learning and behavioral properties have evolved in the same environment following same rules as all other organisms. Alexander noted the inheritance of genetic code for form and function. Sexual recombination and random mutations produce endless variation in succeeding generations. Natural selection, the demands of environment select those organisms which are best suited to survive and reproduce with endless genetic variety. Alexander states that interaction of these phenomena and successions of environments account for traits and history of all life forms. Natural selection with directional component is the driving force in evolution. Traits is adaptive to explicit circumstances and selected. The brain promotes the survival and multiplication of genes. Conrad Waddington compared the relationship of inherited and **acquired behavior** in humans with topography of landscape. For traits like eye color, the topography is a single deep channel. Once egg and sperm meet only one eye color is possible. Many behaviors similar in form across culture, suggest that behavior follows a defined path. Paul Ekman showed pictures of facial expressions to children of American and of Stone Age cultures; 80% of the children in both recognized the emotion being expressed in the face of the other culture.

## 11.1 Biocultural Evolution or Gene-Culture-Co evolution

Humanity's sojourn on the earth is significant as origin of photosynthesis before some 2 billion years. Our evolution allows for intentional culturally-acquired adaptations and their cultural transmission. **Biocultural evolution** dominates all life and places the values and intentions of humans as driving force of future evolution. The religion, broadly conceived as the DNA of cultural replication, would take center stage in biocultural evolution. Humans are linked with rest of nature. All biological creatures depend on biological processes to sustain our lives. All species adapt to an environment and change the environment to some extent. We have changed the environment drastically from the beginning of agriculture over last 10,000 years. The relationship between our biology and our culture, encoded in genetic and neural evolution, project outwards onto the environment

Intensification of human interferences on environments is found in mining, fishery, trade, commerce, travel construction, energy consumption, forestry, communications, and agriculture. **Human population** has increased 4 fold between 1890 and 1990 with 13 fold increase in urban population. Now there are over 6 billion humans. Birth rates are declining in most regions. However, we can expect growth over the next few decades (McNeill, 2000).Sustenance for growing human population require increased food production , clean water, sanitation ,energy consumption, and prevention and cure of diseases. The world economy grew 14-fold between 1890 to 1990, with increase in industrial output by 40-fold fueled by 16-fold increase in energy use and 9-fold increase in water consumption. The domestic cattle population and domestic pig population grew 4-fold and 9-fold respectively. Land under **cultivation** doubled during this period with 20 percent decrease in forest area. Marine fish catch increased 35 fold in the last century (McNeill, 2000).

## 11.2 Complex Distributed Systems

The whole is more than sum of its parts and emergent property of whole need to be studied for their properties that manifest themselves at integrated level. Distributed parts contribute to holistic dynamics of system (Bunge, 2003). Kelly seeks a meta theory of complex creativity in nature, human consciousness, computer viruses, economic markets, technology, and culture. He examines phenomena as diverse as telephone system, robotics, ecological restoration, evolution and social insects (Kelly 1994). Biologists view the genome as a complex distributed system. The common view of genome is a series of programmable on-and-off switches. The genome could be viewed like a bureaucracy in which genes, groups of genes are interlinked in multiple and dynamic interactive patterns. Gene expression in production of proteins is hardly a one-way street, as cytoplasm and environment can alter result. Some 100,000 varieties of proteins fold and wrap around themselves in ways which are poorly understood.

**Ecosystems** are characterized by distributed complexity. Species and environments compete and cooperate in many ways to create phenomena of an interactive ecosystem. The whole evolution is a complex distributed system in which gene sequences and phenotypic traits are shared. Dinosaurs are extinct, but parts of their genes and phenotypic traits is still replicating in mammals, reptiles, birds. Brain is an example of complex distributed system. A single neuron does not do much. The 100 billion neurons in our brain, each with more than 1000 synaptic connections, may be the most complex entity. A brain requires co-evolution with our vocal cords, rest of central nervous system and body, intricacies of nature, and relationships to other humans in culture. We now recognize that highly creative processes in life are best understood as complex distributed systems.

## 11.3 Complexity Horizons

It is believed that $CO_2$, $CH_4$, and other emissions in atmosphere are rising that increase the atmospheric temperature. Effects of such gases on climate are yet to be clearly understood. Regional and global increases or decreases in precipitation levels, decreases or increases reflective albedo of cloud cover, and circulation flows in ocean currents and atmospheric jet streams need to be studied. Change in ocean phytoplankton, possible thawing of artic tundra, and increased carbon fixing by plants in response to increased atmospheric carbon level are also note-worthy. All these processes create feedback loops that could change the effects of increased greenhouse gases that need to be modeled. Complex distributed systems can be robust, or sensitive with non-linear and emergent properties which confound us with complexity horizons. There is a profound moral ambiguity about biocultural evolution in the 21st century.

## 11.4 Lamarckian Wild Cards

Pattern of evolution exhibited by humans is more Lamarckian than Darwinian. Acquired features are passed to future generations through education. In human cultural evolution, selection operates on background of intentional adaptation. **Environmental engineering** changes the background of natural selection, which becomes domesticated in new environments. As we move towards large-scale genetic engineering, we end up with Lamarckism, in which phenotype directs genotype. In the 20th century we live longer, eat better, and have homes, better education, more scopes for travel and entertainment. We are safe in clean environments than our ancestors 100 or 1000 years ago. Impressive data support this claim (Templeton, 1997). We fly around the world at 30,000 feet. One hundred years ago trains, steam ships, and walking were the modes of transportation. Human life expectancy has doubled in the last century. New technologies may extend life span beyond 120 years. By resetting the telomeres at the end of our chromosomes, we can reverse cell senescence. Chemical, nuclear, biological, and conventional weapons have proliferated and likely to be used in the next century. Anthropogenic changes in environment have the potential of spinning out of control, whether it is through soil salination, depleted aquifers, climate changes, barren oceans, voracious weed species, and industrial pollution. The growing resistance to antibiotics could cause pandemic as happened 1918-1919 influenza pandemic that killed about 40 million people.

Fifty-years ago about half of the population in the US was involved in agriculture. Now it is less than 3 percent. On one hand, we have changed little over the millennia as we are still biologically and psychologically identical to our ancestors 10,000 years ago. We are conceived in passion, born in pain, with long period of childhood dependency, after which we initiate adulthood. We work, love, raise children, and grow old. New medical technologies are going to revolutionize human reproduction. It would be the free market that regulates new reprogenetic technologies. Wealthy parents want to send their kids to expensive private schools, or to pay for their orthodontics. If the technology enables to engineer a healthier, attractive, and intelligent baby, then rich parents will avail such technologies. A **speciation horizon** for humanity in the next century is predicted , in which genetically-enriched humans will become a separate breeding population from natural humans, and the "Gen-rich" will also dominate (Silver 1997). Haraway popularizes and philosophically examines the notion of cyborgs, i.e., cybernetic organisms. Cyborgs are about specific historical machines and people in interaction that often turns out to be painfully counterintuitive for analyst of techno science. **Cyborg** attempts to refigure provocatively the border relations among specific humans, other organisms, and machines (Haraway, 1996). E.O. Wilson envisions humans primarily through retrospective lens of genetic and evolutionary pasts. **Sociobiology**, now called evolutionary psychology attempts to understand human behavior based on survival and reproduction. Our nature encoded over millennia in our genes gives us some freedom in behavior, but control us from wandering far away from our fundamental nature.

## 11.5 Technology Matters

Traditional ethics would evoke the precautionary principle - do no harm. We should desist those technological and cultural changes that cause harm. Genetically modified food could cause irreparable damage to ecosystems. Without applying biotechnology, it is difficult to feed the world with reduced cropland .It may also difficult to restore the ecosystems, solve the energy crisis, or cure the next pandemic. Human values, beliefs, and practices are also involved in the equation. Religions as the DNA of cultural evolution glue us together in the whole. Religions represent the collective wisdom of the ages which transcends us. Strict environmental, or cultural preservationism would freeze history in principal and in any case would not work in practice. Life is a thermodynamically disequilibrious process. Equilibrium could result in death, but much disequilibrium also results in death.

## 11.6  Evolution of Infectious Disease: An Analysis of AIDS

The interaction between human populations and disease show the pathogen's adaptation and the biocultural responses of humans to pathogen. The cultural system could influence the disease, and disease could alter cultural adaptation. The **social system** organizes to control the disease (Fabrega, 1975). The ecological model derived from epidemiology originally concerned with the interaction between the host, pathogen, and environment often failed to consider relevant socio-cultural factors that might underlie the disease. Later some variables have added/ modified for better understanding of disease and its impact on population. The modifications include the population as the unit of study to consider disease in broader biological and social context. Secondly, the ecological model is no longer restricted to study of pathogens as the only source of disease. Broader category of insults is now considered as the source of disease (Audy and Dunn, 1974). Insults are pathogens, toxins, physical forces causing trauma, chemical pollutants, and psychological factors which adversely affect the host population. **Disease** is thus defined as the lowering of an individual's or population's ability to cope with its environment. Health is defined as an individual's or population's ability to rally from the effects of these insults. Health and disease are considered to be a continuum and not an either-or condition (Audy and Dunn, 1974).The significant change in ecological model is the transformation of our perception of environment. Originally the environment was restricted to consider the biotic, climatic, topographic, and geographic factors influencing disease. This concept of the environment now includes the cultural system as a part of human disease interaction (Marston Bates, 1953 and Jacque May, 1960). Culture is comprised of technology, social organization, and ideology. Ideology and social organization include an ethno medical perspective which analyzes the process by which contemporary societies define disease and their attempt to deal with a disease threat.

The **cultural system** buffer the population from the insults that emanate from the environment. The technology, social organization, and ideology often create insults that affect the health of population. Changes in technology, social organization, and ideology can produce new insults. Human have faced a struggle with disease that threatens their adaptation. During the last 4 million years, human **disease ecology** has changed significantly due to changes in environment, evolution of species, and cultural adaptation. These created different environments for pathogens and altered their interaction with human populations. There has been dramatic change in pattern of disease and human response, especially within last 10,000 years (Armelagos and McArdle, 1975). When gathering and hunting were the sole means of human subsistence, population size was small, and density was low. Population size and density remained low throughout the Paleolithic. Fertility and **mortality** rate in such small gathering hunting populations were perhaps balanced and population growth was low and stable. Some argue that gatherer-hunters were at their maximum fertility which was balanced by high mortality. Other argues that gatherer hunter maintained a stable population with controlled moderate fertility balanced by moderate mortality. The Neolithic caused a major shift in subsistence resulting in dramatic increase in population size and density. Some argue that the Neolithic economy generated food surpluses which aided in population growth. The abundance of food would led to a better nourished and healthier population with reduced mortality. Since populations attained maximum fertility, a rapid increase in population was expected. The biological consequence of shift from gathering and hunting to **agriculture** presents a serious picture of health and disease. Instead of experiencing improved health, there is evidence of an increase in infectious and nutritional disease.

## 11.7 Disease in Gatherer-hunters

Gatherer hunters would have 2 types of disease (Polgar, 1964). One class of disease would be those organisms that had adapted to prehominid ancestors and persisted with mass which evolved into hominids. Head and body lice (*Pediculus humanus*), pinworms, and malaria are included in this group. Cockburn (1967b) adds to this list most internal protozoa found in modern humans and bacteria as *Salmonella*, and *Staphylococci*. The malaria in early hominids was dismissed because of their small population size and adaptation to savannah, an environment lacking mosquitoes (Livingstone, 1958). The second classes of diseases are the zoonotic, with nonhuman animals as their primary host and only incidentally infect humans. Humans could be infected by *zoonoses* through insect bites, by eating contaminated flesh, and from wounds inflicted by animals. Sleeping sickness, tetanus, trichinosis, tularemia, leptospirosis, and schistosomiasis, scrub typhus, relapsing fever, are the zoonotic diseases which afflicted earlier gatherer-hunters.

The range of earliest hominids was limited to tropical savannah which limited the spread of potential disease agents. There was eventually an expansion of habitat into temperate and eventually the tundra zones. The hominids would have avoided large areas of the **African landscape** because of tsetse flies and thus avoided the trypanosomas

(Lambrecht, 1985). The evolution of human and its expansion into new **ecological niches** would have led to a change in pattern of trypanosome infection. The contagious diseases like influenza, smallpox, measles, and mumps would have been missing. There would have been few viruses infecting these early hominids (Burnet, 1962). The viral diseases found in nonhuman primates would have been easily transmitted to hominids Cockburn. (1967a,b).

## 11.8 Disease in Agricultural Populations

A shift to agriculture would increase the number and impact of disease in sedentary populations which would increase parasitic disease spread by contact with human excreta. Frequent movement of the base camp and forays away from base camp would decrease human contact with excreta in gathering hunting groups. The proximity of living area and their waste deposit sites to water supply is a source of contamination in sedentary populations. While sediments could and did occur before the Neolithic period in those areas with abundant resources, the shift to agriculture necessitated sedentary living. The herding of animals increased the scope of contact with zoonotic diseases. Domestication of animals in the Neolithic caused a steady supply of disease vectors. **Zoonotic infections** increased because of domesticated animals like goats, cattle, pigs, sheep, and fowl. Animal products like skin, milk, hair and the dust raised by the animals, could transmit anthrax, tuberculosis, Q fever, and brucellosis (Polgar, 1964). Cultivation exposes workers to insect bites and diseases such as scrub typhys (Audy, 1961). The slash-and burn agriculture in West Africa exposed populations to *Anopheles gambiae* causing malaria (Livingstone, 1958).

The development of urban centers is recent. Cities with 50,000 people were established by 3000 B.C in the Near East. In the New World, urban settlements of 200,000 people were in existence by 600 A.D. Settlements of this size increase the already difficult problem of removing human wastes and delivering uncontaminated water to people. Cholera transmitted by contaminated water was a potential problem. Diseases like typhus (carried by lice) and plague bacillus (transmitted by fleas, or by the respiratory route) could be spread from person to person. Viral diseases like measles, mumps, chicken - and smallpox could be spread in similar fashion. During **urbanization**, populations were large enough to maintain endemic disease. Population of one million would be necessary to maintain measles as an endemic disease (Cockburn, 1967b). Urban development with expansion of populations into new areas results in introduction of novel diseases (McNeill, 1976). The exploration of the New World may have been the source of treponemal infection that was transmitted to the Old World (Baker and Armelagos, 1988). The treponemal infection in New World was endemic and not sexually transmitted.

In the Old World, the sexual transmission of treponeme created a different environment for pathogen and it resulted in severe and acute infection. Crowding in the urban centers, prostitution, and increase sexual promiscuity may have been factors in venereal transmission (Hudson, 1965). Agriculturalists faced greater infectious disease stress than gatherer-hunters. Zoonotic diseases in gatherer-hunters would likely have the greatest impact on segment of the society that contains the producers (those between ages 20 and 40). This segment, in its daily rounds, is more likely to come into contact with the animals that are the vector of disease.

The occurrence of **endemic diseases** in larger urban agriculturalist areas would likely kills infants, children, and olds. Predictability

of mortality allows them to reduce birth spacing to meet the increase in mortality. Sedentary societies can wean infants earlier, allowing women to become pregnant again. Mortality may not be as great as impact of zoonotic diseases on gatherer-hunters. Those who survive (because of acquired immunity) will be protected from pathogens. Protected producers segment would be able to reproduce and continue to extract the resources for survival. Industrialization would lead to an even greater environmental and social transformation. City dwellers would have contend with industrial wastes and polluted water and air. Slums would become the focal point for poverty and spread of disease. Epidemics of smallpox measles, typhus, typhoid, diphtheria, and yellow fever in urban settings are well documented (Polgar, 1964). Tuberculosis, pneumonia and bronchitis are serious problems. The rapid transportation system led to rapid transmission of HIV from one continent to another.

The patterns of AIDS transmission in the Americas and Africa are not unexpected. AIDS is extracting a large toll in both developing and developed nations. Nations like the United States experience new outbreaks of infectious disease. The ecological context explicitly sets health and disease in a system of mutually interacting organic, inorganic, and cultural environments. The cultural environment receives special emphasis since it is within culture that much of the behavior surrounding health and disease is played out. The host, the target of insults, can be studied at several levels. The impact of insult can be analyzed at level of population, individual, or even at sub-individual level. Individuals are actors making choices that may or may not be environmentally constrained. In an actor-oriented behavioral model, we can focus on process of coping and adapting. The rationale for this approach follows from observation that health, disease, and illness are not simply objective, verifiable concepts but are informed by perceptions and social relations of patient, healer, family, and community. All aspects of cultural environment, and host response to insults entails modification, or use of resources within all these environmental subsystems. The goals of individuals, options available and appropriate for response, and multiple constraints in environment that influence response are a part of understanding the adaptation of a group. Options for response to insult are not infinite. They may be constrained by wealth, ongoing and future social relations, class position, and ideology. Rarely do individuals or at-risk groups have one insult with which to contend; rather they must cope with multiple constraints which may augment a disease process, limit treatment options (poverty), and even alter host goals. By focusing on goals, options, and constraints of affected individuals and/or groups, we can better evaluate the relative benefit of treatment, and thus have a more thorough measure of efficacy in response to illness. The relative benefit of treatment must be assessed.

Adaptation is a continuous dynamic process. Past response shapes future expectations and options and capacity for coping. Coping with one set of conditions shapes constraints and options for coping with additional stressors. Given constraints and context in which they occur, immediate responses are inextricably linked to long-term adaptive processes. In employing an actor-based model of coping, no arbitrary distinction is drawn between notions of health, disease, and human behavioral response. Both ethno medicine and **biomedicine** in medical systems offer appropriate options for response to insult. It is insult which is inextricably a part of larger ecological context in which actor operates. The implications of this approach differ from other ecological adaptive models. A model of adaptive process conceptually locates both **proximate** and **ultimate causation** of illness within limitations and contradictions of actor goals, options, and constraints. Rather than creating a model in service of larger social, political, and economic powers, it is interesting to examine how people cope with multiple constraints and limited options, and the social and biological consequences of this process. We identify the individuals and the groups at-risk, locate the contradictions, and thus begin to recognize the needs of those at risk and the action needed to rectify this situation. The implications for change are 2 fold; change emanates from collective action, and when implementing health projects, existing strategies need to be considered for so as not to remove the power, control, and predictability which already rest within the populations affected. This approach would serves as a theoretical framework for health and disease in general, and **AIDS** specifically as reported elsewhere.

Starting from the diets humans chosed the tools they began to use, the languages they spoke and the cultural, musical and religious traditions they devised. Many complex interactions of biological and cultural evolution(s) finally have come into focus. A prominent example is the ability to digest lactose (milk sugar), which normally ceases in mammals after their time spans of weaning. But today, most Europeans inherited lifelong lactose tolerance in contrast to many Asian or African people. Because formerly rare mutations among those populations who lived in cultural settings including milk-giving cattle spread within a few thousand years. Another prominent example is the Skin color, which usually tended to lighten in colder climates with less threat by **UV-rays** but the danger of shortages in certain vitamins. Some Northern people as the Inuit are defying this rule, having retained comparably darker skins by their diet in fishes and meat rich on those vitamins. Darwin assumed that evolution included biocultural traits of tool-making, speech, music and religion. All of them evolved by bestowing (on average) reproductive advantages on those phenotypes to wield them. Many adamant evolutionists are accepting multitudes of findings to almost any traits - but are trying to exclude religiosity from their logic. They are (increasingly desperate) trying to erect another bastion of dualism as a kind of a last stand. But from the perspective of evolutionary studies on religiosity and religions, it is clear that we are in no need of any special theories circumventing universal evolution. Religiosity and religious traditions have been part of our biocultural nature since the Upper Paleolithic, and successful religions are bestowing huge reproductive benefits on those humans developing religious practice and affiliation.

Topography of language is complex. Humans have an innate drive to acquire speech. This "**first language**" is universal in humans across culture. The upper slope of topography of language is well-defined; though the lower slope of second language is broad and ill-defined, shifting with culture. Primary mental abilities and perceptual and motor skills are mostly influenced by heredity, while personality traits are the least influenced. It seems that ability to cope with relatively unvarying problems in physical environment develop along narrow channels, while qualities of personality, which represent adjustments to rapidly shifting social environment, are more malleable. Actions that must be induced rapidly are guided by emotion, for example childhood phobias, fears of snakes, spiders, rats, heights, dislike of different tastes. In early human history, fears and **phobias** might have provided an extra margin of security needed to ensure survival. If humans evolved following evolutionary laws, then even the human

propensity to create ritual, both religious and secular, may well be in response to biological messages. The universal human tendency to dichotomize, (to classify) other human beings into 2 artificial categories, kin verses non-kin, members versus non-members, makes complete sense in terms of genetic advantage. Individuals seek to perpetuate their own genes at the expense of stranger. Intraspecific competition is a basic tenet of evolutionary theory. The key to many aspects of human culture is the hypertrophy, the exaggerated growth of pre-existing structures. The basic social responses of hunter-gatherers have metamorphosed from relatively modest environmental adaptations into elaborate, sometimes monstrous forms in more advanced societies. Nationalism and racism are the culturally nurtured outgrowths of simple tribalism. Within Kungtribe in Africa, the Nyae Nyae Kung speak of themselves as perfect and clean, and other Kung people as alien murderers who use deadly poisons. More developed civilizations have raised this mechanism of self aggrandizement to an art form, exalting them by divine allow and denigrating others with false histories. Often the origins of these cultural elaborations are hidden.

Marvin Harris has suggested that chronic meat shortage may be the underlying mechanism in evolution of certain religious beliefs in some cultures. Ancient Mexico was deficient in large game. Thus as Aztec civilization flourished in Valley of Mexico, acquisition of meat protein grew increasingly difficult. When Cortez entered the Aztec capital of Tenochititlan in 1521, he found a hundred thousand skulls stacked in neat rows in the city center. The Aztecs believed that human sacrifice was necessary to please the high Gods. In fact, immediately after their hearts had been cut out, the victims were systematically butchered, and their parts were distributed and eaten. Genetic coding for quality protein had caused elaborate religious ritual, serving largely to conceal the cannibalism adopted to fulfill the genetic imperative for protein. Biologically determined behavior pervades all aspects of human life. Four areas of human behavior will be discussed for biological roots; aggression, sexuality, religion, and learning.

## 11.9 Aggression

Warfare, an organized aggression, has been endemic throughout history. Virtually all societies have invented elaborate rules and laws designed to minimize internal and authorize external conflict in an effort to exonerate the genes, to the tiny minority of societies that appear to be nearly or entirely pacific. Innateness refers to measurable probability that a trait will develop, not to certainty that trait will develop in all environments. Most kinds of **aggressive behavior** among members of the same species are responses to crowding in environment. Animals use aggression to gain control over necessities that are scarce. As population grows denser, aggression increases. Some species seldom or never run short of basic necessities of life which are typically pacific toward one another.

If aggression confers no advantage, it is unlikely to be encoded into innate behavior. Humans, while markedly predisposed towards aggression, are not the most violent animal. Recent studies of hyenas, lions, and langur monkeys have shown that individual engage in lethal fighting, infanticide, and **cannibalism** at a rate far above that of humans. Alongside of ants, which conduct assassinations, skirmishes and pitched battles on their way to work, men are all but tranquillized pacifists (Figure 11.1).

**Figure 11.1** Fighting in animal seal

Aggression in humans is not an inborn drive that grows inexorably toward discharge. It is an interaction of genetic potential and learning. It has been found in one study of ten warlike and ten pacifist societies that the practice of war is accompanied with a greater development of combatant sports and other lesser forms of violent aggression. Many organisms exhibit territoriality. **Territories** seem to contain an almost invincible center. The resident animal defends the territory more vigorously when intruders attempt to us it, and defender usually wins. It has a moral advantage over trespassers. The biological formula of territorialism translates easily into territorial practices of human societies. Areas defended by hunter-gatherers are those that appear to be the economically defensible. The Western Shoshoni of the Great Basin, whose resource-poor land was too vast to defend, had no concept of ownership of land. In contrast the Owens Valley Paiute occupied fertile land in areas of abundant game. The valley was finely divided into resource units, each owned by a different band. These territories defended by social and religious allowed an occasional threats and attacks.

The force behind most warlike policies is **ethnocentrism**, the irrationally exaggerated allegiance of individuals to their kin and fellow tribesmen. This powerful drive to favor one's own social group which in ancient times was always one's kin group makes complete sense in terms of evolutionary theory, in which selection for kin's genes may augment or even supplant selection for one's own genes. Thus modern man walks in shadow of primitive man, dividing the world into 2 tangible parts, the near environment of home, kin, nation-state, and the more distant universe of outside villages, wild animals and enemies. The cultural evolution of aggression appears to be guided by 3 forces, 1) genetic predisposition toward learning some form of group aggression, 2) necessities imposed by the environment, and 3) history of the group, which biases toward the adoption of particular modes of behavior. To use the landscape metaphor, the channels of societal aggression are deep.

A society is not likely to avoid them altogether. A society is influenced to take a particular direction by physical and cultural realities in which it exists. Evidence suggests that culturally entrenched warfare can be reversed. In pre-European times, the Maori of New Zealand were the most aggressive people. Insults, hostility and retribution were carefully tallied by the 40-odd tribes, and victory by force of arms was the highest achievement. The introduction of European weapons lead to an escalating arm race, but the price paid even by victors proved unbearable. By 1830, the dominant tribe had begun to question the use of fighting for revenge; the old values crumbled soon afterward. Now societies have come within one step of nuclear annihilation. Human beings are strongly predisposed to respond with unreasoning hostility to external threats, and to escalate this hostility to overwhelm the source of the threat by a wide margin of safety. We seem to be programmed to partition other people into friends and aliens, in the same sense that bird is inclined to learn **territorial songs**. These rules are likely to have developed over the millions of years of human evolution, and have conferred a advantage on those who conformed to them most fully. The rules of aggression are largely obsolete.

## 11.10  Memes, and Evolution of Morals, Culture, and Religion

The moral codes, culture and religion of human society are explained via memetic evolution. R. Dawkins coined the term meme to explain the replicating ideas. **Memes** have gained acceptance. Law is linked to the social interests of man. Justice, truthfulness and self-control are desirable social qualities. Evolution, the only explanation of trusted moral law considers all aspects of the moral life. The approval, or censure of an act is proportionate to the social value of the act. We get morally serious rules about washing, sneezing, coughing, excreting, and wearing hats. **Morals** are ideas to provide some social value. Using memes to understand morals provides a **model** for the development of the moral ideas and explain many nonsensical and repressive morals. Memes can be viewed as ideas and genes that code for the proteins in the brain that store ideas. Memes propagate by communication including spoken language, writing, art, music, and simple observation. Memes arise from material conditions, from other memes, or from the brain's observation of the environment, which produce new memes. Memes develop through the directed or undirected mutation and are selected for by natural selection. The mind is the vector for memes. Memes that convey beneficial quality to the vectors have the increased chance of survival and propagation.

Memes as morals convey socially beneficial ideas, like opposition to murder and theft, and are likely to be retained in a society and spread. Societies lacking such memes would have less selective advantage, and would die out. If memes are similar to **genes**, then one meme would simply be one gene. In the biological evolution, genetic drift plays major role in the propagation of genes. **Memetic drift** play a major role in propagation of memes, a part of meme complexes. Many deleterious memes or memes lacking effect become bundled into meme complexes which contain few primary advantageous memes and thereby piggyback on the propagation of the meme complex. Religions incorporating a variety of memes are the extensive meme complexes.

Some memes make up the religion and may produce beneficial traits for society. Other memes may have no impact. The negative effects in the meme complex may be outweighed by the beneficial effects of the other positive memes. Memes with detrimental impact may reinforce the retention and spread of the beneficial memes. Comparison between social humans and social insects give insight into the effect of memes on behavior. All insects in a hive, or colony produced by a single queen share the same genetic code. Social insect is unique with their self-sacrifice behavior, fighting and dying for the colony. Self sacrifice of workers is selected for through evolution. The workers, the genetic extensions of the queen cannot replicate, but increase the spreading of their own genes by sacrifice. As their genes don't replicate through them, they replicate through the queen. **Sacrificial behavior** among humans has genetic and memetic components (Figure 11.2).

Human's genetic coding promotes sacrificial behavior to protect family members which increase the likelihood of the propagation of the individual's genes shared by other family members. Sacrificial behavior is genetically selected for among social insects as the actual genes are not propagated by the worker insects, but by the queen. Humans engage in sacrificial behavior have zero genetic benefit as they are being driven to sacrifice by memes, not genes. Humans sacrifice their genes to defend or propagate ideas (memes), showing that memes are more powerful than genes in controlling human behavior. All organisms share the fundamental genes like the genes that code for transcription factor needed to transcribe RNA to make proteins. Such genes are highly conserved and shared by all life for their usefulness. Memes that are highly conserved, like memes for morals against murder, are common in different meme complexes. They are highly conserved and are useful across

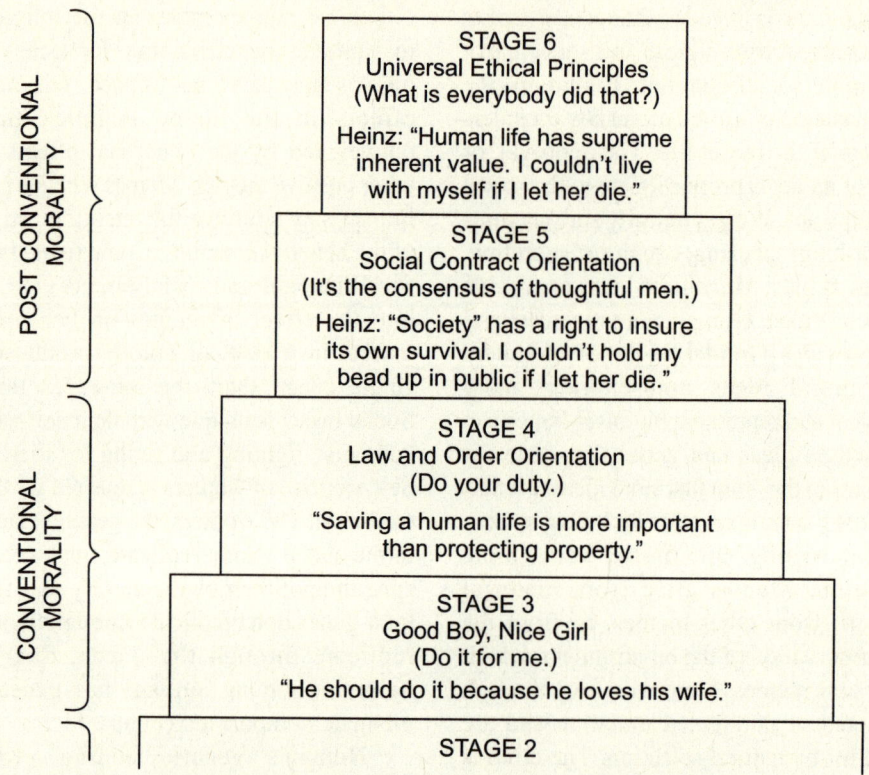

**Figure 11.2**   Kohlberg model of moral development

populations. Meme complexes follow the same general rules of Darwinian evolution, descent with modification, competition, and natural selection. The societies, the collections of vectors that express the traits of memes act similar to organismic life.

Life is a highly competitive struggle for survival, in which aggression and conquest exist. Memes that induce their vectors (humans) to be aggressive, militant, and evangelical naturally out-compete .Other memes induce their vectors to be peaceful and forgiving. Meme complexes that exhibited aggression, militancy, strong devotion, engendered the willingness to fight and die onto spread the beliefs, and has been highly evangelical. The natural selection dictates that aggressive memes out compete passive memes. However, memes can be aggressive without inducing physically aggressive behavior. Aggressive memes compel their own replication. Physical structures of organisms evolved over time. Behaviors, the products of evolution, provide explanations for various aspects of human behavior and cognition. Understanding evolutionary explanations for human behavior require first understanding about the material basis of reality and mechanisms of perception of reality by the human brain.

## 11.11   Perception and the Senses

Our sensory organs detect various aspects of the material world; create signals that are sent to the brain. The brain uses to construct a model based on the signals. The model that the brain constructs is in sync with the external reality.

We interact with the external world through our **mental model**. Color, sound, smell, taste, and sensation are not real. They do not exist in the material world outside the mind. No object has color, or smell, or taste. There is nothing as sound. Compression waves travels in a medium like air or water based on the laws of nature in deterministic way. The waves do not have a sound. Our brains create the sound. Different people and different organisms can model such waves in different ways.

Dolphins and bats visually model their sonar, to perceive visual images from the reception of their sonar sound waves. This holds true for all our other senses. Color is a mental construct. Many different electromagnetic waves travel throughout the universe and bombard the material world. Our eyes detect a small rage of these waves. Certain range of waves, called the visible spectrum, triggers a response in the cells of our eyes, which then send electronic signals to the brain. Such signals are used to create a model. Every individual organism detects slightly different range of electromagnetic waves. The brain uses color to represent different wavelengths. However, what really exists is not color; it is different frequencies of electromagnetic waves bouncing off objects. All color we visualize is a mental representation of a wavelength. Taste and smell is similarly our brain's model for the chemical properties of the environment. Taste and smell are fabricated by the mind. Nothing has an inherent taste, or smell. Taste and smell are not real. They are literally figments of our imagination (Figure 11.3).

## 11.12  Olfactory System

Humans differentiate about 10,000 chemicals using smell. Molecules in the air enter our nose and bind to receptors of the nasal cavity. Signals transmit to the brain showing that receptors are triggered. The brain creates a

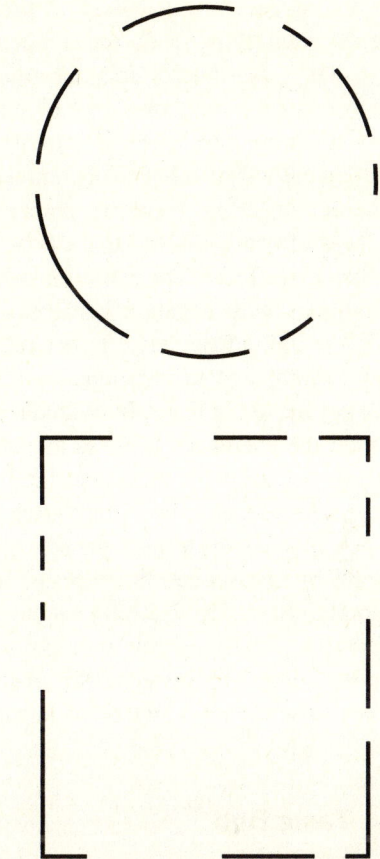

**Figure 11.3**   Human brain perceives complete shapes evern in incomplete form

sensation to register the signal. Perception of smell is instinctive and learned. Perception of chemicals changes through experiences. Nothing has inherent smell. The olfactory system informs us about chemicals that come into contact. Different people may perceive different smells in response to the same chemicals. High degree of similarity in perception of the same chemicals exists in all people which show a common genetic basis for **perception** of the smell. Cigarettes initially smell bad to a person. Once he gets addicted smell of cigarettes is perceived as pleasant. Perception of smell is the product of evolution. Our brains have evolved to perceive certain chemicals in feces negatively. Volatile chemicals emanating from feces become airborne. The nose detects such chemicals. Human feces are the common carrier of human diseases. Coming into contact with feces increases an individual's chance of contracting diseases and suffering. Negative perception of the feces affects an individual's behavior so that they avoid feces. The evolution selects for individuals with negative perception of feces. These individuals have higher rate of survival as compared to individuals without negative perception of feces. Individuals who don't smell the chemicals in feces are likely to come in contact with feces and likely to contract a disease and die. Comparing the human perception of the feces to the perception of the same by flies suggest that feces perhaps smell good to flies. The feces are a source of food for flies. Flies are not vulnerable to mammalian diseases. Instead, the organic molecules in feces are a source of nutrition for flies.

## 11.13   Taste Bud

Five basic types of chemical receptors called taste buds in our tongue can taste sweet, sour, salt, bitter, and umami (savory). Taste buds detect the presence of chemicals, not really detect these flavors. The brain creates the flavor in imagination. The sour **taste buds**, detect the acidity by detecting free H+ ions which in contact with the sour taste buds, send signal to the brain. Umami triggered by meats is a receptor for amino acids. We refer this taste as savory. The basis of perception is critical for understanding the evolution of thought, behavior, and perception. The recognized qualities are not products of the external world, but the products of our minds. We come into contact with acidity, either in fruits, or in spoiled meat. The detection of acidity generates 2 different responses in people depending on the context. When acidity is detected by itself, without any sugar present, then we have a negative reaction to sour. We react positively to sour when it is combined by sweet, as this combination is associated with fruits, but sour (acidity) without sugars, is an indicator of rotten food. When meats rot, the acidity increases. Sour is an indicator of unripe fruits as the sugars in the fruit have not yet developed. Fruits do not become less **acidic** as they ripen but develop more sugars, which change the perception of sour (Figure 11.4).

We react negatively to bitter as the compounds triggering the bitter taste buds are found in poisonous plants. Thus, a negative perception of bitter was selected for. Sweet tastes buds detect sugars, and the main sources of sugars in nature are fruits. Sugars, an energy storing compound, are also found in honey. The fruit surrounds the seeds of a plant are a plant placenta which provide nutrients and energy for the seeds and allow them to germinate and grow faster than the bare seed. The trees that produce fruits get selective advantage over trees that do not produce fruit. There are so many tropical fruits, and seedlings in the tropics that grow very fast to survive. Honey produced from plant products is an energy and nutrient rich substance. The bee feed it to their offspring to promote growth. Thus, the sugars and nutrients in the honey originate in the plants contain sugar, many vitamins and minerals. Taste buds cannot detect the vitamins and minerals. Sugary substance is produced in the wild by an organism as a food source for its offspring. Sweet products in nature are equivalent of milk. Fruit, the milk of the plant, is highly nutritious and is given to offspring that are unable to collect their nutrients. Brains, through years of evolution, have associated sweetness with nutrition.

**Sweetness** is only the detection of sugar which indirectly guided to consume nutritious

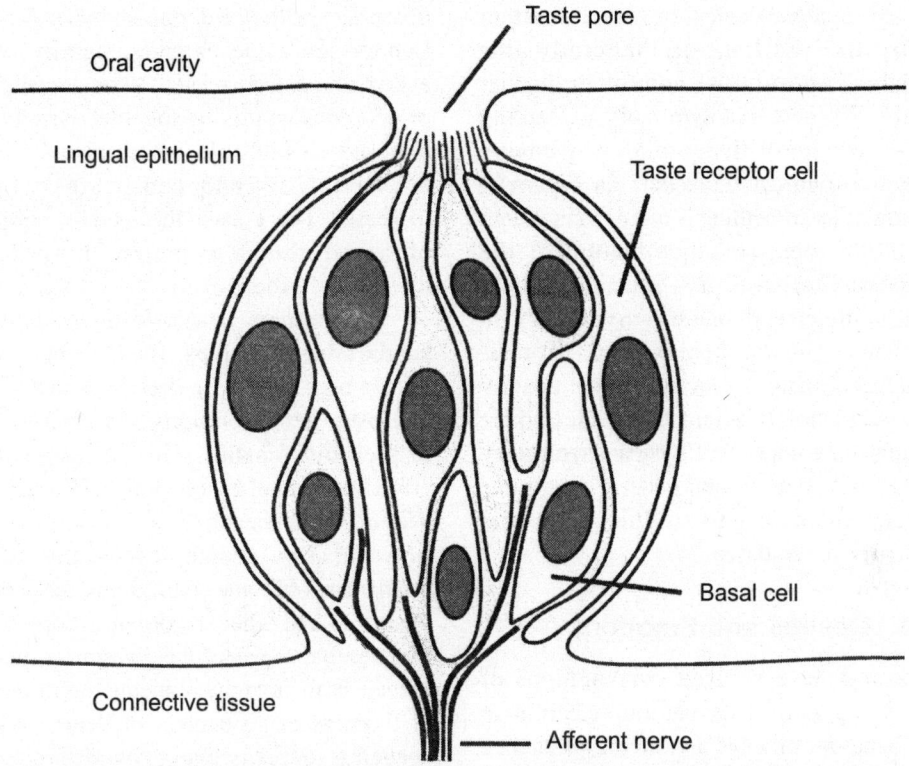

**Figure 11.4** Taste bud

substances, like vitamins and minerals. In the modern world, sugar is used to make non-nutritious foods more attractive. Our taste mechanisms are crude, so we can't really tell if something is actually good for us, or not, although all we can detect sugar. As far as our brains are concerned, a sweet food is a healthy. Our instincts work on the premise that the sweeter something is the better for us, but that is not true when people engineer sweet foods that have no nutritional value. Certain things taste good through undirected mutation. Some of our ancestors had desires for things that happened to be healthy for them. These ancestors survived the best and were thus selected for, passing on their taste preferences

Perceptions drive behavior. The ways through which material world is perceived have produced and selected for through the evolution. **Negative perceptions** discourage individuals from interacting with an object or encourage individuals to take action to remove an object from the vicinity. If something smells bad, tastes bad, sounds bad, or looks ugly, then an individual will tend to remove him, or herself from its presence, kill the offending animal, or spit the object out and not eat it. If an object looks appealing and tastes good, then an individual seek that object to eat. Perceptions and tastes have been shaped through natural selection. On-going variation of some individual's tastes is due to recent mutations, which have not yet been selected for against. Tastes can be impacted by experience to some extent.

People view **symmetry** as aesthetically pleasing and asymmetry as ugly. Animals are generally symmetrically constructed either radially like starfish, or bilaterally like chordates, arthropods, and other higher animals. We perceive symmetry as healthy and as a pattern of living things. Symmetry in the environment represents an organism and shows that something is more than random object like rock, or other feature of the landscape. The brain pays special attention to the symmetrical pattern. Symmetry in organisms shows health and fitness. Symmetrical mates in fact show the healthy mates. Attraction to symmetry is selected for through natural selection. **Physical asymmetry** is linked with detrimental genetic conditions and infectious diseases. Thus, physical asymmetry is repulsive.

## 11.14  Desires and Emotions

Our desires have evolved over millions of years, being shaped by natural selection to guide behavior in ways that increases chance of survival and procreation. Human **desires** have evolved in a completely different context. The civilized human has lived in for the past 10,000 years. Our strongest desires relate to food and sex. Animals desire to eat to sustain the body. Sex is the essential aspect of life. Reproduction continues the chain of life and passed on genes. Humans evolved in small family groups and remained in these groups until the dawn of civilization. One of the earliest events of civilization was the coming together of larger groups of people, who spanned more than single family units. When multiple family units began living in close contact with one another then sex became complicated. When people only lived in small family group sex created conflicts as potential mates were rare and people didn't live in long term mate ownership among other individuals who had different ownership. Early human groups were likely similar to lion prides today, large groups

of females with one dominant male. An entire lion pride is one extended family, where everyone is closely related. When larger groups of people began living together, people were exposed to multiple potential mates on a constant basis and conflict over mates increased. This created the need for regulation of sexual desires to protect the collective interests of group.

The primary sexual features of women are the breasts and hips. The male brain views female breasts, release chemicals in brain and a pleasure sensation occurs which cause men to seek out females. This mechanism was effective in small people groups in wilderness. When people began living in large groups surrounded by people always, this became problematic. In some civilizations, they covered the breast and other sex organs, removing the features that triggered the pleasure stimulators in male brain, thereby reducing the stimulation of those pleasure centers in **brain**. When a needed resource is scarce, the desire for that resource increases. Covering the **sex organs** in some civilizations increased the desires for sex, leading to social problems and causing a never ending escalating loop in societies.

**Emotions** are the most basic types of hormone based behavioral mechanisms that creates strong desires indirectly affecting behavior. When organisms act on emotions, they are not consciously acting in direct relation to outside circumstances, they act relative to internal hormones, which are triggered via crude mechanisms in response to stimulus. Fear is a basic emotion. When an organism goes out into an open space in daylight, their brain detects the surrounding conditions, and certain hormone is produced that causes the animal to have an unpleasant feeling. In reaction to the unpleasant feeling, the animal moves to a location where they no longer have the unpleasant feeling. Mice have now been genetically engineered to be fearless. The genes that produce the **hormone** for fear reaction were disrupted and mice no longer have any fear. Instead of avoiding open spaces, they happily go into open spaces and feed, or sleep there, with none of the normal fear. In the wild, such mice would quickly be eaten, i.e. naturally selected against.

**Love** is another emotion. Feeling of love is linked to amount of oxytocin. The administration of oxytocin via nasal spray increases trust and confidence in others. Oxytocin is released to brain after having sex and after giving birth. It is an important for emotional bonding. Hormones that induce individuals to find pleasure in the company of others love, have been selected for because they increase the chance of sexual intercourse and increase the chance that sexual partners will work together to raise successfully their progeny. Love for progeny increases the care for the progeny, leading to successful propagation of their genes. Love for other members of the group evolved, when our ancestors lived in family units so that forming love bonds (Figure 11.5) to others in group increased the likelihood of spreading genes by helping others who shared the same genes. Now we induce that same love response for non-family members, through several social tricks. In the past, individuals identified members of their genetic group subconsciously by their appearance, sound, and behavior. People recognized their family members as the people who looked the most like them, sounded the most like them, and acted the most like them. Now people trust naturally those people who look, sound, and act like them. Some techniques that can stimulate trust and harmony are to use uniforms, so that everyone looks similar, to speak with a similar accent, and to share common activities. These things trigger the evolved mechanisms that identify people genetically close to ourselves, whom we then naturally trust and love. This

**Figure 11.5** Love bonds

is one reason why uniforms are so effective in developing highly cooperative units, because the uniforms create a natural sense of family since everyone looks similar. Trust can be beneficial as there are advantages from cooperation which requires trust so that one is not taken advantage. The brain subconsciously detects various behaviors of others and releases **oxytocin** based on certain evolved criteria and patterns like eye movements, perspiration, pulse rate, and body language. People trust individuals who are more similar to themselves than individuals who are less similar, because individuals who were more similar are more closely related genetically. From the lowest order to the highest order, life is fundamentally based on killing and consuming other organisms. Plants are the only organisms that don't depend on killing others for their survival, but even plants are in fierce competition. Different types of plants produce toxins, not only for self-defense against animals and fungus, but also to kill other plants. Many types of trees have toxins in their leaves that retard the growth of plants that are of a different species.

## 11.15   Intentionality

Ernst Mayr pointed out to assume that evolutionary adaptations have functions, as their purposefulness is contained in the underlying genetic and environmental programs that bring about such adaptations. The origin and evolution of such programs are not purposeful. Human free will does not exist. We can and do make choices as first posited by Democritus. The fruit of chance and necessity are random or are determined by a combination of genetic and evolutionary heritage and personal histories. William Provine has asserted that **evolutionary biology** has eliminated the possibility of human free will. Nearly all economists, psychologists, and sociologists asserted that the idea of human free is irrelevant. The demise of free is by far the most difficult implication of evolutionary theory for most people to accept. Even most atheists cannot accept the existence of human free will. The basic human desires for blame and revenge need consideration. If free will does not exist, then we cannot blame people for their actions, nor extract revenge. The concept of free will is itself logically unintelligible. It cannot be defined in a way that avoids logical circularity, or which does not contradict what we understand about nature and evolution. The feeling of free is an illusion, and may itself be a product of natural selection.

## 11.16   Assumptions of Optimality Theory

To treat life history evolution as an optimality problem, one assumes a definition of fitness which defines a relationship between traits and fitness, describes trade offs between traits, then finds the combination of traits that maximizes fitness. **Optimality theory** assumes that mortality and fecundity rates are constant and that age structure of population is stable. It also assumes that fitness of a phenotype is adequately measured by estimating its reproductive success in population composed exclusively of identical phenotypes. Fitness is measured as rate at which population of identical phenotypes would grow. Assumption of identical phenotypes implies asexual reproduction or perfect heritability (Stearns, 2000). Success of classical theory suggests that the things on which it chose to concentrate - age and size specific impacts on mortality and reproduction in a stable, asexual population are more important for its purposes .

## 11.17   Conceptual Advantages of Reaction Norms

A reaction norm is a property of genotype. It describes the set of phenotypes produced by that genotype across a range of (Stearns, 2000) environmental conditions. **Reaction norms** for life history traits have at least 2 important characteristics:

1. They clearly distinguish between effects of nature and nurture, between component of reaction to environment that has evolved over many generations and component that is due to developmental reaction of this particular organism to this particular environment in this generation.
2. Theory can predict them. This extends the range of things that theory can predict and thus makes the theory easier to test and to improve.

The classical theory deals with problem of risk of reproductive failure. Evolutionary risk varies with fitness: reduced fitness increases variance. Measuring fitness, as geometric mean of per-generation reproductive success, accounts for long-term risk, as this measure takes into account the effects of variance. This fitness measure makes clear that in risky environment, it can pay to evolve traits that reduce variance in fitness, even if this results in some reduction in arithmetic mean fitness.

There are 3 methods to deal with risk

1. Spread the risk by increasing the number of independent reproductive events. Spreading risk is always preferable to not spreading risk unless there is an associated cost, in which case there will be an intermediate best degree of risk spreading.

2. If several traits contribute to fitness through a mean and a variance, then a given fitness - a long-term geometric mean growth rate can result from many combinations of means and variances of traits. Some components might contribute low mean and low variance, others high mean and high variance. The combinations that yield the same fitness value form mean-variance fitness are isocline. Investment should be distributed across components to yield maximum fitness given intrinsic constraints.

3. The connection of traits to fitness is usually non-linear, as the connection of wealth to utility in economics. If the relationship of fitness to trait is concave down, then reducing variance in trait will increase fitness. If fitness-trait relation is concave up, then increasing variance in trait will increase fitness. A concave-down relationship between a trait and fitness implies risk-averse, variance-reducing behavior. A concave-up relationship implies risk-prone, variance-increasing behavior. With those concepts

as background now we can proceed to achievements of classical theory (Figure 11.6).

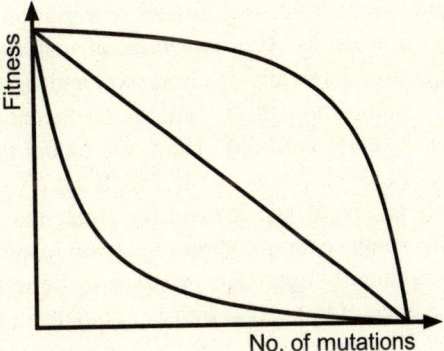

**Figure 11.6**  Evolution of sex. The effects of synergistic and antagonistic epistasis

## 11.18   Age and Size at Maturity

Age and size at maturity are optimized when positive difference between benefits and costs of maturation at different ages and sizes is the greatest. Benefits of early maturation are costs of later maturation, and vice versa. Benefits of early maturation include shorter generation time and shorter period of exposure to juvenile mortality before the first reproductive event. One benefit of later maturation is longer period in which to grow, leading to larger size at maturity and greater fecundity, as fecundity often increases with body size. Another benefit of later maturation can be production of high quality offspring, in sense that offspring have better survival per time unit as juveniles. If this per-time-unit effect is large enough, it can more than compensate for risks of a longer juvenile period.

### *Example of a Simple Optimum*

For many organisms, fecundity can be estimated in field, as growth rate of body size, and if one assumes the population is stable, per-time-unit juvenile mortality rate can be inferred. That is enough to model optimal age and size at maturity. When this was done for eastern fence lizard, the result was a curve, a fitness profile relating fitness to variation in age at maturity. Consequences of making 1 or both of 2 assumptions were explored effects of variation in age at maturity on fecundity and juvenile mortality. Here, the model that best fit the data was one in which there were only effects of age at maturity on fecundity. This simple example shows the main features of optimality approach– **costs and benefits** of variation in a target trait are modeled, and result of model is a curve relating fitness to variation in target trait. Because of assumption of costs and benefits of trait variation, fitness profile has an intermediate optimum, and theory predicts that trait should have the value corresponding with peak in fitness profile. The peak can be sharp, or flat. Prediction is more robust if peak is sharp. It is less likely that effects not taken into account in model could make big difference to **prediction**. When the peak is flat, it is easy for some other factor to make difference to prediction. Whether the peak is sharp or flat, optimality models predict that traits involved in trade offs contribute most to fitness at intermediate values. Thus connections among traits provide a cause of stabilizing selection intrinsic to organism. This is a feature of trait evolution. An important component of environment of a trait consists of other traits out of which organism are built. Its interactions with other traits generate selection pressures on trait as interaction of whole organism with external environment generate selection pressures on organism.

### Example of Optimal Reaction Norm

Optimality theory predicts a single optimal point and trait response to environmental variation. A population of genetically identical organisms is thoroughly mixed and offspring of all individuals encounter all kinds of microenvironments present and all microenvironments contribute equally to reproduction .In some environments, organisms grow rapidly; in others they grow slowly. Take 2 simple maturation rules: always mature at the same size, and always mature at same age. Each entails a cost when environment varies. An **organism** following first rule always mature at same size must wait a long time to mature in poor environments, where growth is slow and while it waits, it runs the risk of dying. The cost of this rule is the risk of mortality. An organism following second rule always mature at same age avoid the mortality risk, but encounters a different problem. In poor environments, it matures at small size with decreased fecundity resulting in few offspring. Evolution reach a compromise that can be predicted by a model incorporating the costs and benefits of variation in age and size at maturity in heterogeneous environments .The optimal reaction norm depicted is only one of many possibilities, perhaps the simplest. It suggests that organism should mature young and large when growth conditions are good and old, and small when growth conditions are poor. Such models, more thoroughly explored by Berrigan and Koella explain some variation in reaction norms found in nature. Their predictions have attractive features that better approaches should attempt to preserve. These include:

1. The clear separation of nature - the shape and position of reaction norm in phenotype space - from nurture - the particular point on the reaction norm at which an individual organism matures.
2. Prediction of a kind of phenotypic response that is relatively easy to measure.
3. Prediction of a kind of response that connects to population dynamics and

inter-specific interactions when growth, fecundity, and mortality rates depend on population densities.

### Relaxing the Implausible Assumptions of Simple Model

This simple model makes a critical assumption that each point on the reaction norm can be treated as though it was associated with an independent, isolated, stable population. In any heterogeneous environment in which growth rates and reproductive rates vary, the contributions of different environments to growth rate of whole population also vary. In a simple extreme case, instead of having a continuous range of environments, there are a good environment in which most of **reproduction** occurs and a poor environment in which organisms can survive but contribute little to reproduction. A net flow of individuals from good environment into poor environment thus occur. Most genes have ancestors that reproduced in source environment. As there is little evolutionary experience of sink environment, population is primarily adapted to source, not to sink, and in extreme case the reaction norm is simply be a flat line with value of trait that is optimal in the source. In more general case, where there are many types of environment rather than just 2, the reaction norm is shaped according to the frequency with which successful reproduction occurs in each environment. Environments that contribute most to reproduction strongly shape the reaction norm, and contribute little to reproduction having little effect on the reaction norm. This insight was arrived at simultaneously and independently by Houston and McNamara and Kawecki and Stearns.

**Frequency dependence:** One limitation of optimality theory that it gives no scope to consider what happens when 2 or more life history phenotypes compete with one another in a population. Phenotypes that evolved in the past are continually tested by mutants, or immigrants that attempt to invade and take over the population. The theory that handles such situations is evolutionary game theory. Kawecki worked out a case to show how the change in approach makes difference to predictions. He modeled the age and size at maturity of 2 individuals competing within a patch for limited resource. In that circumstance **evolutionarily stable strategy** (ESS) is to mature later and at a larger size than age and size that maximize fitness. An analogous result for plant height under competition for light was derived earlier by Mirmirani and Oster. Sinervo et al. found 2 color morphs of a lizard associated with life history phenotypes through genetic correlations. The 2 morphs compete with one another in a way that generates short-term population cycles. Neither morph is optimal; each through its increase in frequency decreases its own fitness and prepares the way for other to increase in frequency. Both density- and frequency-dependence are present and are needed to explain the behavior of system, in which life history phenotypes cycle persistently.

## SUMMARY

1. Every organism is a package of assembled organic molecules. Organisms mate, produce offspring and their traits get passed to next generation. Non-photosynthetic organisms steal the products from the plants. Physical structures of organisms evolved over time.

2. Behaviors of organisms are products of evolution which provide explanations for various aspects of human behavior and cognition. Understanding evolutionary explanations for human behavior require understanding about the material basis and mechanisms of perception of reality.

3. Evolution of languages is the non-biological and more complex than biological evolution. Memes can be viewed as ideas Genes code for the proteins in brain that store ideas. Memes propagate by communication including spoken language, writing, art, music, and simple observation. Memes arise from material conditions from other memes or from brain's observation of the environment which produce new memes.

4. Memes that convey beneficial quality to the vectors have the increased chance of survival and propagation. Sacrificial behavior in humans has genetic and memetic components. Human's genetic coding promotes sacrificial behavior to protect family members which increase the likelihood of the propagation of the individual's genes that are shared by other family members.

5. Color, sound, smell, taste, and sensation are not real. No object has color or smell or taste. Perception of smell is the product of evolution.

6. Volatile chemicals emanating from feces become airborne. Nose detects such chemicals. Human feces are the carrier of human diseases. Coming into contact with feces increases an individual's chance of contracting diseases and suffering. Negative perception of the feces affects an individual's behavior so that they avoid feces. Evolution selects for individuals with negative perception of feces in individuals who have higher rate of survival as compared to individuals without negative perception of feces.

7. People in small family group sex created conflicts as potential mates were rare and people didn't live in long term mate ownership among other individuals who had different ownership. Early human groups were likely similar to lion prides today, large groups of females with one dominant male.

8. The male brain views female breasts, release chemicals in brain and pleasure sensation occurs which cause men to seek out females. This was effective in small people groups in wilderness.

9. Sexual selection is important evolutionary mechanism, moving certain alleles into future just as natural selection does. Females who choose showy mates have showy male offspring, who will in turn be more attractive to females in the future, so genes that promote the preference are passed on.

10. In any heterogeneous environment in which growth rates and reproductive rates vary, the contributions of different environments to the growth rate of whole population also vary. There are 2 kinds of environment: a good environment in which most of reproduction occurs, and a poor environment in which organisms can survive but contribute little to reproduction.

# REVIEW QUESTIONS

## Short Answer Questions

1. What do you understand by the term "bio-cultural evolution"?
2. What do you understand by the anthropogenic changes?
3. Comment on the reprogenetic technologies.
4. Mention the function of the cultural system.
5. Define the zoonotic disease
6. What is the purpose of aggression in animals?
7. Define memetic evolution.
8. Mention examples of sacrificial behaviour.

## Long Answer Questions

1. Write a note on desires and emotion in evolutionary context.
2. Critically examine the evolution of morals, culture and religion.
3. Describe the evolution of disease in agricultural population.
4. Write an essay on the gene-culture co evolution.

# REFERENCES

Armelagos, G.J., and McArdle, A. 1975. Population, disease, and evolution. In : Swedlund, A.C. (ed.): Population Studies in Archaeology and Biological Anthropology: A Symposium. Memoir of the Society of American Archaeology ,No. 30.

Audy, J.R. 1961. The ecology of scrub typhus. In : May, J.M. (ed.): Studies in Disease Ecology. New York: Hafner Publishing.

Audy, J.R., and Dunn, F.L.1974. Health and disease. In : Sargent, F . I1 (ed.): Human Ecology. Amsterdam: North Holland Publishing Co.

Baker, B, and Armelagos, G.J.1988 Origin and antiquity of syphilis: A dilemma in paleopathological diagnosisand interpretation. Curr. Anthropol. 29 (51) :703-737.

Bates, M. 1953. Human Ecology. In Kroeber, A.L. (ed.):Anthropology Today. Chicago: University of Chicago Press.

Bunge, M. A. 2003. Emergence and Convergence: Qualitative Novelty and the Unity of Knowledge. Buffalo, NY: University of Toronto Press.

Burnet, F.M .1962. Natural History of Infectious Disease. Cambridge: Cambridge University Press.

Cockburn, T. A. 1967a. Infections of the order primates. In :Cockburn, T.A. (ed.): Infectious Diseases: Their Evolution and Eradication. Springfield.

Cockburn, T.A.1967b. The evolution of human infectious diseases. In : Cockburn, T. A. (ed.): Infectious Diseases: Their Evolution and Eradication. Springfield.

Cockburn, T.A.1971. Infectious disease in ancient populations. Curr. Anthropol. 12(11):4542.

Gould, S. J. 1996. Full House. New York: Harmony Books

Haraway, D. 1996. Modest witness @ Second Millennium. The Female Man meets OncoMouse™. New York: Routledge.

Hudson, E.H.1965. Treponematosis and man's social evolution. Am. Anthropol. 67:885-901.

Kelly, K. 1994. Out of Control: The New Biology of Machines, Social Systems, and the Economic World. New York: Addison-Wesley.

Lambrecht, F.L. 1985. Trypanosomas and hominid evolution. Bioscience ,64: 646.

Livingstone, F.B. 1958. Anthropological implication of sickle-cell gene distribution in West Africa. Am. Anthropol, 60: 533-552.

May, J. M.1960. The Ecology of Human Disease. Ann.N.Y. Acad. Sci., 84:789-794.

McNeill, J.R. 2000. Something New Under the Sun: An Environmental History of the Twentieth-Century World. New York: W.W. Norton & Company.

McNeill, W.H.1976. Plagues and People. Garden City: Anchor/Doubleday.

McShea, Daniel, W., and Brandon, R . 2010. Biology's first Law: The tendency to diversity and complexity to increase in organic systems. Chicago: University of Chicago Press,

Morgan, Lewis H. 1877 .Ancient Society. New York: Holt.

Polgar, S . 1964. Evolution and the ills of mankind. In STax (ed.): Horizons of Anthropology. Chicago: Aldine.

Sanderson, Stephen K. 2007. Evolutionaism and its Critics: Deconstructing and Reconstructing an Evolutionary Interpretation of Human Society. Boulder: Paradigm Publishers,

Silver, Lee. M. 1997. Remaking Eden: Cloning and Beyond in a Brave New World. First Edition ed. New York: Avon Books.

Stearns, S.C. 2000. Life history evolution: successes, limitations, and prospects .Naturwissenschaften , 87: 476-486

Templeton, John Marks. 1997. Is Progress Speeding Up? Our Multiplying Multitudes of Blessings. Radnor, PA: Templeton Foundation Press.

# Glossary

**Acquired trait:** Phenotypic characteristic without genetic basis acquired during growth and development and do not pass to the next generation.

**Adaptation:** Adjustment to the environment through the natural selection acting on genotypes. Heritable character that improves organism's ability to survive and reproduce, or genetic change influenced by natural selection in the population

**Adaptive landscape:** Graph of the average fitness of population in relation to its genotype frequencies. Peaks of the landscape correspond to genotypic frequencies at which the average fitness is high, while in valleys to genotypic frequencies the average fitness is low.

**Adaptive radiation:** The diversification of a species, or group of species into the different species, or subspecies typically adapted to various ecological niches.

Event of rapid cladogenesis occurs under conditions where new feature permits a lineage to move into a new niche or new habitat, and is then called an adaptive radiation.

**Adaptive strategies:** Strategy to cope with competition, or environmental conditions.

**Albinism:** The genetically inherited condition with a marked deficiency of pigmentation. An individual with such traits is an albino. The gene for albinism is recessive and it only shows up in the phenotype of homozygous recessive people.

**Algae:** Group of various simple organisms containing chlorophyll and live in aquatic habitats, or in moist situations on land. They range from macroscopic seaweeds such as giant kelp, which often exceeds 30 m in length, to microscopic filamentous and single-celled *Spirogyra* and *Chlorella*.

**Allee effect:** The benefit individuals get due to the presence of conspecific.

**Allele:** The known variation of a specific gene.

**Allelic exclusion:** Expression of only one of the two homologous alleles at a locus in the heterozygosity.

**Allelopathy:** The influence of a living plant on other plants, or microorganisms through production of chemicals like carbohydrates and lipids, alkaloids, nitrogen-containing compounds, phenolics and terpenoids.

**Allen's Rule:** Warm-blooded animals living in colder environments tend to have shorter appendages than animals in warmer areas.

**Allogeneic:** Two genetically dissimilar individuals of the same species.

**Allometry:** Method that compares differential growth rates of parts of an organism.

**Allopatric speciation:** Speciation followed by the geographical isolation of subpopulations of the species.

**Allopatry:** Living in separate places.

**Alternation of generations:** The alternation of sexual and asexual form in the life cycle.

**Altruism:** Helping others without direct benefit which sometimes cause harm to one.

**Amino acids:** Building blocks of proteins.

**Amniotes:** The group of reptiles, birds, and mammals. Their embryo is enclosed within a membrane called an amnion.

**Amphipathic:** Molecule with a hydrophobic and a hydrophilic part.

**Anagensis:** Evolutionary change along an unbranching lineage; change without speciation.

**Analogous structures:** Structures in different species that look alike or perform similar functions that have evolved convergently but not from the common ancestor.

**Analogy:** Similarity due to convergent evolution but not due to common ancestry.

**Anatomy:** The structure of an organism, or one of its parts.

**Ancestor:** Any organism, population, or species from which other organism, population, or species is descended.

**Ancestral homology:** Homology that evolved before the common ancestor of a set of species, and is present in other species outside that set of species.

**Angiosperm:** The recently evolved largest group of plants with reproductive organs in their flowers.

**Anisogamy:** Sexual reproduction in which one sex produces larger sex cells than those of the other.

**Anthropoid:** A member of the group of primates made up of monkeys, apes, and humans.

**Antibacterial:** Having the ability to kill bacteria.

**Antibiotic resistance:** Heritable trait in microbes that enables them to survive in the presence of an antibiotic.

**Antibiotics:** Substances that destroy or inhibit the growth of microbes, particularly the disease-causing bacteria.

**Apes:** Species belonging to the Family Pongidae of the Order Primates. They have no tails.

**Apoptosis:** The genetically programmed cell death at specific times during embryonic morphogenesis and development, metamorphosis, and during cell turnover including the maturation of T and B cells of the immune system in adult.

**Aposematic colouring:** Colouration that warns the predators about poisonous, or distasteful nature of the organism. Such colour of poisonous organisms is sometimes mimicked by others to take advantage against predators.

**Aquatic:** Living in water.

**Arboreal:** Tree-living.

**Archaea:** A prokaryote kingdom that has not diverged much from the ancestral prokaryote. Archeabacteria live in extreme conditions. The 3 major groups are Halobacteria, Sulphobacteria and Methanogens. All other prokaryotes are grouped in Eubacteria.

**Archeology:** The study of human history and prehistory through the excavation of site and the analysis of physical remains.

**Archetype:** Original form, or body plan from which a group of organisms develops.

**Archezoa:** One of the kingdom level taxa consisting of the ancient unicellular eukaryotes with a nucleus and rod shaped chromosome but no mitochondria or plastid, and supposed to be the intermediate stage between prokaryotes and eukaryotes.

**Artifact:** A human made preserved object that is studied to learn about a particular time period.

**Artificial selection:** Selective evolutionary pressure imposed by humans to obtain selective breeds.

**Asexual reproduction:** Reproduction that do not depend on a sexual process.

**Assortative matings:** Reproduction in which mate selection is based on physical, cultural, or religious grounds and is not random

**Asteroid:** A small rocky or metallic body orbiting the Sun. About 20,000 have been observed, ranging in size from several hundred kilometers across down to dust particles.

**Atomistic:** Inheritance in which the entities controlling heredity are relatively distinct, permanent, and capable of independent action.

**Australopithecine:** A group of bipedal hominid species belonging to the *Australopithecus*.

**Australopithecus afarensis:** An early australopithecine species that was bipedal.

**Australopithecus:** The extinct Genus of Plio-Pleistocene hominids found in South and East Africa and the evolutionary link between apes and humans.

**Autosome:** Any chromosome other than the sex chromosome.

**Automorphy:** A trait present in one member of a lineage, or in one lineage among many.

**Background extinction:** The normal occurrence of extinction.

**Bacteria:** Tiny, single-celled, prokaryotic organisms. Some cause serious infectious diseases.

**Bacteriophage:** The virus that infects a bacterium.

**Balanced polymorphism:** The maintenance of 2 or more alleles in a population due to selective advantage of the heterozygote.

**Balancing selection:** Selection involving opposing forces in which selective advantages and disadvantages cancel each other out. In heterozygote advantage, an allele selected against the homozygous state is retained because of the superiority of heterozygote.

**Barr body:** The inactivated X chromosome in the nucleus of somatic mammalian cells. Normally only seen in female cells. It is the result of the dosage compensation.

**Basal group:** The earliest diverging group within a clade; for instance, to hypothesize that sponges are basal animals is to suggest that the lineage(s) leading to sponges diverged from the lineage that gave rise to all other animals.

**Base:** A compound, usually containing nitrogen that can accept an $H^+$. It is used to show the non-sugar components of nucleotides. The 5 bases–adenine, guanine, cytosine, thymine and uracil form the nucleic acids.

**Bateman's Principle:** Males gain fitness by increasing their mating success and females maximize their fitness by investing in longevity as their reproductive effort is much higher.

**Batesian mimicry:** Mimicry in which one non-poisonous species mimics another poisonous species.

**Bergmann's Rule:** Northern races of mammals and birds tend to be larger than Southern races of the same species.

**Big bang theory:** The universe began in a state of compression to infinite density, and that in one instant all matter and energy began expanding and have continued expanding ever since.

**Binary fission:** Mode of reproduction not involving any sex but division of a parent cell into 2 equal sized offspring.

**Biodiversity:** Diversity of species, genes and ecosystems.

**Biogenetic law:** Ontogeny recapitulates phylogeny.

**Biogeography:** The study of changing pattern of geographical distribution of plants and animals across the earth.

**Biological species concept:** A species is a set of organisms that can interbreed among each other.

**Biome:** Plant group occupying a major terrestrial region.

**Biosphere:** The boundary of the earth where life is found.

**Bipedal:** Two-footed posture and locomotion.

**Bipedalism:** Walking upright on 2 hind legs; using 2 legs for locomotion.

**Blending inheritance:** The historically influential but factually erroneous theory that organisms contain a blend of their parents' hereditary factors that pass on to their offspring.

**Bony fish:** Fishes having jaws, bony skeleton and overlapping scales.

**Bottleneck effect:** Reduction in genetic diversity of a population resulting from an ecological crisis that wipes out most of its members.

**Bottleneck:** Drastic reduction in the population size followed by an expansion which alters gene pool as a result of subsequent genetic drift.

**Brachiopod:** Marine invertebrates resembling bivalve molluscs because of their hinged shells.

**C value:** The amount of DNA comprising the haploid genome of a species.

**Cambrian:** The first period in the Palaeozoic era characterized by many invertebrates. The sudden appearance of the major animal phyla during the period is called the Cambrian explosion.

**Carnivorous:** Feeding on meat or other animal tissue.

**Carrier:** A healthy person bearing a recessive trait.

**Catarrhini:** The divisions of Primates containing the old world monkeys and apes.

**Cell:** The basic structural and functional unit of the living organisms.

**Cenozoic:** The era from 65 mya to the present, when the modern continents formed and modern animals and plants evolved.

**Character:** Heritable trait of an organism.

**Chordate:** A member of the Phylum Chordata with a hollow dorsal nerve cord; a rod like notochord; and paired gill slits, although in some, these are apparent in early embryonic stages.

**Chromosome:** A structure that carries DNA along with various proteins, particularly the histones.

**Chronology:** The linear sequence of events according to time.

**Clade:** All the descendants of any given species.

**Cladism:** The members of a group share recent common ancestor with one another than with the members of any other group.

**Cladistic species concept:** Species is the lineage of populations between 2 phylogenetic branch points.

**Cladogenesis:** The development of a new clade.

**Cladogram:** A diagram depicting a hypothetical branching sequence of lineages leading to the taxa under consideration.

**Class:** A category of taxonomic classification between Order and Phylum.

**Classification:** The arrangement of organisms into hierarchical groups.

**Cline:** A geographic gradient in the frequency of a gene.

**Clone:** The cells derived from a single cell by repeated cell division and having the same genetic constitution.

**Co adaptation:** Interaction of genes at the genotypic level. Natural selection acts on the complex product of such interactions. The correlated variation and adaptation are present in 2 mutually dependent organisms.

**Co evolution:** Evolution in 2 or more species, such as predator and its prey, or a parasite and its host, in which evolutionary changes in one species influence the evolution of the other species.

**Coalescence theory:** The theory that estimates the time for divergence from the last common ancestor.

**Coalescence:** Growing into each other, uniting into one whole.

**Coding sequence:** The portion of a gene that is transcribed into mRNA and translated into protein.

**Codominance:** Equal effect on the phenotype of 2 alleles of the same locus.

**Codon bias:** Although several codons code for a single amino acid, an organism may have preferred codon for each amino acid which is called codon bias.

**Codon:** Triplet bases in the DNA coding for one amino acid.

**Coefficient of relatedness:** $r = n (0.5) L$, where n is the alternative routes between the related individuals along which a particular allele can be inherited; L is the number of meiosis or generation links.

**Coelacanth:** one of the deep-water, lung less fish which thought to have gone extinct about 65 million years ago.

**Common ancestor:** The most recent ancestral form or species from which 2 different species evolved.

**Comparative biology:** Study of patterns among more than one species.

**Comparative method:** Study of adaptation by comparing many species.

**Concerted evolution:** Preservation of sequence homology among members of a multigene family within the same species.

**Conjugation:** In unicellular organisms, temporary cell contact between complementary genders and exchange of genetic material, as in *Paramecium aurelia*, or one-way transfer of genes as in *Spirogyra*.

**Conodont:** A jawless fish that had tiny, tooth-like pieces that is abundant in the fossil record. These were the earliest known vertebrates.

**Consanguineous mating:** Matings between 2 individuals who share a common ancestor in the preceding 2, or 3 generations.

**Contiguous gene syndromes:** Disorders caused by micro deletions or micro duplications in neighbouring functional genes. Inheritance is usually sporadic but recurrences are possible.

**Continental drift:** The process by which the continents move as part of large plates floating on the earth's mantle.

**Contrivance:** An object or characteristic used or modified to do something different from its usual use.

**Convergence:** Similarities which have arisen independently in 2 or more organisms that are not closely related. Contrast with homology.

**Convergent evolution:** Evolution of 2 or more different lineages towards similar morphology due to similar adaptive pressures. Examples of convergence are fins or fin-like structures in fish, cuttlefish and whales.

**Coral:** These tiny animals make calcium carbonate skeletons that are well known as a key part of tropical reefs. The skeletons of the extinct rugose and tabulate corals are known from fossils.

**Cranium:** The part of the skull that protects the brain in vertebrates.

**Creationism:** The religious doctrine that all living things on earth were created separately by a supernatural creator, as stated in the Cretaceous:

**Cretaceous:** The youngest of the three geologic periods of the Mesozoic era.

**Crossing-over:** The exchange of genetic material between non-sister chromatids of homologous chromosomes during meiosis. This results in a new and unique combination of genes on the daughter chromosome which pass on to the offspring.

**Crown group:** All the taxa descended from a major cladogenesis event, recognized by possessing the clade's synapomorphy.

**Crustacean:** A zoological class that has chitinous /or calcareous exoskeleton.

**Cryptic female choice:** Besides the pre-copulatory female sexual selection, the post-copulatory selection processes going on in the female reproductive system. This mechanism is the basis for differential fertilization which includes sperm selection as opposed to pollen selection in plants.

**Cyanobacteria:** Unicellular, photosynthetic prokaryote, formerly known as blue-green algae. It contains chlorophyll but not chloroplast. They reproduce by fission.

**Cytogenetics:** The study of the structure, function, and abnormalities of chromosomes.

**Cytoplasm:** The region of a eukaryotic cell outside the nucleus.

**Darwinian evolution:** Evolution by the process of natural selection acting on random variation.

**Darwinism:** Darwin's theory that species originated by evolution from other species and that evolution is mainly driven by natural selection. It differs from neo-Darwinism mainly in that Darwin did not know about Mendelian inheritance.

**De facto:** In fact; in reality.

**De novo:** From new as opposed to inherited.

**Degeneracy:** A feature of the genetic code in which more than one nucleotide triplet code for the same amino acid.

**Denaturation:** Reversible disruption of hydrogen bonds between nucleotides converting a double-stranded DNA molecule to single-stranded molecules as caused by heating, or strong alkali treatment.

**Derived homology:** Homology that first evolved in the common ancestor of a set of species and is unique to those species.

**Derived:** Describes a character that is present in one or more sub-clades, but not all of a clade. A derived character is inferred to be a modified version of the primitive condition of that character, and have arisen later in the evolution of the clade.

**Devonian Period:** A geologic period in the Phanerozoic Eon.

**Diatom:** Single-celled algae common among the marine phytoplankton. Their 2-part shells have intricate patterns and fit together like the 2 parts of a shirt box.

**Dicentric chromosome:** One chromosome having 2 centromeres.

**Diffusion:** The process in which molecules move passively from a region of high concentration to a region of low concentration.

**Dikaryotic:** A cell that contains 2 separate haploid nuclei which is different from being haploid or diploid as seen in fungal heterokaryons.

**Dinoflagellate:** Single-celled algae possessing 2 tail-like extensions called flagella that are used for movement. These live freely in other organisms like corals.

**Diploid:** Having 2 sets of genes and 2 sets of chromosomes. Many common species, including humans are diploid.

**Directional selection:** Selection causing a consistent directional change in the population through time.

**Disruptive selection:** Selection against the middle range of variation causing an increase in the frequency of a trait showing the extreme ranges of its variation. Such selection might cause one species to evolve into two.

**Divergent evolution:** Evolutionary change that results in increasing morphological difference between initially more similar lineages.

**Diversity:** The numbers of taxa, or variation in morphology.

**DNA base sequence:** A chain of repeating

units of deoxyribonucleotides arranged in a particular pattern.

**DNA:** The large double-stranded molecule carrying the genetic code. It consists of four bases, phosphate and sugar.

**Domain:** Region of a protein with a distinct tertiary structure and characteristic activity.

**Dominance:** The property of some alleles to determine the phenotype for any particular gene by masking the effects of the other allele. Homozygosity or heterozygosity for the dominant allele results in the same genotype in complete dominance. Incomplete dominance appears as a blend of the phenotypes corresponding to the 2 alleles. In co-dominance, both alleles equally contribute to the phenotype.

**Dominant Allele:** An allele that masks the presence of a recessive allele.

**Double helix:** The twisted ladder shape characteristic of DNA molecules.

**Drift evolution:** A high rate of immunologically significant mutations in certain viruses which results in drifting away from recognition by the immune system by antigenic change. Influenza virus, HIV and HCV change their antigenic structure through drift evolution.

***Drosophila melanogester*:** Common fruit fly containing only 4 pairs of chromosomes.

**Duplication:** The occurrence of a second copy of a particular sequence of DNA. The duplicate sequence may appear next to the original, or be copied elsewhere into the genome. When the duplicated sequence is a gene, this is called gene duplication.

**Ecogenetics:** The branch of genetics that studies influence of genetic factors on human susceptibility to environmental health risks. It deals with the genetic basis of environmental toxicity, prevention and control of environment-related disease. Ecogenetics interacts with ecology, molecular genetics, toxicology, public health medicine and environmental epidemiology.

**Ecological niches:** The specific micro-habitats to which populations or organisms adapt.

**Ecological species concept:** A species is a set of organisms adapted to a particular, discrete set of resources in the environment.

**Ecology:** The study of the interrelationships among living organisms and their environment.

**Ecosystem:** A community of organisms interacting with a particular environment.

**Ediacaran fauna:** The oldest fauna known.

**Effective population size (N or Ne):** Number of individuals contributing 'unique' chromosomes to the next generation which is always less then or equal to the actual population size. Inbreeding effectively reduces Ne because of the identity of most chromosomes in the population.

**Electrophoresis:** The method of separation of molecules according to their motility in an electric field which has been mainly used to distinguish different forms of proteins. The electrophoretic motility is influenced by size and electric charge of the molecule.

**ELISA:** An assay for quantifying the presence of an antigen by using an enzyme linked to an antibody to the antigen.

**Ellis-van Creveld syndrome:** A genetic disorder characterized by dwarfism, extra fingers, and malformation of the arms, wrist, and heart. The majority of this rare syndrome has been found among the Amish.

**Emigration:** Movement of organism out of an area.

**Empirical:** Determined by experiment.

**Endonuclease:** An enzyme which cuts a nucleic acid by cleaving the phosphodiester bonds between 2 internal residues.

**Endosymbiosis:** When one organism takes up permanent residence within another, such that the 2 become a single functional organism. Mitochondria and plastids are known to be resulted from endosymbiosis.

**Enzyme:** A protein that acts as a catalyst for chemical reactions.

**Eocene epoch:** A temporal subdivision (epoch) of the Tertiary period.

**Epidemiology:** Study of the distribution and determinants of health-related events to trace their cause and control.

**Epigenesis:** Theory that the development of an embryo consists of the gradual production and organization of parts.

**Epigenetics:** Study of heritable changes in gene expression without a change in DNA sequence. Epigenetic phenomena like imprinting and paramutation violate Mendelian principles of heredity.

***Escherichia coli*:** A gram-negative bacterium whose genome has been sequenced entirety. It is the model for studying the prokaryotes.

**Ethology:** Study of animal behaviour under normal conditions.

**Eugenics:** The field that attempts for improving the quality of human species by selective breeding.

**Eukaryotic cell:** The cell where DNA lies within a true nucleus.

**Evolution:** The process that results in heritable changes in a population and the changes must be passed on to the next generations.

**Evolutionarily stable strategy:** A strategy such that if most members of a population adopt it would generate reproductive fitness greater than any other strategy.

**Evolutionary classification:** Method of classification using both cladistic and phenetic classificatory principles which permits paraphyletic groups and monophyletic groups, but excludes polyphyletic groups.

**Evolutionary Developmental Biology (Evo-Devo):** Studying the relationships between changes in developmental pathways due to mutations in regulatory parts of genes and evolutionary changes. The emphasis is on the changes in expression patterns of the genes involved in developmental pathways rather than structural changes in genes.

**Evolutionary Distance:** The quantitatively measured difference between the phenetic appearance of 2 groups of individuals, like populations or species, or the difference in their gene frequencies.

**Evolutionary tree:** A diagram which depicts the hypothetical phylogeny of the taxa. The points at which lineages split represent ancestor taxa to the descendant taxa appearing at the terminal points of the cladogram.

**Exon:** The nucleotide sequences of some genes consist of parts that code for amino acids, with other parts that do not code for amino acids interspersed among them. The coding parts which are translated are called exons; the interspersed non-coding parts are called introns.

**Extinction:** The disappearance of a species or a population.

**Extra-chromosomal inheritance:** Non-Mendelian inheritance due to extra-nuclear DNA. The transmission of the trait only occurs from mothers.

**F1:** First filial hybrids arising from a first cross. Subsequent generations are denoted by F2, F3 etc.

**Family:** The category of taxonomic classification between Order and Genus. The members of family share a close similarity.

**Fauna:** Certain species of animals occurring in a particular region or period; grouping of animal taxa.

**Fermentation:** A series of reactions occurring under anaerobic conditions in some microorganisms in which glucose is converted into simpler substances with the release of energy.

**Fertilization:** Fusion of female and male gametes to form a zygote from which a new individual develops.

**Fetus:** Final development stage before birth.

**First cousin:** Someone who is related as a result of being a child of one's uncle or aunt.

**Fisher's Fundamental Theorem:** The rate of increase in fitness is equal to the additive genetic variance in fitness. There is a lot of variation in the population when the value is large.

**Fitness:** Lifetime reproductive success of an individual. It has 2 components, survival and reproductive success. Variation in fitness is the major driving force in evolution.

**Fixation:** When a new mutant allele appears at a locus, it may lost in the following generations or remain in population at varying frequency from one generation to another. If it replaces, the other already existing allele in diallelic locus, it is fixed as the only allele at that locus. This process is called fixation.

**Fossil:** An organism, or a physical part of an organism, or imprint of an organism that is preserved from ancient times in rock, amber, or by other means. New techniques have revealed the presence of cellular and molecular fossils.

**Founder effect:** A type of genetic drift in which allele frequencies are altered in small population, which is a nonrandom sample of a larger population.

**Founder principle:** Effect of a small population size in which the genes of few people are inherited over time by a large number of descendents. This is known as the founder effect and the Sewall Wright effect.

**Frame shift mutations:** Mutations, usually deletions, or insertions that change the reading frame of the codon triplets.

**Frequency:** The number of times that something happens.

**Frequency-dependent selection:** Selection in which the fitness of a genotype depends on its frequency in the population.

**Fungi:** A group of organisms comprising molds and mushrooms which exist as single cells, or make up a multicellular body called a mycelium. Fungi lack chlorophyll and secrete digestive enzymes for decomposing other biological tissues.

**Gaia hypothesis:** A hypothesis developed by Lovelock, Watson, Margulis and others which proposes that the physical and chemical condition of the earth is actively made fit and comfortable by the presence of life itself. This is in contrast to the conventional wisdom which held that life adapted to the planetary conditions and they evolved their separate ways.

**Gamete:** A haploid reproductive cell such as sperm (or pollen) and egg (oocyte).

**Gastropoda:** The class of molluscs that contains the most species. Gastropods include snails and slugs that are marine, freshwater, and terrestrial.

**Gause's Rule:** Two species cannot live the same way in the same place at the same time which is only possible through evolution of niche differentiation.

**Gender:** Differences between any 2 complementary organisms of the same species that render them enable of mating.

**Gene conversion:** Partial sequence transfer from one allele to another changes one gene or allele to another one. It is the common mechanism, especially for the HLA-B locus, in the generation of new MHC alleles. Less common is conversion between alleles of different MHC loci.

**Gene expression:** The conversion of a gene's coded information into the structures operating in the cell. Expressed genes are transcribed and translated all the way to proteins, and those that are transcribed into RNA but not translated into protein.

**Gene family:** A set of related genes occupying various loci in the DNA, certainly formed by duplication of an ancestral gene and having a recognizably similar sequence. Members of a gene family may be functionally similar or differ widely.

**Gene flow:** Movement of genes within a population, or between 2 populations following genetic admixture which creates new combinations of genes or alleles in individuals. This is one source of variation in the natural selection.

**Gene frequency:** Frequency in the population of a particular gene relative to other genes at its locus which is expressed as a percentage.

**Gene pool:** All the genes in a population at a particular time.

**Gene:** The unit of heredity that carries information from one generation to the next. In addition to the coding regions (exons), a gene may have non-coding intervening sequences (introns) and transcription-control regions.

**Genetic anticipation:** Progressive shift of the age of start of a hereditary disease to earlier ages in successive generations. It may occur when a parent is a mosaic, and the child has the full mutation in all cells. Triplet repeat expansion may show anticipation, when the number of repeats increases with each generation.

**Genetic code:** The code relating nucleotide triplets in the mRNA to amino acids in the proteins.

**Genetic determinism:** The belief that genes alone form all characteristics of an individual organism.

**Genetic distance:** Measurement of genetic relatedness of populations based on the number of allelic substitutions per locus that occur during the separate evolution of 2 populations.

**Genetic drift:** Evolutionary change over generations due to random events in small populations. It operates unless overcome by strong selective forces. Widely different HLA allele frequencies among South Amerindian tribes are the result of probable genetic drift in each small tribe.

**Genetic engineering:** Removing genes from the DNA of one species and incorporating them into the DNA of another species.

**Genetic equilibrium:** When a population is not evolving from generation to generation and the population's gene pool frequencies remain unchanged.

**Genetic fitness:** Classic genetic fitness is the average direct reproductive success of an individual possessing a specific genotype in comparison to others in the population. Inclusive fitness is the classic fitness plus the probability that an individual's genotype may be passed on through relatives.

**Genetic heterogeneity:** Presence of several different genotypes contributing to the genetic component of a disease on their own.

**Genetic linkage:** Segregation of 2 or more genes together as a unit which reflects a lack of meiotic crossovers between 2 genes.

**Genetic load:** Average number of lethal equivalents per individual in a population, which are propagated by heterozygote in a masked state.

**Gene Mapping:** Determining the physical location of a gene on a chromosome.

**Genetic relatedness (r):** A quantitative measure of genetic relatedness between individuals. In diploid species, $r=1/2$ between full siblings, or parent and child; $r=1/4$ for half siblings or aunt/uncle versus niece/nephew, or for grandparents versus grandchildren; $r=1/8$ for first cousins; $r=0$ for non-relatives.

**Genetic variance:** Phenotypic variance in a population due to genetic heterogeneity.

**Genetics:** Study of variation, heredity and their physical basis in DNA.

**Genocopy:** A gene/genotype causing the same phenotype as another gene/genotype. Genocopies are the basis of genetic heterogeneity.

**Genome:** The full genetic complement of an individual. In humans, each individual possesses about 2.9 billion nucleotides in the entire DNA that makes up his or her genome.

**Genomic instability:** The first phenomena in the formation of malignancies due to defects in DNA repair and cell cycle controls. This can occur by gain-of-function mutations in proto-oncogenes or loss-of-function mutations in tumor suppressor genes.

**Genotype:** The genetic constitution of an individual.

**Genotype-environment (GxE) interaction:** The modification of genetic risk factors by environmental risk and protective factors and the role of specific genetic risk factors in determining individual differences in vulnerability to environmental risk factors.

**Geologic time:** The time scale describing events in the history of the earth.

**Germ line:** Genetic material transmitted from one generation to the next through the gametes.

**Germ plasm:** The reproductive cells in an organism, or the cells that produce the gametes.

**Germination:** The initial stages in the growth of a seed to form a seedling. The embryonic shoot and embryonic root emerge and grow upward and downward.

**Gestation:** The period in animals bearing live young from the fertilization of the egg and its implantation into the wall of the uterus until the birth of the young.

**Glaciation:** The formation of large sheets of ice across land which marks the beginning of ice ages, when the makeup of the earth and organisms on it changes.

**Glucose-6-phosphate dehydrogenase deficiency:** A genetically inherited X-linked error in metabolism caused by an inadequate amount of the enzyme glucose-6-phosphate dehydrogenase in rbcs. When exposed to certain environmental influences, such as fava beans, the red cells with this deficiency burst, resulting in severe anemia.

**Graptolite:** A small, colonial, often planktonic marine animal that was very abundant in the oceans 300 to 500 million years ago; now extinct.

**Great Apes:** Chimpanzees, gorillas, and orangutans.

**Group selection:** The selection operating between groups of individuals produce attributes beneficial to a group in competition with other groups rather than attributes beneficial to individuals.

**Gymnosperm:** Woody plants whose life histories include alternation of generations and ovules are not enclosed in a carpel. The pollen typically germinates on the surface of the ovule.

**Haeckel's 'Biogenetic Law':** Proposed by Ernst Haeckel in 1874 as an attempt to explain the relationship between ontogeny and phylogeny. It states that ontogeny recapitulates phylogeny, i.e., an embryo repeats in its development the evolutionary history of its species as it passes through stages in which it resembles its past ancestors.

**Haldane's Hypothesis:** Selection to lower recombination on the Y-chromosome causing pleiotropic reduction in recombination rates on other chromosomes.

**Half-life:** The time it takes for one-half of the atoms of radioactive material to decay to stable form. The half-life of carbon-14 is 5,568 years.

**Haploid:** The condition with only one set of chromosomes.

**Haplotype:** A set of genes at more than one locus inherited by an individual from one of its parents.

**Hardy-Weinberg equilibrium:** In an infinitely large population, gene and genotype frequencies remain stable as long as there is no selection, mutation, or migration. For a bi-allelic locus where the gene frequencies are p and q: $p^2+2pq+q^2 = 1$.

**Hemizygous:** As in any X-linked trait in males, absence of a homologous counterpart for an allele which may also result from deletion.

**Hemoglobin:** A protein that carries oxygen from the lungs throughout the body.

**Heredity:** The process by which characteristics are passed from one generation to the next.

**Heritability:** The proportion of the total phenotypic variance that is attributable to hereditary variance.

**Heritable:** Partly or wholly determined by genes.

**Hermaphroditism:** Having both male and female sex organs in one individual.

**Heterogametic sex:** The sex with 2 different

sex chromosomes (XY). Human and *Drosophila* males are the heterogametic sex, but in birds, moths, some fish and amphibians, females are the heterogametic sex (ZW).

**Heterogeneous nuclear RNA:** RNA products immediately synthesized from the DNA template in the nucleus has a short half-life and are very heterogeneous and very large.

**Heterokaryon:** The cell containing more than one genetically different nucleus.

**Heterozygosity:** Presence of 2 different alleles at a locus in a diploid organism which is the result of inheritance of different alleles from parents.

**Heterozygote advantage:** Also called over dominance. Genome-wide heterozygosity conferring advantage for the common diseases.

**Heterozygote:** An individual with 2 different alleles at a genetic locus.

**Heterozygous:** Having 2 different alleles for a particular trait.

**Hierarchy:** A series in which each element is categorized into successive ranks with each level subordinate to the one above.

**Homeo box:** Conserved protein sequence, which forms a DNA-binding domain in a class of transcription factors.

**Homeostasis:** A self-regulating process in development, such that the organism grows up to have much the same form independent of the external influences it experiences while growing up.

**Homeotic mutation:** A mutation causing one structure of an organism to grow in the place suitable to another. In the mutation called antennapedia in the fruit fly, a foot grows in the antennal socket.

**Hominid:** A member of the Hominidae family.

***Homo erectus*:** A species of hominid that lived between 1.8 mya and 300,000 years ago.

***Homo habilis*:** A species of hominid that lived between 1.9 and 1.8 mya.

***Homo neanderthalensis*:** A species of hominid that lived between 150,000 and 30,000 years ago in Europe and Western Asia.

***Homo sapiens*:** Modern humans evolved to their present form about 100,000 years ago.

**Homogametic:** The sex with 2 of the same kind of chromosomes.

**Homologous chromosomes:** Chromosomes that occur in pairs; one from the male parent and the other from the female parent. They participate in crossing-over and contain the same array of genes but may contain different alleles at those loci.

**Homologous structure:** The structures shared by a set of related species as they are inherited, with or without modification from the common ancestor. The bones that support a bat's wing are similar to those of a human arm.

**Homology:** Similarity due to inheritance from a common ancestor. An example is mammals' back legs.

**Homozygosity mapping:** Recessive diseases require 2 copies of an allele for expression. Bacuse of linkage disequilibrium, loci surrounding the disease locus tend to be homozygous in affected individuals. Searching for homozygous segments in diseased individuals help to locate the disease gene. This is called homozygosity mapping.

**Homozygosity:** Presence of 2 identical alleles at a locus in diploid organism which is the result of inheritance of identical alleles from both parents.

**Homozygote:** An individual having 2 copies of the same allele at a genetic locus.

**Homozygous:** Having identical alleles for a particular trait.

**Hox genes:** A particular subgroup of homeobox genes that function to pattern the axis of an organism's body and determine where limbs and other body segments will grow in the developing embryo.

**Huntington's disease:** Severe genetically inherited fatal degenerative nerve disorder resulting in a progressive loss of muscle control that inevitably leads to paralysis and death.

**Hybrid vigor:** Unusual growth, strength, and health of heterozygote offspring from 2 less vigorous homozygote parents.

**Hybrid:** The offspring of 2 distinct species.

**Hybridization:** The specific reassociation of complementary strands of nucleic acids.

**Hypothesis:** An unproven but testable scientific proposition.

**Immigration:** The movement of organisms into an area.

**Immutability:** The ability to withstand change.

***In vitro*:** Refers to experiments carried out outside living body.

***In vivo*:** Refers to experiments taking place in living organisms or natural settings.

**Inbreeding depression:** Reduction in offspring fitness resulting from mating between blood relatives.

**Inbreeding:** Production of offspring by (blood) related parents.

**Incest:** Sexual relationships between parents and children, or between brothers and sisters.

**Induction:** The process of deriving general principles from particular facts.

**Inference:** A conclusion drawn from evidence.

**In-group:** In a cladistic analysis, the set of taxa which are hypothesized to be closely related to each other than any are to the out-group.

**Inheritance of acquired characters:** The theory that an individual inherits characters that its parents acquired during their lifetimes.

**Insectivorous:** Feeding largely or exclusively on insects.

**Intelligent design:** The argument that complex biological structures have been designed by supernatural or extra-terrestrial intelligence.

**Invertebrate:** All animals other than those in the phylum Chordata; lower metazoans. They do not possess a notochord or vertebral column.

**Isolating mechanism:** Any mechanism, like a difference between species in courtship behavior or breeding season that results in reproductive isolation between the species.

**Karyotype:** A photomicrograph of metaphase chromosomes. Normal human karyotype consists of 46 chromosomes; of which 44 are somatic and 2 are sex chromosomes.

**Kin recognition:** Ability to distinguish the incest from unrelated members of a species.

**Kin selection:** Natural selection that lowers an individual's own chance of survival but raises that of a relative.

**Kingdom:** The major taxonomic group in the classification of living organisms with the exception of informal division of prokaryotic and eukaryotic empires. The five Kingdoms are Monera, Protoctista, Fungi, Plants and Animals.

**Land bridge:** A connection between 2 land masses, especially the continents for migration of plants and animals from one land mass to the other.

**Larva:** The immature stage of an animal that has a different form from the adult. The caterpillar is the larval stage of a butterfly or moth.

**Law:** A description of how a natural phenomenon will occur under given circumstances.

**Lek:** A special site where communal courtship display takes place by swarms of animals. Lekking is best known in male birds.

**Lemur:** A small, tree-dwelling primate belonging to the prosimians.

**Lethal recessive:** The case in which inheriting 2 recessive alleles of a gene causes the death of the organism.

**Lineage:** Any continuous line of descent; any series of organisms connected by reproduction by parent of offspring.

**Linnaean classification:** A hierarchical method of naming classificatory groups, invented by the Linnaeus. Each individual is assigned to a species, genus, family, order, class, phylum, and kingdom, and some intermediate classificatory levels. Species are referred to by a Linnaean binomial of its genus and species.

**Living fossil:** An extant species which is morphologically similar to a species from the ancient past. Coelacanth, Horseshoe Crab, Amazon River Dolphin are examples.

**Locus:** The position on a chromosome occupied by a particular gene.

**Macro mutation:** Mutation of large phenotypic effect, one that produces a phenotype well outside the range of variation previously existing in the population.

**Macroevolution:** Evolution that occurs at the higher levels of taxa.

**Mammals:** One of the classes in the Phylum Chordata which contains about 4500 species in 15 Orders.

**Mammary glands:** These are specialized glands that produce milk for feeding young in mammals.

**Mandible:** A part of the bony structure of a jaw. It is the lower jaw in vertebrates; one of the paired appendages closest to the mouth in arthropods.

**Marsupial mammals:** A group of mammals whose females give birth to young at very early stage of development. The newborns complete their development while sucking in a pouch, a permanent feature of the female.

**Mastodon:** An extinct elephant-like mammal.

**Maternal inheritance:** Diseases due to mutations in mtDNA are transmitted only by mothers as all mitochondria are inherited via the egg. All offspring of an affected female are at risk of inheriting the abnormality, but no offspring of an affected male are at risk. Clinical manifestations may be due to variable mixtures of mutant and normal mitochondrial genomes within cells and tissues.

**Mating type:** Genetically determined characteristics of bacteria, ciliates, fungi and algae, determine their ability to conjugate and undergo sexual reproduction with other members of the species. The mushroom *Schizophyllum* has more than 50,000 mating types (genders) encoded in 2 separate loci.

**Meiosis:** Cell division with two phases resulting in four haploid cells from a diploid cell.

**Meme:** Richard Dawkins coined the term for a unit of culture, like an idea, skill, story, or custom, passed from person to person by imitation or teaching. Some argue that memes are the cultural equivalent of genes, and reproduce, mutate, selected, and evolve in a similar way.

**Mendelian inheritance:** Inheritance of traits mediated by nuclear genes according to the laws of Gregor Mendel.

**Messenger RNA:** RNA produced by transcription from the DNA which acts as the message that is decoded to form proteins.

**Metabolism:** The processes that occur in a living organism to maintain life. There are 2 kinds of metabolism: anabolism, and catabolism.

**Metamorphosis:** One or more changes in form during the life cycle of an organism in which the juvenile stages differ from the adult as found in the transition from a tadpole to an adult frog.

**Metazoa:** Major division in the Animal Kingdom consisting of multicellular animals.

**Microbe:** Used to refer to microscopic organisms like bacteria, protozoa or viruses that cause disease or infection.

**Microcephaly:** An extremely rare genetically inherited condition in which babies are born with small brains and heads as found among the children of 23 Old Order Amish families in Lancaster County, Pennsylvania.

**Microevolution:** Evolution of species over relatively short times.

**Mimicry:** The resemblance of one kind of organism to another to make the organism difficult to find, to discourage the potential predators, or to attract potential prey.

**Missense mutation:** A mutation that causes the substitution of one amino acid for another.

**Missing link:** The absent member needed to complete an evolutionary lineage.

**Mitochondrial DNA:** The maternally inherited nucleic acid found in cytoplasm whose homologue in plants is chloroplastic DNA which mostly codes for tRNAs, rRNAs and ATP synthases. It is more related to bacterial DNA than to eukaryotic nuclear DNA.

**Mitochondrion:** An organelle in eukaryotic cells which produce enzymes to convert food to energy. They contain DNA coding for some mitochondrial proteins.

**Mitosis:** Cell division into 2 identical daughter cells with the same chromosome number as the mother cell. Replicated chromosomes separate and each chromatid go to a daughter cell.

**Molecular clock:** The theory that molecules evolve at about constant rate. The difference between the form of a molecule in 2 species is then proportional to the time since the species diverged from a common ancestor, and molecules have great value in the inference of phylogeny.

**Molecular mimicry:** Resemblance of a DNA sequence or a polypeptide by an unrelated sequence at the nucleotide or amino acid level. Mimicry of MHC proteins is an immunoevasion mechanism used by pathogens.

**Mollusc:** An invertebrate that has a fleshy, muscular body including snails, bivalves, squids, and octopuses.

**Monera:** One of the 5 Kingdoms containing all prokaryotes. It contains archaebacteria, eubacteria and cyanobacteria.

**Monogamy:** A reproductive strategy in which one male and one female mate and reproduce with each other.

**Monophyletic group:** A set of species containing a common ancestor and all of its descendants, and not containing any organisms that are not the descendants of that common ancestor.

**Monophyletic:** A group of organisms which includes the most recent common ancestor of all of its members and all of the descendants of that most recent common ancestor.

**Monotremes:** The mammals whose females lay eggs. The young hatch and develop in the mother's pouch, which is present only when needed. Two species of spiny anteater and the duck-billed platypus are the living monotremes.

**Morphology:** Study of the form, shape, and structure of organisms.

**Mosaicism:** Mosaicism is the presence of more than one cell lines differing in genotype or karyotype but derived from one zygote. Post-zygotic new mutations result in mosaic individuals who may not be affected themselves, but are at risk of bearing multiple affected offspring.

**mRNA expression profile:** The identities and absolute or relative levels of mRNAs in a specific cell/tissue type in a given physiological, developmental or pathological state.

**Muller's Ratchet:** The continual decrease in fitness due to accumulation of mutations without compensating mutations and recombination in an asexual lineage. Recombination is more common than mutation, so it can take care of mutations as they arise. This is one of the reasons for the evolution of sex.

**Müllerian mimicry:** A kind of mimicry in which 2 poisonous species evolve to look like one another.

**Multivariate analysis:** A statistical analysis of several variables assessed simultaneously and it should be preferred over individual analyses of pairs of variables.

**Mutation rate:** Number of mutations at a particular locus, which occur per gene per cell generation. This is the source of variation in asexual organisms.

**Mutation:** Any heritable change brought about by an alteration in the genetic material including gene conversion, deletion, duplication, and insertion .

**Mya:** Million years ago.

**Natural selection:** Many individuals of each species are born than can possibly survive; and there is a frequent recurrent struggle for existence, thus any being, if it varies slightly in any manner profitable to itself, under the complex and varying conditions of life, will have a better chance of surviving, and thus be naturally selected.

**Nature-nurture debate:** The debate on the relative contributions of gene (nature) and environment (nurture) to the characteristics of an organism.

**Neanderthal:** A hominid, similar to but distinct from modern humans that lived in Europe and Western Asia about 150,000 to 30,000 years ago.

**Negative assortative mating:** Type of non-random mating in which individuals of unlike phenotype mate often than predicted under random mating conditions.

**Neo-Darwinism:** Darwin's theory of natural selection plus Mendelian inheritance.

*Neurospora crassa*: Haploid, heterothallic, filamentous Ascomycete fungus.

**Neutral mutation:** The mutation with the same fitness as the other allele or alleles at its locus.

**Neutral theory:** The theory suggests that much variation at the molecular level is due to the interaction between drift and mutation rather than being actively maintained by selection.

**Niche:** The ecological role of a species; the resources it consumes and habitats it occupies.

**Nomadic:** People who live in no fixed place but move in search of food or grazing land for their animals.

**Notochord:** A flexible skeletal rod running the length of the body in the embryos of the chordates.

**Nucleotide:** A building block of DNA and RNA consisting of a sugar and phosphate backbone with a base attached.

**Nucleus:** A region of eukaryotic cells, enclosed within a membrane, containing the DNA.

**Numerical taxonomy:** Any method of taxonomy using numerical measurements which often refers to phenetic classification using large numbers of quantitatively measured characters.

**Order:** The taxonomic level between class and family.

**Organelle:** Any of a number of distinct small structures found in the cytoplasm of eukaryotic cells.

**Organisms:** Living things.

**Orthogenesis:** The wrong idea that species tend to evolve in a fixed direction because of some inherent force driving them to do so.

**Out-group:** In a cladistic analysis, any taxon used to help resolve the polarity of characters, and which is hypothesized to be less closely related to each of the taxa under consideration than any are to each other.

**Paleoanthropologist:** Scientist who uses fossil evidence to study early human ancestors.

**Paleobiology:** Biological study of fossils.

**Paleontologist:** Scientist who studies fossils to understand life in prehistoric times.

**Paleontology:** Study of fossils.

**Pangaea:** A super continent which began to break apart into the modern continents about 260 million years ago, causing the isolation of various groups of organisms from each other.

**Panmixis:** Random mating throughout a population.

**Parallel evolution:** Evolution of roughly similar changes in 2 or more closely related lineages.

*Paramecium*: A unicellular Protozoa belonging to the group Ciliates. Although normally reproduces asexually, they also undergo sexual conjugation.

**Parapatric speciation:** Speciation in which new species forms from a population contiguous with the ancestral species' geographic range.

Speciation that occurs as a result of 2 populations diverging in adjacent geographical areas.

**Paraphyletic group:** A set of species containing an ancestral species together with some of its descendants. The species included in the group are those that have continued to resemble the ancestor. The excluded species have evolved rapidly and no longer resemble their ancestor.

**Paraphyletic:** A group of organisms which includes the most common ancestor of all of its members, but not all the descendants of that most recent common ancestor.

**Parasite:** An organism that lives on or in a plant or animal of a different species, taking food without providing any benefit to the host.

**Parsimony:** A rule used to choose among possible cladograms, which states that the cladogram implying the least number of changes in character states is the best.

**Parthenogenesis:** Reproduction involving unfertilized eggs.

**Paternity:** The identity of the father of an offspring.

**Pathogen:** A microbe that causes disease.

**Pathological:** Related to or caused by disease.

**PCR:** A technique that allows amplification of specific DNA segments in a short time.

**Peripatric speciation:** Synonym of peripheral isolate speciation.

**Peripheral isolate speciation:** A form of allopatric speciation in which the new species is formed from a small population isolated at the edge of the ancestral population's geographic range.

**Pesticide-resistant insects:** Insects with the ability to survive and reproduce in the presence of pesticides. These resistant variants increase in frequency over time if pesticides remain present in their environment.

**Phenetic classification:** Classification in which species are grouped together with other species that they most closely resemble phenotypically.

**Phenetic species concept:** Species is a set of organisms that is phenotypically similar to one another.

**Phenocopy:** A condition resembling a genetically determined one. Some teratogens may cause congenital anomalies mimicking genetically caused anomalies. Deafness is another example of phenocopy which may be genetic or non-genetic.

**Phenotype:** The visible characteristics of an organism.

**Phenotypic characters:** Individual traits that can be observed in an organism and that result from the interaction between the organism's genetic makeup and its environment.

**Pheromone:** Species-specific chemical produced by an animal to communicate with an effect on their behaviour without being consciously perceived as smell. .

**Photoreceptor cell:** A cell of the nervous system that reacts to the light. It contains a pigment that undergoes a chemical change when light is absorbed. This chemical change stimulates electrical changes in the photoreceptor.

**Phyletic gradualism:** The evolutionary mode characterized by slow and gradual modifications of biological structures leading to speciation as opposed to punctuated equilibrium.

**Phylogenetic foot printing:** The use of phylogenetic comparisons to know the conserved functional elements.

**Phylogenetics:** Study of reconstructing evolutionary genealogical ties between taxa and line of descent of species.

**Phylogeny:** The evolutionary tree showing the inferred relationships of descent and common ancestry of any taxa.

**Phylum:** The highest levels of taxonomic classification.

**Phytoplankton:** Microscopic aquatic organisms that use photosynthesis to capture and harness solar energy.

**Placental mammals:** The group of mammals in which the young develop inside the mother, nourished by a placenta. The young are born in an advanced stage of development.

**Placoderm:** An extinct bottom-dwelling fish that first developed jaws and paired fins.

**Plankton:** Minute animals (zooplankton) and plants (phytoplankton) that float and drift in water, near the surface. Both in the sea and in freshwater, small plants can photosynthesize. Many sessile adult organisms disperse by means of planktonic larval stage.

**Plasmid:** A genetic element that exists freely of the main DNA in the cell. In bacteria, plasmids exist as small loops of DNA and are passed between cells independently.

**Plesiomorphy:** A primitive character state of the taxa .

**Polarity of characters:** The nature of characters used in a cladistic analysis, either original or derived. Original characters are acquired by an ancestor deeper in the phylogeny than the recent common ancestor of the taxa under consideration. Derived characters are acquired by the recent common ancestor of the taxa under consideration.

**Polyandry:** A mating system in which one female mate with many males. Seahorses and jacanas are examples.

**Polygenic:** Traits controlled by 2 or more genetic loci and are influenced by environment.

**Polygyny:** Mating strategy in which one male mate with several females. Lions, peacocks, and gorillas are examples

**Polymorphism:** Presence of discreetly different forms of a gene. It is a Mendelian trait that exists in the population in at least 2 phenotypes, neither of which occurs at a frequency of less than 1%.

**Polyphyletic group:** A set of species descended from more than one common ancestor. The common ancestor of all species in the group is not a member of the polyphyletic group.

**Polyphyletic:** A group of organisms which does not include the recent common ancestor of those organisms; the ancestor does not possess the character shared by members of the group.

**Polyploidy:** The situation in which the organism possess more than 2 (2n) sets of chromosomes. Polyploidy is a common method for sympatric speciation.

**Population biology:** The study of the patterns in which organisms are related in time and space which combines disciplines like population genetics, ecology, and taxonomy.

**Population genetics:** The branch of genetics that deals with frequencies of alleles and genotypes in breeding populations and with selective influences on the genetic composition of the population.

**Position effect:** A difference in phenotype based on the position of a gene, often caused by heterochromatin nearby. The change in a gene's location may cause a change in its expression.

**Post-zygotic isolation:** A type of reproductive isolation in which a zygote is formed but then either fails to develop or develops into a sterile adult. Donkeys and horses are post-zygotically isolated from one another; a male donkey and a female horse mate to produce a mule which is sterile.

**Pre-Cambrian Eon:** The entire geological time before the Cambrian period.

**Pre-mRNA (precursor mRNA):** The primary transcript and intermediates in RNA processing that yield functional (mature) mRNA.

**Pre-zygotic isolation:** A type of reproductive isolation in which 2 species never reach the stage of successful mating, and no zygote is formed. Species that have different breeding seasons or courtship displays, never recognize one another as potential mates.

**Primary sex ratio:** The estimated male-to-female sex ratio at fertilization which is generally more than 1 in most mammals.

**Primates:** One of the mammalian Orders which includes Lemurs, old world and new world Monkeys, Apes and Humans.

**Primitive:** A character state that is present in the common ancestor of a clade which is the original condition of that character within the clade.

**Prion:** An infectious agent, which does not have any nucleic acid but causes scrapie in sheep, kuru and Creutzfeldt-Jakob disease in humans.

**Prokaryotic cell:** The cell in which the DNA is not enclosed in a nucleus. It consists of Eubacteria and Archaebacteria.

**Prosimians (Prosimii):** The suborder of Primates including Lemurs, Lorises and Tarsiers.

**Protein clock hypothesis:** The idea that amino acid replacements occur at a constant rate in a protein family and the degree of divergence between 2 species can be used to estimate the time elapsed since their divergence.

**Protein:** The biomolecule made up of a sequence of amino acids.

**Pseudo-extinction:** The apparent disappearance of a taxon which is not due to the death of all members, but the evolution of novel features in one or more lineages, so that the new clades are not recognized as belonging to the paraphyletic ancestral group, whose members have ceased to exist.

**Pseudoalleles:** Genes that behave like alleles but can be separated by crossing over.

**Pseudogenes:** A gene with a nonsense mutation and no transcription ability.

**Punnett square:** A graphical method of showing all the potential combinations of offspring genotypes that can occur and their probability give the parent genotypes. Punnett squares are used by genetics counselors to predict the odds of a couple passing on particular inherited traits.

**Purine:** A type of base in the DNA. Adenine and guanine are purines.

**Pyrimidine:** A kind of base. Cytosine and thymine are pyrimidines.

**Quantitative character:** A character displaying a continuous phenotypic range rather than discrete classes. A gene affecting a quantitative character is a quantitative trait locus, or QTL.

**Quantitative genetics:** The statistical study of the genetics of quantitative characters. Quantitative genetic characters do not assort in a simple way in crosses. Examples are physiological activity and reproductive rate.

**Quasidominance:** Direct transmission of a recessive trait giving the impression of dominance as happens if the recessive gene is frequent or inbreeding is intense.

**Quasispecies:** The whole population of phylogenetically related variants observed within a single individual. Viruses with high mutation rates like HIV and hepatitis C virus occur like this.

**Race:** A geographic subdivision of a species distinguished from others by the allele frequencies of a number of genes.

**Radiometric dating:** Dating technique that uses the decay rate of radioactive isotopes to estimate the age of an object.

**Random mating:** Mating without preference for mates.

**Rank:** In traditional taxonomy, taxa are ranked according to their level of inclusiveness.

**Receptors:** Proteins that bind to other specific molecules. Usually on the surface of a cell, receptors bind to antibodies or hormones.

**Recessive allele:** Allele that is masked in the phenotype by the presence of a dominant allele and is expressed in the phenotype when the genotype is homozygous recessive.

**Recessive:** A trait that is not expressed in heterozygotes.

**Recognition species concept:** A species is a set of organisms that recognize one another as potential mates; they have a shared mate recognition system.

**Recombination:** The exchange of genetic material between a homologous pair of chromosomes during meiosis.

**Red Queen Theory:** An organism's biotic environment evolves to the detriment of the organism. Sex and recombination result in progeny genetically different from the previous generations and less susceptible to the antagonistic advances made during the previous generations, particularly by their parasites.

**Reinforcement:** An increase in reproductive isolation between incipient species by natural selection. Reinforcement amounts to selection for assortative mating between the incipiently speciating forms.

**Relatedness:** Two clades are more closely related when they share a recent common ancestor between them than they do with any other clade.

**Relative dating:** The process of ordering fossils, rocks, and geologic events from old to young.

**Repetitive DNA:** Non-coding DNA sequence blocks repeatedly occurring in chromosomal DNA which do not have any function but those capping the chromosomes prevent the loss of genetic information after each replication.

**Reproductive cells:**   Sperm or unfertilized ovum produced in the testes and ovaries of animals with half the number of chromosomes found in somatic cells.

**Reproductive isolation:**   Two populations of opposite sex are considered reproductively isolated from one another if they cannot together produce fertile offspring.

**RFLP (restriction fragment length polymorphism):**   Genetic polymorphism as revealed by the sizes of fragments produced with a particular restriction endonuclease enzyme.

**Ribosomal RNA:**   The RNA that constitutes the ribosomes and provides the site for translation.

**Ribosome:**   The site of protein synthesis in the cell.

**Ring species:**   A situation in which 2 reproductively isolated populations living in the same region are connected by a geographic ring of populations that can interbreed.

**RNA:**   A single-stranded nucleic acid found both in nucleus and cytoplasm. It contains uracil instead of thymine, and its sugar molecule is ribose. Ribonucleic acid act as the intermediaries by which the hereditary code of DNA is converted into proteins. In some viruses, RNA is the hereditary molecule.

**Selection:**   Process which favors one feature of organisms over another feature found in the population.

**Selectionism:**   The theory that some class of evolutionary events, like molecular or phenotypic changes, have been caused by natural selection.

**Selective pressures:**   Environmental forces such as scarcity of food or extreme temperatures that result in the survival of only certain organisms with characteristics that provide resistance.

**Self-fertilization:**   The fusion of male and female gametes produced by the same individual. Self-fertilization allows an individual to create a local population, but it fails to provide variability within a population and limits the possibilities for adaptation to environmental change. Some plants reproduce by self-fertilization but most hermaphroditic animals rarely use self-fertilization, since many of them have adaptations encouraging cross-fertilization.

**Sex cell:**   A gamete, either a sperm or an ovum.

**Sex chromosome:**   The chromosome or chromosomes that influence sex determination.

**Sex:**   Formation of a new organism containing genetic material from more than a single parent.

**Sexual dimorphism:**   The existence within a species, of differences in morphology between the sexes.

**Sexual reproduction:**   Reproduction requiring the union of sex cells, which are the products of meiotic division.

**Sexual selection:**   Natural selection operating on factors that contributes to an organism's mating success.

**Sexually dimorphic:**   When males and females of a species have different appearances, which may include size, coloration, or other features.

**Sibling species:** Two species evolved from a common ancestor and are genetically distinct but morphologically similar.

**Sickle cell anemia:** A disease in which poorly formed red blood cells cannot bind correctly to oxygen, resulting in low iron, blood clotting, and joint pain.

**Silent mutation:** Base-pair substitution, which alters a codon but does not alter phenotype due to the degeneracy of the genetic code.

**Single Nucleotide Polymorphism (SNP):** A single nucleotide change in the DNA code.

**Sister group:** The two clades resulting from the splitting of the single lineage.

**Small population size effect:** Rapid changes in gene pool frequencies that occur in small populations.

**Social Darwinism:** Applies the principles of selection to the structure of society, asserting that social structure is determined by how well people are suited to living conditions.

**Sperm:** A male sex cell or gamete.

**Splicing:** An event which takes place within the nucleus whereby introns are removed from the precursor mRNA and the exons are joined together as a post-transcriptional modification.

**Sponges:** Sponges are the phylogenetic oldest Metazoa still extant .They branched off first from the common ancestor of all metazoans, the Urmetazoa.

**Spore:** In plants, small reproductive bodies capable of giving rise to a new offspring either immediately, or after a period of dormancy and can be produced asexually or sexually.

**Stabilizing selection:** Natural selection against the extreme deviation from the average.

**Stepped cline:** A cline with a sudden change in gene frequency.

**Stop codon:** Codons that signal the end of a growing polypeptide chain.

**Sub-Saharan Africa:** The region of Africa south of the Sahara desert.

**Substitution:** The evolutionary replacement of one allele by another in a population.

**Supernatural:** Relating to phenomena that cannot be described by natural laws, cannot be tested by scientific methodology, and are outside the realm of science.

**Susceptibility gene:** A gene that is neither necessary nor sufficient to cause a disease but increases the risk of development for it.

**Symbiosis:** Two organisms living together and both benefit from this. Examples include the coral polyps and algae Zooxanthellae.

**Sympatric speciation:** Speciation via populations with overlapping geographic ranges.

**Sympatry:** Living in the same geographic region.

**Synapomorphy:** A character which is derived, shared by the taxa under consideration and is used to infer common ancestry.

**Syngamy:** The union of the nuclei of 2 gametes following fertilization to form a single nucleus for the zygote.

**Syngeneic:** Genetically identical (isogeneic) members of the same species like monozygotic twins.

**Synonymous (silent) base change:** A change in the nucleotide sequence that does not cause an amino acid change.

**Synthetic theory of evolution:** Proposed to explain the transformation of a species by natural selection, mutation, etc. and for the splitting of a species into reproductively isolated subgroups.

**Systematics:** Classification of living things with regard to their evolutionary relationship.

**Tarsier:** One of three species of small nocturnal primate belonging to the genus *Tarsius*, found in Sumatra, Borneo, Celebes, and the Philippines.

**TATA box (Goldberg-Hogness box):** A short nucleotide sequence in the promoter 25 to 35 bp upstream to the transcription initiation (cap) site of eukaryotic genes to which RNA polymerase II binds.

**Taxon:** Any group of organisms to which a rank of taxonomic name is applied.

**Taxonomic hierarchy:** Taxa are classified within the kingdom, phylum, class, order, family, genus, species, and subspecies.

**Taxonomy:** The theory and practice of biological classification.

**Teleology:** It asserts that there is an element of purpose or design behind the workings of nature.

**Teleost:** Bony fishes with well-developed bone structure.

**Telocentric:** A chromosome with a terminal centromere.

**Terrestrial:** Living on land.

**Tetrapods:** Vertebrate animals other than fishes.

**Thecodont:** A diverse group of Triassic reptiles that included large 4-legged carnivores, armored herbivores.

**Theory:** A well-substantiated explanation of some aspect of the natural world that incorporates many confirmed observations, laws, and verified hypotheses.

**Topoisomerase:** A class of enzymes that convert DNA from one topological form to another.

**Trait:** A characteristic or condition.

**Trans-acting gene:** A gene acting on or co-operating with another gene on a different chromosome.

**Transcription:** The process by which messenger RNA is read from the DNA forming a gene.

**Transfer RNA:** A type of RNA that brings the amino acids to the ribosomes to make proteins.

**Transformism:** The evolutionary theory of Lamarck in which changes occur within a lineage of populations, but in which lineages do not split and do not go extinct.

**Transgenic:** An organism that contains genes from another species.

**Transition:** A mutation changing one purine into the other purine, or one pyrimidine into the other pyrimidine.

**Transitional fossil:** A fossil or group of fossils representing a series of similar species, genera, or families that link an older group of organisms to a younger group.

**Translation:** The process by which a protein is manufactured at a ribosome, using mRNA code and transfer RNA to supply the amino acids.

**Translocation:** Transfer of chromosomal material between chromosomes.

**Transversion:** A mutation caused by the substitution of a purine for a pyrimidine or vice versa in DNA or RNA.

**Trilobite:** An extinct marine arthropod common from the Cambrian to the Permian eras. Trilobite fossils are abundant in rocks of this period.

**Trisomy:** The presence of 3 copies of a specific chromosome.

**Trivers-Willard Hypothesis:** In species with a long period of parental investment after birth of young, biases in parental behaviour toward offspring of different sex, according to the parental condition is expected ; parents in better condition would be expected to show a bias toward male offspring.

**Typology:** The definition of classificatory groups by phenetic similarity to a type specimen.

**Unequal crossing-over:** A crossing-over in which the 2 chromosomes do not exchange equal lengths of DNA; one receives more than the other.

**Variance:** A measure of how variable a set of numbers are. It is the sum of squared deviations from the mean divided by (n-1) .

**Variant:** Because of the ambiguity in the definitions of mutation and polymorphism, any genetic change is called a sequence variation and such alleles are called variants.

**Vector:** A plasmid, phage, or cosmid into which foreign DNA may be inserted for cloning.

**Vertebrates:** A subphylum in the Phylum Chordata of the Kingdom Animalia. All members have a notochord and a cranium.

**Vestigial:** Any structures that have been greatly reduced in size and function over time to the extent that they now appear to have little or no current function.

**Vicariance:** Speciation which occurs as a result of the separation and isolation of portions of an orginal population.

**Viroid:** A disease-causing agent consisting of only a single-stranded, short RNA molecule.

**Virulence:** The disease-producing ability of a microorganism.

**Virus:** An entity that is capable of reproducing only by infecting a bacterial or eukaryotic cell.

**Weismann's hypothesis:** Evolutionary function of sex is to provide variation for natural selection to act on.

**Wright-Fisher model:** The most widely used population genetics model for reproduction which assumes a finite and constant size (N) and non-overlapping population and random mating. One of the results is that if a new allele appears in the population, its fixation probability is its frequency (1/2N).

**X-chromosome:** One of the sex chromosomes in humans. Females have 2 copies and males have one copy, which is invariably maternal in origin.

**Xenolog:** Homologs resulting from horizontal transfer of a gene between two organisms.

***Xenopus*:** An amphibian who shared a common ancestor with mammals about 350 million years ago. Its eggs are large and have front-to-back orientation even before they are fertilized.

**Zygote:** A cell formed by the fusion of sperm and egg.

# Index